Praise for Nate Silver and *The Signal and the Noise*

A *Wall Street Journal* Nonfiction Book of the Year

"Mr. Silver, just thirty-four, is an expert at finding signal in noise. . . . Lively prose—from energetic to outraged . . . illustrates his dos and don'ts through a series of interesting essays that examine how predictions are made in fields including chess, baseball, weather forecasting, earthquake analysis, and politics. . . . [The] chapter on global warming is one of the most objective and honest analyses I've seen. . . . even the noise makes for a good read."

—*The New York Times*

"Not so different in spirit from the way public intellectuals like John Kenneth Galbraith once shaped discussions of economic policy and public figures like Walter Cronkite helped sway opinion on the Vietnam War . . . could turn out to be one of the more momentous books of the decade."

—*The New York Times Book Review*

"A serious treatise about the craft of prediction—without academic mathematics—cheerily aimed at lay readers. Silver's coverage is polymathic, ranging from poker and earthquakes to climate change and terrorism . . . an homage to Thomas Bayes, a long-neglected statistical scholar."

—*The New York Review of Books*

"Mr. Silver's breezy style makes even the most difficult statistical material accessible. What is more, his arguments and examples are painstakingly researched."

—*The Wall Street Journal*

"Nate Silver is the Kurt Cobain of statistics. . . . His ambitious new book, *The Signal and the Noise*, is a practical handbook and a philosophical manifesto in one, following the theme of prediction through a series of case studies ranging from hurricane tracking to professional poker to counterterrorism. It will be a supremely valuable resource for anyone who wants to make good guesses about the future, or who wants to assess the guesses made by others. In other words, everyone."

—*The Boston Globe*

"Silver delivers an improbably breezy read on what is essentially a primer on making predictions by interlacing theory and numerical nitty-gritty with a series of stories about people who have to know numbers."

"*The Signal and the Noise* is many things—an introduction to the Bayesian theory of probability, a meditation on luck and character, a commentary on poker's insights into life—but its most important function is its most basic and absolutely necessary one right now: a guide to detecting and avoiding bullshit dressed up as data. . . . What is most refreshing . . . is its humility. Sometimes we have to deal with not knowing, and we need somebody to tell us that." —*Esquire*

"[An] entertaining popularization of a subject that scares many people off . . . Silver's journey from consulting to baseball analytics to professional poker to political prognosticating is very much that of a restless and curious mind. And this, more than number crunching, is where real forecasting prowess comes from." —*Slate*

"Nate Silver serves as a sort of Zen master to American election watchers. . . . In the spirit of Nassim Nicholas Taleb's widely read *The Black Swan*, Mr. Silver asserts that humans are overconfident in their predictive abilities, that they struggle to think in probabilistic terms and build models that do not allow for uncertainty." —*The Economist*

"Silver explores our attempts at forecasting stocks, storms, sports, and anything else not set in stone." —*Wired*

"*The Signal and the Noise* is essential reading in the era of Big Data that touches every business, every sports event, and every policy maker." —Forbes.com

"Laser sharp. Surprisingly, statistics in Silver's hands is not without some fun." —*Smithsonian Magazine*

"A substantial, wide-ranging, and potentially important gauntlet of probabilistic thinking based on actual data thrown at the feet of a culture determined to sweep away silly liberal notions like 'facts.'" —*The Village Voice*

"Silver shines a light on six hundred years of human intelligence-gathering—from the advent of the printing press all the way through the Industrial Revolution and up to the current day—and he finds that it's been an inspiring climb. We've learned so much, and we still have so much left to learn." —MLB.com

"From the housing bubble to political science, the best and perhaps the brightest routinely blow the biggest calls because they can't separate the signal (truth) from the noise (distractions). We'll risk one prediction, though: Silver's book will be hard to put down." —*Mother Jones*

"Lord and god of the algorithm." —Jon Stewart, *The Daily Show*

"Nate Silver's *The Signal and the Noise* is *The Soul of a New Machine* for the twenty-first century (a century we thought we'd be a lot better at predicting than we actually are). Our political discourse is already better informed and more data-driven because of Nate's influence. But here he shows us what *he* has always been able to see in the numbers—the heart and the ethical imperative of getting the quantitative questions right. A wonderful read—totally engrossing."
—Rachel Maddow, author of *Drift*

"Yogi Berra was right: 'forecasting is hard, especially about the future.' In this important book, Nate Silver explains why the performance of experts varies from prescient to useless and why we must plan for the unexpected. Must reading for anyone who cares about what might happen next."
—Richard H. Thaler, coauthor of *Nudge*

"Making predictions in the era of 'big data' is not what you might imagine. Nate Silver's refreshing and original book provides unpredictably illuminating insights differentiating objective and subjective realities in forecasting our future. He reminds us that the human element is still essential in predicting advances in science, technology, and even politics . . . if we were only wise enough to learn from our mistakes." —Governor Jon Huntsman

"Here's a prediction: after you read *The Signal and the Noise*, you'll have much more insight into why some models work well—and also why many don't. You'll learn to pay more attention to weather forecasts for the coming week—and none at all for weather forecasts beyond that. Nate Silver takes a complex, difficult subject and makes it fun, interesting, and relevant."
—Peter Orszag, former director of the Office of Management and Budget

"Projection, prediction, assumption, trepidation, anticipation, expectation, estimation . . . we wouldn't have eighty words like this in the English language if it wasn't central to our lives. We tend not to take prediction seriously because, on some level, we know that we don't know. Silver shows us how this inevitable part of life goes awry when projected on a grand scale into the murky worlds of politics, science, and economics. Dancing through chess, sports, snowstorms, global warming, and the *McLaughlin Group*, he makes a serious and systematic effort to show us how to clean the noise off the signal."
—Bill James, author of *The Bill James Baseball Abstract*

PENGUIN BOOKS

THE SIGNAL AND THE NOISE

Nate Silver is the founder and editor in chief of FiveThirtyEight.com.

The Signal and the Noise

Why So Many Predictions Fail—but Some Don't

NATE SILVER

PENGUIN BOOKS

PENGUIN BOOKS
Published by the Penguin Group
Penguin Group (USA) LLC
375 Hudson Street
New York, New York 10014

USA | Canada | UK | Ireland | Australia | New Zealand | India | South Africa | China
penguin.com
A Penguin Random House Company

First published in the United States of America by The Penguin Press,
a member of Penguin Group (USA) Inc., 2012
Published with a new preface in Penguin Books 2015

Illustrations credits:
Figure 4-2: Courtesy of Dr. Tim Parker, University of Oxford
Figure 7-1: From "1918 Influenza: The Mother of All Pandemics" by Jeffery Taubenberger and David Morens, *Emerging Infectious Disease Journal*, vol. 12, no. 1, January 2006, Centers for Disease Control and Prevention
Figures 9-2, 9-3C, 9-4, 9-5, 9-6, and 9-7: by Cburnett, Wikimedia Commons
Figure 12-2: Courtesy of Dr. J. Scott Armstrong, The Warton School, University of Pennsylvania

THE LIBRARY OF CONGRESS HAS CATALOGED THE HARDCOVER EDITION AS FOLLOWS:
Silver, Nate
The signal and the noise : why so many predictions fail—but some don't / Nate Silver
p. cm.
Includes bibliographic references and index.
ISBN 978-1-59420-411-1 (hc.)
ISBN 978-0-14-312508-2 (pbk.)
1. Forecasting. 2. Forecasting—Methodology. 3. Forecasting—history.
4. Bayesian statistical decision theory. 5. Knowledge, Theory of. I. Title.
CB158.S54 2012
519.5'42—dc23 2012027308

Printed in the United States of America
3 5 7 9 10 8 6 4 2

DESIGNED BY AMANDA DEWEY

To Mom and Dad

CONTENTS

PREFACE TO THE PAPERBACK EDITION

At about the time *The Signal and the Noise* was first published in September 2012, "Big Data" was on its way becoming a Big Idea. Google searches for the term doubled over the course of a year,[1] as did mentions of it in the news media.[2] Hundreds of books were published on the subject. If you picked up any business periodical in 2013, advertisements for Big Data were as ubiquitous as cigarettes in an episode of *Mad Men*.

But by late 2014, there was evidence that trend had reached its apex. The frequency with which Big Data was mentioned in corporate press releases had slowed down and possibly begun to decline.[3] The technology research firm Gartner even declared that Big Data had passed the peak of its "hype cycle."[4]

I hope that Gartner is right. Coming to a better understanding of data and statistics is essential to help us navigate our lives. But as with most emerging technologies, the widespread benefits to science, industry, and human welfare will come only after the hype has died down.

FIGURE P-1: BIG DATA MENTIONS IN CORPORATE PRESS RELEASES

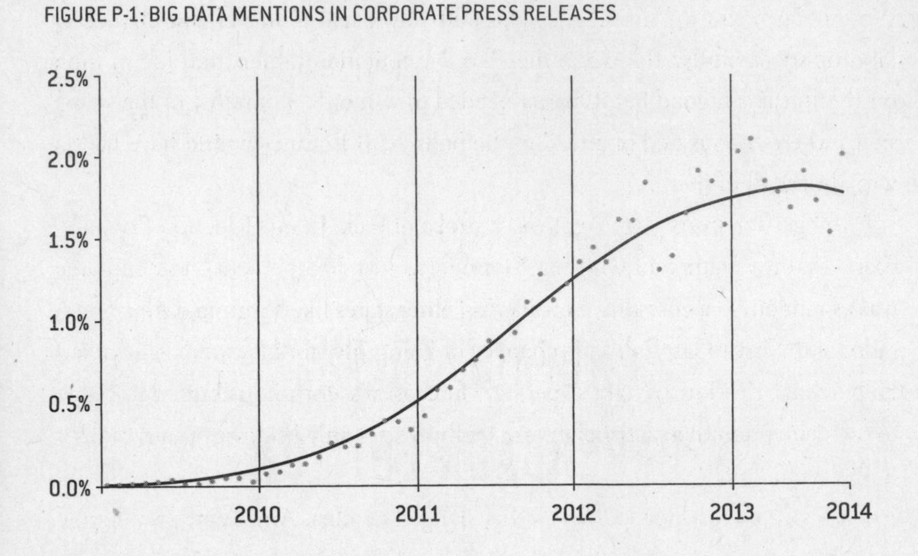

I worry that certain events in my life have contributed to the hype cycle. On November 6, 2012, the statistical model at my Web site FiveThirtyEight "called" the winner of the American presidential election correctly in all fifty states. I received a congratulatory phone call from the White House. I was hailed as "lord and god of the algorithm" by *The Daily Show*'s Jon Stewart. My name briefly received more Google search traffic than the vice president of the United States.

I enjoyed some of the attention, but I felt like an outlier—even a fluke. Mostly I was getting credit for having pointed out the obvious—and most of the rest was luck.*

To be sure, it was reasonably clear by Election Day that President Obama was poised to win reelection. When voters went to the polls on election morning, FiveThirtyEight's statistical model put his chances of winning the Electoral College at about 90 percent.† A 90 percent chance is not quite a sure thing:

* Polling-based forecasts like FiveThirtyEight's are also not the best example of "big" data. Although some election models are modestly more complex than others, accounting for factors like the state of the economy along with the polls, they use relatively little data and make relatively few assumptions about it (the most important assumption is simply that the polls will be about as accurate as they have been historically). Most of these models can be run in anywhere from a few microseconds to a few minutes with an ordinary laptop.

† It was not alone: a half-dozen competing models also had Obama as a heavy favorite, as did bookmakers and betting markets.

Would you board a plane if the pilot told you it had a 90 percent chance of landing successfully? But when there's only reputation rather than life or limb on the line, it's a good bet. Obama needed to win only a handful of the swing states where he was tied or ahead in the polls; Mitt Romney would have had to win almost all of them.

But getting every state right was a stroke of luck. In our Election Day forecast, Obama's chance of winning Florida was just 50.3 percent—the outcome was as random as a coin flip. Considering other states like Virginia, Ohio, Colorado, and North Carolina, our chances of going fifty-for-fifty were only about 20 percent.[5] FiveThirtyEight's "perfect" forecast was fortuitous but contributed to the perception that statisticians are soothsayers—only using computers rather than crystal balls.

This is a wrongheaded and rather dangerous idea. American presidential elections are the exception to the rule—one of the few examples of a complex system in which outcomes are usually more certain than the conventional wisdom implies. (There are a number of reasons for this, not least that the conventional wisdom is often not very wise when it comes to politics.) Far more often, as this book will explain, we overrate our ability to predict the world around us. With some regularity, events that are said to be certain fail to come to fruition—or those that are deemed impossible turn out to occur.

If all of this is so simple, why did so many pundits get the 2012 election wrong? It wasn't just on the fringe of the blogosphere that conservatives insisted that the polls were "skewed" toward President Obama. Thoughtful conservatives like George F. Will[6] and Michael Barone[7] also predicted a Romney win, sometimes by near-landslide proportions.

One part of the answer is obvious: the pundits didn't have much incentive to make the right call. You can get invited back on television with a far worse track record than Barone's or Will's—provided you speak with some conviction and have a viewpoint that matches the producer's goals.

An alternative interpretation is slightly less cynical but potentially harder to swallow: human judgment is intrinsically fallible. It's hard for any of us (myself included) to recognize how much our relatively narrow range of experience can color our interpretation of the evidence. There's so much information out there today that none of us can plausibly consume all of it. We're constantly making decisions about what Web site to read, which television channel to watch, and where to focus our attention.

Having a better understanding of statistics almost certainly helps. Over the past decade, the number of people employed as statisticians in the United States has increased by 35 percent[8] even as the overall job market has stagnated. But it's a necessary rather than sufficient part of the solution. Some of the examples of failed predictions in this book concern people with exceptional intelligence and exemplary statistical training—but whose biases still got in the way.

These problems are not so simple and so this book does not promote simple answers to them. It makes some recommendations but they are philosophical as much as technical. Once we're getting the big stuff right—coming to a better understanding of probably and uncertainty; learning to recognize our biases; appreciating the value of diversity, incentives, and experimentation—we'll have the luxury of worrying about the finer points of technique.

Gartner's hype cycle ultimately has a happy ending. After the peak of inflated expectations there's a "trough of disillusionment"—what happens when people come to recognize that the new technology will still require a lot of hard work.

FIGURE P-2: GARTNER'S HYPE CYCLE

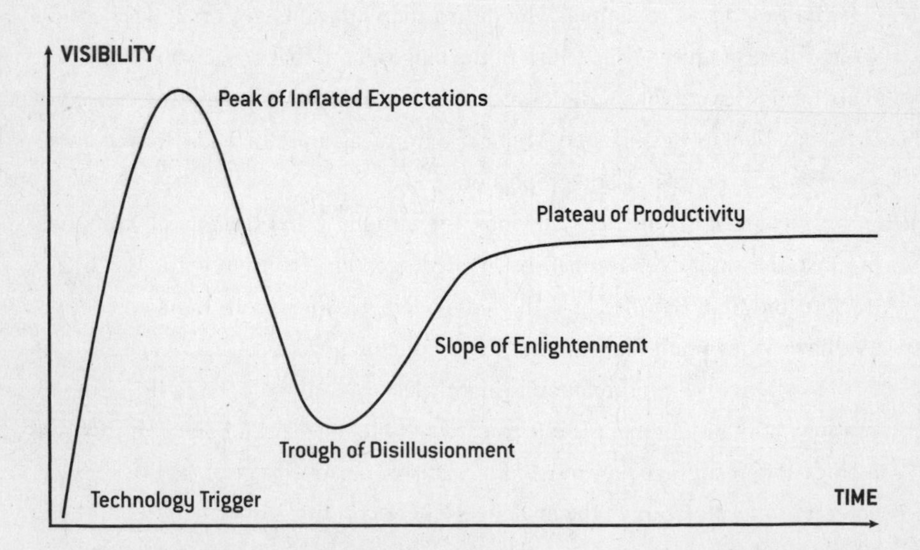

But right when views of the new technology have begun to lapse from healthy skepticism into overt cynicism, that technology can begin to pay some dividends. (We've been through this before: after the computer boom in the 1970s and the Internet commerce boom of the late 1990s, among other examples.) Eventually it matures to the point when there are fewer glossy advertisements but more gains in productivity—it may even have become so commonplace that we take it for granted. I hope this book can accelerate the process, however slightly.

The Signal and the Noise

INTRODUCTION

This is a book about information, technology, and scientific progress. This is a book about competition, free markets, and the evolution of ideas. This is a book about the things that make us smarter than any computer, and a book about human error. This is a book about how we learn, one step at a time, to come to knowledge of the objective world, and why we sometimes take a step back.

This is a book about prediction, which sits at the intersection of all these things. It is a study of why some predictions succeed and why some fail. My hope is that we might gain a little more insight into planning our futures and become a little less likely to repeat our mistakes.

More Information, More Problems

The original revolution in information technology came not with the microchip, but with the printing press. Johannes Gutenberg's invention in 1440 made information available to the masses, and the explosion of ideas it produced had unintended consequences and unpredictable effects. It was a spark for the Industrial Revolution in 1775,[1] a tipping point in which civilization suddenly went from having made almost no scientific or economic progress for most of its existence to the exponential rates of growth and change that are familiar to us today. It set in motion the events that would produce the European Enlightenment and the founding of the American Republic.

But the printing press would first produce something else: hundreds of years of holy war. As mankind came to believe it could predict its fate and choose its destiny, the bloodiest epoch in human history followed.[2]

Books had existed prior to Gutenberg, but they were not widely written and they were not widely read. Instead, they were luxury items for the nobility, produced one copy at a time by scribes.[3] The going rate for reproducing a single manuscript was about one florin (a gold coin worth about $200 in today's dollars) per five pages,[4] so a book like the one you're reading now would cost around $20,000. It would probably also come with a litany of transcription errors, since it would be a copy of a copy of a copy, the mistakes having multiplied and mutated through each generation.

This made the accumulation of knowledge extremely difficult. It required heroic effort to prevent the volume of recorded knowledge from actually *decreasing,* since the books might decay faster than they could be reproduced. Various editions of the Bible survived, along with a small number of canonical texts, like from Plato and Aristotle. But an untold amount of wisdom was lost to the ages,[5] and there was little incentive to record more of it to the page.

The pursuit of knowledge seemed inherently futile, if not altogether vain. If today we feel a sense of impermanence because things are changing so rapidly, impermanence was a far more literal concern for the generations before us. There was "nothing new under the sun," as the beautiful Bible verses in Ecclesiastes put it—not so much because everything had been discovered but because everything would be forgotten.[6]

The printing press changed that, and did so permanently and profoundly. Almost overnight, the cost of producing a book decreased by about three hundred times,[7] so a book that might have cost $20,000 in today's dollars instead cost $70. Printing presses spread very rapidly throughout Europe; from Gutenberg's Germany to Rome, Seville, Paris, and Basel by 1470, and then to almost all other major European cities within another ten years.[8] The number of books being produced grew exponentially, increasing by about thirty times in the first century after the printing press was invented.[9] The store of human knowledge had begun to accumulate, and rapidly.

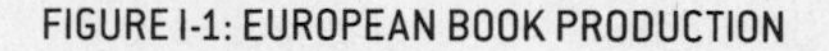
FIGURE I-1: EUROPEAN BOOK PRODUCTION

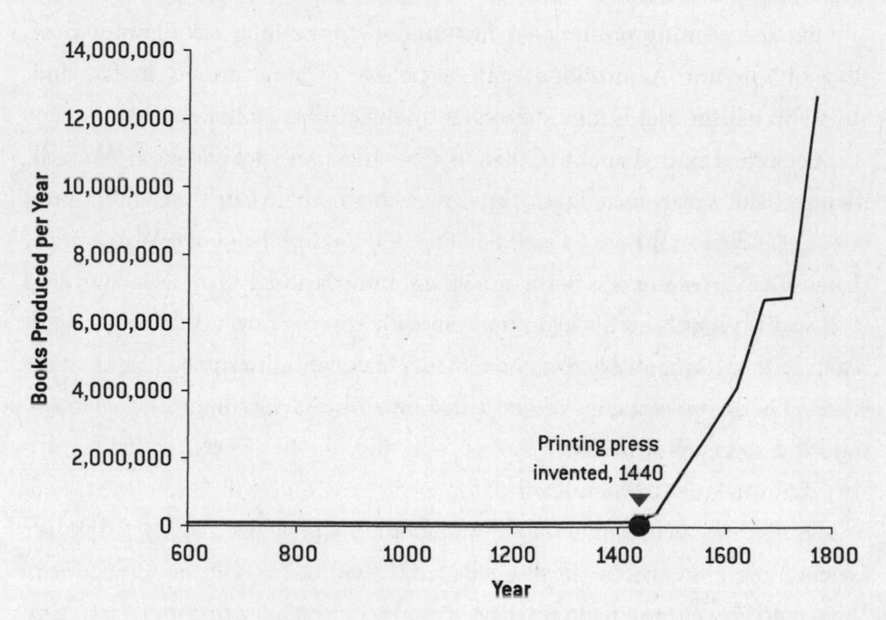

As was the case during the early days of the World Wide Web, however, the quality of the information was highly varied. While the printing press paid almost immediate dividends in the production of higher quality maps,[10] the bestseller list soon came to be dominated by heretical religious texts and pseudoscientific ones.[11] Errors could now be mass-produced, like in the so-called Wicked Bible, which committed the most unfortunate typo in history to the page: thou *shalt* commit adultery.[12] Meanwhile, exposure to so many new ideas was producing mass confusion. The amount of information was increasing much more rapidly than our understanding of what to do with it, or our ability to differentiate the useful information from the mistruths.[13] Paradoxically, the result of having so much more shared knowledge was increasing isolation along national and religious lines. The instinctual shortcut that we take when we have "too much information" is to engage with it selectively, picking out the parts we like and ignoring the remainder, making allies with those who have made the same choices and enemies of the rest.

The most enthusiastic early customers of the printing press were those who used it to evangelize. Martin Luther's *Ninety-five Theses* were not that radical; similar sentiments had been debated many times over. What was revolutionary, as Elizabeth Eisenstein writes, is that Luther's theses "did not stay tacked to the church door."[14] Instead, they were reproduced at least three hundred thousand times by Gutenberg's printing press[15]—a runaway hit even by modern standards.

The schism that Luther's Protestant Reformation produced soon plunged Europe into war. From 1524 to 1648, there was the German Peasants' War, the Schmalkaldic War, the Eighty Years' War, the Thirty Years' War, the French Wars of Religion, the Irish Confederate Wars, the Scottish Civil War, and the English Civil War—many of them raging simultaneously. This is not to neglect the Spanish Inquisition, which began in 1480, or the War of the Holy League from 1508 to 1516, although those had less to do with the spread of Protestantism. The Thirty Years' War alone killed one-third of Germany's population,[16] and the seventeenth century was possibly the bloodiest ever, with the early twentieth staking the main rival claim.[17]

But somehow in the midst of this, the printing press was starting to produce scientific and literary progress. Galileo was sharing his (censored) ideas, and Shakespeare was producing his plays.

Shakespeare's plays often turn on the idea of fate, as much drama does. What makes them so tragic is the gap between what his characters might like to accomplish and what fate provides to them. The idea of controlling one's fate seemed to have become part of the human consciousness by Shakespeare's time—but not yet the competencies to achieve that end. Instead, those who tested fate usually wound up dead.[18]

These themes are explored most vividly in *The Tragedy of Julius Caesar*. Throughout the first half of the play Caesar receives all sorts of apparent warning signs—what he calls predictions[19] ("beware the ides of March")—that his coronation could turn into a slaughter. Caesar of course ignores these signs, quite proudly insisting that they point to someone else's death—or otherwise reading the evidence selectively. Then Caesar is assassinated.

"[But] men may construe things after their fashion / Clean from the purpose of the things themselves," Shakespeare warns us through the voice of Cicero—good advice for anyone seeking to pluck through their newfound wealth of information. It was hard to tell the signal from the noise. The story the data tells us is often the one we'd like to hear, and we usually make sure that it has a happy ending.

And yet if *The Tragedy of Julius Caesar* turned on an ancient idea of prediction—associating it with fatalism, fortune-telling, and superstition—it also introduced a more modern and altogether more radical idea: that we might interpret these signs so as to gain an advantage from them. "Men at some time are masters of their fates," says Cassius, hoping to persuade Brutus to partake in the conspiracy against Caesar.

The idea of man as master of his fate was gaining currency. The words *predict* and *forecast* are largely used interchangeably today, but in Shakespeare's time, they meant different things. A prediction was what the soothsayer told you; a forecast was something more like Cassius's idea.

The term *forecast* came from English's Germanic roots,[20] unlike *predict*, which is from Latin.[21] Forecasting reflected the new Protestant worldliness rather than the otherworldliness of the Holy Roman Empire. Making a forecast typically implied planning under conditions of uncertainty. It suggested having prudence, wisdom, and industriousness, more like the way we now use the word *foresight*.[22]

The theological implications of this idea are complicated.[23] But they were less so for those hoping to make a gainful existence in the terrestrial world. These qualities were strongly associated with the Protestant work ethic, which Max Weber saw as bringing about capitalism and the Industrial Revolution.[24] This notion of *forecasting* was very much tied in to the notion of progress. All that information in all those books ought to have helped us to plan our lives and profitably predict the world's course.

The Protestants who ushered in centuries of holy war were learning how to use their accumulated knowledge to change society. The Industrial Revolution largely began in Protestant countries and largely in those with a free press, where both religious and scientific ideas could flow without fear of censorship.[25]

The importance of the Industrial Revolution is hard to overstate. Throughout essentially all of human history, economic growth had proceeded at a rate of perhaps 0.1 percent per year, enough to allow for a very gradual increase in population, but not *any* growth in per capita living standards.[26] And then, suddenly, there was progress when there had been none. Economic growth began to zoom upward much faster than the growth rate of the population, as it has continued to do through to the present day, the occasional global financial meltdown notwithstanding.[27]

FIGURE I-2: GLOBAL PER CAPITA GDP, 1000–2010

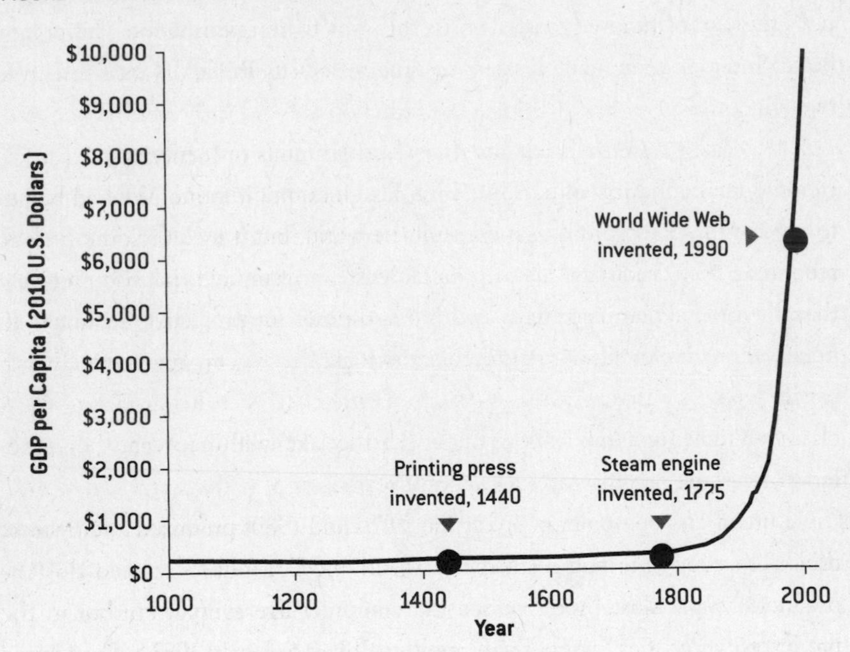

The explosion of information produced by the printing press had done us a world of good, it turned out. It had just taken 330 years—and millions dead in battlefields around Europe—for those advantages to take hold.

The Productivity Paradox

We face danger whenever information growth outpaces our understanding of how to process it. The last forty years of human history imply that it can still take a long time to translate information into useful knowledge, and that if we are not careful, we may take a step back in the meantime.

The term "information age" is not particularly new. It started to come into more widespread use in the late 1970s. The related term "computer age" was used earlier still, starting in about 1970.[28] It was at around this time that computers began to be used more commonly in laboratories and academic settings, even if they had not yet become common as home appliances. This time it did not take three hundred years before the growth in information technology began to produce tangible benefits to human society. But it did take fifteen to twenty.

The 1970s were the high point for "vast amounts of theory applied to extremely small amounts of data," as Paul Krugman put it to me. We had begun to use computers to produce models of the world, but it took us some time to recognize how crude and assumption laden they were, and that the precision that computers were capable of was no substitute for predictive accuracy. In fields ranging from economics to epidemiology, this was an era in which bold predictions were made, and equally often failed. In 1971, for instance, it was claimed that we would be able to predict earthquakes within a decade,[29] a problem that we are no closer to solving forty years later.

Instead, the computer boom of the 1970s and 1980s produced a temporary *decline* in economic and scientific productivity. Economists termed this the productivity paradox. "You can see the computer age everywhere but in the productivity statistics," wrote the economist Robert Solow in 1987.[30] The United States experienced four distinct recessions between 1969 and 1982.[31] The late 1980s were a stronger period for our economy, but less so for countries elsewhere in the world.

Scientific progress is harder to measure than economic progress.[32] But one mark of it is the number of patents produced, especially relative to the investment in research and development. If it has become cheaper to produce a new invention, this suggests that we are using our information wisely and are forging it into knowledge. If it is becoming more expensive, this suggests that we are seeing signals in the noise and wasting our time on false leads.

In the 1960s the United States spent about $1.5 million (adjusted for inflation[33]) per patent application[34] by an American inventor. That figure *rose* rather than fell at the dawn of the information age, however, doubling to a peak of about $3 million in 1986.[35]

FIGURE I-3: RESEARCH AND DEVELOPMENT EXPENDITURES PER PATENT APPLICATION

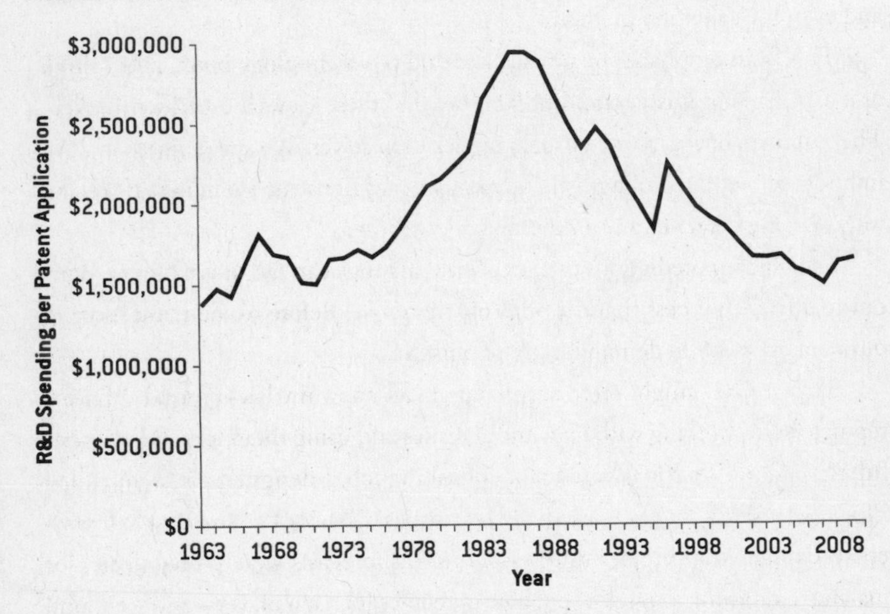

As we came to more realistic views of what that new technology could accomplish for us, our research productivity began to improve again in the 1990s. We wandered up fewer blind alleys; computers began to improve our everyday lives and help our economy. Stories of prediction are often those of long-term progress but short-term regress. Many things that seem predictable over the long run foil our best-laid plans in the meanwhile.

The Promise and Pitfalls of "Big Data"

The fashionable term now is "Big Data." IBM estimates that we are generating 2.5 quintillion bytes of data each day, more than 90 percent of which was created in the last two years.[36]

This exponential growth in information is sometimes seen as a cure-all, as computers were in the 1970s. Chris Anderson, the editor of *Wired* magazine, wrote in 2008 that the sheer volume of data would obviate the need for theory, and even the scientific method.[37]

This is an emphatically pro-science and pro-technology book, and I think of it as a very optimistic one. But it argues that these views are badly mistaken. The numbers have no way of speaking for themselves. We speak for them. We imbue them with meaning. Like Caesar, we may construe them in self-serving ways that are detached from their objective reality.

Data-driven predictions can succeed—and they can fail. It is when we deny our role in the process that the odds of failure rise. Before we demand more of our data, we need to demand more of ourselves.

This attitude might seem surprising if you know my background. I have a reputation for working with data and statistics and using them to make successful predictions. In 2003, bored at a consulting job, I designed a system called PECOTA, which sought to predict the statistics of Major League Baseball players. It contained a number of innovations—its forecasts were probabilistic, for instance, outlining a range of possible outcomes for each player—and we found that it outperformed competing systems when we compared their results. In 2008, I founded the Web site FiveThirtyEight, which sought to forecast the upcoming election. The FiveThirtyEight forecasts correctly predicted the winner of the presidential contest in forty-nine of fifty states as well as the winner of all thirty-five U.S. Senate races.

After the election, I was approached by a number of publishers who wanted to capitalize on the success of books such as *Moneyball* and *Freakonomics* that told the story of nerds conquering the world. This book was conceived of along those lines—as an investigation of data-driven predictions in fields ranging from baseball to finance to national security.

But in speaking with well more than one hundred experts in more than a dozen fields over the course of four years, reading hundreds of journal articles and books, and traveling everywhere from Las Vegas to Copenhagen in pursuit of my investigation, I came to realize that prediction in the era of Big Data was not going very well. I had been lucky on a few levels: first, in having achieved success despite having made many of the mistakes that I will describe, and second, in having chosen my battles well.

Baseball, for instance, is an exceptional case. It happens to be an especially rich and revealing exception, and the book considers why this is so—why a decade after *Moneyball*, stat geeks and scouts are now working in harmony.

The book offers some other hopeful examples. Weather forecasting, which also involves a melding of human judgment and computer power, is one of them. Meteorologists have a bad reputation, but they have made remarkable progress, being able to forecast the landfall position of a hurricane three times more accurately than they were a quarter century ago. Meanwhile, I met poker players and sports bettors who really were beating Las Vegas, and the computer programmers who built IBM's Deep Blue and took down a world chess champion.

But these cases of progress in forecasting must be weighed against a series of failures.

If there is one thing that defines Americans—one thing that makes us exceptional—it is our belief in Cassius's idea that we are in control of our own fates. Our country was founded at the dawn of the Industrial Revolution by religious rebels who had seen that the free flow of ideas had helped to spread not just their religious beliefs, but also those of science and commerce. Most of our strengths and weaknesses as a nation—our ingenuity and our industriousness, our arrogance and our impatience—stem from our unshakable belief in the idea that we choose our own course.

But the new millennium got off to a terrible start for Americans. We had not seen the September 11 attacks coming. The problem was not want of information. As had been the case in the Pearl Harbor attacks six decades earlier, all the signals were there. But we had not put them together. Lacking a proper theory for how terrorists might behave, we were blind to the data and the attacks were an "unknown unknown" to us.

There also were the widespread failures of prediction that accompanied the recent global financial crisis. Our naïve trust in models, and our failure to realize how fragile they were to our choice of assumptions, yielded disastrous results. On a more routine basis, meanwhile, I discovered that we are unable to predict recessions more than a few months in advance, and not for lack of trying. While there has been considerable progress made in controlling inflation, our economic policy makers are otherwise flying blind.

The forecasting models published by political scientists in advance of the 2000 presidential election predicted a landslide 11-point victory for Al Gore.[38] George W. Bush won instead. Rather than being an anomalous result, failures like these have been fairly common in political prediction. A long-term study by Philip E. Tetlock of the University of Pennsylvania found that when political scientists claimed that a political outcome had absolutely *no* chance of occurring, it nevertheless happened about 15 percent of the time. (The political scientists are probably better than television pundits, however.)

There has recently been, as in the 1970s, a revival of attempts to predict earthquakes, most of them using highly mathematical and data-driven techniques. But these predictions envisaged earthquakes that never happened and failed to prepare us for those that did. The Fukushima nuclear reactor had been designed to handle a magnitude 8.6 earthquake, in part because some seismologists concluded that anything larger was impossible. Then came Japan's horrible magnitude 9.1 earthquake in March 2011.

There are entire disciplines in which predictions have been failing, often at great cost to society. Consider something like biomedical research. In 2005, an Athens-raised medical researcher named John P. Ioannidis published a controversial paper titled "Why Most Published Research Findings Are False."[39] The paper studied positive findings documented in peer-reviewed journals: descriptions of successful predictions of medical hypotheses carried out in laboratory experiments. It concluded that most of these findings were likely to fail when applied in the real world. Bayer Laboratories recently confirmed Ioannidis's hypothesis. They could not replicate about *two-thirds* of the positive findings claimed in medical journals when they attempted the experiments themselves.[40]

Big Data *will* produce progress—eventually. How quickly it does, and whether we regress in the meantime, will depend on us.

Why the Future Shocks Us

Biologically, we are not very different from our ancestors. But some stone-age strengths have become information-age weaknesses.

Human beings do not have very many natural defenses. We are not all that fast, and we are not all that strong. We do not have claws or fangs or body armor. We cannot spit venom. We cannot camouflage ourselves. And we cannot fly. Instead, we survive by means of our wits. Our minds are quick. We are wired to detect patterns and respond to opportunities and threats without much hesitation.

"This need of finding patterns, humans have this more than other animals," I was told by Tomaso Poggio, an MIT neuroscientist who studies how our brains process information. "Recognizing objects in difficult situations means generalizing. A newborn baby can recognize the basic pattern of a face. It has been learned by evolution, not by the individual."

The problem, Poggio says, is that these evolutionary instincts sometimes lead us to see patterns when there are none there. "People have been doing that all the time," Poggio said. "Finding patterns in random noise."

The human brain is quite remarkable; it can store perhaps three terabytes of information.[41] And yet that is only about one one-millionth of the information that IBM says is now produced in the world *each day*. So we have to be terribly selective about the information we choose to remember.

Alvin Toffler, writing in the book *Future Shock* in 1970, predicted some of the consequences of what he called "information overload." He thought our defense mechanism would be to simplify the world in ways that confirmed our biases, even as the world itself was growing more diverse and more complex.[42]

Our biological instincts are not always very well adapted to the information-rich modern world. Unless we work *actively* to become aware of the biases we introduce, the returns to additional information may be minimal—or diminishing.

The information overload after the birth of the printing press produced greater sectarianism. Now those different religious ideas could be testified to with more information, more conviction, more "proof"—and less tolerance for dissenting opinion. The same phenomenon seems to be occurring today. Political partisanship began to increase very rapidly in the United States beginning at about the time that Tofller wrote *Future Shock* and it may be accelerating even faster with the advent of the Internet.[43]

These partisan beliefs can upset the equation in which more information will bring us closer to the truth. A recent study in *Nature* found that the *more* informed that strong political partisans were about global warming, the *less* they agreed with one another.[44]

Meanwhile, if the quantity of information is increasing by 2.5 quintillion bytes per day, the amount of *useful* information almost certainly isn't. Most of it is just noise, and the noise is increasing faster than the signal. There are so many hypotheses to test, so many data sets to mine—but a relatively constant amount of objective truth.

The printing press changed the way in which we made mistakes. Routine errors of transcription became less common. But when there was a mistake, it would be reproduced many times over, as in the case of the Wicked Bible.

Complex systems like the World Wide Web have this property. They may not fail as often as simpler ones, but when they fail they fail badly. Capitalism and the Internet, both of which are incredibly efficient at propagating information, create the potential for bad ideas as well as good ones to spread. The bad ideas may produce disproportionate effects. In advance of the financial crisis, the system was so highly levered that a single lax assumption in the credit ratings agencies' models played a huge role in bringing down the whole global financial system.

Regulation is one approach to solving these problems. But I am suspicious that it is an excuse to avoid looking within ourselves for answers. We need to stop, and admit it: we have a prediction problem. We love to predict things—and we aren't very good at it.

The Prediction Solution

If prediction is the central problem of this book, it is also its solution.

Prediction is indispensable to our lives. Every time we choose a route to work, decide whether to go on a second date, or set money aside for a rainy day, we are making a forecast about how the future will proceed—and how our plans will affect the odds for a favorable outcome.

Not all of these day-to-day problems require strenuous thought; we can budget only so much time to each decision. Nevertheless, you are making predictions many times every day, whether or not you realize it.

For this reason, this book views prediction as a shared enterprise rather than as a function that a select group of experts or practitioners perform. It is amusing to poke fun at the experts when their predictions fail. However, we should be careful with our Schadenfreude. To say our predictions are no worse than the experts' is to damn ourselves with some awfully faint praise.

Prediction does play a particularly important role in science, however. Some of you may be uncomfortable with a premise that I have been hinting at and will now state explicitly: we can *never* make perfectly objective predictions. They will *always* be tainted by our subjective point of view.

But this book is emphatically against the nihilistic viewpoint that there is no objective truth. It asserts, rather, that a belief in the objective truth—and a commitment to pursuing it—is the first prerequisite of making better predictions. The forecaster's next commitment is to realize that she perceives it imperfectly.

Prediction is important because it connects subjective and objective reality. Karl Popper, the philosopher of science, recognized this view.[45] For Popper, a hypothesis was not scientific unless it was falsifiable—meaning that it could be tested in the real world by means of a prediction.

What should give us pause is that the few ideas we have tested aren't doing so well, and many of our ideas have not or cannot be tested at all. In economics, it is much easier to test an unemployment rate forecast than a claim about the effectiveness of stimulus spending. In political science, we can test models that are used to predict the outcome of elections, but a theory about how changes to political institutions might affect policy outcomes could take decades to verify.

I do not go as far as Popper in asserting that such theories are therefore unscientific or that they lack any value. However, the fact that the few theories we *can* test have produced quite poor results suggests that many of the ideas we *haven't* tested are very wrong as well. We are undoubtedly living with many delusions that we do not even realize.

But there is a way forward. It is not a solution that relies on half-baked policy ideas—particularly given that I have come to view our political system as a big part of the problem. Rather, the solution requires an attitudinal change.

This attitude is embodied by something called Bayes's theorem, which I introduce in chapter 8. Bayes's theorem is nominally a mathematical formula. But it is really much more than that. It implies that we must think differently about our ideas—and how to test them. We must become more comfortable with probability and uncertainty. We must think more carefully about the assumptions and beliefs that we bring to a problem.

The book divides roughly into halves. The first seven chapters diagnose the prediction problem while the final six explore and apply Bayes's solution.

Each chapter is oriented around a particular subject and describes it in some depth. There is no denying that this is a detailed book—in part because that is often where the devil lies, and in part because my view is that a certain amount of immersion in a topic will provide disproportionately more insight than an executive summary.

The subjects I have chosen are usually those in which there is some publicly shared information. There are fewer examples of forecasters making predictions based on private information (for instance, how a company uses its customer records to forecast demand for a new product). My preference is for topics where you can check out the results for yourself rather than having to take my word for it.

A Short Road Map to the Book

The book weaves between examples from the natural sciences, the social sciences, and from sports and games. It builds from relatively straightforward cases, where the successes and failures of prediction are more easily demarcated, into others that require slightly more finesse.

Chapters 1 through 3 consider the failures of prediction surrounding the recent financial crisis, the successes in baseball, and the realm of political prediction—where some approaches have worked well and others haven't. They should get you thinking about some of the most fundamental questions that underlie the prediction problem. How can we apply our judgment to the data—without succumbing to our biases? When does market competition make forecasts better—and how can it make them worse? How do we reconcile the need to use the past as a guide with our recognition that the future may be different?

Chapters 4 through 7 focus on *dynamic* systems: the behavior of the earth's atmosphere, which brings about the weather; the movement of its tectonic plates, which can cause earthquakes; the complex human interactions that account for the behavior of the American economy; and the spread of infectious diseases. These systems are being studied by some of our best scientists. But dynamic systems make forecasting more difficult, and predictions in these fields have not always gone very well.

Chapters 8 through 10 turn toward solutions—first by introducing you to a sports bettor who applies Bayes's theorem more expertly than many economists or scientists do, and then by considering two other games, chess and poker. Sports and games, because they follow well-defined rules, represent good laboratories for testing our predictive skills. They help us to a better understanding of randomness and uncertainty and provide insight about how we might forge information into knowledge.

Bayes's theorem, however, can also be applied to more existential types of problems. Chapters 11 through 13 consider three of these cases: global warming, terrorism, and bubbles in financial markets. These are hard problems for forecasters and for society. But if we are up to the challenge, we can make our country, our economy, and our planet a little safer.

. . .

The world has come a long way since the days of the printing press. Information is no longer a scarce commodity; we have more of it than we know what to do with. But relatively little of it is useful. We perceive it selectively, subjectively, and without much self-regard for the distortions that this causes. We think we want information when we really want knowledge.

The signal is the truth. The noise is what distracts us from the truth. This is a book about the signal and the noise.

1

A CATASTROPHIC FAILURE OF PREDICTION

It was October 23, 2008. The stock market was in free fall, having plummeted almost 30 percent over the previous five weeks. Once-esteemed companies like Lehman Brothers had gone bankrupt. Credit markets had all but ceased to function. Houses in Las Vegas had lost 40 percent of their value.[1] Unemployment was skyrocketing. Hundreds of billions of dollars had been committed to failing financial firms. Confidence in government was the lowest that pollsters had ever measured.[2] The presidential election was less than two weeks away.

Congress, normally dormant so close to an election, was abuzz with activity. The bailout bills it had passed were sure to be unpopular[3] and it needed to create every impression that the wrongdoers would be punished. The House Oversight Committee had called the heads of the three major credit-rating agencies, Standard & Poor's (S&P), Moody's, and Fitch Ratings, to testify before them. The ratings agencies were charged with assessing the likelihood that trillions of dollars in mortgage-backed securities would go into default. To put it mildly, it appeared they had blown the call.

The Worst Prediction of a Sorry Lot

The crisis of the late 2000s is often thought of as a failure of our political and financial institutions. It was obviously an economic failure of massive proportions. By 2011, four years after the Great Recession officially began, the American economy was still almost $800 billion below its productive potential.[4]

I am convinced, however, that the best way to view the financial crisis is as a failure of judgment—a catastrophic failure of prediction. These predictive failures were widespread, occurring at virtually every stage during, before, and after the crisis and involving everyone from the mortgage brokers to the White House.

The most calamitous failures of prediction usually have a lot in common. We focus on those signals that tell a story about the world as we would like it to be, not how it really is. We ignore the risks that are hardest to measure, even when they pose the greatest threats to our well-being. We make approximations and assumptions about the world that are much cruder than we realize. We abhor uncertainty, even when it is an irreducible part of the problem we are trying to solve. If we want to get at the heart of the financial crisis, we should begin by identifying the greatest predictive failure of all, a prediction that committed all these mistakes.

The ratings agencies had given their AAA rating, normally reserved for a handful of the world's most solvent governments and best-run businesses, to thousands of mortgage-backed securities, financial instruments that allowed investors to bet on the likelihood of someone else defaulting on their home. The ratings issued by these companies are quite explicitly meant to be predictions: estimates of the likelihood that a piece of debt will go into default.[5] Standard & Poor's told investors, for instance, that when it rated a particularly complex type of security known as a collateralized debt obligation (CDO) at AAA, there was only a 0.12 percent probability—about 1 chance in 850—that it would fail to pay out over the next five years.[6] This supposedly made it as safe as a AAA-rated corporate bond[7] and *safer* than S&P now assumes U.S. Treasury bonds to be.[8] The ratings agencies do not grade on a curve.

In fact, around 28 percent of the AAA-rated CDOs defaulted, according to

S&P's internal figures.[9] (Some independent estimates are even higher.[10]) That means that the actual default rates for CDOs were more than *two hundred times higher* than S&P had predicted.[11]

This is just about as complete a failure as it is possible to make in a prediction: trillions of dollars in investments that were rated as being almost completely safe instead turned out to be almost completely unsafe. It was as if the weather forecast had been 86 degrees and sunny, and instead there was a blizzard.

FIGURE 1-1: FORECASTED AND ACTUAL 5-YEAR DEFAULT RATES FOR AAA-RATED CDO TRANCHES

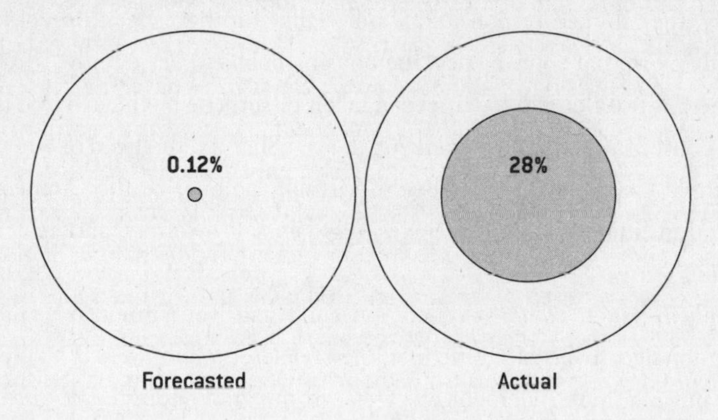

When you make a prediction that goes so badly, you have a choice of how to explain it. One path is to blame external circumstances—what we might think of as "bad luck." Sometimes this is a reasonable choice, or even the correct one. When the National Weather Service says there is a 90 percent chance of clear skies, but it rains instead and spoils your golf outing, you can't really blame them. Decades of historical data show that when the Weather Service says there is a 1 in 10 chance of rain, it really does rain about 10 percent of the time over the long run.*

This explanation becomes less credible, however, when the forecaster does not have a history of successful predictions and when the magnitude of his error

* This property is true of National Weather Service forecasts—but not of your local TV meteorologist. Their incentive may be to exaggerate the probability of rain for the sake of better ratings. More about this in chapter 4.

is larger. In these cases, it is much more likely that the fault lies with the forecaster's model of the world and not with the world itself.

In the instance of CDOs, the ratings agencies had no track record at all: these were new and highly novel securities, and the default rates claimed by S&P were not derived from historical data but instead were assumptions based on a faulty statistical model. Meanwhile, the magnitude of their error was enormous: AAA-rated CDOs were two hundred times more likely to default in practice than they were in theory.

The ratings agencies' shot at redemption would be to admit that the models had been flawed and the mistake had been theirs. But at the congressional hearing, they shirked responsibility and claimed to have been unlucky. They blamed an external contingency: the housing bubble.

"S&P is not alone in having been taken by surprise by the extreme decline in the housing and mortgage markets," Deven Sharma, the head of Standard & Poor's, told Congress that October.[12] "Virtually no one, be they homeowners, financial institutions, rating agencies, regulators or investors, anticipated what is coming."

Nobody saw it coming. When you can't state your innocence, proclaim your ignorance: this is often the first line of defense when there is a failed forecast.[13] But Sharma's statement was a lie, in the grand congressional tradition of "I did not have sexual relations with that woman" and "I have never used steroids."

What is remarkable about the housing bubble is the number of people who did see it coming—and who said so well in advance. Robert Shiller, the Yale economist, had noted its beginnings as early as 2000 in his book *Irrational Exuberance.*[14] Dean Baker, a caustic economist at the Center for Economic and Policy Research, had written about the bubble in August 2002.[15] A correspondent at the *Economist* magazine, normally known for its staid prose, had spoken of the "biggest bubble in history" in June 2005.[16] Paul Krugman, the Nobel Prize–winning economist, wrote of the bubble and its inevitable end in August 2005.[17] "This was baked into the system," Krugman later told me. "The housing crash was not a black swan. The housing crash was the elephant in the room."

Ordinary Americans were also concerned. Google searches on the term "housing bubble" increased roughly tenfold from January 2004 through summer 2005.[18] Interest in the term was heaviest in those states, like California, that had seen the largest run-up in housing prices[19]—and which were about to experience the largest decline. In fact, discussion of the bubble was remarkably widespread. Instances of the two-word phrase "housing bubble" had appeared in just eight news accounts in 2001[20] but jumped to 3,447 references by 2005. The housing bubble was discussed about *ten times per day* in reputable newspapers and periodicals.[21]

And yet, the ratings agencies—whose job it is to measure risk in financial markets—say that they missed it. It should tell you something that they seem to think of this as their *best* line of defense. The problems with their predictions ran very deep.

"I Don't Think They Wanted the Music to Stop"

None of the economists and investors I spoke with for this chapter had a favorable view of the ratings agencies. But they were divided on whether their bad ratings reflected avarice or ignorance—did they know any better?

Jules Kroll is perhaps uniquely qualified to pass judgment on this question: he runs a ratings agency himself. Founded in 2009, Kroll Bond Ratings had just issued its first rating—on a mortgage loan made to the builders of a gigantic shopping center in Arlington, Virginia—when I met him at his office in New York in 2011.

Kroll faults the ratings agencies most of all for their lack of "surveillance." It is an ironic term coming from Kroll, who before getting into the ratings game had become modestly famous (and somewhat immodestly rich) from his original company, Kroll Inc., which acted as a sort of detective agency to patrol corporate fraud. They knew how to sniff out a scam—such as the case of the kidnappers who took a hedge-fund billionaire hostage but foiled themselves by charging a pizza to his credit card.[22] Kroll was sixty-nine years old when I met him, but his bloodhound instincts are keen—and they were triggered when he began to examine what the ratings agencies were doing.

"Surveillance is a term of art in the ratings industry," Kroll told me. "It means keeping investors informed as to what you're seeing. Every month you get a tape* of things like defaults on mortgages, prepayment of mortgages—you get a lot of data. That is the early warning—are things getting better or worse? The world expects you to keep them posted."

The ratings agencies ought to have been just about the first ones to detect problems in the housing market, in other words. They had better information than anyone else: fresh data on whether thousands of borrowers were making their mortgage payments on time. But they did not begin to downgrade large batches of mortgage-backed securities until 2007—at which point the problems had become manifest and foreclosure rates had already doubled.[23]

"These are not stupid people," Kroll told me. "They knew. I don't think they wanted the music to stop."

Kroll Bond Ratings is one of ten registered NRSROs, or nationally recognized statistical rating organizations, firms that are licensed by the Securities and Exchange Commission to rate debt-backed securities. But Moody's, S&P, and Fitch are three of the others, and they have had almost all the market share; S&P and Moody's each rated almost 97 percent of the CDOs that were issued prior to the financial collapse.[24]

One reason that S&P and Moody's enjoyed such a dominant market presence is simply that they had been a part of the club for a long time. They are part of a legal oligopoly; entry into the industry is limited by the government. Meanwhile, a seal of approval from S&P and Moody's is often mandated by the bylaws of large pension funds,[25] about two-thirds of which[26] mention S&P, Moody's, or both by name, requiring that they rate a piece of debt before the pension fund can purchase it.[27]

S&P and Moody's had taken advantage of their select status to build up exceptional profits despite picking résumés out of Wall Street's reject pile.† Moody's[28] revenue from so-called structured-finance ratings increased by more than 800 percent between 1997 and 2007 and came to represent the majority of their ratings business during the bubble years.[29] These products helped Moody's to the highest profit margin of any company in the S&P 500 for five consecutive years during the housing bubble.[30] (In 2010, even after the bubble

---

* *Tape* is also a term of art in the ratings industry, referring to new and fresh data on individual mortgages.

† In 2005, the average Moody's employee made $185,000, compared with the $520,000 received by the average Goldman Sachs employee that same year.

burst and the problems with the ratings agencies had become obvious, Moody's still made a 25 percent profit.[31])

With large profits locked in so long as new CDOs continued to be issued, and no way for investors to verify the accuracy of their ratings until it was too late, the agencies had little incentive to compete on the basis of quality. The CEO of Moody's, Raymond McDaniel, explicitly told his board that ratings quality was the least important factor driving the company's profits.[32]

Instead their equation was simple. The ratings agencies were paid by the issuer of the CDO every time they rated one: the more CDOs, the more profit. A virtually unlimited number of CDOs could be created by combining different types of mortgages—or when that got boring, combining different types of CDOs into derivatives of one another. Rarely did the ratings agencies turn down the opportunity to rate one. A government investigation later uncovered an instant-message exchange between two senior Moody's employees in which one claimed that a security "could be structured by cows" and Moody's would rate it.[33] In some cases, the ratings agencies went further still and abetted debt issuers in manipulating the ratings. In what it claimed was a nod to transparency,[34] S&P provided the issuers with copies of their ratings software. This made it easy for the issuers to determine exactly how many bad mortgages they could add to the pool without seeing its rating decline.[35]

The possibility of a housing bubble, and that it might burst, thus represented a threat to the ratings agencies' gravy train. Human beings have an extraordinary capacity to ignore risks that threaten their livelihood, as though this will make them go away. So perhaps Deven Sharma's claim isn't so implausible—perhaps the ratings agencies really had missed the housing bubble, even if others hadn't.

In fact, however, the ratings agencies quite explicitly considered the possibility that there was a housing bubble. They concluded, remarkably, that it would be no big deal. A memo provided to me by an S&P spokeswoman, Catherine Mathis, detailed how S&P had conducted a simulation in 2005 that anticipated a 20 percent decline in national housing prices over a two-year period—not far from the roughly 30 percent decline in housing prices that actually occurred between 2006 and 2008. The memo concluded that S&P's existing models "captured the risk of a downturn" adequately and that its highly rated securities would "weather a housing downturn without suffering a credit-rating downgrade."[36]

In some ways this is even more troubling than if the ratings agencies had missed the housing bubble entirely. In this book, I'll discuss the danger of "unknown unknowns"—the risks that we are not even aware of. Perhaps the only greater threat is the risks we think we have a handle on, but don't.* In these cases we not only fool ourselves, but our false confidence may be contagious. In the case of the ratings agencies, it helped to infect the entire financial system. "The major difference between a thing that might go wrong and a thing that cannot possibly go wrong is that when a thing that cannot possibly go wrong goes wrong it usually turns out to be impossible to get at or repair," wrote Douglas Adams in *The Hitchhiker's Guide to the Galaxy* series.[37]

But how did the ratings agencies' models, which had all the auspices of scientific precision, do such a poor job of describing reality?

How the Ratings Agencies Got It Wrong

We have to dig a bit deeper to find the source of the problem. The answer requires a little bit of detail about how financial instruments like CDOs are structured, and a little bit about the distinction between *uncertainty* and *risk*.

CDOs are collections of mortgage debt that are broken into different pools, or "tranches," some of which are supposed to be quite risky and others of which are rated as almost completely safe. My friend Anil Kashyap, who teaches a course on the financial crisis to students at the University of Chicago, has come up with a simplified example of a CDO, and I'll use a version of this example here.

Imagine you have a set of five mortgages, each of which you assume has a 5 percent chance of defaulting. You can create a number of bets based on the status of these mortgages, each of which is progressively more risky.

The safest of these bets, what I'll call the Alpha Pool, pays out unless *all five*

---

* These types of risks might fall into the category "unknown knowns."

of the mortgages default. The riskiest, the Epsilon Pool, leaves you on the hook if *any* of the five mortgages defaults. Then there are other steps along the way.

Why might an investor prefer making a bet on the Epsilon Pool to the Alpha Pool? That's easy—because it will be priced more cheaply to account for the greater risk. But say you're a risk-averse investor, such as a pension fund, and that your bylaws prohibit you from investing in poorly rated securities. If you're going to buy anything, it will be the Alpha Pool, which will assuredly be rated AAA.

The Alpha Pool consists of five mortgages, each of which has only a 5 percent chance of defaulting. You lose the bet only if *all five* actually do default. What is the risk of that happening?

Actually, that is not an easy question—and therein lies the problem. The assumptions and approximations you choose will yield profoundly different answers. If you make the wrong assumptions, your model may be extraordinarily wrong.

One assumption is that each mortgage is independent of the others. In this scenario, your risks are well diversified: if a carpenter in Cleveland defaults on his mortgage, this will have no bearing on whether a dentist in Denver does. Under this scenario, the risk of losing your bet would be exceptionally small—the equivalent of rolling snake eyes five times in a row. Specifically, it would be 5 percent taken to the fifth power, which is just one chance in 3,200,000. This supposed miracle of diversification is how the ratings agencies claimed that a group of subprime mortgages that had just a B+ credit rating on average[38]—which would ordinarily imply[39] more than a 20 percent chance of default[40]—had almost no chance of defaulting when pooled together.

The other extreme is to assume that the mortgages, instead of being entirely independent of one another, will all behave exactly alike. That is, either all five mortgages will default or none will. Instead of getting five separate rolls of the dice, you're now staking your bet on the outcome of just one. There's a 5 percent chance that you will roll snake eyes and all the mortgages will default—making your bet *160,000 times riskier* than you had thought originally.[41]

FIGURE 1-2: SIMPLIFIED CDO STRUCTURE

		PROBABILITY OF LOSING BET		
Bet	**Rules**	**If defaults are perfectly uncorrelated**	**If defaults are perfectly correlated**	**Risk multiple**
Alpha Pool	Bet wins unless all 5 mortgages default	0.00003%	5.0%	160,000x
Beta Pool	Bet wins unless 4 of 5 mortgages default	0.003%	5.0%	1,684x
Gamma Pool	Bet wins unless 3 of 5 mortgages default	0.1%	5.0%	44x
Delta Pool	Bet wins unless 2 of 5 mortgages default	2.1%	5.0%	2.3x
Epsilon Pool	Bet wins unless any of 5 mortgages default	20.4%	5.0%	0.2x

Which of these assumptions is more valid will depend on economic conditions. If the economy and the housing market are healthy, the first scenario—the five mortgages have nothing to do with one another—might be a reasonable approximation. Defaults are going to happen from time to time because of unfortunate rolls of the dice: someone gets hit with a huge medical bill, or they lose their job. However, one person's default risk won't have much to do with another's.

But suppose instead that there is some common factor that ties the fate of these homeowners together. For instance: there is a massive housing bubble that has caused home prices to rise by 80 percent without any tangible improvement in the fundamentals. Now you've got trouble: if one borrower defaults, the rest might succumb to the same problems. The risk of losing your bet has increased by orders of magnitude.

The latter scenario was what came into being in the United States beginning in 2007 (we'll conduct a short autopsy on the housing bubble later in this chapter). But it was the former assumption of largely uncorrelated risks that the ratings agencies had bet on. Although the problems with this assumption were understood in the academic literature[42] and by whistle-blowers at the ratings agencies[43] long before the housing bubble burst, the efforts the ratings agencies made to account for it were feeble.

Moody's, for instance, went through a period of making ad hoc adjustments to its model[44] in which it increased the default probability assigned to AAA-rated securities by 50 percent. That might seem like a very prudent attitude: surely a 50 percent buffer will suffice to account for any slack in one's assumptions?

It might have been fine had the potential for error in their forecasts been linear and arithmetic. But leverage, or investments financed by debt, can make the error in a forecast compound many times over, and introduces the potential of highly geometric and nonlinear mistakes. Moody's 50 percent adjustment was like applying sunscreen and claiming it protected you from a nuclear meltdown—wholly inadequate to the scale of the problem. It wasn't just a possibility that their estimates of default risk could be 50 percent too low: they might just as easily have underestimated it by 500 percent or 5,000 percent. In practice, defaults were two hundred times more likely than the ratings agencies claimed, meaning that their model was off by a mere 20,000 percent.

In a broader sense, the ratings agencies' problem was in being unable or uninterested in appreciating the distinction between *risk* and *uncertainty*.

Risk, as first articulated by the economist Frank H. Knight in 1921,[45] is something that you can put a price on. Say that you'll win a poker hand unless your opponent draws to an inside straight: the chances of that happening are *exactly* 1 chance in 11.[46] This is risk. It is not pleasant when you take a "bad beat" in poker, but at least you know the odds of it and can account for it ahead of time. In the long run, you'll make a profit from your opponents making desperate draws with insufficient odds.

Uncertainty, on the other hand, is risk that is hard to measure. You might have some vague awareness of the demons lurking out there. You might even be acutely concerned about them. But you have no real idea how many of them there are or when they might strike. Your back-of-the-envelope estimate might be off by a factor of 100 or by a factor of 1,000; there is no good way to know. This is uncertainty. Risk greases the wheels of a free-market economy; uncertainty grinds them to a halt.

The alchemy that the ratings agencies performed was to spin uncertainty into what looked and felt like risk. They took highly novel securities, subject to an enormous amount of systemic uncertainty, and claimed the ability to quantify just how risky they were. Not only that, but of all possible conclusions, they came to the astounding one that these investments were almost risk-free.

Too many investors mistook these confident conclusions for accurate ones, and too few made backup plans in case things went wrong.

And yet, while the ratings agencies bear substantial responsibility for the financial crisis, they were not alone in making mistakes. The story of the financial crisis as a failure of prediction can be told in three acts.

Act I: The Housing Bubble

An American home has not, historically speaking, been a lucrative investment. In fact, according to an index developed by Robert Shiller and his colleague Karl Case, the market price of an American home has barely increased at all over the long run. After adjusting for inflation, a $10,000 investment made in a home in 1896 would be worth just $10,600 in 1996. The rate of return had been less in a century than the stock market typically produces in a single year.[47]

But if a home was not a profitable investment it had at least been a safe one. Prior to the 2000s, the most significant shift in American housing prices had

FIGURE 1-3: CASE-SHILLER INDEX, PRICES OF U.S. HOMES, 1890–2006
100 = average price in 1890, adjusted for inflation

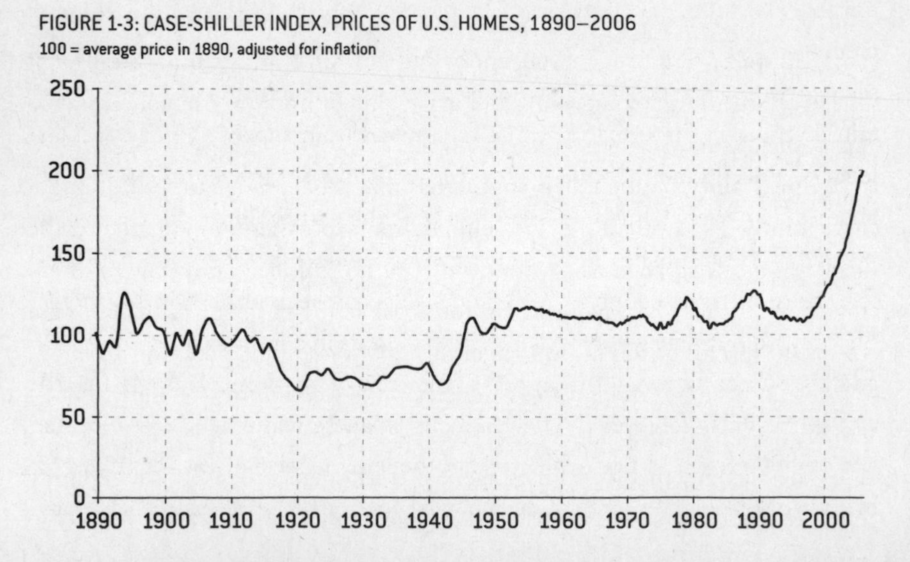

come in the years immediately following World War II, when they increased by about 60 percent relative to their nadir in 1942.

The housing boom of the 1950s, however, had almost nothing in common with the housing bubble of the 2000s. The comparison helps to reveal why the 2000s became such a mess.

The postwar years were associated with a substantial shift in living patterns. Americans had emerged from the war with a glut of savings[48] and into an age of prosperity. There was a great demand for larger living spaces. Between 1940 and 1960, the homeownership rate surged to 62 percent from 44 percent,[49] with most of the growth concentrated in the suburbs.[50] Furthermore, the housing boom was accompanied by the baby boom: the U.S. population was growing at a rate of about 20 percent per decade after the war, about twice its rate of growth during the 2000s. This meant that the number of homeowners increased by about 80 percent during the decade—meeting or exceeding the increase in housing prices.

In the 2000s, by contrast, homeownership rates increased only modestly: to a peak of about 69 percent in 2005 from 65 percent a decade earlier.[51] Few Americans who hadn't already bought a home were in a position to afford one. The 40th percentile of household incomes increased by a nominal 15 percent between 2000 and 2006[52]—not enough to cover inflation, let alone a new home.

Instead, the housing boom had been artificially enhanced—through speculators looking to flip homes and through ever more dubious loans to ever less creditworthy consumers. The 2000s were associated with record-low rates of savings: barely above 1 percent in some years. But a mortgage was easier to obtain than ever.[53] Prices had become untethered from supply and demand, as lenders, brokers, and the ratings agencies—all of whom profited in one way or another from every home sale—strove to keep the party going.

If the United States had never experienced such a housing bubble before, however, other countries had—and results had been uniformly disastrous. Shiller, studying data going back hundreds of years in countries from the Netherlands to Norway, found that as real estate grew to unaffordable levels a crash almost inevitably followed.[54] The infamous Japanese real estate bubble of the early 1990s forms a particularly eerie precedent to the recent U.S. housing bubble, for instance. The price of commercial real estate in Japan increased by about 76 percent over the ten-year period between 1981 and 1991 but then declined by 31 percent over the next five years, a close fit for the trajectory that American home prices took during and after the bubble[55] (figure 1-4).

FIGURE 1-4: JAPANESE COMMERCIAL REAL ESTATE BUBBLE (1981–2001) AND U.S. HOME PRICE BUBBLE (1996–2011)

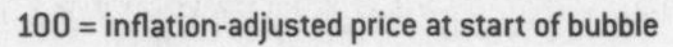
100 = inflation-adjusted price at start of bubble

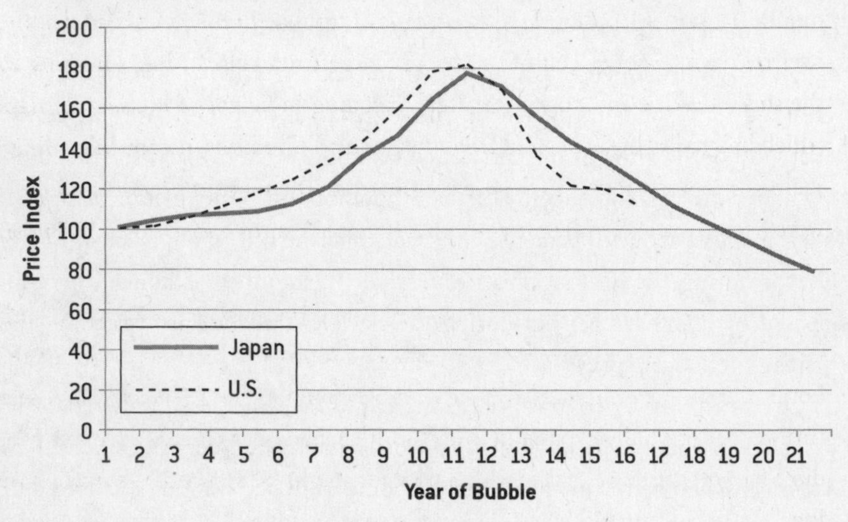

Shiller uncovered another key piece of evidence for the bubble: the people buying the homes had completely unrealistic assumptions about what their investments might return. A survey commissioned by Case and Schiller in 2003 found that homeowners expected their properties to appreciate at a rate of about 13 percent per year.[56] In practice, over that one-hundred-year period from 1896 through 1996[57] to which I referred earlier, sale prices of houses had increased by just 6 percent *total* after inflation, or about 0.06 percent annually.

These homeowners can perhaps be excused for their overconfidence in the housing market. The housing bubble had seeped into the culture to the point where two separate TV programs—one named *Flip This House* and the other named *Flip That House*—were launched within ten days of each other in 2005. Even home buyers who weren't counting on a huge return on investment may have been concerned about keeping up with the Joneses. "I can remember twenty years ago, on the road to Sacramento, there were no traffic jams," I was told by George Akerlof, a frequent colleague of Shiller's, whose office at the

University of California at Berkeley sits at the epicenter of some of the worst declines in housing prices. "Now there tend to be traffic stoppages a good share of the way. That's what people were thinking—if I don't buy now then I'm gonna pay the same price in five years for a house that's ten miles up the road."

Whether homeowners believed that they couldn't lose on a home or couldn't choose to defer the purchase, conditions were growing grimmer by the month. By late 2007 there were clear signs of trouble: home prices had declined over the year in seventeen of the twenty largest markets.[58] More ominous was the sharp decline in housing permits, a leading indicator of housing demand, which had fallen by 50 percent from their peak.[59] Creditors, meanwhile—finally seeing the consequences of their lax standards in the subprime lending market—were becoming less willing to make loans. Foreclosures had doubled by the end of 2007.[60]

Policy makers' first instinct was to reinflate the bubble. Governor Charlie Crist of Florida, one of the worst-hit states, proposed a $10,000 credit for new home buyers.[61] A bill passed by the U.S. Congress in February 2008 went further, substantially expanding the lending capacity of Fannie Mae and Freddie Mac in that hope that more home sales might be spurred.[62] Instead, housing prices continued their inexorable decline, falling a further 20 percent during 2008.

Act II: Leverage, Leverage, Leverage

While quite a few economists identified the housing bubble as it occurred, fewer grasped the consequences of a housing-price collapse for the broader economy. In December 2007, economists in the *Wall Street Journal* forecasting panel predicted only a 38 percent likelihood of a recession over the next year. This was remarkable because, the data would later reveal, the economy was *already in recession* at the time. The economists in another panel, the Survey of Professional Forecasters, thought there was less than a 1 in 500 chance that the economy would crash as badly as it did.[63]

There were two major factors that the economists missed. The first was simply the effect that a drop in housing prices might have on the finances of the average American. As of 2007, middle-class Americans[64] had more than 65 percent of their wealth tied up in their homes.[65] Otherwise they had been getting poorer—they had been using their household equity as ATMs.[66] Nonhousehold wealth—meaning the sum total of things like savings, stocks, pensions, cash, and equity in small businesses—declined by 14 percent[67] for the median family between 2001 and 2007.[68] When the collapse of the housing bubble wiped essentially all their housing equity off the books, middle-class Americans found they were considerably worse off than they had been a few years earlier.

The decline in consumer spending that resulted as consumers came to take a more realistic view of their finances—what economists call a "wealth effect"—is variously estimated at between about 1.5 percent[69] and 3.5 percent[70] of GDP per year, potentially enough to turn average growth into a recession. But a garden-variety recession is one thing. A global financial crisis is another, and the wealth effect does not suffice to explain how the housing bubble triggered one.

In fact, the housing market is a fairly small part of the financial system. In 2007, the total volume of home sales in the United States was about $1.7 trillion—paltry when compared with the $40 trillion in stocks that are traded every year. But in contrast to the activity that was taking place on Main Street, Wall Street was making bets on housing at furious rates. In 2007, the total volume of trades in mortgage-backed securities was about $80 trillion.[71] That meant that for every dollar that someone was willing to put in a mortgage, Wall Street was making almost $50 worth of bets on the side.[72]

Now we have the makings of a financial crisis: home buyers' bets were multiplied fifty times over. The problem can be summed up in a single word: leverage.

If you borrow $20 to wager on the Redskins to beat the Cowboys, that is a leveraged bet.* Likewise, it's leverage when you borrow money to take out a mortgage—or when you borrow money to bet on a mortgage-backed security.

Lehman Brothers, in 2007, had a leverage ratio of about 33 to 1,[73] meaning that it had about $1 in capital for every $33 in financial positions that it held. This meant that if there was just a 3 to 4 percent decline in the value of its

---

* And probably also a stupid bet if you've ever seen the Redskins play.

FIGURE 1-5: HOME SALES VERSUS MORTGAGE-BACKED BETS

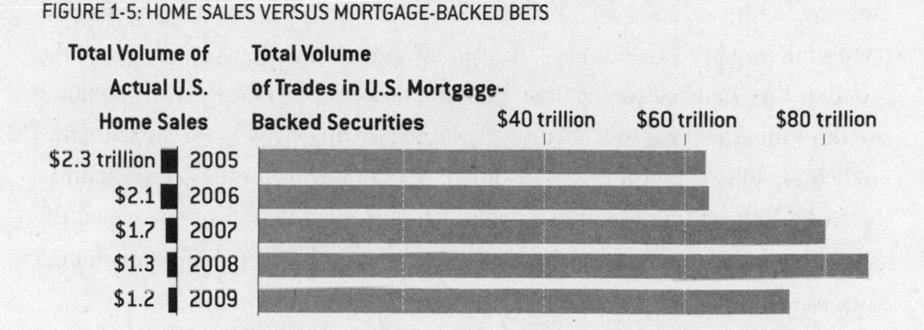

portfolio, Lehman Brothers would have negative equity and would potentially face bankruptcy.[74]

Lehman was not alone in being highly levered: the leverage ratio for other major U.S. banks was about 30 and had been increasing steadily in the run-up to the financial crisis.[75] Although historical data on leverage ratios for U.S. banks is spotty, an analysis by the Bank of England on United Kingdom banks suggests that the overall degree of leverage in the system was either near its historical highs in 2007 or was perhaps altogether unprecedented.[76]

What particularly distinguished Lehman Brothers, however, was its voracious appetite for mortgage-backed securities. The $85 billion it held in mortgage-backed securities in 2007 was about four times more than the underlying value of its capital, meaning that a 25 percent decline in their value would likely be enough to bankrupt the company.[77]

Ordinarily, investors would have been extremely reluctant to purchase assets like these—or at least they would have hedged their bets very carefully.

"If you're in a market and someone's trying to sell you something which you don't understand," George Akerlof told me, "you should think that they're selling you a lemon."

Akerlof wrote a famous paper on this subject called "The Market for Lemons"[78]—it won him a Nobel Prize. In the paper, he demonstrated that in a market plagued by asymmetries of information, the quality of goods will decrease and the market will come to be dominated by crooked sellers and gullible or desperate buyers.

Imagine that a stranger walked up to you on the street and asked if you were interested in buying his used car. He showed you the Blue Book value but was not willing to let you take a test-drive. Wouldn't you be a little suspicious? The core problem in this case is that the stranger knows much more about the car—its repair history, its mileage—than you do. Sensible buyers will avoid transacting in a market like this one at any price. It is a case of *uncertainty* trumping *risk*. You know that you'd need a discount to buy from him—but it's hard to know how much exactly it ought to be. And the lower the man is willing to go on the price, the more convinced you may become that the offer is too good to be true. There may be no such thing as a fair price.

But now imagine that the stranger selling you the car has someone else to vouch for him. Someone who seems credible and trustworthy—a close friend of yours, or someone with whom you have done business previously. Now you might reconsider. This is the role that the ratings agencies played. They vouched for mortgage-backed securities with lots of AAA ratings and helped to enable a market for them that might not otherwise have existed. The market was counting on them to be the Debbie Downer of the mortgage party—but they were acting more like Robert Downey Jr.

Lehman Brothers, in particular, could have used a designated driver. In a conference call in March 2007, Lehman CFO Christopher O'Meara told investors that the recent "hiccup" in the markets did not concern him and that Lehman hoped to do some "bottom fishing" from others who were liquidating their positions prematurely.[79] He explained that the credit quality in the mortgage market was "very strong"—a conclusion that could only have been reached by looking at the AAA ratings for the securities and not at the subprime quality of the collateral. Lehman had bought a lemon.

One year later, as the housing bubble began to burst, Lehman was desperately trying to sell its position. But with the skyrocketing premiums that investors were demanding for credit default swaps—investments that pay you out in the event of a default and which therefore provide the primary means of insurance against one—they were only able to reduce their exposure by about 20 percent.[80] It was too little and too late, and Lehman went bankrupt on September 14, 2008.

Intermission: Fear Is the New Greed

The precise sequence of events that followed the Lehman bankruptcy could fill its own book (and has been described in some excellent ones, like *Too Big to Fail*). It should suffice to remember that when a financial company dies, it can continue to haunt the economy through an afterlife of unmet obligations. If Lehman Brothers was no longer able to pay out on the losing bets that it had made, this meant that somebody else suddenly had a huge hole in his portfolio. Their problems, in turn, might affect yet other companies, with the effects cascading throughout the financial system. Investors and lenders, gawking at the accident but unsure about who owed what to whom, might become unable to distinguish the solvent companies from the zombies and unwilling to lend money at any price, preventing even healthy companies from functioning effectively.

It is for this reason that governments—at great cost to taxpayers as well as to their popularity—sometimes bail out failing financial firms. But the Federal Reserve, which did bail out Bear Stearns and AIG, elected not to do so for Lehman Brothers, defying the expectations of investors and causing the Dow to crash by 500 points when it opened for business the next morning.

Why the government bailed out Bear Stearns and AIG but not Lehman remains unclear. One explanation is that Lehman had been so irresponsible, and its financial position had become so decrepit, that the government wasn't sure what could be accomplished at what price and didn't want to chase good money after bad.[81]

Larry Summers, who was the director of the National Economic Council at the time that I met him in the White House in December 2009,[82] told me that the United States might have had a modestly better outcome had it bailed out Lehman Brothers. But with the excess of leverage in the system, some degree of pain was inevitable.

"It was a self-denying prophecy," Summers told me of the financial crisis. "Everybody leveraged substantially, and when everybody leverages substantially, there's substantial fragility, and their complacency proves to be unwarranted."

"Lehman was a burning cigarette in a very dry forest," he continued a little later. "If that hadn't happened, it's quite likely that something else would have."

Summers thinks of the American economy as consisting of a series of feedback loops. One simple feedback is between supply and demand. Imagine that you are running a lemonade stand.[83] You lower the price of lemonade and sales go up; raise it and they go down. If you're making lots of profit because it's 100 degrees outside and you're the only lemonade stand on the block, the annoying kid across the street opens his own lemonade stand and undercuts your price.

Supply and demand is an example of a **negative feedback**: as prices go up, sales go down. Despite their name, negative feedbacks are a good thing for a market economy. Imagine if the opposite were true and as prices went up, sales went up. You raise the price of lemonade from 25 cents to $2.50—but instead of declining, sales double.[84] Now you raise the price from $2.50 to $25 and they double again. Eventually, you're charging $46,000 for a glass of lemonade—the average income in the United States each year—and all 300 million Americans are lined up around the block to get their fix.

This would be an example of a **positive feedback**. And while it might seem pretty cool at first, you'd soon discover that everyone in the country had gone broke on lemonade. There would be nobody left to manufacture all the video games you were hoping to buy with your profits.

Usually, in Summers's view, negative feedbacks predominate in the American economy, behaving as a sort of thermostat that prevents it from going into recession or becoming overheated. Summers thinks one of the most important feedbacks is between what he calls *fear* and *greed*. Some investors have little appetite for risk and some have plenty, but their preferences balance out: if the price of a stock goes down because a company's financial position deteriorates, the fearful investor sells his shares to a greedy one who is hoping to bottom-feed.

Greed and fear are volatile quantities, however, and the balance can get out of whack. When there is an excess of greed in the system, there is a bubble. When there is an excess of fear, there is a panic.

Ordinarily, we benefit from consulting our friends and neighbors before making a decision. But when their judgment is compromised, this means that ours will be too. People tend to estimate the prices of houses by making comparisons to other houses[85]—if the three-bedroom home in the new subdivision across town is selling for $400,000, the colonial home around the block suddenly looks like steal at $350,000. Under these circumstances, if the price of one house increases, it may make the other houses seem more attractive rather than less.

Or say that you are considering buying another type of asset: a mortgage-backed security. This type of commodity may be even harder to value. But the more investors buy them—and the more the ratings agencies vouch for them—the more confidence you might have that they are safe and worthwhile investments. Hence, you have a positive feedback—and the potential for a bubble.

A negative feedback did eventually rein in the housing market: there weren't any Americans left who could afford homes at their current prices. For that matter, many Americans who had bought homes couldn't really afford them in the first place, and soon their mortgages were underwater. But this was not until trillions of dollars in bets, highly leveraged and impossible to unwind without substantial damage to the economy, had been made on the premise that all the people buying these assets couldn't possibly be wrong.

"We had too much greed and too little fear," Summers told me in 2009. "Now we have too much fear and too little greed."

Act III: This Time Wasn't Different

Once the housing bubble had burst, greedy investors became fearful ones who found uncertainty lurking around every corner. The process of disentangling a financial crisis—everyone trying to figure out who owes what to whom—can produce hangovers that persist for a very long time. The economists Carmen Reinhart and Kenneth Rogoff, studying volumes of financial history for their book *This Time Is Different: Eight Centuries of Financial Folly*, found that financial crises typically produce rises in unemployment that persist for four to six years.[86] Another study by Reinhart, which focused on more recent financial crises, found that ten of the last fifteen countries to endure one had *never* seen their unemployment rates recover to their precrisis levels.[87] This stands in contrast to normal recessions, in which there is typically above-average growth in the year or so following the recession[88] as the economy reverts to the mean, allowing employment to catch up quickly. Yet despite its importance, many economic models made no distinction between the financial system and other parts of the economy.

Reinhart and Rogoff's history lesson was one that the White House might have done more to heed. Soon, they would be responsible for their own notoriously bad prediction.

In January 2009, as Barack Obama was about to take the oath of office, the White House's incoming economic team—led by Summers and Christina Romer, the chair of the Council of Economic Advisers—were charged with preparing the blueprint for a massive stimulus package that was supposed to make up for the lack of demand among consumers and businesses. Romer thought that $1.2 trillion in stimulus was called for.[89] Eventually, the figure was revised downward to about $800 billion after objections from the White House's political team that a trillion-dollar price would be difficult to sell to Congress.

To help pitch the Congress and the country on the stimulus, Romer and her colleagues prepared a memo[90] outlining the depth of the crisis and what the stimulus might do to ameliorate it. The memo prominently featured a graphic predicting how the unemployment rate would track with and without the stimulus. Without the stimulus, the memo said, the unemployment rate, which had been 7.3 percent when last reported in December 2008, would peak at about 9 percent in early 2010. But with the stimulus, employment would never rise above 8 percent and would begin to turn downward as early as July 2009.

Congress passed the stimulus on a party-line vote in February 2009. But unemployment continued to rise—to 9.5 percent in July and then to a peak of 10.1 percent in October 2009. This was much worse than the White House had projected even under the "no stimulus" scenario. Conservative bloggers cheekily updated Romer's graphic every month—but with the actual unemployment rate superimposed on the too-cheery projections (figure 1-6).

People see this graphic now and come to different—and indeed entirely opposite—conclusions about it. Paul Krugman, who had argued from the start that the stimulus was too small,[91] sees it as proof that the White House had dramatically underestimated the shortfall in demand. "The fact that unemployment didn't come down much in the wake of this particular stimulus means that we knew we were facing one hell of a shock from the financial crisis," he

FIGURE 1-6: WHITE HOUSE ECONOMIC PROJECTIONS, JANUARY 2009

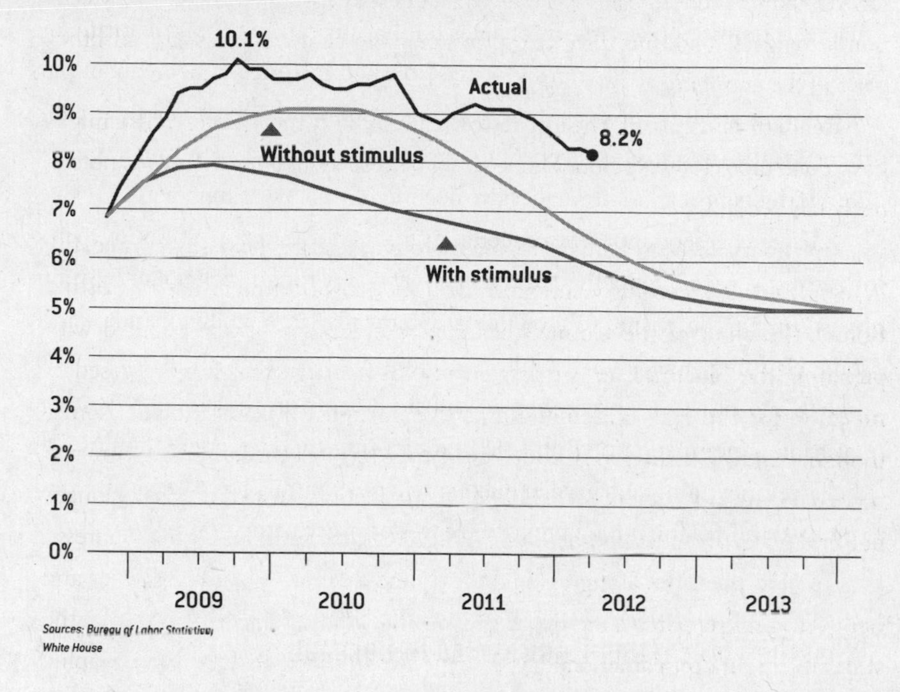

told me. Other economists, of course, take the graph as evidence that the stimulus had completely failed.[92]

The White House can offer its version of S&P's "everyone else made the same mistake" defense. Its forecasts were largely in line with those issued by independent economists at the time.[93] Meanwhile, the initial economic statistics had significantly underestimated the magnitude of the crisis.[94] The government's first estimate—the one available to Romer and Summers at the time the stimulus was being sold—was that GDP had declined at a rate of 3.8 percent in the fall of 2008.[95] In fact, the financial crisis had taken more than twice as large a bite out of the economy. The actual rate of GDP decline had been closer to 9 percent,[96] meaning that the country was about $200 billion poorer than the government first estimated.

Perhaps the White House's more inexcusable error was in making such a precise-seeming forecast—and in failing to prepare the public for the eventuality that it might be wrong. No economist, whether in the White House or elsewhere, has been able to predict the progress of major economic indicators like the unemployment rate with much success. (I take a more detailed look at macroeconomic forecasting in chapter 6.) The uncertainty in an unemployment rate forecast[97] made during a recession had historically been about plus or minus 2 percent.[98] So even if the White House thought 8 percent unemployment was the most likely outcome, it might easily enough have wound up in the double digits instead (or it might have declined to as low as 6 percent).

There is also considerable uncertainty about how effective stimulus spending really is. Estimates of the multiplier effect—how much each dollar in stimulus spending contributes to growth—vary radically from study to study,[99] with some claiming that $1 in stimulus spending returns as much as $4 in GDP growth and others saying the return is just 60 cents on the dollar. When you layer the large uncertainty intrinsic to measuring the effects of stimulus atop the large uncertainty intrinsic to making macroeconomic forecasts of any kind, you have the potential for a prediction that goes very badly.

What the Forecasting Failures Had in Common

There were at least four major failures of prediction that accompanied the financial crisis.

- The housing bubble can be thought of as a poor prediction. Homeowners and investors thought that rising prices implied that home values would continue to rise, when in fact history suggested this made them prone to decline.
- There was a failure on the part of the ratings agencies, as well as by banks like Lehman Brothers, to understand how risky mortgage-backed securities were. Contrary to the assertions they made before Congress, the problem was not that the ratings agencies failed to see the housing bubble. Instead, their forecasting models were full of faulty assumptions and false confidence about the risk that a collapse in housing prices might present.
- There was a widespread failure to anticipate how a housing crisis could trigger a global financial crisis. It had resulted from the high degree of leverage in the market, with $50 in side bets staked on every $1 that an American was willing to invest in a new home.

- Finally, in the immediate aftermath of the financial crisis, there was a failure to predict the scope of the economic problems that it might create. Economists and policy makers did not heed Reinhart and Rogoff's finding that financial crises typically produce very deep and long-lasting recessions.

There is a common thread among these failures of prediction. In each case, as people evaluated the data, they ignored a key piece of context:

- The confidence that homeowners had about housing prices may have stemmed from the fact that there had not been a substantial decline in U.S. housing prices in the recent past. *However,* there had never before been such a widespread increase in U.S. housing prices like the one that preceded the collapse.
- The confidence that the banks had in Moody's and S&P's ability to rate mortgage-backed securities may have been based on the fact that the agencies had generally performed competently in rating other types of financial assets. *However,* the ratings agencies had never before rated securities as novel and complex as credit default options.
- The confidence that economists had in the ability of the financial system to withstand a housing crisis may have arisen because housing price fluctuations had generally not had large effects on the financial system in the past. *However,* the financial system had probably never been so highly leveraged, and it had certainly never made so many side bets on housing before.
- The confidence that policy makers had in the ability of the economy to recuperate quickly from the financial crisis may have come from their experience of recent recessions, most of which had been associated with rapid, "V-shaped" recoveries. *However,* those recessions had not been associated with financial crises, and financial crises are different.

There is a technical term for this type of problem: the events these forecasters were considering were *out of sample.* When there is a major failure of prediction, this problem usually has its fingerprints all over the crime scene.

What does the term mean? A simple example should help to explain it.

Out of Sample, Out of Mind: A Formula for a Failed Prediction

Suppose that you're a very good driver. Almost everyone thinks they're a good driver,[100] but you really have the track record to prove it: just two minor fender benders in thirty years behind the wheel, during which time you have made 20,000 car trips.

You're also not much of a drinker, and one of the things you've absolutely never done is driven drunk. But one year you get a little carried away at your office Christmas party. A good friend of yours is leaving the company, and you've been under a lot of stress: one vodka tonic turns into about twelve. You're blitzed, three sheets to the wind. Should you drive home or call a cab?

That sure seems like an easy question to answer: take the taxi. And cancel your morning meeting.

But you could construct a facetious argument for driving yourself home that went like this: out of a sample of 20,000 car trips, you'd gotten into just two minor accidents, and gotten to your destination safely the other 19,998 times. Those seem like pretty favorable odds. Why go through the inconvenience of calling a cab in the face of such overwhelming evidence?

The problem, of course, is that of those 20,000 car trips, none occurred when you were anywhere near this drunk. Your sample size for drunk driving is not 20,000 trips but zero, and you have no way to use your past experience to forecast your accident risk. This is an example of an out-of-sample problem.

As easy as it might seem to avoid this sort of problem, the ratings agencies made just this mistake. Moody's estimated the extent to which mortgage defaults were correlated with one another by building a model from past data—specifically, they looked at American housing data going back to about the 1980s.[101] The problem is that from the 1980s through the mid-2000s, home prices were always steady or increasing in the United States. Under these circumstances, the assumption that one homeowner's mortgage has little relationship to another's was probably good enough. But nothing in that past data would have described what happened when home prices began to decline in tandem. The housing collapse was an out-of-sample event, and their models were worthless for evaluating default risk under those conditions.

The Mistakes That Were Made—and What We Can Learn from Them

Moody's was not completely helpless, however. They could have come to some more plausible estimates by expanding their horizons. The United States had never experienced such a housing crash before—but other countries had, and the results had been ugly. Perhaps if Moody's had looked at default rates after the Japanese real estate bubble, they could have had some more realistic idea about the precariousness of mortgage-backed securities—and they would not have stamped their AAA rating on them.

But forecasters often resist considering these out-of-sample problems. When we expand our sample to include events further apart from us in time and space, it often means that we will encounter cases in which the relationships we are studying did *not* hold up as well as we are accustomed to. The model will seem to be less powerful. It will look less impressive in a PowerPoint presentation (or a journal article or a blog post). We will be forced to acknowledge that we know less about the world than we thought we did. Our personal and professional incentives almost always discourage us from doing this.

We forget—or we willfully ignore—that our models are simplifications of the world. We figure that if we make a mistake, it will be at the margin.

In complex systems, however, mistakes are not measured in degrees but in whole orders of magnitude. S&P and Moody's underestimated the default risk associated with CDOs by a factor of two hundred. Economists thought there was just a 1 in 500 chance of a recession as severe as what actually occurred.

One of the pervasive risks that we face in the information age, as I wrote in the introduction, is that even if the amount of knowledge in the world is increasing, the gap between what we know and what we think we know may be widening. This syndrome is often associated with very precise-seeming predictions that are not at all accurate. Moody's carried out their calculations to the second decimal place—but they were utterly divorced from reality. This is like claiming you are a good shot because your bullets always end up in about the same place—even though they are nowhere near the target (figure 1-7).

FIGURE 1-7: ACCURACY VERSUS PRECISION

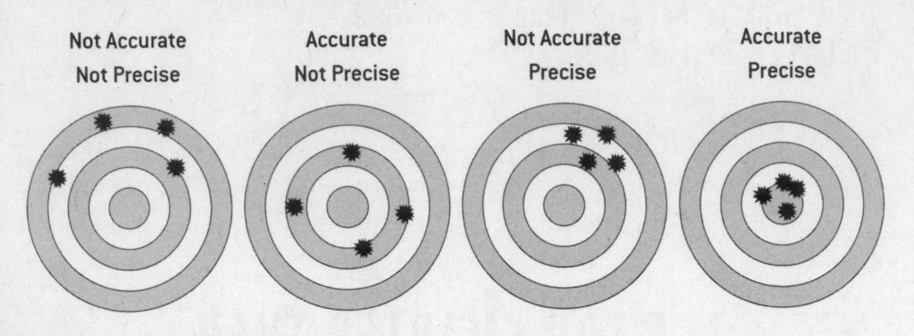

Financial crises—and most other failures of prediction—stem from this false sense of confidence. Precise forecasts masquerade as accurate ones, and some of us get fooled and double-down our bets. It's exactly when we think we have overcome the flaws in our judgment that something as powerful as the American economy can be brought to a screeching halt.

2

ARE YOU SMARTER THAN A TELEVISION PUNDIT?

For many people, political prediction is synonymous with the television program *The McLaughlin Group,* a political roundtable that has been broadcast continually each Sunday since 1982 and parodied by *Saturday Night Live* for nearly as long. The show, hosted by John McLaughlin, a cantankerous octogenarian who ran a failed bid for the United States Senate in 1970, treats political punditry as sport, cycling through four or five subjects in the half hour, with McLaughlin barking at his panelists for answers on subjects from Australian politics to the prospects for extraterrestrial intelligence.

At the end of each edition of *The McLaughlin Group,* the program has a final segment called "Predictions," in which the panelists are given a few seconds to weigh in on some matter of the day. Sometimes, the panelists are permitted to pick a topic and make a prediction about anything even vaguely related to politics. At other times, McLaughlin calls for a "forced prediction," a sort of pop quiz that asks them their take on a specific issue.

Some of McLaughlin's questions—say, to name the next Supreme Court nominee from among several plausible candidates—are difficult to answer. But others are softballs. On the weekend before the 2008 presidential election, for instance, McLaughlin asked his panelists whether John McCain or Barack Obama was going to win.[1]

That one ought not to have required very much thought. Barack Obama had led John McCain in almost every national poll since September 15, 2008, when the collapse of Lehman Brothers had ushered in the worst economic slump since the Great Depression. Obama also led in almost every poll of almost every swing state: in Ohio and Florida and Pennsylvania and New Hampshire—and even in a few states that Democrats don't normally win, like Colorado and Virginia. Statistical models like the one I developed for FiveThirtyEight suggested that Obama had in excess of a 95 percent chance of winning the election. Betting markets were slightly more equivocal, but still had him as a 7 to 1 favorite.[2]

But McLaughlin's first panelist, Pat Buchanan, dodged the question. "The undecideds will decide this weekend," he remarked, drawing guffaws from the rest of the panel. Another guest, the *Chicago Tribune*'s Clarence Page, said the election was "too close to call." Fox News' Monica Crowley was bolder, predicting a McCain win by "half a point." Only *Newsweek*'s Eleanor Clift stated the obvious, predicting a win for the Obama-Biden ticket.

The following Tuesday, Obama became the president-elect with 365 electoral votes to John McCain's 173—almost exactly as polls and statistical models had anticipated. While not a landslide of historic proportions, it certainly hadn't been "too close to call": Obama had beaten John McCain by nearly ten million votes. Anyone who had rendered a prediction to the contrary had some explaining to do.

There would be none of that on *The McLaughlin Group* when the same four panelists gathered again the following week.[3] The panel discussed the statistical minutiae of Obama's win, his selection of Rahm Emanuel as his chief of staff, and his relations with Russian president Dmitry Medvedev. There was no mention of the failed prediction—made on national television in contradiction to essentially all available evidence. In fact, the panelists made it sound as though the outcome had been inevitable all along; Crowley explained that it had been a "change election year" and that McCain had run a terrible campaign—neglecting to mention that she had been willing to bet on that campaign just a week earlier.

Rarely should a forecaster be judged on the basis of a single prediction—but this case may warrant an exception. By the weekend before the election, perhaps the only plausible hypothesis to explain why McCain could still win was if there was massive racial animus against Obama that had gone undetected in the polls.[4] None of the panelists offered this hypothesis, however. Instead they seemed to be operating in an alternate universe in which the polls didn't exist, the economy hadn't collapsed, and President Bush was still reasonably popular rather than dragging down McCain.

Nevertheless, I decided to check to see whether this was some sort of anomaly. Do the panelists on *The McLaughlin Group*—who are paid to talk about politics for a living—have any real skill at forecasting?

I evaluated nearly 1,000 predictions that were made on the final segment of the show by McLaughlin and the rest of the panelists. About a quarter of the predictions were too vague to be analyzed or concerned events in the far future. But I scored the others on a five-point scale ranging from completely false to completely true.

The panel may as well have been flipping coins. I determined 338 of their predictions to be either mostly or completely false. The exact same number—338—were either mostly or completely true.[5]

FIGURE 2-1: *MCLAUGHLIN GROUP* PREDICTIONS ANALYSIS

Completely true	285	39%
Mostly true	53	7%
Partly true, partly false	57	8%
Mostly false	70	10%
Completely false	268	37%
Total predictions evaluated	**733**	**100%**
Predictions not evaluated[6]	249	

Nor were any of the panelists—including Clift, who at least got the 2008 election right—much better than the others. For each panelist, I calculated a percentage score, essentially reflecting the number of predictions they got right. Clift and the three other most frequent panelists—Buchanan, the late Tony Blankley, and McLaughlin himself—each received almost identical scores ranging from 49 percent to 52 percent, meaning that they were about as likely to get a prediction right as wrong.[7] They displayed about as much political acumen as a barbershop quartet.

The McLaughlin Group, of course, is more or less explicitly intended as slapstick entertainment for political junkies. It is a holdover from the shouting match era of programs, such as CNN's *Crossfire*, that featured liberals and conservatives endlessly bickering with one another. Our current echo chamber era isn't much different from the shouting match era, except that the liberals and conservatives are confined to their own channels, separated in your cable lineup by a demilitarized zone demarcated by the Food Network or the Golf Channel.* This arrangement seems to produce higher ratings if not necessarily more reliable analysis.

But what about those who are paid for the accuracy and thoroughness of their scholarship—rather than the volume of their opinions? Are political scientists, or analysts at Washington think tanks, any better at making predictions?

Are Political Scientists Better Than Pundits?

The disintegration of the Soviet Union and other countries of the Eastern bloc occurred at a remarkably fast pace—and all things considered, in a remarkably orderly way.†

On June 12, 1987, Ronald Reagan stood at the Brandenburg Gate and implored Mikhail Gorbachev to tear down the Berlin Wall—an applause line that seemed as audacious as John F. Kennedy's pledge to send a man to the moon. Reagan was prescient; less than two years later, the wall had fallen.

On November 16, 1988, the parliament of the Republic of Estonia, a nation about the size of the state of Maine, declared its independence from the

---

* Most major cable providers keep Fox News and MSNBC at least a couple of channels apart in their lineups.

† Of the series of revolutions in the Eastern bloc in 1989, only the one in Romania entailed substantial bloodshed.

mighty USSR. Less than three years later, Gorbachev parried a coup attempt from hard-liners in Moscow and the Soviet flag was lowered for the last time before the Kremlin; Estonia and the other Soviet Republics would soon become independent nations.

If the fall of the Soviet empire seemed predictable after the fact, however, almost no mainstream political scientist had seen it coming. The few exceptions were often the subject of ridicule.[8] If political scientists couldn't predict the downfall of the Soviet Union—perhaps the most important event in the latter half of the twentieth century—then what exactly were they good for?

Philip Tetlock, a professor of psychology and political science, then at the University of California at Berkeley,[9] was asking some of the same questions. As it happened, he had undertaken an ambitious and unprecedented experiment at the time of the USSR's collapse. Beginning in 1987, Tetlock started collecting predictions from a broad array of experts in academia and government on a variety of topics in domestic politics, economics, and international relations.[10]

Political experts had difficulty anticipating the USSR's collapse, Tetlock found, because a prediction that not only forecast the regime's demise but also understood the reasons for it required different strands of argument to be woven together. There was nothing inherently contradictory about these ideas, but they tended to emanate from people on different sides of the political spectrum,[11] and scholars firmly entrenched in one ideological camp were unlikely to have embraced them both.

On the one hand, Gorbachev was clearly a major part of the story—his desire for reform had been sincere. Had Gorbachev chosen to become an accountant or a poet instead of entering politics, the Soviet Union might have survived at least a few years longer. Liberals were more likely to hold this sympathetic view of Gorbachev. Conservatives were less trusting of him, and some regarded his talk of glasnost as little more than posturing.

Conservatives, on the other hand, were more instinctually critical of communism. They were quicker to understand that the USSR's economy was failing and that life was becoming increasingly difficult for the average citizen. As late as 1990, the CIA estimated—quite wrongly[12]—that the Soviet Union's GDP was about half that of the United States[13] (on a per capita basis, tantamount to where stable democracies like South Korea and Portugal are today). In fact, more recent evidence has found that the Soviet economy—weakened by its long war with Afghanistan and the central government's inattention to a variety of social problems—was roughly $1 trillion poorer than the CIA had thought and was shrinking by as much as 5 percent annually, with inflation well into the double digits.

Take these two factors together, and the Soviet Union's collapse is fairly easy to envision. By opening the country's media and its markets and giving his citizens greater democratic authority, Gorbachev had provided his people with the mechanism to catalyze a regime change. And because of the dilapidated state of the country's economy, they were happy to take him up on his offer. The center was too weak to hold: not only were Estonians sick of Russians, but Russians were nearly as sick of Estonians, since the satellite republics contributed less to the Soviet economy than they received in subsidies from Moscow.[14] Once the dominoes began falling in Eastern Europe—Czechoslovakia, Poland, Romania, Bulgaria, Hungary, and East Germany were all in the midst of revolution by the end of 1989—there was little Gorbachev or anyone else could do to prevent them from caving the country in. A lot of Soviet scholars understood parts of the problem, but few experts had put all the puzzle pieces together, and almost no one had forecast the USSR's sudden collapse.

Tetlock, inspired by the example of the Soviet Union, began to take surveys of expert opinion in other areas—asking the experts to make predictions about the Gulf War, the Japanese real-estate bubble, the potential secession of Quebec from Canada, and almost every other major event of the 1980s and 1990s. Was the failure to predict the collapse of the Soviet Union an anomaly, or does "expert" political analysis rarely live up to its billing? His studies, which spanned more than fifteen years, were eventually published in the 2005 book *Expert Political Judgment*.

Tetlock's conclusion was damning. The experts in his survey—regardless of their occupation, experience, or subfield—had done barely any better than random chance, and they had done worse than even rudimentary statistical methods at predicting future political events. They were grossly overconfident and terrible at calculating probabilities: about 15 percent of events that they claimed had *no chance* of occurring in fact happened, while about 25 percent of those that they said were *absolutely sure things* in fact failed to occur.[15] It didn't matter whether the experts were making predictions about economics, domestic politics, or international affairs; their judgment was equally bad across the board.

The Right Attitude for Making Better Predictions: Be Foxy

While the experts' performance was poor in the aggregate, however, Tetlock found that some had done better than others. On the losing side were those experts whose predictions were cited most frequently in the media. The more interviews that an expert had done with the press, Tetlock found, the *worse* his predictions tended to be.

Another subgroup of experts had done relatively well, however. Tetlock, with his training as a psychologist, had been interested in the experts' cognitive styles—how they thought about the world. So he administered some questions lifted from personality tests to all the experts.

On the basis of their responses to these questions, Tetlock was able to classify his experts along a spectrum between what he called *hedgehogs* and *foxes*. The reference to hedgehogs and foxes comes from the title of an Isaiah Berlin essay on the Russian novelist Leo Tolstoy—*The Hedgehog and the Fox*. Berlin had in turn borrowed his title from a passage attributed to the Greek poet Archilochus: "The fox knows many little things, but the hedgehog knows one big thing."

Unless you are a fan of Tolstoy—or of flowery prose—you'll have no particular reason to read Berlin's essay. But the basic idea is that writers and thinkers can be divided into two broad categories:

- **Hedgehogs** are type A personalities who believe in Big Ideas—in governing principles about the world that behave as though they were physical laws and undergird virtually every interaction in society. Think Karl Marx and class struggle, or Sigmund Freud and the unconscious. Or Malcolm Gladwell and the "tipping point."
- **Foxes**, on the other hand, are scrappy creatures who believe in a plethora of little ideas and in taking a multitude of approaches toward a problem. They

tend to be more tolerant of nuance, uncertainty, complexity, and dissenting opinion. If hedgehogs are hunters, always looking out for the big kill, then foxes are gatherers.

Foxes, Tetlock found, are considerably better at forecasting than hedgehogs. They had come closer to the mark on the Soviet Union, for instance. Rather than seeing the USSR in highly ideological terms—as an intrinsically "evil empire," or as a relatively successful (and perhaps even admirable) example of a Marxist economic system—they instead saw it for what it was: an increasingly dysfunctional nation that was in danger of coming apart at the seams. Whereas the hedgehogs' forecasts were barely any better than random chance, the foxes' demonstrated predictive skill.

FIGURE 2-2: ATTITUDES OF FOXES AND HEDGEHOGS

How Foxes Think	How Hedgehogs Think
Multidisciplinary: Incorporate ideas from different disciplines and regardless of their origin on the political spectrum.	**Specialized:** Often have spent the bulk of their careers on one or two great problems. May view the opinions of "outsiders" skeptically.
Adaptable: Find a new approach—or pursue multiple approaches at the same time—if they aren't sure the original one is working.	**Stalwart:** Stick to the same "all-in" approach—new data is used to refine the original model.
Self-critical: Sometimes willing (if rarely happy) to acknowledge mistakes in their predictions and accept the blame for them.	**Stubborn:** Mistakes are blamed on bad luck or on idiosyncratic circumstances—a good model had a bad day.
Tolerant of complexity: See the universe as complicated, perhaps to the point of many fundamental problems being irresolvable or inherently unpredictable.	**Order-seeking:** Expect that the world will be found to abide by relatively simple governing relationships once the signal is identified through the noise.
Cautious: Express their predictions in probabilistic terms and qualify their opinions.	**Confident:** Rarely hedge their predictions and are reluctant to change them.
Empirical: Rely more on observation than theory.	**Ideological:** Expect that solutions to many day-to-day problems are manifestations of some grander theory or struggle.
Foxes are better forecasters.	**Hedgehogs are weaker forecasters.**

Why Hedgehogs Make Better Television Guests

I met Tetlock for lunch one winter afternoon at the Hotel Durant, a stately and sunlit property just off the Berkeley campus. Naturally enough, Tetlock revealed himself to be a fox: soft-spoken and studious, with a habit of pausing for twenty or thirty seconds before answering my questions (lest he provide me with too incautiously considered a response).

"What are the incentives for a public intellectual?" Tetlock asked me. "There are some academics who are quite content to be relatively anonymous. But there are other people who aspire to be public intellectuals, to be pretty bold and to attach nonnegligible probabilities to fairly dramatic change. That's much more likely to bring you attention."

Big, bold, hedgehog-like predictions, in other words, are more likely to get you on television. Consider the case of Dick Morris, a former adviser to Bill Clinton who now serves as a commentator for Fox News. Morris is a classic hedgehog, and his strategy seems to be to make as dramatic a prediction as possible when given the chance. In 2005, Morris proclaimed that George W. Bush's handling of Hurricane Katrina would help Bush to regain his standing with the public.[16] On the eve of the 2008 elections, he predicted that Barack Obama would win Tennessee and Arkansas.[17] In 2010, Morris predicted that the Republicans could easily win one hundred seats in the U.S. House of Representatives.[18] In 2011, he said that Donald Trump would run for the Republican nomination—and had a "damn good" chance of winning it.[19]

All those predictions turned out to be horribly wrong. Katrina was the beginning of the end for Bush—not the start of a rebound. Obama lost Tennessee and Arkansas badly—in fact, they were among the only states in which he performed worse than John Kerry had four years earlier. Republicans had a good night in November 2010, but they gained sixty-three seats, not one hundred. Trump officially declined to run for president just two weeks after Morris insisted he would do so.

But Morris is quick on his feet, entertaining, and successful at marketing himself—he remains in the regular rotation at Fox News and has sold his books to hundreds of thousands of people.

Foxes sometimes have more trouble fitting into type A cultures like television, business, and politics. Their belief that many problems are hard to forecast—and that we should be explicit about accounting for these uncertainties—may be mistaken for a lack of self-confidence. Their pluralistic approach may be mistaken for a lack of conviction; Harry Truman famously demanded a "one-handed economist," frustrated that the foxes in his administration couldn't give him an unqualified answer.

But foxes happen to make much better predictions. They are quicker to recognize how noisy the data can be, and they are less inclined to chase false signals. They know more about what they don't know.

If you're looking for a doctor to predict the course of a medical condition or an investment adviser to maximize the return on your retirement savings, you may want to entrust a fox. She might make more modest claims about what she is able to achieve—but she is much more likely to actually realize them.

Why Political Predictions Tend to Fail

Fox-like attitudes may be especially important when it comes to making predictions about politics. There are some particular traps that can make suckers of hedgehogs in the arena of political prediction and which foxes are more careful to avoid.

One of these is simply partisan ideology. Morris, despite having advised Bill Clinton, identifies as a Republican and raises funds for their candidates—and his conservative views fit in with those of his network, Fox News. But liberals are not immune from the propensity to be hedgehogs. In my study of the accuracy of predictions made by *McLaughlin Group* members, Eleanor Clift—who is usually the most liberal member of the panel—almost never issued a prediction that would imply a more favorable outcome for Republicans than the consensus of the group. That may have served her well in predicting the outcome of the 2008 election, but she was no more accurate than her conservative counterparts over the long run.

Academic experts like the ones that Tetlock studied can suffer from the same problem. In fact, a little knowledge may be a dangerous thing in the hands of a hedgehog with a Ph.D. One of Tetlock's more remarkable findings is that, while foxes tend to get better at forecasting with experience, the opposite is true of hedgehogs: their performance tends to *worsen* as they pick up additional credentials. Tetlock believes the more facts hedgehogs have at their command, the more opportunities they have to permute and manipulate them in ways that confirm their biases. The situation is analogous to what might happen if you put a hypochondriac in a dark room with an Internet connection. The more time that you give him, the more information he has at his disposal, the more ridiculous the self-diagnosis he'll come up with; before long he'll be mistaking a common cold for the bubonic plague.

But while Tetlock found that left-wing and right-wing hedgehogs made especially poor predictions, he also found that foxes of all political persuasions were more immune from these effects.[20] Foxes may have emphatic convictions about the way the world *ought* to be. But they can usually separate that from their analysis of the way that the world actually is and how it is likely to be in the near future.

Hedgehogs, by contrast, have more trouble distinguishing their rooting interest from their analysis. Instead, in Tetlock's words, they create "a blurry fusion between facts and values all lumped together." They take a prejudicial view toward the evidence, seeing what they want to see and not what is really there.

You can apply Tetlock's test to diagnose whether you are a hedgehog: Do your predictions improve when you have access to more information? In theory, more information should give your predictions a wind at their back—you can always ignore the information if it doesn't seem to be helpful. But hedgehogs often trap themselves in the briar patch.

Consider the case of the *National Journal* Political Insiders' Poll, a survey of roughly 180 politicians, political consultants, pollsters, and pundits. The survey is divided between Democratic and Republican partisans, but both groups are asked the same questions. Regardless of their political persuasions, this group leans hedgehog: political operatives are proud of their battle scars, and see themselves as locked in a perpetual struggle against the other side of the cocktail party.

A few days ahead of the 2010 midterm elections, *National Journal* asked its panelists whether Democrats were likely to retain control of both the House and the Senate.[21] There was near-universal agreement on these questions: Democrats would keep the Senate but Republicans would take control of the House (the panel was right on both accounts). Both the Democratic and Republican insiders were also almost agreed on the overall magnitude of Republican gains in the House; the Democratic experts called for them to pick up 47 seats, while Republicans predicted a 53-seat gain—a trivial difference considering that there are 435 House seats.

National Journal, however, also asked its panelists to predict the outcome of eleven individual elections, a mix of Senate, House, and gubernatorial races. Here, the differences were much greater. The panel split on the winners they expected in the Senate races in Nevada, Illinois, and Pennsylvania, the governor's race in Florida, and a key House race in Iowa. Overall, Republican panelists expected Democrats to win just one of the eleven races, while Democratic panelists expected them to win 6 of the 11. (The actual outcome, predictably enough, was somewhere in the middle—Democrats won three of the eleven races that *National Journal* had asked about.[22])

Obviously, partisanship plays some role here: Democrats and Republicans were each rooting for the home team. That does not suffice to explain, however, the unusual divide in the way that the panel answered the different types of questions. When asked in *general* terms about how well Republicans were likely to do, there was almost no difference between the panelists. They differed profoundly, however, when asked about specific cases—these brought the partisan differences to the surface.[23]

Too much information can be a bad thing in the hands of a hedgehog. The question of how many seats Republicans were likely to gain on Democrats overall is an abstract one: unless you'd studied all 435 races, there was little additional detail that could help you to resolve it. By contrast, when asked about any one particular race—say, the Senate race in Nevada—the panelists had all kinds of information at their disposal: not just the polls there, but also news accounts they'd read about the race, gossip they'd heard from their friends, or what they thought about the candidates when they saw them on television. They might even know the candidates or the people who work for them personally.

Hedgehogs who have lots of information construct stories—stories that are neater and tidier than the real world, with protagonists and villains, winners and losers, climaxes and dénouements—and, usually, a happy ending for the home team. The candidate who is down ten points in the polls is going to win, goddamnit, because I know the candidate and I know the voters in her state, and maybe I heard something from her press secretary about how the polls are tightening—and have you seen her latest commercial?

When we construct these stories, we can lose the ability to think about the evidence critically. Elections typically present compelling narratives. Whatever you thought about the politics of Barack Obama or Sarah Palin or John McCain or Hillary Clinton in 2008, they had persuasive life stories: reported books on the campaign, like *Game Change*, read like tightly bestselling novels. The candidates who ran in 2012 were a less appealing lot but still more than sufficed to provide for the usual ensemble of dramatic clichés from tragedy (Herman Cain?) to farce (Rick Perry).

You can get lost in the narrative. Politics may be especially susceptible to poor predictions precisely because of its human elements: a good election engages our dramatic sensibilities. This does not mean that you must feel totally dispassionate about a political event in order to make a good prediction about it. But it does mean that a fox's aloof attitude can pay dividends.

A Fox-Like Approach to Forecasting

I had the idea for FiveThirtyEight* while waiting out a delayed flight at Louis Armstrong New Orleans International Airport in February 2008. For some reason—possibly the Cajun martinis had stirred something up—it suddenly seemed obvious that someone needed to build a Web site that predicted how well Hillary Clinton and Barack Obama, then still in heated contention for the Democratic nomination, would fare against John McCain.

My interest in electoral politics had begun slightly earlier, however—and had been mostly the result of frustration rather any affection for the political process. I had carefully monitored the Congress's attempt to ban Internet poker

* FiveThirtyEight's name refers to the number of votes in the Electoral College (538).

in 2006, which was then one of my main sources of income. I found political coverage wanting even as compared with something like sports, where the "Moneyball revolution" had significantly improved analysis.

During the run-up to the primary I found myself watching more and more political TV, mostly MSNBC and CNN and Fox News. A lot of the coverage was vapid. Despite the election being many months away, commentary focused on the inevitability of Clinton's nomination, ignoring the uncertainty intrinsic to such early polls. There seemed to be too much focus on Clinton's gender and Obama's race.[24] There was an obsession with determining which candidate had "won the day" by making some clever quip at a press conference or getting some no-name senator to endorse them—things that 99 percent of voters did not care about.

Political news, and especially the important news that really affects the campaign, proceeds at an irregular pace. But news coverage is produced every day. Most of it is filler, packaged in the form of stories that are designed to obscure its unimportance.* Not only does political coverage often lose the signal—it frequently accentuates the noise. If there are a number of polls in a state that show the Republican ahead, it won't make news when another one says the same thing. But if a new poll comes out showing the Democrat with the lead, it will grab headlines—even though the poll is probably an outlier and won't predict the outcome accurately.

The bar set by the competition, in other words, was invitingly low. Someone could look like a genius simply by doing some fairly basic research into what really has predictive power in a political campaign. So I began blogging at the Web site Daily Kos, posting detailed and data-driven analyses on issues like polls and fundraising numbers. I studied which polling firms had been most accurate in the past, and how much winning one state—Iowa, for instance—tended to shift the numbers in another. The articles quickly gained a following, even though the commentary at sites like Daily Kos is usually more qualitative (and partisan) than quantitative. In March 2008, I spun my analysis out to my

* The classic form of media bias is "rooting for the story"—hoping for a more dramatic outcome that might increase newspaper sales.

own Web site, FiveThirtyEight, which sought to make predictions about the general election.

The FiveThirtyEight forecasting model started out pretty simple—basically, it took an average of polls but weighted them according to their past accuracy—then gradually became more intricate. But it abided by three broad principles, all of which are very fox-like.

Principle I: Think Probabilistically

Almost all the forecasts that I publish, in politics and other fields, are probabilistic. Instead of spitting out just one number and claiming to know exactly what will happen, I instead articulate a range of possible outcomes. On November 2, 2010, for instance, my forecast for how many seats Republicans might gain in the U.S. House looks like what you see in figure 2-3.

The most likely range of outcomes—enough to cover about half of all possible cases—was a Republican gain of between 45 and 65 seats (their actual gain was 63 seats). But there was also the possibility that Republicans might win 70 or 80 seats—if almost certainly not the 100 that Dick Morris had predicted. Conversely, there was also the chance that Democrats would hold just enough seats to keep the House.

The wide distribution of outcomes represented the most honest expression of the uncertainty in the real world. The forecast was built from forecasts of each of the 435 House seats individually—and an exceptionally large number of those races looked to be extremely close. As it happened, a remarkable 77 seats were decided by a single-digit margin.[25] Had the Democrats beaten their forecasts by just a couple of points in most of the competitive districts, they could easily have retained the House. Had the Republicans done the opposite, they could have run their gains into truly astonishing numbers. A small change in the political tides could have produced a dramatically different result; it would have been foolish to pin things down to an exact number.

This probabilistic principle also holds when I am forecasting the outcome in an individual race. How likely is a candidate to win, for instance, if he's

FIGURE 2-3: FIVETHIRTYEIGHT HOUSE FORECAST, NOVEMBER 2, 2010

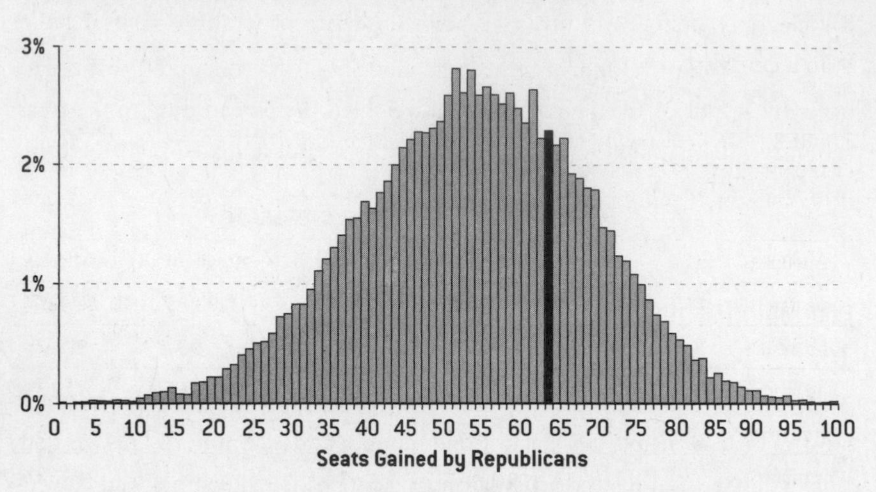

ahead by five points in the polls? This is the sort of question that FiveThirtyEight's models are trying to address.

The answer depends significantly on the type of race that he's involved in. The further down the ballot you go, the more volatile the polls tend to be: polls of House races are less accurate than polls of Senate races, which are in turn less accurate than polls of presidential races. Polls of primaries, also, are considerably less accurate than general election polls. During the 2008 Democratic primaries, the average poll missed by about eight points, far more than implied by its margin of error. The problems in polls of the Republican primaries of 2012 may have been even worse.[26] In many of the major states, in fact—including Iowa, South Carolina, Florida, Michigan, Washington, Colorado, Ohio, Alabama, and Mississippi—the candidate ahead in the polls a week before the election *lost*.

But polls do become more accurate the closer you get to Election Day. Figure 2-4 presents some results from a simplified version of the FiveThirtyEight Senate forecasting model, which uses data from 1998 through 2008 to infer the probability that a candidate will win on the basis of the size of his lead in the polling average. A Senate candidate with a five-point lead on the day before the election, for instance, has historically won his race about 95 percent of the time—almost a sure thing, even though news accounts are sure to describe the race as "too close to call." By contrast, a five-point lead a year before the election translates to just a 59 percent chance of winning—barely better than a coin flip.

FIGURE 2-4: PROBABILITY OF SENATE CANDIDATE WINNING, BASED ON SIZE OF LEAD IN POLLING AVERAGE

	SIZE OF LEAD			
Time Until Election	**1 Point**	**5 Points**	**10 Points**	**20 Points**
One day	64%	95%	99.7%	99.999%
One week	60%	89%	98%	99.97%
One month	57%	81%	95%	99.7%
Three months	55%	72%	87%	98%
Six months	53%	66%	79%	93%
One year	52%	59%	67%	81%

The FiveThirtyEight models provide much of their value in this way. It's very easy to look at an election, see that one candidate is ahead in all or most of the polls, and determine that he's the favorite to win. (With some exceptions, this assumption will be correct.) What becomes much trickier is determining exactly how much of a favorite he is. Our brains, wired to detect patterns, are always looking for a signal, when instead we should appreciate how noisy the data is.

I've grown accustomed to this type of thinking because my background consists of experience in two disciplines, sports and poker, in which you'll see pretty much everything at least once. Play enough poker hands, and you'll make your share of royal flushes. Play a few more, and you'll find that your opponent has made a royal flush when you have a full house. Sports, especially baseball, also provide for plenty of opportunity for low-probability events to occur. The Boston Red Sox failed to make the playoffs in 2011 despite having a 99.7 percent chance of doing so at one point[27]—although I wouldn't question anyone who says the normal laws of probability don't apply when it comes to the Red Sox or the Chicago Cubs.

Politicians and political observers, however, find this lack of clarity upsetting. In 2010, a Democratic congressman called me a few weeks in advance of the election. He represented a safely Democratic district on the West Coast. But given how well Republicans were doing that year, he was nevertheless concerned about losing his seat. What he wanted to know was exactly how much uncertainty there was in our forecast. Our numbers gave him, to the nearest approximation, a 100 percent chance of winning. But did 100 percent really mean 99 percent, or 99.99 percent, or 99.9999 percent? If the latter—a 1 in 100,000 chance of losing—he was prepared to donate his campaign funds to other candidates in more vulnerable districts. But he wasn't willing to take a 1 in 100 risk.

Political partisans, meanwhile, may misinterpret the role of uncertainty in a forecast; they will think of it as hedging your bets and building in an excuse for yourself in case you get the prediction wrong. That is not really the idea. If you forecast that a particular incumbent congressman will win his race 90 percent of the time, you're also forecasting that he should lose it 10 percent of the time.[28] The signature of a good forecast is that each of these probabilities turns out to be about right over the long run.

Tetlock's hedgehogs were especially bad at understanding these probabilities. When you say that an event has a 90 percent chance of happening, that has a *very* specific and objective meaning. But our brains translate it into something more subjective. Evidence from the psychologists Daniel Kahneman and Amos Tversky suggests that these subjective estimates don't always match up with the reality. We have trouble distinguishing a 90 percent chance that the plane will land safely from a 99 percent chance or a 99.9999 percent chance, even though these imply vastly different things about whether we ought to book our ticket.

With practice, our estimates can get better. What distinguished Tetlock's hedgehogs is that they were too stubborn to learn from their mistakes. Acknowledging the real-world uncertainty in their forecasts would require them to acknowledge to the imperfections in their theories about how the world was supposed to behave—the last thing that an ideologue wants to do.

Principle 2: Today's Forecast Is the First Forecast of the Rest of Your Life

Another misconception is that a good prediction shouldn't change. Certainly, if there are wild gyrations in your forecast from day to day, that may be a bad sign—either of a badly designed model, or that the phenomenon you are attempting to predict isn't very predictable at all. In 2012, when I published forecasts of the Republican primaries in advance of each state, solely according to the polls there, the probabilities often shifted substantially just as the polls did.

When the outcome is more predictable—as a general election is in the late stages of the race—the forecasts will normally be more stable. The comment that I heard most frequently from Democrats after the 2008 election was that they turned to FiveThirtyEight to help keep them calm.* By the end of a presidential race, as many as thirty or forty polls might be released every day from different states, and some of these results will inevitably fall outside the margin of error. Candidates, strategists, and television commentators—who have some vested interest in making the race seem closer than it really is—might focus on these outlier polls, but the FiveThirtyEight model found that they usually didn't make much difference.

Ultimately, the right attitude is that you should make the best forecast possible *today*—regardless of what you said last week, last month, or last year. Making a new forecast does not mean that the old forecast just disappears. (Ideally, you should keep a record of it and let people evaluate how well you did over the whole course of predicting an event.) But if you have reason to think that yesterday's forecast was wrong, there is no glory in sticking to it. "When the facts change, I change my mind," the economist John Maynard Keynes famously said. "What do you do, sir?"

Some people don't like this type of course-correcting analysis and mistake it for a sign of weakness. It seems like cheating to change your mind—the equivalent of sticking your finger out and seeing which way the wind is blowing.[29]

---

* Not surprisingly, no Democrats told me this after the 2010 campaign—when our models consistently showed them losing badly.

The critiques usually rely, implicitly or explicitly, on the notion that politics is analogous to something like physics or biology, abiding by fundamental laws that are intrinsically knowable and predicable. (One of my most frequent critics is a professor of neuroscience at Princeton.[30]) Under those circumstances, new information doesn't matter very much; elections should follow a predictable orbit, like a comet hurtling toward Earth.

Instead of physics or biology, however, electoral forecasting resembles something like poker: we can observe our opponent's behavior and pick up a few clues, but we can't see his cards. Making the most of that limited information requires a willingness to update one's forecast as newer and better information becomes available. It is the alternative—failing to change our forecast because we risk embarrassment by doing so—that reveals a lack of courage.

Principle 3: Look for Consensus

Every hedgehog fantasizes that they will make a daring, audacious, outside-the-box prediction—one that differs radically from the consensus view on a subject. Their colleagues ostracize them; even their golden retrievers start to look at them a bit funny. But then the prediction turns out to be exactly, profoundly, indubitably right. Two days later, they are on the front page of the *Wall Street Journal* and sitting on Jay Leno's couch, singled out as a bold and brave pioneer.

Every now and then, it might be correct to make a forecast like this. The expert consensus can be wrong—someone who had forecasted the collapse of the Soviet Union would have deserved most of the kudos that came to him. But the fantasy scenario is hugely unlikely. Even though foxes, myself included, aren't really a conformist lot, we get worried anytime our forecasts differ radically from those being produced by our competitors.

Quite a lot of evidence suggests that aggregate or group forecasts are more accurate than individual ones, often somewhere between 15 and 20 percent more accurate depending on the discipline. That doesn't necessarily mean the group forecasts are *good*. (We'll explore this subject in more depth later in the book.) But it does mean that you can benefit from applying multiple perspectives toward a problem.

"Foxes often manage to do inside their heads what you'd do with a whole group of hedgehogs," Tetlock told me. What he means is that foxes have developed an ability to emulate this consensus process. Instead of asking questions of a whole group of experts, they are constantly asking questions of themselves. Often this implies that they will aggregate different types of information together—as a group of people with different ideas about the world naturally would—instead of treating any one piece of evidence as though it is the Holy Grail. (FiveThirtyEight's forecasts, for instance, typically combine polling data with information about the economy, the demographics of a state, and so forth.) Forecasters who have failed to heed Tetlock's guidance have often paid the price for it.

Beware Magic-Bullet Forecasts

In advance of the 2000 election, the economist Douglas Hibbs published a forecasting model that claimed to produce remarkably accurate predictions about how presidential elections would turn out, based on just two variables, one related to economic growth and the other to the number of military casualties.[31] Hibbs made some very audacious and hedgehogish claims. He said accounting for a president's approval rating (historically a very reliable indicator of his likelihood to be reelected) would not improve his forecasts at all. Nor did the inflation rate or the unemployment rate matter. And the identity of the candidates made no difference: a party may as well nominate a highly ideological senator like George McGovern as a centrist and war hero like Dwight D. Eisenhower. The key instead, Hibbs asserted, was a relatively obscure economic variable called real disposable income per capita.

So how did the model do? It forecasted a landslide victory for Al Gore, predicting him to win the election by 9 percentage points. But George W. Bush won instead after the recount in Florida. Gore did win the nationwide popular vote, but the model had implied that the election would be nowhere near close, attributing only about a 1 in 80 chance to such a tight finish.[32]

There were several other models that took a similar approach, claiming they had boiled down something as complex as a presidential election to a two-variable formula. (Strangely, none of them used the *same* two variables.) Some of them, in fact, have a far worse track record than Hibbs's method. In 2000, one of these models projected a nineteen-point victory for Gore and would have given billions-to-one odds against the actual outcome.[33]

These models had come into vogue after the 1988 election, in which the fundamentals seemed to favor George H. W. Bush—the economy was good and Bush's Republican predecessor Reagan was popular—but the polls had favored Michael Dukakis until late in the race.[34] Bush wound up winning easily.

Since these models came to be more widely published, however, their track record has been quite poor. On average, in the five presidential elections since 1992, the typical "fundamentals-based" model—one that ignored the polls and claimed to discern exactly how voters would behave without them—has missed the final margin between the major candidates by almost 7 percentage points.[35] Models that take a more fox-like approach, combining economic data with polling data and other types of information, have produced more reliable results.

Weighing Qualitative Information

The failure of these magic-bullet forecasting models came even though they were quantitative, relying on published economic statistics. In fact, some of the very worst forecasts that I document in this book are quantitative. The ratings agencies, for instance, had models that came to precise, "data-driven" estimates of how likely different types of mortgages were to default. These models were dangerously wrong because they relied on a self-serving assumption—that the default risk for different mortgages had little to do with one another—that made no sense in the midst of a housing and credit bubble. To be certain, I have a strong preference for more quantitative approaches in my own forecasts. But hedgehogs can take any type of information and have it reinforce their biases, while foxes who have practice in weighing different types of information together can sometimes benefit from accounting for qualitative along with quantitative factors.

Few political analysts have a longer track record of success than the tight-knit team that runs the Cook Political Report. The group, founded in 1984 by a genial, round-faced Louisianan named Charlie Cook, is relatively little known outside the Beltway. But political junkies have relied on Cook's forecasts for years and have rarely had reason to be disappointed with their results.

Cook and his team have one specific mission: to predict the outcome of U.S. elections, particularly to the Congress. This means issuing forecasts for all 435 races for the U.S. House, as well as the 35 or so races for the U.S. Senate that take place every other year.

Predicting the outcome of Senate or gubernatorial races is relatively easy. The candidates are generally well known to voters, and the most important races attract widespread attention and are polled routinely by reputable firms. Under these circumstances, it is hard to improve on a good method for aggregating polls, like the one I use at FiveThirtyEight.

House races are another matter, however. The candidates often rise from relative obscurity—city councilmen or small-business owners who decide to take their shot at national politics—and in some cases are barely known to voters until just days before the election. Congressional districts, meanwhile, are spread throughout literally every corner of the country, giving rise to any number of demographic idiosyncrasies. The polling in House districts tends to be erratic at best[36] when it is available at all, which it often isn't.

But this does not mean there is no information available to analysts like Cook. Indeed, there is an abundance of it: in addition to polls, there is data on the demographics of the district and on how it has voted in past elections. There is data on overall partisan trends throughout the country, such as approval ratings for the incumbent president. There is data on fund-raising, which must be scrupulously reported to the Federal Elections Commission.

Other types of information are more qualitative, but are nonetheless potentially useful. Is the candidate a good public speaker? How in tune is her platform with the peculiarities of the district? What type of ads is she running? A political campaign is essentially a small business: How well does she manage people?

Of course, all of that information could just get you into trouble if you were a hedgehog who wasn't weighing it carefully. But Cook Political has a lot of experience in making forecasts, and they have an impressive track record of accuracy.

Cook Political classifies races along a seven-point scale ranging from Solid Republican—a race that the Republican candidate is almost certain to win—to Solid Democrat (just the opposite). Between 1998 and 2010, the races that Cook described as Solid Republican were in fact won by the Republican candidate on 1,205 out of 1,207 occasions—well over 99 percent of the time. Likewise, races that they described as Solid Democrat were won by the Democrat in 1,226 out of 1,229 instances.

Many of the races that Cook places into the Solid Democrat or Solid Republican categories occur in districts where the same party wins every year by landslide margins—these are not that hard to call. But Cook Political has done just about as well in races that require considerably more skill to forecast. Elections they've classified as merely "leaning" toward the Republican candidate, for instance, have in fact been won by the Republican about 95 percent of the time. Likewise, races they've characterized as leaning to the Democrat have been won by the Democrat 92 percent of the time.[37] Furthermore, the Cook forecasts have a good track record even when they disagree with quantitative indicators like polls.[38]

I visited the Cook Political team in Washington one day in September 2010, about five weeks ahead of that November's elections, and spent the afternoon with David Wasserman, a curly-haired thirtysomething who manages their House forecasts.

The most unique feature of Cook's process is their candidate interviews. At election time, the entryway to the fifth floor of the Watergate complex, where the Cook offices are located, becomes a literal revolving door, with candidates dropping by for hourlong chats in between fund-raising and strategy sessions. Wasserman had three interviews scheduled on the day that I visited. He offered to let me sit in on one of them with a Republican candidate named Dan Kapanke. Kapanke was hoping to unseat the incumbent Democrat Ron Kind in Wisconsin's Third Congressional District, which encompasses a number of small communities in the southwestern corner of the state. Cook Political had the race rated as Likely Democrat, which means they assigned Kapanke only a small chance of victory, but they were considering moving it into a more favorable category, Lean Democrat.

Kapanke, a state senator who ran a farm supply business, had the gruff demeanor of a high-school gym teacher. He also had a thick Wisconsin accent: when he spoke about the La Crosse Loggers, the minor-league baseball team that he owns, I wasn't certain whether he was referring to "logger" (as in timber cutter), or "lager" (as in beer)—either one of which would have been an apropos nickname for a ball club from Wisconsin. At the same time, his plain-spokenness helped to overcome what he might have lacked in charm—and he had consistently won his State Senate seat in a district that ordinarily voted Democratic.[39]

Wasserman, however, takes something of a poker player's approach to his interviews. He is stone-faced and unfailingly professional, but he is subtly seeking to put the candidate under some stress so that that they might reveal more information to him.

"My basic technique," he told me, "is to try to establish a comfortable and friendly rapport with a candidate early on in an interview, mostly by getting them to talk about the fuzzy details of where they are from. Then I try to ask more pointed questions. *Name an issue where you disagree with your party's leadership*. The goal isn't so much to get them to unravel as it is to get a feel for their style and approach."

His interview with Kapanke followed this template. Wasserman's knowledge of the nooks and crannies of political geography can make him seem like a local, and Kapanke was happy to talk shop about the intricacies of his district—just how many voters he needed to win in La Crosse to make up for the ones he'd lose in Eau Claire. But he stumbled over a series of questions on allegations that he had used contributions from lobbyists to buy a new set of lights for the Loggers' ballpark.[40]

It was small-bore stuff; it wasn't like Kapanke had been accused of cheating on his wife or his taxes. But it was enough to dissuade Wasserman from changing the rating.[41] Indeed, Kapanke lost his election that November by about 9,500 votes, even though Republicans won their races throughout most of the similar districts in the Midwest.

This is, in fact, the more common occurrence; Wasserman will usually maintain the same rating after the interview. As hard as he works to glean new information from the candidates, it is often not important enough to override his prior take on the race.

Wasserman's approach works because he is capable of evaluating this information without becoming dazzled by the candidate sitting in front of him. A lot of less-capable analysts would open themselves to being charmed, lied to, spun, or would otherwise get hopelessly lost in the narrative of the campaign. Or they would fall in love with their *own* spin about the candidate's interview skills, neglecting all the other information that was pertinent to the race.

Wasserman instead considers everything in the broader political context. A terrific Democratic candidate who aces her interview might not stand a chance in a district that the Republican normally wins by twenty points.

So why bother with the candidate interviews at all? Mostly, Wasserman is looking for red flags—like the time when the Democratic congressman Eric Massa (who would later abruptly resign from Congress after accusations that he sexually harassed a male staffer) kept asking Wasserman how old he was. The psychologist Paul Meehl called these "broken leg" cases—situations where there is something so glaring that it would be foolish not to account for it.[42]

Catching a few of these each year helps Wasserman to call a few extra races right. He is able to weigh the information from his interviews without *overweighing* it, which might actually make his forecasts worse. Whether information comes in a quantitative or qualitative flavor is not as important as how you use it.

It Isn't Easy to Be Objective

In this book, I use the terms *objective* and *subjective* carefully. The word *objective* is sometimes taken to be synonymous with *quantitative*, but it isn't. Instead it means seeing beyond our personal biases and prejudices and toward the truth of a problem.[43]

Pure objectivity is desirable but unattainable in this world. When we make a forecast, we have a choice from among many different methods. Some of these might rely solely on quantitative variables like polls, while approaches like

Wasserman's may consider qualitative factors as well. All of them, however, introduce decisions and assumptions that have to be made by the forecaster. Wherever there is human judgment there is the potential for bias. The way to become more objective is to recognize the influence that our assumptions play in our forecasts and to question ourselves about them. In politics, between our ideological predispositions and our propensity to weave tidy narratives from noisy data, this can be especially difficult.

So you will need to adopt some different habits from the pundits you see on TV. You will need to learn how to express—and quantify—the uncertainty in your predictions. You will need to update your forecast as facts and circumstances change. You will need to recognize that there is wisdom in seeing the world from a different viewpoint. The more you are willing to do these things, the more capable you will be of evaluating a wide variety of information without abusing it.

In short, you will need to learn how to think like a fox. The foxy forecaster recognizes the limitations that human judgment imposes in predicting the world's course. Knowing those limits can help her to get a few more predictions right.

3

ALL I CARE ABOUT IS W'S AND L'S

The Red Sox were in a very bad mood. They had just returned from New York, where they had lost all three games of a weekend series to the hated Yankees, ending their chances to win the 2009 American League East title. With only seven games left in the regular season, the Red Sox were almost certain to make the playoffs as the American League's wild card,* but this was not how the organization wanted to go into the postseason. Statistical studies have shown that the way a team finishes the regular season has little bearing on how they perform in the playoffs,[1] but the Red Sox were starting to sense that it was not their year.

I was at Fenway Park to speak to one person: the Red Sox's star second baseman, Dustin Pedroia. Pedroia had been one of my favorite players since 2006, when PECOTA, the projection system that I developed for the organization

* The Red Sox's truly epic collapse—when they lost sixteen of their last twenty-one games and missed the playoffs after seeming almost certain to qualify for them—would come two years later in 2011.

Baseball Prospectus, had predicted that he would become one of the best players in baseball. PECOTA's prediction stood sharply in contrast to the position of many scouts, who dismissed Pedroia as "not physically gifted,"[2] critiquing his short stature and his loopy swing and concluding that he would be a marginal player. Whereas PECOTA ranked Pedroia as the fourth best prospect in baseball in 2006,[3] the publication *Baseball America,* which traditionally gives more emphasis to the views of scouts, put him at seventy-seventh. Instead, reports like this one (filed by ESPN's Keith Law[4] early in Pedroia's rookie season) were typical.[5]

> Dustin Pedroia doesn't have the strength or bat speed to hit major-league pitching consistently, and he has no power. If he can continue to hit .260 or so, he'll be useful, and he probably has a future as a backup infielder if he can stop rolling over to third base and shortstop.

Law published that comment on May 12, 2007, at which point Pedroia was hitting .247 and had just one home run.[6] Truth be told, I was losing my faith too; I had watched most of his at-bats and Pedroia looked overmatched at the plate.*

But almost as though he wanted to prove his doubters wrong, Pedroia started hitting the tar out of the baseball. Over the course of his next fifteen games, he hit a remarkable .472, bringing his batting average, which had dropped to as low as .158 in April, all the way up to .336.

In July, two months after Law's report, Pedroia made the American League All-Star Team. In October, he helped the Red Sox win only their second World Series since 1918. That November he was named Rookie of the Year. The next season, at the age of twenty-four, Pedroia took the Most Valuable Player award as the American League's best all-around performer. He wasn't a backup infielder any longer but a superstar. The scouts had seriously underestimated him.

I had come to Fenway because I wanted to understand what made Pedroia tick. I had prepared a whole list of questions, and the Red Sox had arranged a

---

* I traded Pedroia away in one of my fantasy leagues.

press credential for me and given me field-level access. It wasn't going to be easy, I knew. A major-league playing field, which players regard as their sanctuary, is not the best place to conduct an interview. The Red Sox, coming off their losing weekend, were grumpy and tense.

As I watched Pedroia take infield practice, grabbing throws from Kevin Youkilis, the team's hulking third baseman, and relaying them to his new first baseman, Casey Kotchman, it was clear that there was something different about him. Pedroia's actions were precise, whereas Youkilis botched a few plays and Kotchman's attention seemed to wander. But mostly there was his attitude: Pedroia whipped the ball around the infield, looking annoyed whenever he perceived a lack of concentration from his teammates.

After about fifteen minutes of practice, the Red Sox left the infield to the Toronto Blue Jays, their opponents that evening. Pedroia walked past me as I stood on the first-base side of the infield, just a couple of yards from the Red Sox's dugout. The scouts were right about his stature: Pedroia is officially listed at five feet nine—my height if you're rounding up—but I had a good two inches on him. They are also right about his decidedly nonathletic appearance. Balding at age twenty-five, Pedroia had as much hair on his chin as on his head, and a little paunch showed through his home whites. If you saw him on the street, you might take him for a video rental clerk.

Pedroia turned to enter the dugout, where he sat all by himself. This seemed like the perfect moment to catch him, so I mustered up my courage.

"Hey, Dustin, ya got a minute?"

Pedroia stared at me suspiciously for a couple of seconds, and then declared—in as condescending a manner as possible, every syllable spaced out for emphasis: "No. I don't. I'm trying to get ready for the big-league-base-ball-game."

I hung around the field for a few minutes trying to recover my dignity before ambling up to the press box to watch the game.

The next day, after my credential had expired and I'd headed back to New York, I sent my friend David Laurila, a former colleague of mine at Baseball Prospectus and a veteran interviewer, on a reconnaissance mission to see if he could get something more useful out of Pedroia. But Pedroia wasn't much more talkative, giving Laurila about the blandest quote imaginable. "You know what? I'm a guy who doesn't care about numbers and stats," he told Laurila. "All I care about is W's and L's. I care about wins and losses. Nothing else matters to me."

Pedroia had learned to speak in this kind of cliché after getting himself into all kinds of trouble when straying from the party line. Like the time he called his hometown of Woodland, California, a "dump."[7] "You can quote me on that," Pedroia told *Boston* magazine, "I don't give a shit."

He doesn't give a shit. I would come to realize that without that attitude, Pedroia might have let the scouting reports go to his head and never have made the big leagues.

Building a Baseball Forecasting System

I have been a fan of baseball—and baseball statistics—for as long as I can remember. My hometown team, the Detroit Tigers, won the World Series in 1984 when I was six. As an annoying little math prodigy, I was attracted to all the numbers in the game, buying my first baseball card at seven, reading my first *Elias Baseball Analyst* at ten, and creating my own statistic at twelve. (It somewhat implausibly concluded that the obscure Red Sox infielder Tim Naehring was one of the best players in the game.)

My interest peaked, however, in 2002. At the time Michael Lewis was busy writing *Moneyball*, the soon-to-be national bestseller that chronicled the rise of the Oakland Athletics and their statistically savvy general manager Billy Beane. Bill James, who twenty-five years earlier had ushered in the Sabermetric era* by publishing a book called *The Bill James Baseball Abstract*, was soon to be hired as a consultant by the Red Sox. An unhealthy obsession with baseball statistics suddenly seemed like it could be more than just a hobby—and as it happened, I was looking for a new job.

---

* James coined the term *sabermetrics* to describe the systematic study of baseball, especially through statistics. The term is derived from the acronym SABR, for Society of American Baseball Research, which was instrumental in the popularization of James's work.

Two years out of college, I was living in Chicago and working as something called a transfer pricing consultant for the accounting firm KPMG. The job wasn't so bad. My bosses and coworkers were friendly and professional. The pay was honest and I felt secure.

But telling a company how to set prices at its cell phone factory in Malaysia so as to minimize its tax exposure, or hopping a 6 A.M. flight to St. Louis to value contracts for a coal company, was not exactly my idea of stimulating work. It was all too riskless, too prudent, and too routine for a restless twenty-four-year-old, and I was as bored as I'd ever been. The one advantage, however, was that I had a lot of extra time on my hands. So in my empty hours I started building a colorful spreadsheet full of baseball statistics that would later become the basis for PECOTA.

While in college, I had started reading an annual publication called *Baseball Prospectus*. The series was founded in 1996 by Gary Huckabay, an ebullient and sarcastic redhead who had recruited a team of writers from the early Internet newsgroup rec.sport.baseball, then at the vanguard of the statistical analysis of the sport. Huckabay had sensed a market opportunity: Bill James had stopped publishing his *Abstracts* in 1988, and most of the products that endeavored to replace it were not as good, or had folded during the prolonged baseball strike of 1994 and 1995. The first *Baseball Prospectus*, published in 1996, was produced one copy at a time on a laser printer, accidentally omitted the St. Louis Cardinals chapter, and sold just seventy-five copies. But the book quickly developed a cult following, with sales increasing exponentially each year.

Baseball Prospectus was a stat geek's wet dream. There were the reams and reams of numbers, not just for major-league players but also for minor-league prospects whose performance had been "translated" to the major-league level. The writing was sharp if sometimes esoteric, full of *Simpsons* references, jokes about obscure '80s porn films, and sarcastic asides about the group's least favorite general managers.

But most important were its predictions about how each player would perform in the next season, in the form of a Huckabay-developed projection system called Vladimir. The system seemed to be the next step in the revolution that James had begun.

A good baseball projection system must accomplish three basic tasks:

1. Account for the context of a player's statistics
2. Separate out skill from luck
3. Understand how a player's performance evolves as he ages—what is known as the *aging curve*

The first task is relatively easy. Baseball, uniquely among the major American sports, has always been played on fields with nonstandard dimensions. It's much easier to put up a high batting average in snug and boxy Fenway Park, whose contours are shaped by compact New England street grids, than in the cavernous environs of Dodger Stadium, which is surrounded by a moat of parking lot. By observing how players perform both at home and on the road, we can develop "park factors" to account for the degree of difficulty that a player faces. (For example, Fred Lynn, an MVP with the Red Sox during the 1970s, hit .347 over the course of his career at Fenway Park but just .264 at every other stadium.) Likewise, by observing what happens to players who switch from the National League to the American League, we can tell quite a bit about which league is better and account for the strength of a player's competition.

The World's Richest Data Set

The second chore—separating skill from luck—requires more work. Baseball is designed in such a way that luck tends to predominate in the near term: even the best teams lose about one-third of their ball games, and even the best hitters fail to get on base three out of every five times. Sometimes luck will obscure a player's real skill level even over the course of a whole year. During a given season, a true .275 hitter has about a 10 percent chance of hitting .300 and a 10 percent chance of hitting .250 on the basis of luck alone.[8]

What a well-designed forecasting system can do is sort out which statistics are relatively more susceptible to luck; batting average, for instance, is more erratic than home runs. This is even more important for pitchers, whose statistics are notoriously inconsistent. If you want to predict a pitcher's win-loss record, looking at the number of strikeouts he recorded and the number of walks he yielded is more informative than looking at his W's and L's from the previous season, because the former statistics are much more consistent from year to year.

The goal, as in formulating any prediction, is to weed out the root cause: striking batters out prevents them from getting on base, preventing them from getting on base prevents them from scoring runs, and preventing them from scoring runs prevents them from winning games. However, the further downstream you go, the more noise will be introduced into the system: a pitcher's win-loss record is affected as much by how many runs his offense scores, something that he has essentially no control over, as by how well he pitches. The Seattle Mariners' star pitcher Felix Hernandez went 19-5 in 2009 but 13-12 in 2010 despite pitching roughly as well in both years, because the Mariners had some epically terrible hitters in 2010.

Cases like these are not at all uncommon and tend to make themselves known if you spend any real effort to sort through the data. Baseball offers perhaps the world's richest data set: pretty much everything that has happened on a major-league playing field in the past 140 years has been dutifully and accurately recorded, and hundreds of players play in the big leagues every year. Meanwhile, although baseball is a team sport, it proceeds in a highly orderly way: pitchers take their turn in the rotation, hitters take their turn in the batting order, and they are largely responsible for their own statistics.* There are relatively few problems involving complexity and nonlinearity. The causality is easy to sort out.

That makes life easy for a baseball forecaster. A hypothesis can usually be tested empirically, and proven or disproven to a high degree of statistical satisfaction. In fields like economic or political forecasting where the data is much sparser—one presidential election every four years, not hundreds of new data points ever year—you won't have that luxury and your prediction is more likely to go astray.

---

* Contrast this with football, in which a great offensive line can make an All-Pro out of a mediocre running back, or basketball, in which the synergy between a point guard and a power forward can make them more than the sum of their parts.

Behold: The Aging Curve

All this assumes, however, that a player's skill level is constant from year to year—if only we could separate the signal from the noise, we'd know everything that we needed to. In fact, a baseball player's skills are in a constant state of flux, and therein lies the challenge.

By looking at statistics for thousands of players, James had discovered that the typical player[9] continues to improve until he is in his late twenties, at which point his skills usually begin to atrophy, especially once he reaches his midthirties.[10] This gave James one of his most important inventions: the *aging curve.*

Olympic gymnasts peak in their teens; poets in their twenties; chess players in their thirties[11]; applied economists in their forties,[12] and the average age of a Fortune 500 CEO is 55.[13] A baseball player, James found, peaks at age twenty-seven. Of the fifty MVP winners between 1985 and 2009, 60 percent were between the ages of twenty-five and twenty-nine, and 20 percent were aged twenty-seven exactly. This is when the combination of physical attributes and mental attributes needed to play the game well seem to be in the best balance.

FIGURE 3-1: AGING CURVE FOR HITTER

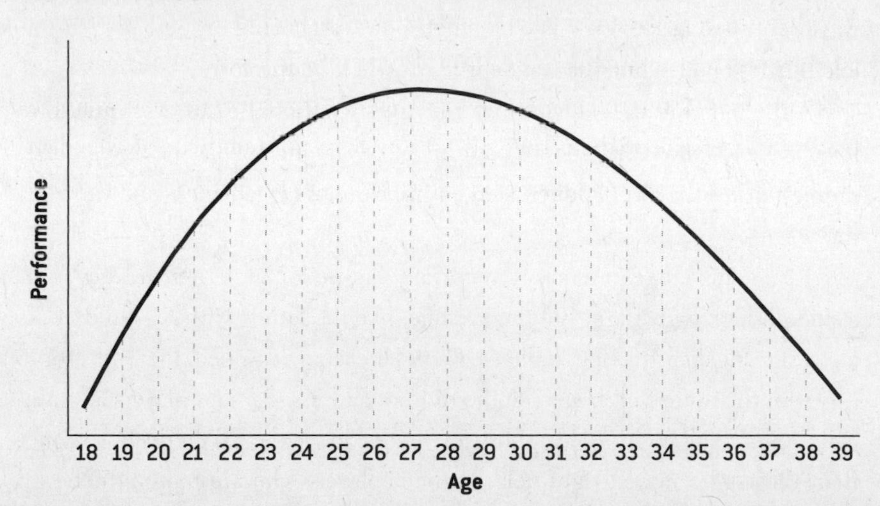

This notion of the aging curve would have been extremely valuable to any team that had read James's work. Under baseball's contract rules, players do not become free agents until fairly late in their careers: after they've played at least six full major-league seasons (before then, they are under the exclusive control of the club that drafted them and cannot command a full market price). Since the typical rookie reaches the big leagues at twenty-three or twenty-four years old, he might not become a free agent until he is thirty—just *after* his window of peak performance has eclipsed. Teams were paying premium dollars for free agents on the assumption that they would replicate in their thirties the production they had exhibited in their twenties; in fact, it usually declined, and since Major League Baseball contracts are guaranteed, the teams had no recourse.

But James's aging curve painted too smooth a picture. Sure, the *average* player might peak at age twenty-seven. As anyone who has paid his dues staring at the backs of baseball cards can tell you, however, players age at different paces. Bob Horner, a third baseman with the Atlanta Braves during the 1980s, won the Rookie of the Year award when he was just twenty and made the All-Star team when he was twenty-four; the common assumption at the time was that he was bound for the Hall of Fame. But by age thirty, set back by injuries and an ill-advised stint with the Yakult Swallows of the Japanese League, he was out of professional baseball entirely. On the other hand, the Seattle Mariner great Edgar Martinez did not have a steady job in the big leagues until he was twenty-seven. He was a late bloomer, however, having his best years in his late thirties and leading the league in RBIs when he was forty.

Although Horner and Martinez may be exceptional cases, it is quite rare for players to follow the smooth patterns of development that the aging curve implies; instead, a sort of punctuated equilibrium of jagged peaks and valleys is the norm.

Real aging curves are noisy—very noisy (figure 3-2). *On average*, they form a smooth-looking pattern. But the average, like the family with 1.7 children, is just a statistical abstraction. Perhaps, Gary Huckabay reasoned, there was some signal in the noise that James's curve did not address. Perhaps players at physically demanding positions like shortstop tended to see their skills decline sooner than those who played right field. Perhaps players who are more athletic all

FIGURE 3-2: NOISY AGING PATTERNS FOR DIFFERENT HITTERS

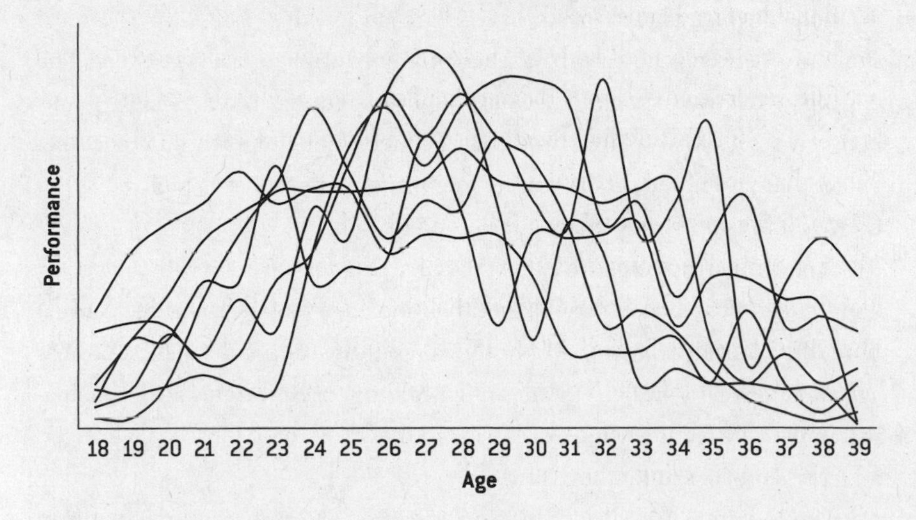

around can be expected to have longer careers than those who have just one or two strong skills.

Huckabay's system hypothesized that there are twenty-six distinct aging curves, each applying to a different type of player.[14] If Huckabay was correct, you could assess which curve was appropriate to which player and could therefore predict how that player's career would track. If a player was on the Bob Horner track, he might have an early peak and an early decline. Or if he was more like Martinez, his best seasons might come later on.

While Huckabay's Vladimir nailed some of its predictions, it ultimately was not much more accurate than the slow-and-steady type of projections developed by James[15] that applied the same aging curve to every player. Some of the issue was that twenty-six was an arbitrary number for Huckabay's categories, and it required as much art as science to figure out which group a player belonged in.

However, a person must have a diverse array of physical and mental skills to play baseball at an elite level: muscle memory, physical strength, hand-eye coordination, bat speed, pitch recognition, and the willpower to stay focused when his team endures a slump. Vladimir's notion of different aging curves seemed like a more natural fit for the complexities inherent in human performance. In developing PECOTA, I tried to borrow some elements from Huckabay and some from Bill James.

In the 1986 *Baseball Abstract*, James introduced something called similarity scores, which as their name implies are designed to assess the statistical similarity between the career statistics of any two major-league players. The concept is relatively simple. They start out by assigning a score of 1,000 points between a set of two players, and then deduct points for each difference between them.[16] Highly similar players might maintain scores as high as 950 or even 975, but the discrepancies quickly add up.

The similarity scores are extremely satisfying for anybody with a working knowledge of baseball history. Rather than look at a player's statistics in a vacuum, they provide some sense of historical context. Pedroia's statistics through age twenty-five, for instance, were similar to those of Rod Carew, the Panamanian great who led the Minnesota Twins in the 1970s, or to Charlie Gehringer, a Depression-era star for the Tigers.

James mainly intended for his similarity scores to be backward looking: to analyze, for instance, how suitable a player's statistics were for the Hall of Fame. If you were trying to make the case that your favorite player belonged in Cooperstown, and you had observed that 9 of the 10 players with the most similar statistics had made it there, you'd have a very strong argument.

But couldn't similarity scores be predictive, too? If we could identify, say, the one hundred players who were most comparable to Pedroia through a given age, might not the performance of those players over the balance of their careers tell us something about how Pedroia was likely to develop?

This was the idea that I set out to work on—and slowly, over the course of those long days at KPMG in 2002, PECOTA began to develop. It took the form of a giant, colorful Excel spreadsheet—fortuitously so, since Excel was one of the main tools that I used in my day job at KPMG. (Every time one of my bosses walked by, they assumed I was diligently working on a highly elaborate model for one of our clients.[17])

Eventually, by stealing an hour or two at a time during slow periods during the workday, and a few more while at home at night, I developed a database consisting of more than 10,000 player-seasons (every major-league season since World War II was waged[18]) as well as an algorithm to compare any one player with another. The algorithm was somewhat more elaborate than James's and sought to take full advantage of baseball's exceptionally rich data set. It used a different method for comparing a set of players—what is technically known as a *nearest neighbor analysis*. It also considered a wider variety of factors—including things like a player's height and weight, that are traditionally more in the domain of scouting.

Like Huckabay's system, PECOTA provided for the possibility that different types of players might age in different ways. But it didn't try to force them onto one of twenty-six development curves; instead, it let this occur naturally by identifying a set of comparable players somewhere in baseball's statistical galaxy. If it turned out, for instance, that a disproportionate number of Dustin Pedroia's comparable players turned into strong major leaguers, that might imply something about Pedroia's chances for success.

More often, however, a player's most comparable players will be a mixed bag; the paths of players who might have similar statistics through a given point in their careers can diverge wildly thereafter. I mentioned that under James's similarity scores, Pedroia was found to be similar to Charlie Gehringer and Rod Carew, two players who had long and illustrious careers and who eventually made the Hall of Fame. But Pedroia's statistics over that period were also similar to Jose Vidro, an undistinguished second baseman for the Montreal Expos.

These differences can be especially dramatic for minor-league players. In 2009, when PECOTA identified the top comparables for Jason Heyward, then a nineteen-year-old prospect in the Atlanta Braves system, you could find everything from Hall of Famer to murder victim. Chipper Jones, Heyward's number-two comparable, is an example of the former case: one of the greatest Atlanta Braves ever, he's played seventeen seasons with the club and is a lifetime .304 hitter with more than 450 home runs. On the other hand, there was Dernell Stenson, a promising prospect whose numbers were also similar to Heyward's. After playing in a developmental-league game in Arizona in 2003, he was tied up, shot, and run over with his own SUV in an apparently random act of violence.

All of Heyward's comparables were big, strong, multitalented athletes who were high draft picks and displayed precocious skills in the minor leagues. But they met radically different fortunes. PECOTA's innovation was to acknowledge this by providing a range of possible outcomes for each player, based on the precedents set by his comparables: essentially, best-case, worst-case, and most-likely scenarios. An endless array of outcomes can and will happen when we are trying to predict human performance.

So far, things have been up and down for Heyward. After a terrific 2009 in which he was named the Minor League Player of the Year, he hit eight home runs in his first thirty major-league games upon making his debut with the Braves in 2010 and made the All-Star team, exceeding all expectations. His sophomore season in 2011 was rougher, however, and he hit just .227. A good statistical forecasting system might have found some reason to be optimistic after Heyward's 2011 season: his numbers were essentially the same except for his batting average, and batting average is subject to more luck than other statistics.

But can statistics tell you everything you'll want to know about a player? Ten years ago, that was the hottest topic in baseball.

Can't We All Just Get Along?

A slipshod but nevertheless commonplace reading of *Moneyball* is that it was a story about the conflict between two rival gangs—"statheads" and "scouts"—that centered on the different paradigms that each group had adopted to evaluate player performance (statistics, of course, for the statheads, and "tools" for the scouts).

In 2003, when *Moneyball* was published, Michael Lewis's readers would not have been wrong to pick up on some animosity between the two groups. (The book itself probably contributed to some of the hostility.) When I attended baseball's Winter Meetings that year at the New Orleans Marriott, it was like being back in high school. On one side were the jocks, who, like water buffaloes at an oasis, could sometimes be found sipping whiskey and exchanging old war stories at the hotel bar. More often they sequestered themselves in their hotel rooms to negotiate trades with one another. These were the baseball lifers: mostly in their forties and fifties, many of them former athletes, they had paid their dues and gradually worked their way up through the organizational hierarchy. On the other side were the nerds: herds of twenty- and thirtysomethings, armed with laptop bags and color-printed position papers, doing lap after lap around the lobby hoping to pin down a jock and beg him for a job. There wasn't much interaction between the two camps, and each side regarded the other as arrogant and closed-minded.

The real source of conflict may simply have been that the jocks perceived the nerds as threats to their jobs, usually citing real or perceived cuts in scouting budgets as their evidence. "It is adversarial right now," Eddie Bane, the scouting director of the Anaheim Angels told *Baseball America* in a contentious roundtable at the conference that focused on *Moneyball*.[19] "Some of our old-time guys are losing jobs that we didn't feel they should be losing. Maybe the cutbacks were due to money or whatever. But we correlate it to the fact that some of the computer stuff is causing that. And we resent it."

How many teams had really cut their scouting budgets is unclear. The Toronto Blue Jays were one team that did so, and they paid the price for it, with a series of poor drafts between 2002 and 2005. But the budget cuts were forced by the peculiarities of their corporate parent, Rogers Communications, which was struggling with a weak Canadian dollar, and not the whim of their then general manager, the Beane disciple J. P. Ricciardi.

It's now been a decade since the publication of *Moneyball*, however, and these brushfires have long since burned themselves out. The success of the Red Sox, who won their first World Series title in eighty-six years in 2004 with a fusion approach that emphasized both statistics and scouting, may have been a key factor in the détente. Organizations that would have been classified as "scouting" organizations in 2003, like the St. Louis Cardinals, have since adopted a more analytic approach and are now among the most innovative in the sport. "Stathead" teams like the Oakland A's have expanded rather than contracted their scouting budgets.[20]

The economic recession of 2007 through 2009 may have further encouraged the use of analytic methods. Although baseball weathered the recession fairly well, suddenly everyone had become a *Moneyball* team, needing to make the best use possible of constrained budgets.[21] There was no shortage of cheap stathead labor: economics and computer science graduates from Harvard and Yale, who might otherwise have planned on a $400,000-a-year job at an investment bank, were willing to move to Tampa or Cleveland and work around the clock for one-tenth of that salary. The $40,000 nerd was a better investment than a $40 million free agent who was doomed to see his performance revert to the mean.

It has not, however, been a unilateral victory for the statheads. If the statheads have proved their worth, so have the scouts.

PECOTA Versus Scouts: Scouts Win

PECOTA originally stood for Pitcher Empirical Comparison and Optimization Test Algorithm: a clunky acronym that was designed to spell out the name of Bill Pecota, a marginal infielder with the Kansas City Royals during the 1980s who was nevertheless a constant thorn in the side of my favorite Detroit Tigers.*

The program had initially been built to project the performance of pitchers rather than hitters. The performance of pitchers is notoriously hard to predict, so much so that after experimenting for a couple of years with a system called WFG—you can guess what the acronym stands for—Baseball Prospectus had given up and left their prediction lines blank. I sensed an opportunity and pitched PECOTA to Huckabay. Somewhat to my surprise, he and the Baseball Prospectus crew were persuaded by it; they offered to purchase PECOTA for me in exchange for stock in Baseball Prospectus on the condition that I developed a similar system for hitters.[22] I did so, and the first set of PECOTA projections were published the next winter in *Baseball Prospectus 2003*.

When the 2003 season was over, we discovered that PECOTA had performed a little better than the other commercial forecasting systems.[23] Indeed, each year from 2003 through 2008, the system matched or bettered the competition every time that we or others tested it,[24] while also beating the Vegas over-under lines.[25] There were also some fortuitous successes that boosted the system's reputation. In 2007, for instance, PECOTA predicted that the Chicago White Sox—just two years removed from winning a World Series title—would finish instead with just seventy-two wins. The forecast was met with howls of protest from the Chicago media and from the White Sox front office.[26] But it turned out to be exactly right: the White Sox went 72-90.

---

* Although Bill Pecota hit just .249 for his career overall, he hit .303 in games against the Tigers.

By 2009 or so, however, the other systems were catching up and sometimes beating PECOTA. As I had borrowed from James and Huckabay, other researchers had borrowed some of PECOTA's innovations while adding new wrinkles of their own. Some of these systems are very good. When you rank the best forecasts each year in terms of how well they predict the performance of major league players, the more advanced ones will now usually come within a percentage point or two of one another.[27]

I had long been interested in another goal for PECOTA, however: projecting the performance of *minor* league players like Pedroia. This is potentially much harder. And because few other systems were doing it until recently, the only real competition was the scouts.

In 2006, I published a list of PECOTA's top 100 prospects for the first time, comparing the rankings against the scouting-based list published at the same time by *Baseball America*. The players in the PECOTA list were ranked by how much value they were expected to contribute over the next six seasons once they matriculated to the major leagues.[28]

The 2011 season marked the sixth year since the forecasts were issued, so I was finally able to open up the time capsule and see how well they performed. Although the players on this list are still fairly young, we should have a pretty good idea by now of whether they are stars, benchwarmers, or burnouts.

This list did have Pedroia ranked as the fourth best prospect in baseball. And there were other successes for PECOTA. The system thought highly of the prospect Ian Kinsler, whom Baseball America did not have ranked at all; he has since made two All-Star teams and has become one of the cogs in the Texas Rangers' offense. PECOTA liked Matt Kemp, the Dodgers superstar who nearly won baseball's elusive Triple Crown in 2011, better than *Baseball America* did.

But have you ever heard of Joel Guzman? Donald Murphy? Yusemiro Petit? Unless you are a baseball junkie, probably not. PECOTA liked those players as well.

Baseball America also had its share of misses: the scouts were much too optimistic about Brandon Wood, Lastings Milledge, and Mark Rogers. But they seemed to have a few more hits. They identified stars like the Red Sox pitcher Jon Lester, the Rockies' shortstop Troy Tulowitzki, and the Baltimore

Orioles outfielder Nick Markakis, all of whom had middling minor-league statistics and whom PECOTA had not ranked at all.

There is enough data to compare the systems statistically. Specifically, we can look at the number of wins the players on each list generated for their major-league teams in the form of a statistic called wins above replacement player, or WARP,[29] which is meant to capture all the ways that a player contributes value on the baseball diamond: hitting, pitching, and defense.

The players in the PECOTA list had generated 546 wins for their major-league teams through 2011 (figure 3-3). But the players in *Baseball America's* list did better, producing 630 wins. Although the scouts' judgment is sometimes flawed, they were adding plenty of value: their forecasts were about 15 percent better than ones that relied on statistics alone. That might not sound like a big difference, but it really adds up. Baseball teams are willing to pay about $4 million per win on the free-agent market.[30] The extra wins the scouts identified were thus worth a total of $336 million over this period.*

FIGURE 3-3: WINS GENERATED THROUGH 2011, PECOTA AND BASEBALL AMERICA TOP 100 PROSPECT LISTS (2006)

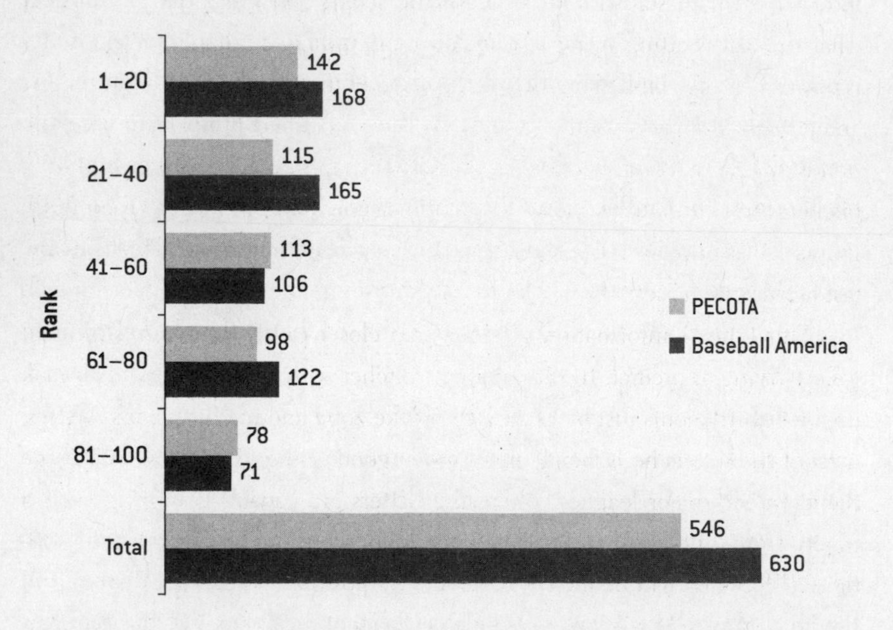

* That works out to about $1.9 million per team per season.

The Biases of Scouts and Statheads

Although it would have been cool if the PECOTA list had gotten the better of the scouts, I didn't expect it to happen. As I wrote shortly after the lists were published:[31]

> As much fun as it is to play up the scouts-versus-stats angle, I don't expect the PECOTA rankings to be as accurate as . . . the rankings you might get from *Baseball America*.
>
> The fuel of any ranking system is information—and being able to look at both scouting and statistical information means that you have more fuel. The only way that a purely stat-based prospect list should be able to beat a hybrid list is if the biases introduced by the process are so strong that they overwhelm the benefit.

In other words, scouts use a hybrid approach. They have access to more information than statistics alone. Both the scouts and PECOTA can look at what a player's batting average or ERA was; an unbiased system like PECOTA is probably a little bit better at removing some of the noise from those numbers and placing them into context. Scouts, however, have access to a *lot* of information that PECOTA has no idea about. Rather than having to infer how hard a pitcher throws from his strikeout total, for instance, they can take out their radar guns and time his fastball velocity. Or they can use their stopwatches to see how fast he runs the bases.

This type of information gets one step closer to the root causes of what we are trying to predict. In the minors, a pitcher with a weak fastball can rack up a lot of strikeouts just by finding the strike zone and mixing up his pitches; most of the hitters he is facing aren't much good, so he may as well challenge them. In the major leagues, where the batters are capable of hitting even a ninety-eight-mile-per-hour fastball out of the park, the odds are against the soft-tosser. PECOTA will be fooled by these false positives while a good scout will not be. Conversely, a scout may be able to identify players who have major-league talent but who have yet to harness it.

To be sure, whenever human judgment is involved, it also introduces the potential for bias. As we saw in chapter 2, more information actually can make matters *worse* for people who take the wrong attitude toward prediction and use it as an excuse to advance a partisan theory about the way that the world is supposed to work—instead of trying to get at the truth.

Perhaps in the pre-*Moneyball* era, these biases were getting the better of the scouts. They may have been more concerned about the aesthetics of a player—did he fill out his uniform in the right way?—than about his talent. If recent *Baseball America* lists have been very good, the ones from the early 1990s[32] were full of notorious busts—highly touted prospects like Todd Van Poppel, Ruben Rivera, and Brien Taylor who never amounted to much.

But statheads can have their biases too. One of the most pernicious ones is to assume that if something cannot easily be quantified, it does not matter. In baseball, for instance, defense has long been much harder to measure than batting or pitching. In the mid-1990s, Beane's Oakland A's teams placed little emphasis on defense, and their outfield was manned by slow and bulky players, like Matt Stairs, who came out of the womb as designated hitters. As analysis of defense advanced, it became apparent that the A's defective defense was costing them as many as eight to ten wins per season,[33] effectively taking them out of contention no matter how good their batting statistics were. Beane got the memo, and his more recent and successful teams have had relatively good defenses.

These blind spots can extract an even larger price when it comes to forecasting the performance of minor-league players. With an established major-league player, the question is essentially whether he can continue to perform as he has in the past. An especially clever statistical forecasting system might be able to divine an upward or downward trend of a few percentage points.[34] But if you simply assume that the player will do about as well next season as he has in his past couple, you won't be too far off. Most likely, his future capability will not differ that much from his present output.

Baseball is unique among the major professional sports, however, for its extremely deep minor-league system. Whereas the National Football League has no officially sanctioned minor league, and the NBA has just a few minor-league teams, baseball has 240, eight for each major-league parent. Moreover, whereas basketball or football players can jump from college or even high school straight into the pros and be impact players immediately upon arrival, this kind of instant stardom is extremely rare in baseball. Even the most talented draft picks may have to bide their time in Billings or Bakersfield or Binghamton before advancing to the major leagues.

It is very challenging to predict the performance of these players because we are hoping that they will eventually be able to do something that they are *not* capable of at present: perform at a high level in the major leagues. Save for a literal once-in-a-generation prospect like Bryce Harper, the best high school hitter in the country would get killed if he had to face off against major-league pitching. He will have to get bigger, stronger, smarter, and more disciplined in order to play in the majors—all of which will require some combination of hard work and good fortune. Imagine if you walked into an average high school classroom, got to observe the students for a few days, and were asked to predict which of them would become doctors, lawyers, and entrepreneurs, and which ones would struggle to make ends meet. I suppose you could look at their grades and SAT scores and who seemed to have more friends, but you'd have to make some pretty wild guesses.

And yet amateur scouts (and any statistical system designed to emulate them) are expected to do exactly this. Although some baseball players are drafted out of college, many others come straight from high school, and the scouting process can begin when they're as young as their midteens. Like any group of young men, these players will be full of hormones and postadolescent angst, still growing into their bodies, dealing with the temptations of booze and the opposite sex. Imagine if you had to entrust the future of your business to a set of entitled nineteen-year-olds.

Beyond the Five Tools

As Lewis described in *Moneyball*, Billy Beane was one of those players who had prodigious talent but failed to realize it; a first-round draft pick in 1980, he played just 148 games in the majors and hit .219 for his career. Beane had a Hall of Fame career, however, compared with prospects like John Sanders, who is now a scout for the Los Angeles Dodgers.

Sanders once played in the major leagues. Exactly once—like Moonlight Graham from *Field of Dreams*. On April, 13, 1965, when Sanders was nineteen, the Kansas City Athletics used him as a pinch-runner in the seventh inning of a game against the Detroit Tigers. Sanders didn't so much as advance a base: the last two hitters popped out, and he was replaced before the next inning began.[35] He would never play in the majors again.

Sanders did not lack talent. He had been a multisport star at Grand Island High School in Nebraska: All-State quarterback in 1963, All-State basketball in 1964, a gold-medal winner in the discus at the state track meet.[36] Baseball might not even have been his best sport. But he was darned good at it, and when he graduated in the summer of 1964, he had a professional contract from the A's to accompany his diploma.

But Sanders's development was stymied by something called the Bonus Baby rule. Before the introduction of the major-league draft, in 1965, all amateur players were free agents and teams could pay them whatever they wanted. To prevent the wealthiest teams from hoarding the talent, the rule extracted a punishment—players who received a large signing bonus were required to spend their first two professional seasons on the major-league roster, even though they were nowhere near ready to play at that level.[37]

The rule really punished bright prospects like Sanders. Most of the Bonus Babies spent their time riding the bench, rarely seeing any big-league action. They were shut out from getting everyday game experience at the very time they needed it the most. Fans and teammates, wondering why some peach-fuzzed nineteen-year-old was being paid thousands to be a glorified batboy, were unlikely to be sympathetic to their plight. Although a few Bonus Babies like Sandy Koufax and Harmon Killebrew went on to have Hall of Fame careers, many other talented prospects of the era never overcame the experience.

Sanders's background—going from perhaps the best amateur athlete in the state of Nebraska to a footnote in the *Baseball Encyclopedia*—gives him unique insight into the psyches of young players. I caught up with him by cell phone one morning while he was driving from North Carolina to Georgia to watch the Braves' top affiliate play.

At the 2003 Winter Meetings in New Orleans, I would have pegged Sanders as one of the "jocks." He devoted his life to the sport after his (brief) playing career ended. But Sanders has never seen the game through the jocks-versus-nerds prism.

"I love to evaluate," he told me. "I've always enjoyed statistical proofs even way back in the day when we did it with calculators and adding machines." Sanders relayed an anecdote: "One of the scouts said, 'Well, let's face it, guys, what's the first thing we do when we go to the ballpark? We go to the press room, we get the stats. We get the stats! What's wrong with that? That's what you do.' "

Statistics, indeed, have been a part of the fabric of baseball since the very beginning. The first newspaper box score, which included five categories of statistics for each player—runs, hits, putouts, assists, and errors—was published by Henry Chadwick in 1859,[38] twelve years before the first professional league was established, in 1871. Many of the *Moneyball*-era debates concerned not *whether* statistics should be used, but *which* ones should be taken into account. On-base percentage (OBP), for instance, as analysts like James had been pointing out for years, is more highly correlated with scoring runs (and winning games) than batting average, a finding which long went underappreciated by traditionalists within the industry.[39]

This type of debate was usually fought on the statheads' turf. That OBP conveys more useful information than batting average, or that a pitcher's ERA is a fairer indicator of his performance than his win-loss record are scientific facts, as much as the earth's revolution around the sun; the statheads were unambiguously right about them. But being on the winning side of these arguments may have led the stathead community to be complacent or dismissive about other points on which there is much more ambiguity.

The further you get away from the majors—the more you are trying to *predict* a player's performance instead of *measure* it—the less useful statistics are. Statistics at the more advanced minor-league levels, like Double-A and Triple-A, have been shown to be *almost* as predictive as major-league numbers. But statistics at the lower minor-league levels are less reliable, and the numbers for college or high school players have very little predictive power.

The scouts' traditional alternative to statistics are the Five Tools: hitting for power, hitting for average, speed, arm strength, and defensive range. It is a list that has drawn, and deserves, much criticism. Plate discipline—which consists of drawing walks and avoiding strikeouts—is not represented. And the invocation of the Five Tools sometimes conveys the impression that they are all equally important, when in fact hitting for power is far more important than arm strength at all positions save for shortstop and catcher.

There is also reason to think that the Five Tools alone won't tell us very much. As a player works his way up the minor-league ladder, the tools should increasingly be reflected in the player's statistics—or he probably isn't bound to advance any higher. Some of the categories, in fact, essentially *are* statistics: "hitting for average" is expressed by a player's batting average; "hitting for power" by doubles and home runs. If a scout tells you that a certain player grades out at a seventy on the eighty-point scouting scale in the hitting-for-power category, but he's struggling to hit ten home runs per year for the Altoona Curve, how much faith would you really put in the scouting report?

Sanders, the industry veteran, is skeptical of the emphasis placed on the Five Tools. "The impact toolbox is obvious to anyone. Runs fast, throws hard, all that stuff. Scouts can walk in and see those immediately," he told me. "I think the question is—are those skills used effectively to [make for] winning ballplayers? Are those tools converted into usable skills? Bat speed—we can see that fairly quickly. But if the person has bat speed but he doesn't trust it—if he's always jumping out at pitches—that's not usable."

Sanders's focus is less on physical tools and more on usable, game-ready skills. The extent to which one can be translated to the other depends on what he calls a player's mental toolbox. The mental tools are often slower to develop than the physical ones; Sanders's wife is a special-needs educator and pointed him toward research suggesting that most of us are still in a state of mental adolescence until about the age of twenty-four.[40] Before that age, Sanders will cut a player some slack if he sees signs that their mental tools are developing. After that, he needs to see performance. Interestingly, twenty-four is right about the age when a player is usually in Double-A and his performance starts to become more predictable from his statistics.

Sanders has no formal definition of what a player's mental toolbox should include, but over the course of our conversation, I identified five different intellectual and psychological abilities that he believes help to predict success at the major-league level.

- **Preparedness and Work Ethic** Baseball is unlike almost all other professional sports in that games are played six or seven times a week. A baseball player can't get "amped up" for game day as a football or basketball player might; he has to be ready to perform at a professional level *every day.* This means he must have a certain amount of discipline. Sanders likes to get to the ballpark early because he contends he can better detect this attribute in a player's pregame rituals than during the game itself. Pedroia, for instance, was clearly more focused than his teammates during infield practice that September night at Fenway Park. He had his routine, and he wasn't putting up with any distractions—and certainly not some reporter he'd never heard of trying to interview him.
- **Concentration and Focus** Although related to preparedness, this category specifically concerns the manner in which a player conducts himself during the course of the game. Baseball is a reflex sport. A hitter has about three tenths of a second to decide whether to swing at a pitch;[41] an infielder has to react to a sharply hit grounder as soon as it comes off the bat. "If a player is not energized, I don't know what we can do with them," Sanders says. "I want a shortstop, a middle infielder, to have pitch-by-pitch focus that's off the charts."
- **Competitiveness and Self-Confidence** While it may seem like a given that any professional athlete would be a natural-born competitor, baseball players must overcome self-doubt and other psychological obstacles in the early stages of their careers. One moment, they were the king of the hill in high school; the next, they are riding busses between Kannapolis and Greensboro, reading about their failures on the Internet each time they go into a slump. When Sanders sees a talented player underachieving, he wonders, "Is there a desire to succeed to the degree that there's a failure mechanism kicking in? Is there a fear of failure? Is the desire to succeed significant enough to overcome the fear of failure?"
- **Stress Management and Humility** In baseball, even the best hitters fail a majority of the time, and every player will enter a slump at certain points

during the season. The ability to cope with this failure requires a short memory and a certain sense of humor. One of Sanders's favorite scouting tactics is to observe how a player reacts after a rough or unlucky play. "I like to see a hitter, when he flails at a pitch, when he takes a big swing and to the fans it looks ridiculous, I like to look down and see a smile on his face. And then the next time—*bam*—four hundred feet!" These skills will only become more critical once a player reaches the majors and has to deal with scrutiny from fans and the media.

- **Adaptiveness and Learning Ability** How successfully is the player able to process new information during a game? Listen to advice from his coaches? How does he adapt when his life situation changes? What if he's traded—or asked to play a new position? The path between amateur ball and the major leagues is rarely linear even for the most talented prospects—and so a great player can't be too rigid in his mental approach. "Players who are successful at this game are people who, when they walk down the hall of the building and there's a turn, they make a round turn. They don't make a sharp turn," Sanders observes. "It's a controlled intensity."

These same habits, of course, are important in many human endeavors. Some of them even have applications for forecasters, especially the one Sanders calls adaptiveness: How do you react to new information when it comes your way? Being too hot under the collar and overreacting to a change in the circumstances or being too cool to change your mind when the evidence dictates otherwise will lead to bad predictions.

Few professions, however, are as competitive as baseball. Among the thousands of professional baseball players, and the hundreds of thousands of amateurs, only 750 are able to play in the major leagues at any given time, and only a few dozen of those will be All-Stars. Sanders's job is to search for those exceptional individuals who defy the odds. He has to work nearly as hard at his job as the players do, and he is still out on the road almost every day in his late sixties.

But Sanders provides the Dodgers with the most valuable kind of information—the kind of information that other people don't have.

Information Is the Name of the Game

Billy Beane, the protagonist of *Moneyball*, sees relentless information gathering as the secret to good scouting.

"What defines a good scout? Finding out information that other people can't," he told me. "Getting to know the kid. Getting to know the family. There's just some things that you have to find out in person."

Beane should know. Much of the success of the A's was a result of the team's statistical aptitude. But their scouting of amateur players was just as critical to their achievements. Most of the team's stars during the early 2000s period that *Moneyball* chronicled—Miguel Tejada, Jason Giambi, Barry Zito, Tim Hudson, Eric Chavez—had been signed and developed by the club.

Beane told me the A's scouting budget is now much higher than it has ever been. Moreover, he said it was the A's fascination with statistical analysis that led them to increase it. As we've seen, baseball players do not become free agents until after six full seasons, which is usually not until they're at least thirty. As Bill James's analysis of the aging curve revealed, this often leads clubs to overspend on free agents—after all, their best years are usually behind them. But there is a flip side to this: *before* a player is thirty, he can provide tremendous value to his club. Moreover, baseball's economics are structured such that younger players can often be had for pennies on the dollar.[42]

If a baseball team is viewed, as with any other business, from a standpoint of profits and losses, almost all the value is created by the scouting and development process. If a team's forecasting system is exceptionally good, perhaps it can pay $10 million a year for a player whose real value is $12 million. But if its *scouting* is really good, it might be paying the same player just $400,000. That is how you compete in a small market like Oakland.

So the A's have no lack of respect for the role of scouting; quite the contrary. Nor, Beane made clear, do they shy away from looking at a player's mental makeup when deciding which ones to bring into their organization.

The organization still very much believes in rigorous analysis. The rigor and discipline is applied, however, in the way the organization processes the information it collects, and not in declaring certain types of information off-limits.

"The proportion of objective versus subjective analysis is weighted more in some organizations than in others," he explained. "From our standpoint in Oakland, we're sort of forced into making objective decisions versus gut-feel decisions. If we in Oakland happen to be right on a gut-feel decision one time, my guess is it would be random. And we're not in a position to be making random decisions and hope we get lucky. If we're playing blackjack, and the dealer's showing a four and we have a six, hitting on the sixteen just doesn't make sense for us."

The key to making a good forecast, as we observed in chapter 2, is not in limiting yourself to quantitative information. Rather, it's having a good process for weighing the information appropriately. This is the essence of Beane's philosophy: collect as much information as possible, but then be as rigorous and disciplined as possible when analyzing it.

The litmus test for whether you are a competent forecaster is if more information makes your predictions better. If you're screwing it up, you have some bad habits and attitudes, like Phil Tetlock's political pundits did. If Prospect A is hitting .300 with twenty home runs and works at a soup kitchen during his off days, and Prospect B is hitting .300 with twenty home runs but hits up nightclubs and snorts coke during his free time, there is probably no way to *quantify* this distinction. But you'd sure as hell want to take it into account.

Many times, in fact, it is possible to translate qualitative information into quantitative information.* Scouts actually do rate players using a hard numerical scale, which ranges from 20 to 80 in each category. There's no reason you couldn't place that into a statistical model alongside a player's batting average[43] and see where it adds value; some teams, like the Cardinals, already attempt versions of this.

Indeed, the line between stats and scouting, and qualitative and quantitative information, has become very blurry in the baseball industry. Take, for example, the introduction of Pitch f/x, a system of three-dimensional cameras that have now been installed at every major-league stadium. Pitch f/x can measure

* In the model that I use to forecast the outcome of U.S. House elections, for instance, I combine quantitative information like polls with the qualitative ratings from experts like the Cook Political Report. If Cook says a particular race is a toss-up, I code that as a 0. If they put the race in the Lean Democrat category, I code that as a +1, and so forth. Combining this information produces a better forecast than either type of information taken alone.

not just how fast a pitch travels—that has been possible for years with radar guns—but how much it moves, horizontally and vertically, before reaching the plate. We can now say statistically, for instance, that Zack Greinke, a young pitcher with the Milwaukee Brewers who won the 2009 Cy Young Award as his league's best pitcher, has baseball's best slider,[44] or that Mariano Rivera's cut fastball is really as good as reputed.[45] Traditionally, these things were considered to be in the domain of scouting; now they're another variable that can be placed into a projection system.

We're not far from a point where we might have a complete three-dimensional recording of everything that takes place on a baseball field. We'll soon be able to measure exactly how good a jump Jacoby Ellsbury gets on a fly ball hit over his head. We'll know exactly how fast Ichiro Suzuki rounds the bases, or exactly how quickly Yadier Molina gets the ball down to second base when he's trying to throw out an opposing base-stealer.

This new technology will not kill scouting any more than *Moneyball* did, but it may change its emphasis toward the things that are even harder to quantify and where the information is more exclusive, like a player's mental tools. Smart scouts like Sanders are already ahead of the curve.

Why Pedroia Was Predicted to Fail . . .

But why were the scouts so wrong about Dustin Pedroia?

All the scouts were in agreement on the basic facts about him. Everyone knew that Pedroia was a very good hitter for average, that he had a smart approach at the plate, and that his mental toolbox was "off the charts." Everyone knew that he had a long swing; that his defense was steady but unspectacular; that his foot speed was no better than average; that he was short and did not have a terrific physique.

It was an idiosyncratic profile for a young player, however, and a lot of scouts didn't know what to make of it. "When you draw up a player, scouts have a feel for what they want to see," Sanders told me. "Prototypical standards. Dustin went against the grain in some of those areas, starting with his size."

When we can't fit a square peg into a round hole, we'll usually blame the peg—when sometimes it's the rigidity of our thinking that accounts for our failure to accommodate it. Our first instinct is to place information into categories—usually a relatively small number of categories since they'll be easier to keep track of. (Think of how the Census Bureau classifies people from hundreds of ethnic groups into just six racial categories or how thousands of artists are placed into a taxonomy of a few musical genres.)

This might work well enough most of the time. But when we have trouble categorizing something, we'll often overlook it or misjudge it. This is one of the reasons that Beane avoids what he calls "gut-feel" decisions. If he relies too heavily on his first impressions, he'll let potentially valuable prospects slip through the cracks—and he can't afford that with a payroll like Oakland's.

A system like PECOTA, which searches through thousands of players to find the ones with similar profiles, has a more rigorous way to categorize players. It could place Pedroia's skills more within their proper context.

PECOTA's search found some favorable precedents. Pedroia's short stature, for example, may actually have been an advantage given the rest of his skills. In baseball, the strike zone is defined as running from a player's shoulders to his knees. The smaller the athlete, the smaller the target the pitcher gets to throw at. A player like Pedroia with a good batting eye can take especial advantage of this.

Meanwhile, being lower to the ground can be an asset to a second baseman's defense. It's a position that relies on agility, having catlike reflexes to ground balls that come hot off the bat. Many of the best second basemen in baseball history have been very short. Of the seventeen in the Hall of Fame, only two—Nap Lajoie and Ryne Sandberg—were over six feet tall.[46] Joe Morgan, perhaps the greatest second baseman of all time, was just five seven.

Scouts are very good at what they do, but this is a case where they had categorized a player too quickly and too prejudicially. Pedroia's diminutive stature was in some ways a strength.

Still, there were no guarantees: PECOTA had not seen Pedroia's success as certain, just that the odds were in his favor. The scouts saw the odds as weighted against him. What made the difference is that the Red Sox believed in Dustin Pedroia. And fortunately for the Red Sox, Pedroia believed in Pedroia, too.

. . . and How He Beat the Odds

I met Bill James for the first time on a panel at the *New Yorker* Festival in October 2009. There was a fancy party afterward and he looked out of place among the fashionistas, wearing an exceedingly colorful sweater and a pair of scuff-marked clogs that looked like they were two sizes too big. While everyone else at the party was chasing down Susan Sarandon, we parked ourselves at the bar and chatted for a while.[47]

James's responsibilities with the Red Sox are varied—and kept close to the vest. (He wasn't able to go into much detail about them on the record.) But after having spent a quarter century writing on baseball as an outside agitator, he's mellowed into late adulthood. The sport looks a little different to him now that he is on the inside; he is quicker to recognize the mental aspects of the game.

"There are a lot of things I wrote in the eighties that weren't right," he told me. "The big change was my having children. I know it's a cliché, but once you have children you start to understand that everyone is somebody's baby. It is an insiders-outsiders thing. You grow up and these people are characters on TV or video games or baseball cards—you don't really think about the fact that these guys are humans and doing the best they can."

I was struck by how similar many of James's comments were to those of Beane and Sanders, even though they traditionally have approached the sport from very different angles. Indeed, if you put the transcripts of my conversations with James, Beane, and Sanders into a lineup, you would scarcely be able to pick them out (except that James's would be much funnier). He's come to recognize the value that the Red Sox scouts provide to the club and believes that it's parallel to his own mission. In baseball, success is measured in a very particular way—by W's and L's—so it's easy to keep everyone on the right track. If more information is making your predictions worse, you'll be out of a job—not entitled to a lifetime appointment to *The McLaughlin Group*.

"On a certain level the way that I see baseball and the way the scouts see baseball is pretty similar," James continued. "Maybe it's one of those things where if you go far enough to the right and far enough to the left on the political spectrum you find people saying the exact same thing. What scouts are trying to see is really the same thing that I am trying to see."

James was assisting with the Red Sox draft process in 2004 when they took Pedroia with the sixty-fifth pick. He had written a favorable report on Pedroia but recommended that they draft someone else. Nevertheless, he was pleased with the selection, and has enjoyed watching Pedroia make his critics look foolish.

There were a few moments early in his career, however, when even some of Pedroia's biggest fans were starting to have their doubts. Pedroia was first called up to the big leagues in August 2006, playing in thirty-one games but compiling just a .198 batting average and just six extra-base hits. Nobody was too concerned about this; the Red Sox, unusually for them, were well out of playoff contention in the final weeks of the season, and New England's attention was already turning to the Celtics and the Patriots. But the next year, entrusted with the full-time second base job, he started out slowly as well, his batting average just .172 after the first month of the season.

A team like the Cubs, who until recently were notorious for their haphazard decision-making process, might have cut Pedroia at this point. For many clubs, every action is met by an equal and opposite overreaction. The Red Sox, on the other hand, are disciplined by their more systematic approach. And when the Red Sox looked at Pedroia at that point in the season, James told me, they actually saw a lot to like. Pedroia was making plenty of contact with the baseball—it just hadn't been falling for hits. The numbers, most likely, would start to trend his way.

"We all have moments of losing confidence in the data," James told me. "You probably know this, but if you look back at the previous year, when Dustin hit .180 or something, if you go back and look at his swing-and-miss percentage, it was maybe about 8 percent, maybe 9 percent. It was the same during that period in the spring when he was struggling. It was always logically apparent—when you swing as hard as he does, there's no way in the world that you can make that much contact and keep hitting .180."

The Red Sox hadn't taken their decision to draft Pedroia lightly. They were still observing him do the same things that had made him successful at every other amateur and professional level. If they were going to bench Pedroia, the decision would have to be just as carefully considered as the one to promote him in the first place. They did not let the data dictate their decision without placing it in a larger context.

Their only concern, James told me, was whether Pedroia would start to doubt himself. And another player might have—but not Pedroia, who suffers neither fools nor critics.

"Fortunately, Dustin is really cocky, because if he was the kind of person who was intimidated—if he had listened to those people—it would have ruined him. He didn't listen to people. He continued to dig in and swing from his heels and eventually things turned around for him."

Pedroia has what John Sanders calls a "major league memory"—which is to say a short one. He isn't troubled by a slump, because he is damned sure that he's playing the game the right way, and in the long run, that's what matters. Indeed, he has very little tolerance for anything that distracts him from doing his job. This doesn't make him the most generous human being, but it is exactly what he needs in order to play second base for the Boston Red Sox, and that's the only thing that Pedroia cares about.

"Our weaknesses and our strengths are always very intimately connected," James said. "Pedroia made strengths out of things that would be weaknesses for other players."

The Real Lessons of *Moneyball*

"As Michael Lewis said, the debate is over," Billy Beane declared when we were discussing *Moneyball*. For a time, *Moneyball* was very threatening to people in the game; it seemed to imply that their jobs and livelihoods were at stake. But the reckoning never came—scouts were never replaced by computers. In fact, the demand to know what the future holds for different types of baseball players—whether couched in terms of scouting reports or statistical systems like PECOTA—still greatly exceeds the supply. Millions of dollars—and the outcome of future World Series—are put at stake each time a team decides which player to draft, whom to trade for, how much they should pay for a free agent.

Teams are increasingly using every tool at their disposal to make these decisions. The information revolution has lived up to its billing in baseball, even though it has been a letdown in so many other fields, because of the sport's unique combination of rapidly developing technology, well-aligned incentives, tough competition, and rich data.

This isn't necessarily making life easier for Beane, who expressed concern that the other teams have copied the A's best tricks. Very few teams, for instance, now fail to understand the importance of OBP or neglect the role played by defense—and what hasn't changed is that those teams still have a lot more money than the A's.

In the most competitive industries, like sports, the best forecasters must constantly innovate. It's easy to adopt a goal of "exploit market inefficiencies." But that doesn't really give you a plan for how to find them and then determine whether they represent fresh dawns or false leads. It's hard to have an idea that nobody else has thought of. It's even harder to have a *good* idea—and when you do, it will soon be duplicated.

That is why this book shies away from promoting quick-fix solutions that imply you can just go about your business in a slightly different way and outpredict the competition. Good innovators typically think very big *and* they think very small. New ideas are sometimes found in the most granular details of a problem where few others bother to look. And they are sometimes found when you are doing your most abstract and philosophical thinking, considering why the world is the way that it is and whether there might be an alternative to the dominant paradigm. Rarely can they be found in the temperate latitudes between these two spaces, where we spend 99 percent of our lives. The categorizations and approximations we make in the normal course of our lives are usually good enough to get by, but sometimes we let information that might give us a competitive advantage slip through the cracks.

The key is to develop tools and habits so that you are more often looking for ideas and information in the right places—and in honing the skills required to harness them into W's and L's once you've found them.

It's hard work. But baseball will remain an unusually fertile proving ground for innovators. There hasn't been a really groundbreaking forecasting system since PECOTA's debut ten years ago. But someone will come along and take advantage of Pitch f/x data in a smart way, or will figure out how to fuse quantitative and qualitative evaluations of player performance. All this will happen, and sooner rather than later—possibly in the time that this book is at the printer.

"The people who are coming into the game, the creativity, the intelligence—it's unparalleled right now," Beane told me. "In ten years if I applied for this job I wouldn't even get an interview."

Moneyball is dead; long live *Moneyball*.

4

FOR YEARS YOU'VE BEEN TELLING US THAT RAIN IS GREEN

On Tuesday, August 23, 2005, an Air Force reconnaissance plane picked up signs of a disturbance over the Bahamas.[1] There were "several small vortices," it reported, spirals of wind rotating in a counterclockwise motion from east to west—away from the expanse of the Atlantic and toward the United States. This disruption in wind patterns was hard to detect from clouds or from satellite data, but cargo ships were beginning to recognize it. The National Hurricane Center thought there was enough evidence to characterize the disturbance as a tropical cyclone, labeling it Tropical Depression Twelve. It was a "tricky" storm that might develop into something more serious or might just as easily dissipate; about half of all tropical depressions in the Atlantic Basin eventually become hurricanes.[2]

The depression strengthened quickly, however, and by Wednesday afternoon one of the Hurricane Center's computer models was already predicting a double landfall in the United States—a first one over southern Florida and a second that might "[take] the cyclone to New Orleans."[3] The storm had gathered enough strength to become a hurricane and it was given a name, Katrina.[4]

Katrina's first landfall—it passed just north of Miami and then zoomed through the Florida Everglades a few hours later as a Category 1 hurricane—had not been prolonged enough to threaten many lives. But it had also not been long enough to take much energy out of the storm. Instead, Katrina was gaining strength in the warm waters of the Gulf of Mexico. In the wee hours of Saturday morning the forecast really took a turn for the worse: Katrina had become a Category 3 hurricane, on its way to being a Category 5. And its forecast track had gradually been moving westward, away from the Florida Panhandle and toward Mississippi and Louisiana. The computer models were now in agreement: the storm seemed bound for New Orleans.[5]

"I think I had five congressional hearings after Katrina." said Max Mayfield, who was director of the National Hurricane Center at the time the storm hit, when I asked him to recall when he first recognized the full magnitude of the threat. "One of them asked me when I first became concerned with New Orleans. I said 'Sixty years ago.'"

A direct strike of a major hurricane on New Orleans had long been every weather forecaster's worst nightmare. The city presented a perfect set of circumstances that might contribute to the death and destruction there. On the one hand there was its geography: New Orleans does not border the Gulf of Mexico as much as sink into it. Much of the population lived below sea level and was counting on protection from an outmoded system of levees and a set of natural barriers that had literally been washing away to sea.[6] On the other hand there was its culture. New Orleans does many things well, but there are two things that it proudly refuses to do. New Orleans does not move quickly, and New Orleans does not place much faith in authority. If it did those things, New Orleans would not really be New Orleans. It would also have been much better prepared to deal with Katrina, since those are the exact two things you need to do when a hurricane threatens to strike.

The National Hurricane Center nailed its forecast of Katrina; it anticipated a potential hit on the city almost five days before the levees were breached, and concluded that some version of the nightmare scenario was probable more than forty-eight hours away. Twenty or thirty years ago, this much advance warning would almost certainly not have been possible, and fewer people would have been evacuated. The Hurricane Center's forecast, and the steady advances made in weather forecasting over the past few decades, undoubtedly saved many lives.

Not everyone listened to the forecast, however. About 80,000 New Orleanians[7]—almost a fifth of the city's population at the time—failed to evacuate the city, and 1,600 of them died. Surveys of the survivors found that about two-thirds of them did not think the storm would be as bad as it was.[8] Others had been confused by a bungled evacuation order; the city's mayor, Ray Nagin, waited almost twenty-four hours to call for a mandatory evacuation, despite pleas from Mayfield and from other public officials. Still other residents—impoverished, elderly, or disconnected from the news—could not have fled even if they had wanted to.

Weather forecasting is one of the success stories in this book, a case of man and machine joining forces to understand and sometimes anticipate the complexities of nature. That we can sometimes predict nature's course, however, does not mean we can alter it. Nor does a forecast do much good if there is no one willing to listen to it. The story of Katrina is one of human ingenuity and human error.

The Weather of Supercomputers

The supercomputer labs at the National Center for Atmospheric Research (NCAR) in Boulder, Colorado, literally produce their own weather. They are hot: the 77 trillion calculations that the IBM Bluefire supercomputer makes every second generate a substantial amount of radiant energy. They are windy: all that heat must be cooled, lest the nation's ability to forecast its weather be placed into jeopardy, and so a series of high-pressure fans blast oxygen on the computers at all times. And they are noisy: the fans are loud enough that hearing protection is standard operating equipment.

The Bluefire is divided into eleven cabinets, each about eight feet tall and two feet wide with a bright green racing stripe running down the side. From the back, they look about how you might expect a supercomputer to look: a mass of crossed cables and blinking blue lights feeding into the machine's brain stem. From the front, they are about the size and shape of a portable toilet, complete with what appears to be a door with a silver handle.

"They look a little bit like Porta-Potties," I tell Dr. Richard Loft, the director of technology development for NCAR, who oversees the supercomputer lab.

Those in the meteorology business are used to being the butt of jokes. Larry David, in the show *Curb Your Enthusiasm*, posits that meterologists sometimes predict rain when there won't be just so they can get a head start on everyone else at the golf course.[9] Political commercials use weather metaphors as a basis to attack their opponents,[10] usually to suggest that they are always flip-flopping on the issues. Most people assume that weather forecasters just aren't very good at what they do.

Indeed, it was tempting to look at the rows of whirring computers and wonder if this was all an exercise in futility: All this to forecast the weather? And they still can't tell us whether it's going to rain tomorrow?

Loft did not look amused. Improved computing power has not really improved earthquake or economic forecasts in any obvious way. But meteorology is a field in which there has been considerable, even remarkable, progress. The power of Loft's supercomputers is a big part of the reason why.

A Very Brief History of Weather Forecasting

"Allow me to deviate from the normal flight plan," Loft said back in his office. He proved to have a sense of humor after all—quirky and offbeat, like a more self-aware version of Dwight Schrute from *The Office*.* From the very beginnings of history, Loft explained, man has tried to predict his environment. "You go back to Chaco Canyon or Stonehenge and people realized they could predict the shortest day of the year and the longest day of the year. That the moon moved in predictable ways. But there are things an ancient man couldn't predict. Ambush from some kind of animal. A flash flood or a thunderstorm."

* Loft bears some physical resemblance to Rainn Wilson, the actor who plays Schrute.

Today we might take it for granted that we can predict where a hurricane will hit days in advance, but meteorology was very late to develop into a successful science. For centuries, progress was halting. The Babylonians, who were advanced astronomers, produced weather prognostications that have been preserved on stone tablets for more than 6,000 years.[11] Ultimately, however, they deferred to their rain god Ningirsu. Aristotle wrote a treatise on meteorology[12] and had a few solid intuitions, but all in all it was one of his feebler attempts. It's only been in the past fifty years or so, as computer power has improved, that any real progress has been made.

You might not think of the weather report as an exercise in metaphysics, but the very idea of predicting the weather evokes age-old debates about predestination and free will. "Is everything written, or do we write it ourselves?" Loft asked. "This has been a basic problem for human beings. And there really were two lines of thought.

"One comes through Saint Augustine and Calvinism," he continued, describing people who believed in predestination. Under this philosophy, humans might have the ability to predict the course they would follow. But there was nothing they could do to alter it. Everything was carried out in accordance with God's plan. "This is against the Jesuits and Thomas Aquinas who said we actually have free will. This question is about whether the world is predictable or unpredictable."

The debate about predictability began to be carried out on different terms during the Age of Enlightenment and the Industrial Revolution. Isaac Newton's mechanics had seemed to suggest that the universe was highly orderly and predictable, abiding by relatively simple physical laws. The idea of scientific, technological, and economic progress—which by no means could be taken for granted in the centuries before then—began to emerge, along with the notion that mankind might learn to control its own fate. Predestination was subsumed by a new idea, that of scientific *determinism*.

The idea takes on various forms, but no one took it further than Pierre-Simon Laplace, a French astronomer and mathematician. In 1814, Laplace made the following postulate, which later came to be known as Laplace's Demon:

> We may regard the present state of the universe as the effect of its past and the cause of its future. An intellect which at a certain moment would know all forces that set nature in motion, and all positions of all items of which nature is composed, if this intellect were also vast enough to submit these data to analysis, it would embrace in a single formula the movements of the greatest bodies of the universe and those of the tiniest atom; for such an intellect nothing would be uncertain and the future just like the past would be present before its eyes.[13]

Given perfect knowledge of present conditions ("all positions of all items of which nature is composed"), and perfect knowledge of the laws that govern the universe ("all forces that set nature in motion"), we ought to be able to make *perfect* predictions ("the future just like the past would be present"). The movement of every particle in the universe should be as predictable as that of the balls on a billiard table. Human beings might not be up to the task, Laplace conceded. But if we were smart enough (and if we had fast enough computers) we could predict the weather and everything else—and we would find that nature itself is perfect.

Laplace's Demon has been controversial for all its two-hundred-year existence. At loggerheads with the determinists are the probabilists, who believe that the conditions of the universe are knowable only with some degree of uncertainty.* Probabilism was, at first, mostly an *epistemological* paradigm: it avowed that there were limits on man's ability to come to grips with the universe. More recently, with the discovery of quantum mechanics, scientists and philosophers have asked whether the universe *itself* behaves probabilistically. The particles Laplace sought to identify begin to behave like waves when you look closely enough—they seem to occupy no fixed position. How can you predict where something is going to go when you don't know where it is in the first place? You can't. This is the basis for the theoretical physicist Werner Heisenberg's famous uncertainty principle.[14] Physicists interpret the uncertainty

* As we will see in chapter 8, it is not so easy to characterize Laplace along this spectrum. He was an eclectic thinker and was instrumental in the development of probability theory, motivated in part by his conviction that human beings were unlikely to live up to the perfection he saw in nature.

principle in different ways, but it suggests that Laplace's postulate cannot literally be true. *Perfect* predictions are impossible if the universe itself is random.

Fortunately, weather does not require quantum mechanics for us to study it. It happens at a molecular (rather than an atomic) level, and molecules are much too large to be discernibly impacted by quantum physics. Moreover, we understand the chemistry and Newtonian physics that govern the weather fairly well, and we have for a long time.

So what about a revised version of Laplace's Demon? If we knew the position of every molecule in the earth's atmosphere—a much humbler request than deigning to know the position of every atomic particle in the universe—could we make perfect weather predictions? Or is there a degree of randomness inherent in the weather as well?

The Matrix

Purely statistical predictions about the weather have long been possible. Given that it rained today, what is the probability that it will rain tomorrow? A meteorologist could look up all the past instances of rain in his database and give us an answer about that. Or he could look toward long-term averages: it rains about 35 percent of the time in London in March.[15]

The problem is that these sorts of predictions aren't very useful—not precise enough to tell you whether to carry an umbrella, let alone to forecast the path of a hurricane. So meteorologists have been after something else. Instead of a statistical model, they wanted a living and breathing one that simulated the physical processes that govern the weather.

Our ability to compute the weather has long lagged behind our theoretical understanding of it, however. We know which equations to solve and roughly what the right answers are, but we just aren't fast enough to calculate them for every molecule in the earth's atmosphere. Instead, we have to make some approximations.

The most intuitive way to do this is to simplify the problem by breaking the atmosphere down into a finite series of pixels—what meteorologists variously refer to as a matrix, a lattice, or a grid. According to Loft, the earliest credible attempt to do this was made in 1916 by Lewis Fry Richardson, a prolific English physicist. Richardson wanted to determine the weather over northern Germany at a particular time: at 1 P.M. on May 20, 1910. This was not, strictly speaking, a *prediction*, the date being some six years in the past. But Richardson had a lot of data: a series of observations of temperature, barometric pressures and wind speeds that had been gathered by the German government. And he had a lot of time: he was serving in northern France as part of a volunteer ambulance unit and had little to do in between rounds of artillery fire. So Richardson broke Germany down into a series of two-dimensional boxes, each measuring three degrees of latitude (about 210 miles) by three degrees of longitude across. Then he went to work attempting to solve the chemical equations that governed the weather in each square and how they might affect weather in the adjacent ones.

FIGURE 4-1: RICHARDSON'S MATRIX: THE BIRTH OF MODERN WEATHER FORECASTING

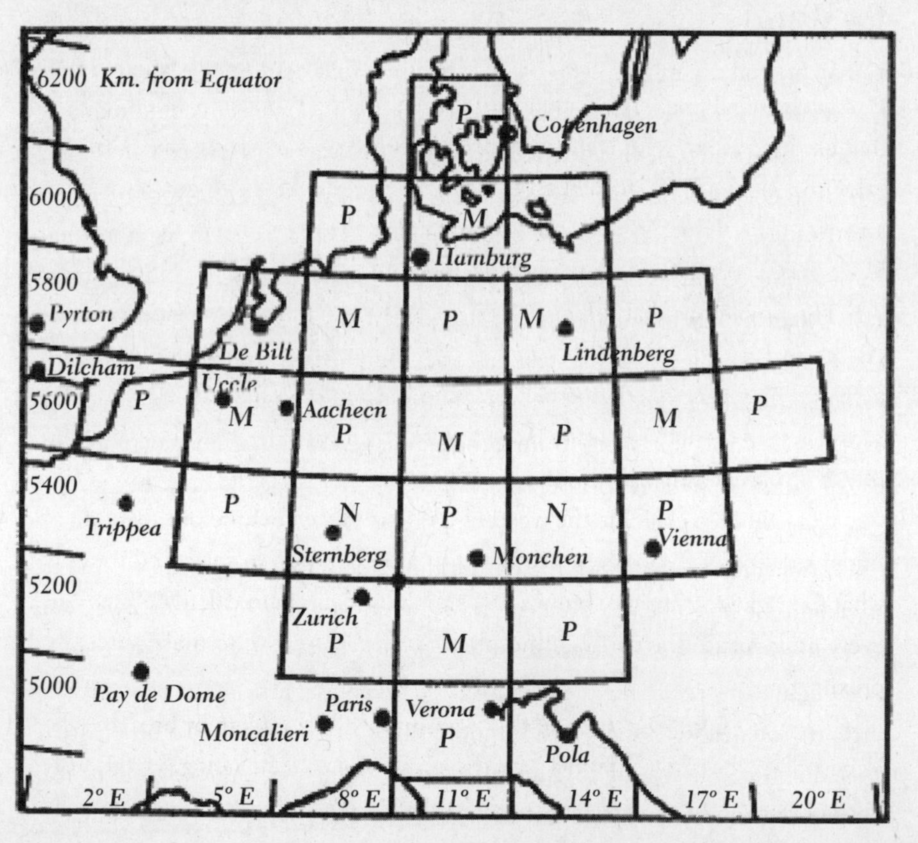

Richardson's experiment, unfortunately, failed miserably[16]—it "predicted" a dramatic rise in barometric pressure that hadn't occurred in the real world on the day in question. But he published his results nevertheless. It certainly *seemed* like the right way to predict the weather—to solve it from first principles, taking advantage of our strong theoretical understanding of how the system behaves, rather than relying on a crude statistical approximation.

The problem was that Richardson's method required an awful lot of work. Computers were more suitable to the paradigm that he had established. As you'll see in chapter 9, computers aren't good at every task we hope they might accomplish and have been far from a panacea for prediction. But computers are very good at computing: at repeating the same arithmetic tasks over and over again and doing so quickly and accurately. Tasks like chess that abide by relatively simple rules, but which are difficult computationally, are right in their wheelhouse. So, potentially, was the weather.

The first computer weather forecast was made in 1950 by the mathematician John von Neumann, who used a machine that could make about 5,000 calculations per second.[17] That was a lot faster than Richardson could manage with a pencil and paper in a French hay field. Still, the forecast wasn't any good, failing to do any better than a more-or-less random guess.

Eventually, by the mid-1960s, computers would start to demonstrate some skill at weather forecasting. And the Bluefire—some 15 *billion* times faster than the first computer forecast and perhaps a *quadrillion* times faster than Richardson—displays quite a bit of acumen because of the speed of computation. Weather forecasting is much better today than it was even fifteen or twenty years ago. But, while computing power has improved exponentially in recent decades, progress in the accuracy of weather forecasts has been steady but slow.

There are essentially two reasons for this. One is that the world isn't one or two dimensional. The most reliable way to improve the accuracy of a weather forecast—getting one step closer to solving for the behavior of each molecule—is to reduce the size of the grid that you use to represent the atmosphere. Richardson's squares were about two hundred miles by two hundred miles across, providing for at best a highly generalized view of the planet (you could nearly squeeze both New York and Boston—which can have very different weather—

into the same two hundred by two hundred square). Suppose you wanted to reduce the diameter of the squares in half, to a resolution of one hundred by one hundred. That improves the precision of your forecast, but it also increases the number of equations you need to solve. In fact, it would increase this number not twofold but fourfold—since you're doubling the magnitude both lengthwise and widthwise. That means, more or less, that you need four times as much computer power to produce a solution.

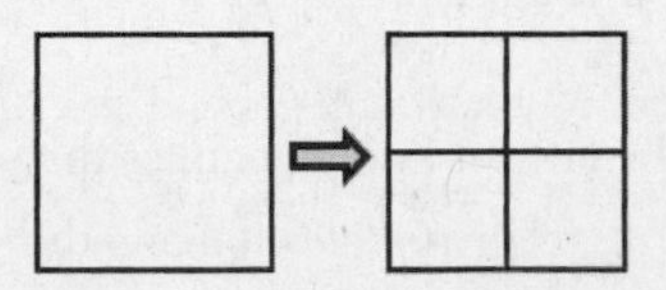

But there are more dimensions to worry about than just two. Different patterns can take hold in the upper atmosphere, in the lower atmosphere, in the oceans, and near the earth's surface. In a three-dimensional universe, a twofold increase in the resolution of our grid will require an *eight*fold increase in computer power:

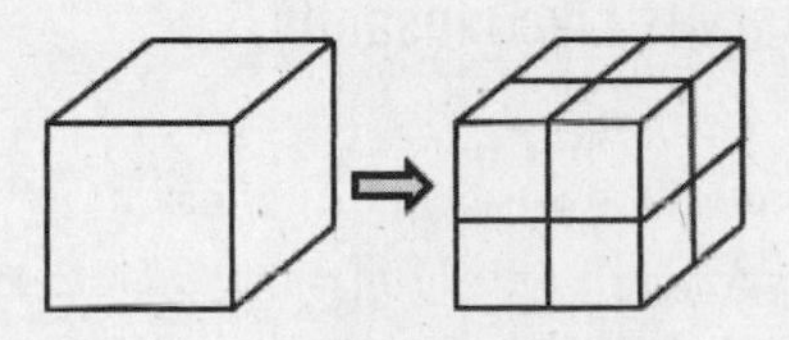

And then there is the fourth dimension: time. A meteorological model is no good if it's static—the idea is to know how the weather is changing from one moment to the next. A thunderstorm moves at about forty miles per hour: if you have a three-dimensional grid that is forty by forty by forty across, you can monitor the storm's movement by collecting one observation every hour. But if you halve the dimensions of the grid to twenty by twenty by twenty, the storm will now pass through one of the boxes every *half* hour. That means you need

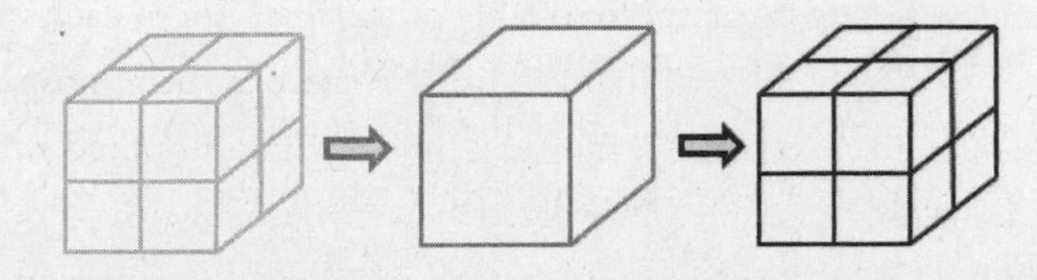

to halve the time parameter as well—again doubling your requirement to sixteen times as much computing power as you had originally.

If this was the only problem it wouldn't be prohibitive. Although you need, roughly speaking, to get ahold of sixteen times more processing power in order to double the resolution of your weather forecast, processing power has been improving exponentially—doubling about once every two years.[18] That means you only need to wait eight years for a forecast that should be twice as powerful; this is about the pace, incidentally, at which NCAR has been upgrading its supercomputers.

Say you've solved the laws of fluid dynamics that govern the movement of weather systems. They're relatively Newtonian: the uncertainty principle—interesting as it might be to physicists—won't bother you much. You've gotten your hands on a state-of-the-art piece of equipment like the Bluefire. You've hired Richard Loft to design the computer's software and to run its simulations. What could possibly go wrong?

How Chaos Theory Is Like Linsanity

What could go wrong? Chaos theory. You may have heard the expression: *the flap of a butterfly's wings in Brazil can set off a tornado in Texas.* It comes from the title of a paper[19] delivered in 1972 by MIT's Edward Lorenz, who began his career as a meteorologist. Chaos theory applies to systems in which each of two properties hold:

1. The systems are *dynamic*, meaning that the behavior of the system at one point in time influences its behavior in the future;
2. And they are *nonlinear*, meaning they abide by exponential rather than additive relationships.

Dynamic systems give forecasters plenty of problems—as I describe in chapter 6, for example, the fact that the American economy is continually evolving in a chain reaction of events is one reason that it is very difficult to predict. So do nonlinear ones: the mortgage-backed securities that triggered the financial crisis were designed in such a way that small changes in macroeconomic conditions could make them exponentially more likely to default.

When you combine these properties, you can have a real mess. Lorenz did not realize just how profound the problems were until, in the tradition of Alexander Fleming and penicillin[20] or the New York Knicks and Jeremy Lin, he made a major discovery purely by accident.

Lorenz and his team were working to develop a weather forecasting program on an early computer known as a Royal McBee LGP-30.[21] They thought they were getting somewhere until the computer started spitting out erratic results. They began with what they thought was *exactly* the same data and ran what they thought was *exactly* the same code—but the program would forecast clear skies over Kansas in one run, and a thunderstorm in the next.

After spending weeks double-checking their hardware and trying to debug their program, Lorenz and his team eventually discovered that their data wasn't *exactly* the same: one of their technicians had truncated it in the third decimal place. Instead of having the barometric pressure in one corner of their grid read 29.5168, for example, it might instead read 29.517. Surely this couldn't make that much difference?

Lorenz realized that it could. The most basic tenet of chaos theory is that a small change in initial conditions—a butterfly flapping its wings in Brazil—can produce a large and unexpected divergence in outcomes—a tornado in Texas. This does not mean that the behavior of the system is *random,* as the term "chaos" might seem to imply. Nor is chaos theory some modern recitation of Murphy's Law ("whatever can go wrong will go wrong"). It just means that certain types of systems are very hard to predict.

The problem begins when there are inaccuracies in our data. (Or inaccuracies in our assumptions, as in the case of mortgage-backed securities). Imagine that we're *supposed* to be taking the sum of 5 and 5, but we keyed in the second number wrong. Instead of adding 5 and 5, we add 5 and 6. That will give us an answer of 11 when what we really want is 10. We'll be wrong, but not by much: addition, as a linear operation, is pretty forgiving. Exponential operations, however, extract a lot more punishment when there are inaccuracies in our data. If instead of taking 5^5—which should be 3,215—we instead take 5^6 (five to the sixth power), we wind up with an answer of 15,625. That's way off: we've missed our target by 500 percent.

This inaccuracy quickly gets worse if the process is *dynamic*, meaning that our outputs at one stage of the process become our inputs in the next. For instance, say that we're supposed to take five to the fifth power, and then take whatever result we get and apply it to the fifth power again. If we'd made the error described above, and substituted a 6 for the second 5, our results will now be off by a factor of more than 3,000.[22] Our small, seemingly trivial mistake keeps getting larger and larger.

The weather is the epitome of a dynamic system, and the equations that govern the movement of atmospheric gases and fluids are nonlinear—mostly differential equations.[23] Chaos theory therefore most definitely applies to weather forecasting, making the forecasts highly vulnerable to inaccuracies in our data.

Sometimes these inaccuracies arise as the result of human error. The more fundamental issue is that we can only observe our surroundings with a certain degree of precision. No thermometer is perfect, and if it's off in even the third or the fourth decimal place, this can have a profound impact on the forecast.

Figure 4-2 shows the output of fifty runs from a European weather model, which was attempting to make a weather forecast for France and Germany on Christmas Eve, 1999. All these simulations are using the same software, and all are making the same assumptions about how the weather behaves. In fact, the models are completely deterministic: they assume that we could forecast the weather perfectly, if only we knew the initial conditions perfectly. But small changes in the input can produce large differences in the output. The European forecast attempted to account for these errors. In one simulation, the barometric pressure in Hanover might be perturbed just slightly. In another, the wind conditions in Stuttgart are permuted by a fraction of a percent. These small changes might be enough for a huge storm system to develop in Paris in some simulations, while it's a calm winter evening in others.

This is the process by which modern weather forecasts are made. These small changes, introduced intentionally in order to represent the inherent uncertainty in the quality of the observational data, turn the deterministic forecast

FIGURE 4-2: DIVERGENT WEATHER FORECASTS WITH SLIGHTLY DIFFERENT INITIAL CONDITIONS

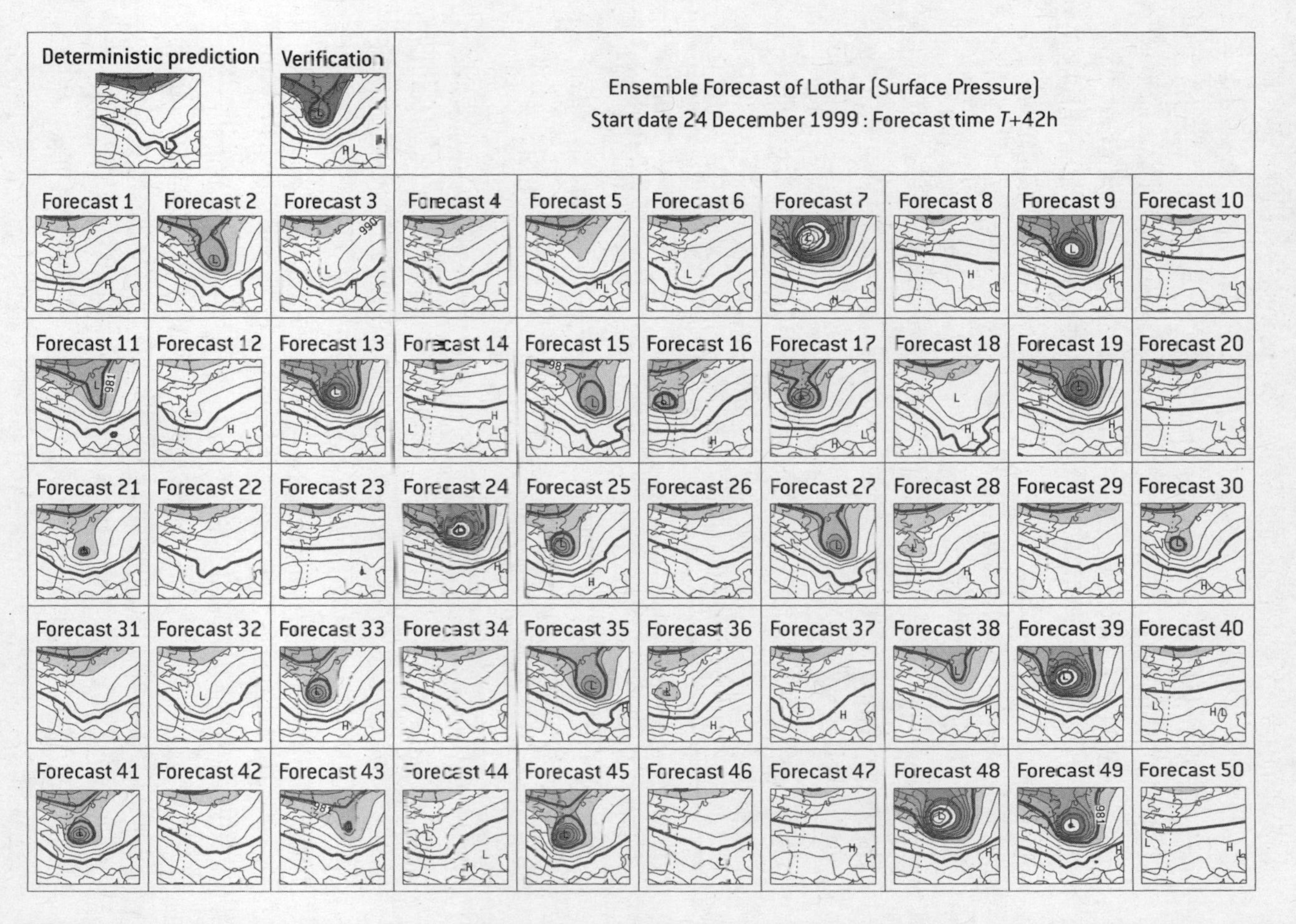

into a probabilistic one. For instance, if your local weatherman tells you that there's a 40 percent chance of rain tomorrow, one way to interpret that is that in 40 percent of his simulations, a storm developed, and in the other 60 percent—using just slightly different initial parameters—it did not.

It is still not quite that simple, however. The programs that meteorologists use to forecast the weather are quite good, but they are not perfect. Instead, the forecasts you actually see reflect a combination of computer and human judgment. Humans can make the computer forecasts better or they can make them worse.

The Vision Thing

The World Weather Building is an ugly, butterscotch-colored, 1970s-era office building in Camp Springs, Maryland, about twenty minutes outside Washington. The building forms the operational headquarters of NOAA—the National Oceanic and Atmospheric Administration—which is the parent of the National Weather Service (NWS) on the government's organization chart.[24] In contrast to NCAR's facilities in Boulder, which provide for sweeping views of the Front Range of the Rocky Mountains, it reminds one of nothing so much as bureaucracy.

The Weather Service was initially organized under the Department of War by President Ulysses S. Grant, who authorized it in 1870. This was partly because President Grant was convinced that only a culture of military discipline could produce the requisite accuracy in forecasting[25] and partly because the whole enterprise was so hopeless that it was only worth bothering with during wartime when you would try almost anything to get an edge.

The public at large became more interested in weather forecasting after the Schoolhouse Blizzard of January 1888. On January 12 that year, initially a relatively warm day in the Great Plains, the temperature dropped almost 30 degrees in a matter of a few hours and a blinding snowstorm came.[26] Hundreds of children, leaving school and caught unaware as the blizzard hit, died of hypothermia on their way home. As crude as early weather forecasts were, it was hoped that they might at least be able to provide some warning about an event so severe. So the National Weather Service was moved to the Department of Agriculture and took on a more civilian-facing mission.*

The Weather Service's origins are still apparent in its culture today. Its weather forecasters work around the clock for middling salaries[27] and they see themselves as public servants. The meteorologists I met in Camp Springs were patriotic people, rarely missing an opportunity to remind me about the importance that weather forecasting plays in keeping the nation's farms, small businesses, airlines, energy sector, military, public services, golf courses, picnic lunches, and schoolchildren up and running, all for pennies on the dollar. (The NWS gets by on just $900 million per year[28]—about $3 per U.S. citizen—even though weather has direct effects on some 20 percent of the nation's economy.[29])

Jim Hoke, one of the meteorologists I met, is the director of the NWS's Hydrometeorological Prediction Center. He is also a thirty-five-year veteran of the field, having taken his turn both on the computational side of the NWS (helping to build the computer models that his forecasters use) and on the operational side (actually making those forecasts and communicating them to the public). As such, he has some perspective on how man and machine intersect in the world of meteorology.

What is it, exactly, that humans can do better than computers that can crunch numbers at seventy-seven teraFLOPS? They can see. Hoke led me onto the forecasting floor, which consisted of a series of workstations marked with blue overhanging signs with such legends as MARITIME FORECAST CENTER and NATIONAL CENTER. Each station was manned by one or two meterologists—accompanied by an armada of flat-screen monitors that displayed full-color maps of every conceivable type of weather data for every corner of the country. The forecasters worked quietly and quickly, with a certain amount of Grant's military precision.[30]

Some of the forecasters were drawing on these maps with what appeared to be a light pen, painstakingly adjusting the contours of temperature gradients produced by the computer models—fifteen miles westward over the Mississippi

* That was not the Weather Service's final move. In 1940, with an eye toward assisting the Civil Aeronautics Administration and the bourgeoning industry of manned flight, the Congress migrated it to the Department of Commerce, where it resides today.

Delta, thirty miles northward into Lake Erie. Gradually, they were bringing them one step closer to the Platonic ideal they were hoping to represent.

The forecasters know the flaws in the computer models. These inevitably arise because—as a consequence of chaos theory—even the most trivial bug in the model can have potentially profound effects. Perhaps the computer tends to be too conservative on forecasting nighttime rainfalls in Seattle when there's a low-pressure system in Puget Sound. Perhaps it doesn't know that the fog in Acadia National Park in Maine will clear up by sunrise if the wind is blowing in one direction, but can linger until midmorning if it's coming from another. These are the sorts of distinctions that forecasters glean over time as they learn to work around the flaws in the model, in the way that a skilled pool player can adjust to the dead spots on the table at his local bar.

The unique resource that these forecasters were contributing was their eyesight. It is a valuable tool for forecaters in any discipline—a visual inspection of a graphic showing the interaction between two variables is often a quicker and more reliable way to detect outliers in your data than a statistical test. It's also one of those areas where computers lag well behind the human brain. Distort a series of letters just slightly—as with the CAPTCHA technology that is often used in spam or password protection—and very "smart" computers get very confused. They are too literal-minded, unable to recognize the pattern once its subjected to even the slightest degree of manipulation. Humans by contrast, out of pure evolutionary necessity, have very powerful visual cortexes. They rapidly parse through any distortions in the data in order to identify abstract qualities like pattern and organization—qualities that happen to be very important in different types of weather systems.

FIGURE 4-3: CAPTCHA

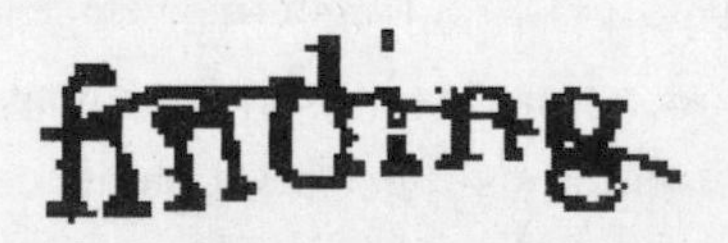

Indeed, back in the old days when meterological computers weren't much help at all, weather forecasting was almost entirely a visual process. Rather than flat screens, weather offices were instead filled with a series of light tables, illuminating maps that meterologists would mark with chalk or drafting pencils, producing a weather forecast fifteen miles at a time. Although the last light table was retired many years ago, the spirit of the technique survives today.

The best forecasters, Hoke explained, need to think visually and abstractly while at the same time being able to sort through the abundance of information the computer provides them with. Moreover, they must understand the dynamic and nonlinear nature of the system they are trying to study. It is not an easy task, requiring vigorous use of both the left and right brain. Many of his forecasters would make for good engineers or good software designers, fields where they could make much higher incomes, but they choose to become meteorologists instead.

The NWS keeps two different sets of books: one that shows how well the computers are doing by themselves and another that accounts for how much value the humans are contributing. According to the agency's statistics, humans improve the accuracy of precipitation forecasts by about 25 percent over the computer guidance alone,[31] and temperature forecasts by about 10 percent.[32] Moreover, according to Hoke, these ratios have been relatively constant over time: as much progress as the computers have made, his forecasters continue to add value on top of it. Vision accounts for a lot.

Being Struck by Lightning Is Increasingly Unlikely

When Hoke began his career, in the mid-'70s, the jokes about weather forecasters had some grounding in truth. On average, for instance, the NWS was missing the high temperature by about 6 degrees when trying to forecast it three days in advance (figure 4-4). That isn't much better than the accuracy you could get just by looking up a table of long-term averages. The partnership between man and machine is paying big dividends, however. Today, the average miss is about 3.5 degrees, meaning that almost half the inaccuracy has been stripped out.

Weather forecasters are also getting better at predicting severe weather. What are your odds of being struck—and killed—by lightning? Actually, this is not a constant number; they depend on how likely you are to be outdoors when

FIGURE 4-4: AVERAGE HIGH TEMPERATURE ERROR IN NWS FORECASTS

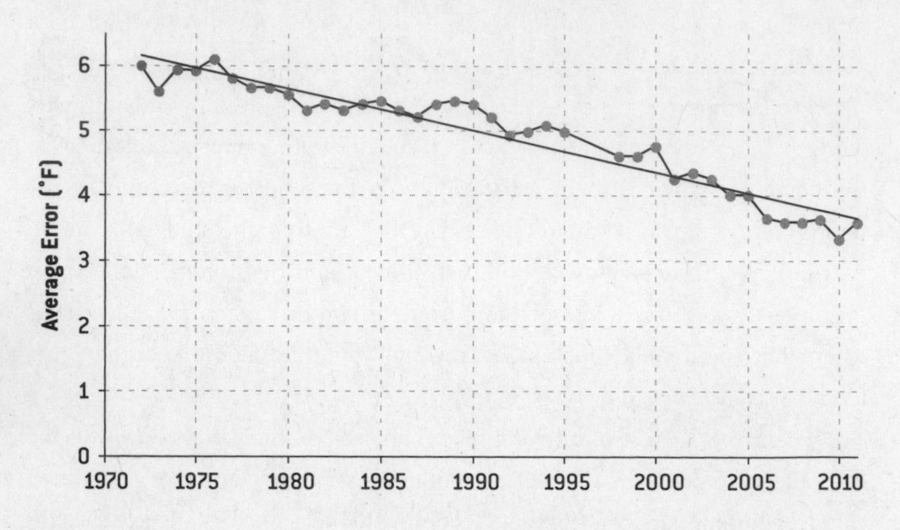

lightning hits and unable to seek shelter in time because you didn't have a good forecast. In 1940, the chance of an American being killed by lightning in a given year was about 1 in 400,000.[33] Today, it's just 1 chance in 11,000,000, making it almost thirty times less likely. Some of this reflects changes in living patterns (more of our work is done indoors now) and improvement in communications technology and medical care, but it's also because of better weather forecasts.

Perhaps the most impressive gains have been in hurricane forecasting. Just twenty-five years ago, when the National Hurricane Center tried to forecast where a hurricane would hit three days in advance of landfall, it missed by an average of 350 miles.[34] That isn't very useful on a human scale. Draw a 350-mile radius outward from New Orleans, for instance, and it covers all points from Houston, Texas, to Tallahassee, Florida (figure 4-5). You can't evacuate an area that large.

Today, however, the average miss is only about one hundred miles, enough to cover only southeastern Louisiana and the southern tip of Mississippi. The hurricane will still hit outside that circle some of the time, but now we are looking at a relatively small area in which an impact is even money or better—small enough that you could plausibly evacuate it seventy-two hours in advance. In 1985, by contrast, it was not until twenty-four hours in advance of landfall that

FIGURE 4-5: IMPROVEMENT IN HURRICANE TRACK FORECASTING

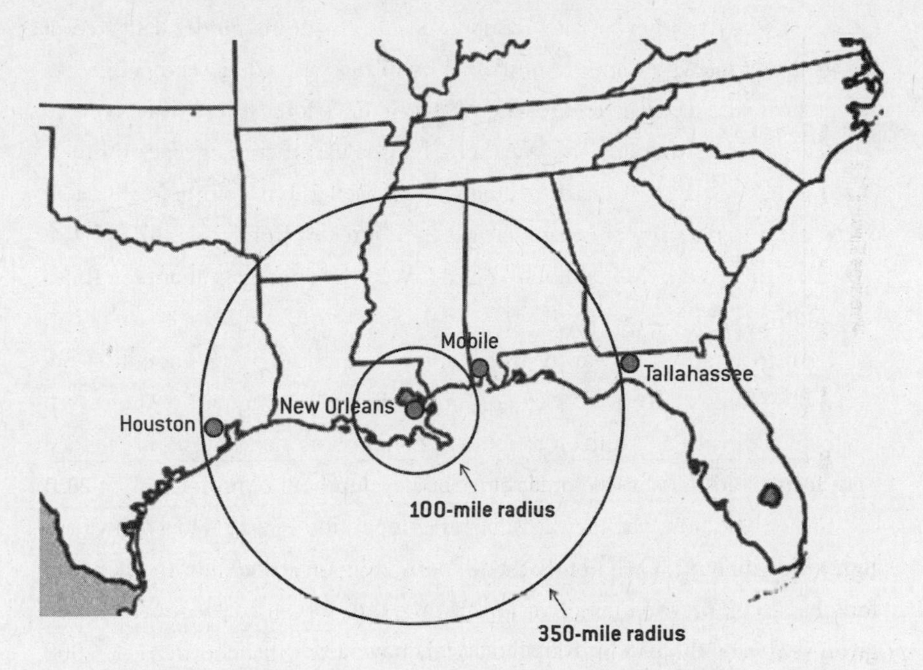

hurricane forecasts displayed the same skill. What this means is that we now have about forty-eight hours of additional warning time before a storm hits—and as we will see later, every hour is critical when it comes to evacuating a city like New Orleans.*

The Weather Service hasn't yet slain Laplace's Demon, but you'd think they might get more credit than they do. The science of weather forecasting is a success story despite the challenges posed by the intricacies of the weather system. As you'll find throughout this book, cases like these are more the exception than the rule when it comes to making forecasts. (Save your jokes for the economists instead.)

Instead, the National Weather Service often goes unappreciated. It faces stiff competition from private industry,[35] competition that occurs on a some-

* Unfortunately, although forecasters have gotten much better at figuring out where a hurricane will hit, they haven't really gotten any better at predicting how powerful it will be when it does. The reason is that the forces that govern the intensity of a storm occur at much smaller scales than the ones that determine its track. This means they require a much finer lattice, and even the Bluefire is not yet up to the task.

what uneven playing field. In contrast to most of its counterparts around the world, the Weather Service is supposed to provide its model data free of charge to anyone who wants it (most other countries with good weather bureaus charge licensing or usage fees for their government's forecasts). Private companies like AccuWeather and the Weather Channel can then piggyback off their handiwork to develop their own products and sell them commercially. The overwhelming majority of consumers get their forecast from one of the private providers; the Weather Channel's Web site, Weather.com, gets about ten times more traffic than Weather.gov.[36]

I am generally a big fan of free-market competition, or competition between the public and private sectors. Competition was a big part of the reason that baseball evolved as quickly as it did to better combine the insights gleaned from scouts and statistics in forecasting the development of prospects.

In baseball, however, the yardstick for competition is clear: How many ballgames did you win? (Or if not that, how many ballgames did you win relative to how much you spent.) In weather forecasting, the story is a little more complicated, and the public and private forecasters have differing agendas.

What Makes a Forecast Good?

"A pure researcher wouldn't be caught dead watching the Weather Channel, but lots of them do behind closed doors," Dr. Bruce Rose, the affable principal scientist and vice president at the Weather Channel (TWC), informed me. Rose wasn't quite willing to say that TWC's forecasts are *better* than those issued by the government, but they are *different*, he claimed, and oriented more toward the needs of a typical consumer.

"The models typically aren't measured on how well they predict practical weather elements," he continued. "It's really important if, in New York City, you get an inch of rain rather than ten inches of snow.[37] That's a huge [distinction] for the average consumer, but scientists just aren't interested in that."

Much of Dr. Rose's time, indeed, is devoted to highly pragmatic and even somewhat banal problems related to how customers interpret his forecasts. For instance: how to develop algorithms that translate raw weather data into everyday verbiage. What does *bitterly cold* mean? A *chance of flurries*? Just where is the dividing line between *partly cloudy* and *mostly cloudy*? The Weather Channel needs to figure this out, and it needs to establish formal rules for doing so, since it issues far too many forecasts for the verbiage to be determined on an ad hoc basis.

Sometimes the need to adapt the forecast to the consumer can take on comical dimensions. For many years, the Weather Channel had indicated rain on their radar maps with green shading (occasionally accompanied by yellow and red for severe storms). At some point in 2001, someone in the marketing department got the bright idea to make rain blue instead—which is, after all, what we think of as the color of water. The Weather Channel was quickly besieged with phone calls from outraged—and occasionally terrified—consumers, some of whom mistook the blue blotches for some kind of heretofore unknown precipitation (plasma storms? radioactive fallout?). "That was a nuclear meltdown," Dr. Rose told me. "Somebody wrote in and said, 'For years you've been telling us that rain is green—and now it's blue? What madness is this?'"

But the Weather Channel also takes its meteorology very seriously. And at least in theory, there is reason to think that they might be able to make a better forecast than the government. The Weather Channel, after all, gets to use *all* of the government's raw data as their starting point and then add whatever value they might be able to contribute on their own.

The question is, what is a "better" forecast? I've been defining it simply as a more accurate one. But there are some competing ideas, and they are pertinent in weather forecasting.

An influential 1993 essay[38] by Allan Murphy, then a meteorologist at Oregon State University, posited that there were three definitions of forecast quality that were commonplace in the weather forecasting community. Murphy wasn't necessarily advocating that one or another definition was better; he was trying to faciliate a more open and honest conversation about them. Versions of these definitions can be applied in almost any field in which forecasts or predictions are made.

One way to judge a forecast, Murphy wrote—perhaps the most obvious one—was through what he called "quality," but which might be better defined as **accuracy**. That is, did the actual weather match the forecast?

A second measure was what Murphy labeled "consistency" but which I think of as **honesty**. However accurate the forecast turned out to be, was it the best one the forecaster was capable of at the time? Did it reflect her best judgment, or was it modified in some way before being presented to the public?

Finally, Murphy said, there was the **economic value** of a forecast. Did it help the public and policy makers to make better decisions?

Murphy's distinction between accuracy and honesty is subtle but important. When I make a forecast that turns out to be wrong, I'll often ask myself whether it was the best forecast I could have made given what I knew at the time. Sometimes I'll conclude that it was: my thought process was sound; I had done my research, built a good model, and carefully specified how much uncertainty there was in the problem. Other times, of course, I'll find that there was something I didn't like about it. Maybe I had too hastily dismissed a key piece of evidence. Maybe I had overestimated the predictability of the problem. Maybe I had been biased in some way, or otherwise had the wrong incentives.

I don't mean to suggest that you should beat yourself up every time your forecast is off the mark. To the contrary, one sign that you have made a good forecast is that you are equally at peace with however things turn out—not all of which is within your immediate control. But there is always room to ask yourself what objectives you had in mind when you made your decision.

In the long run, Murphy's goals of accuracy and honesty should converge when we have the right incentives. But sometimes we do not. The political commentators on *The McLaughlin Group*, for instance, probably cared more about sounding smart on telvision than about making accurate predictions. They may well have been behaving rationally. But if they were deliberately making bad forecasts because they wanted to appeal to a partisan audience, or to be invited back on the show, they failed Murphy's honesty-in-forecasting test.

Murphy's third criterion, the economic value of a forecast, can complicate matters further. One can sympathize with Dr. Rose's position that, for instance, a city's forecast might deserve more attention if it is close to its freezing point, and its precipitation might come down as rain, ice, or snow, each of which would have different effects on the morning commute and residents' safety. This, however, is more a matter of where the Weather Channel focuses its resources and places its emphasis. It does not necessarily impeach the forecast's accuracy or honesty. Newspapers strive to ensure that all their articles are accurate and honest, but they still need to decide which ones to put on the front page. The Weather Channel has to make similar decisions, and the economic impact of a forecast is a reasonable basis for doing so.

There are also times, however, when the goals may come into more conflict, and commercial success takes precedence over accuracy.

When Competition Makes Forecasts Worse

There are two basic tests that any weather forecast must pass to demonstrate its merit:

1. It must do better than what meteorologists call **persistence**: the assumption that the weather will be the same tomorrow (and the next day) as it was today.
2. It must also beat **climatology**, the long-term historical average of conditions on a particular date in a particular area.

These were the methods that were available to our ancestors long before Richardson, Lorenz, and the Bluefire came along; if we can't improve on them, then all that expensive computer power must not be doing much good.

We have lots of data, going back at least to World War II, on past weather *outcomes*: I can go to Wunderground.com, for instance, and tell you that the weather at 7 A.M. in Lansing, Michigan, on January 13, 1978—the date and time when I was born—was 18 degrees with light snow and winds from the northeast.[39] But relatively few people had bothered to collect information on past weather *forecasts*. Was snow expected in Lansing that morning? It was one of the few pieces of information that you might have expected to find on the Internet but couldn't.

In 2002 an entrepeneur named Eric Floehr, a computer science graduate from Ohio State who was working for MCI, changed that. Floehr simply started collecting data on the forecasts issued by the NWS, the Weather Channel, and AccuWeather, to see if the government model or the private-sector forecasts were more accurate. This was mostly for his own edification at first—a sort of very large scale science fair project—but it quickly evolved into a profitable business, ForecastWatch.com, which repackages the data into highly customized reports for clients ranging from energy traders (for whom a fraction of a degree can translate into tens of thousands of dollars) to academics.

Floehr found that there wasn't any one clear overall winner. His data suggests that AccuWeather has the best precipitation forecasts by a small margin, that the Weather Channel has slightly better temperature forecasts, and the government's forecasts are solid all around. They're all pretty good.

But the further out in time these models go, the less accurate they turn out to be (figure 4-6). Forecasts made eight days in advance, for example, demonstrate almost no skill; they beat persistence but are barely better than climatology. And at intervals of nine or more days in advance, the professional forecasts were actually a bit worse than climatology.

FIGURE 4-6: COMPARISON OF HIGH-TEMPERATURE FORECASTS[40]

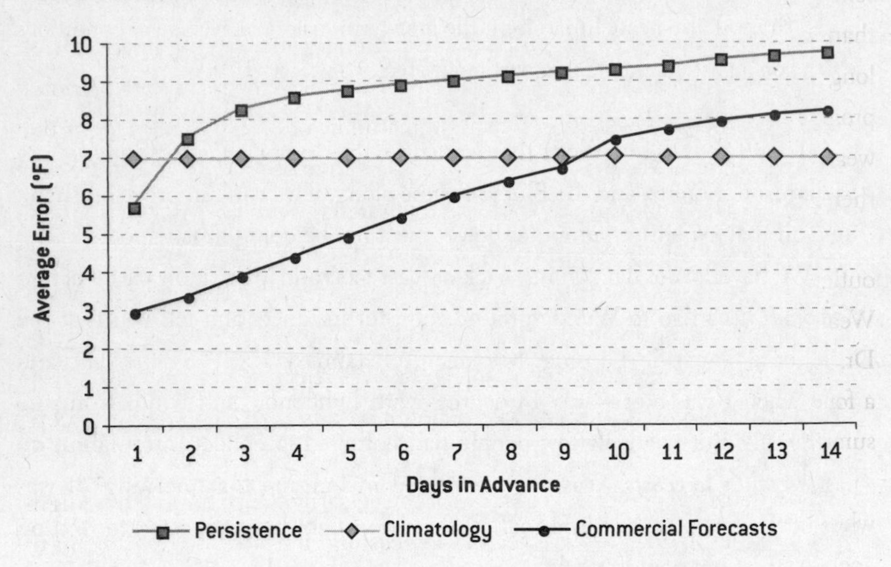

After a little more than a week, Loft told me, chaos theory completely takes over, and the dynamic memory of the atmopshere erases itself. Although the following analogy is somewhat imprecise, it may help to think of the atmosphere as akin to a NASCAR oval, with various weather systems represented by individual cars that are running along the track. For the first couple of dozen laps around the track, knowing the starting order of the cars should allow us to make a pretty good prediction of the order in which they might pass by. Our predictions won't be perfect—there'll be crashes, pit stops, and engine failures that we've failed to account for—but they will be a lot better than random. Soon, however, the faster cars will start to lap the slower ones, and before long the field will be completely jumbled up. Perhaps the second-placed car is running side by side with the sixteenth-placed one (which is about to get lapped), as well as the one in the twenty-eighth place (which has already been lapped once and is in danger of being lapped again). What we knew of the initial conditions of the race is of almost no value to us. Likewise, once the atmosphere has had enough time to circulate, the weather patterns bear so little resemblence to their starting positions that the models don't do any good.

Still, Floehr's finding raises a couple of disturbing questions. It would be one thing if, after seven or eight days, the computer models demonstrated essentially *zero* skill. But instead, they actually display *negative* skill: they are *worse* than what you or I could do sitting around at home and looking up a table of long-term weather averages. How can this be? It is likely because the computer programs, which are hypersensitive to the naturally occurring feedbacks in the weather system, begin to produce feedbacks of their own. It's not merely that there is no longer a signal amid the noise, but that the noise is being amplified.

The bigger question is why, if these longer-term forecasts aren't any good, outlets like the Weather Channel (which publishes ten-day forecasts) and AccuWeather (which ups the ante and goes for fifteen) continue to produce them. Dr. Rose took the position that doing so doesn't really cause any harm; even a forecast based purely on climatology might be of some interest to their consumers.

The statistical reality of accuracy isn't necessarily the governing paradigm when it comes to commercial weather forecasting. It's more the *perception* of accuracy that adds value in the eyes of the consumer.

For instance, the for-profit weather forecasters rarely predict exactly a 50 percent chance of rain, which might seem wishy-washy and indecisive to consumers.[41] Instead, they'll flip a coin and round up to 60, or down to 40, even though this makes the forecasts both less accurate and less honest.[42]

Floehr also uncovered a more flagrant example of fudging the numbers, something that may be the worst-kept secret in the weather industry. Most commercial weather forecasts are *biased*, and probably deliberately so. In particular, they are biased toward forecasting more precipitation than will actually occur[43]—what meteorologists call a "wet bias." The further you get from the government's original data, and the more consumer facing the forecasts, the worse this bias becomes. Forecasts "add value" by subtracting accuracy.

How to Know if Your Forecasts Are All Wet

One of the most important tests of a forecast—I would argue that it is the single most important one[44]—is called *calibration*. Out of all the times you said there was a 40 percent chance of rain, how often did rain actually occur? If, over the long run, it really did rain about 40 percent of the time, that means your forecasts were well calibrated. If it wound up raining just 20 percent of the time instead, or 60 percent of the time, they weren't.

Calibration is difficult to achieve in many fields. It requires you to think probabilistically, something that most of us (including most "expert" forecasters) are not very good at. It really tends to punish overconfidence—a trait that most forecasters have in spades. It also requires a lot of data to evaluate fully—cases where forecasters have issued hundreds of predictions.*

Meteoroloigsts meet this standard. They'll forecast the temperatures, and the probability of rain and other precipitation, in hundreds of cities every day. Over the course of a year, they'll make tens of thousands of forecasts.

This sort of high-frequency forecasting is extremely helpful not just when we want to evaluate a forecast but also to the forecasters themselves—they'll get lots of feedback on whether they're doing something wrong and can change course accordingly. Certain computer models, for instance, tend to come out a little wet[45]—forecasting rain more often than they should. But once you are alert to this bias you can correct for it. Likewise, you will soon learn if your forecasts are overconfident.

* A poorly calibrated forecast can sometimes be detected more quickly. If you said there was a 100 percent chance of an outcome occurring and it didn't, or a 0 percent chance and it did, we do not need any more data to conclude that your forecast was wrong.

The National Weather Service's forecasts are, it turns out, admirably well calibrated[46] (figure 4-7). When they say there is a 20 percent chance of rain, it really does rain 20 percent of the time. They have been making good use of feedback, and their forecasts are honest and accurate.

FIGURE 4-7: NATIONAL WEATHER SERVICE CALIBRATION

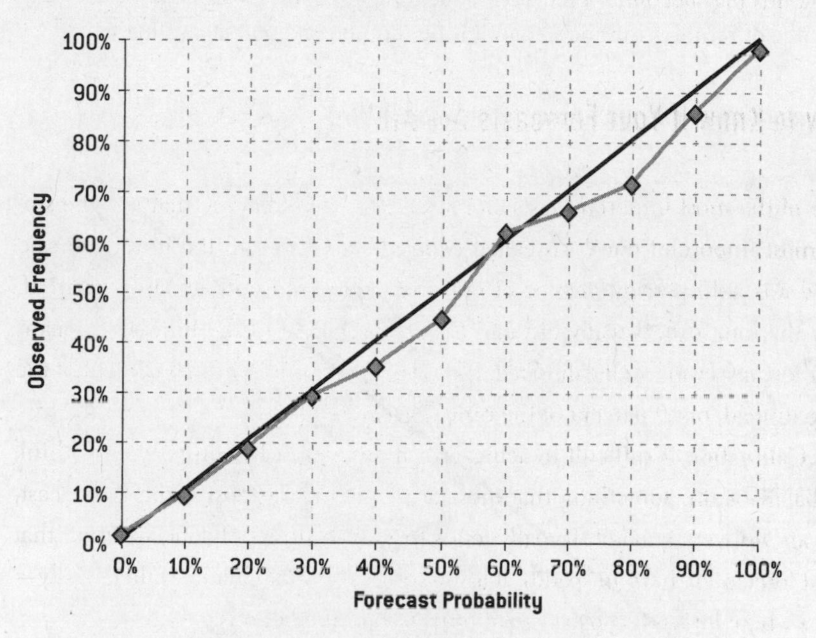

The meteorologists at the Weather Channel will fudge a little bit under certain conditions. Historically, for instance, when they say there is a 20 percent chance of rain, it has actually only rained about 5 percent of the time.[47] In fact, this is deliberate and is something the Weather Channel is willing to admit to. It has to do with their economic incentives.

People notice one type of mistake—the failure to predict rain—more than another kind, false alarms. If it rains when it isn't supposed to, they curse the weatherman for ruining their picnic, whereas an unexpectedly sunny day is taken as a serendipitous bonus. It isn't good science, but as Dr. Rose at the Weather Channel acknolwedged to me: "If the forecast was objective, if it has zero bias in precipitation, we'd probably be in trouble."

Still, the Weather Channel is a relatively buttoned-down organization—

many of their customers mistakenly think they are a government agency—and they play it pretty straight most of the time. Their wet bias is limited to slightly exaggerating the probability of rain when it is unlikely to occur—saying there is a 20 percent chance when they know it is really a 5 or 10 percent chance—covering their butts in the case of an unexpected sprinkle. Otherwise, their forecasts are well calibrated (figure 4-8). When they say there is a 70 percent chance of rain, for instance, that number can be taken at face value.

FIGURE 4-8: THE WEATHER CHANNEL CALIBRATION

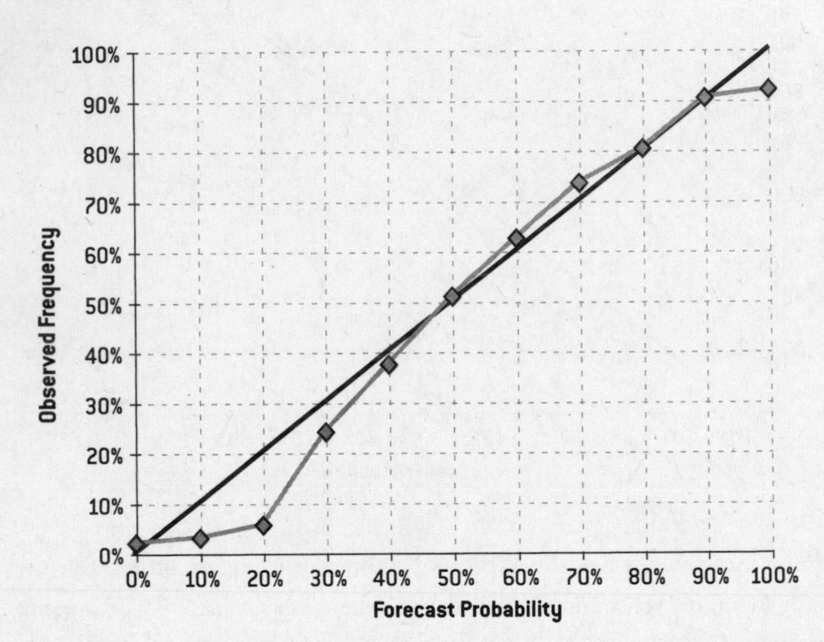

Where things really go haywire is when weather is presented on the local network news. Here, the bias is very pronounced, with accuracy and honesty paying a major price.

Kansas City ought to be a great market for weather forecasting—it has scorching-hot summers, cold winters, tornadoes, and droughts, and it is large enough to be represented by all the major networks. A man there named J. D. Eggleston began tracking local TV forecasts to help his daughter with a fifth-grade classroom project. Eggleston found the analysis so interesting that he continued it for seven months, posting the results to the *Freakonomics* blog.[48]

The TV meteorologists weren't placing much emphasis on accuracy. Instead, their forecasts were quite a bit worse than those issued by the National Weather Service, which they could have taken for free from the Internet and reported on the air. And they weren't remotely well calibrated. In Eggleston's study, when a Kansas City meteorologist said there was a 100 percent chance of rain, it failed to rain about one-third of the time (figure 4-9).

FIGURE 4-9: LOCAL TV METEOROLOGIST CALIBRATION

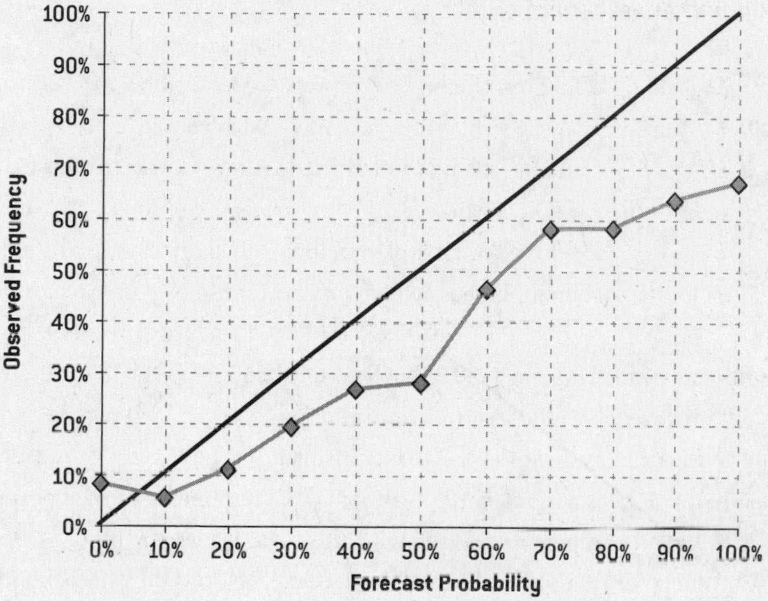

The weather forecasters did not make any apologies for this. "There's not an evaluation of accuracy in hiring meteorologists. Presentation takes precedence over accuracy," one of them told Eggleston. "Accuracy is not a big deal to viewers," said another. The attitude seems to be that this is all in good fun—who cares if there is a little wet bias, especially if it makes for better television? And since the public doesn't think our forecasts are any good anyway, why bother with being accurate?

This logic is a little circular. TV weathermen say they aren't bothering to make accurate forecasts because they figure the public won't believe them anyway. But the public shouldn't believe them, because the forecasts aren't accurate.

This becomes a more serious problem when there is something urgent—something like Hurricane Katrina. Lots of Americans get their weather information from local sources[49] rather than directly from the Hurricane Center, so they will still be relying on the goofball on Channel 7 to provide them with accurate information. If there is a mutual distrust between the weather forecaster and the public, the public may not listen when they need to most.

The Cone of Chaos

As Max Mayfield told Congress, he had been prepared for a storm like Katrina to hit New Orleans for most of his sixty-year life.[50] Mayfield grew up around severe weather—in Oklahoma, the heart of Tornado Alley—and began his forecasting career in the Air Force, where people took risk very seriously and drew up battle plans to prepare for it. What took him longer to learn was how difficult it would be for the National Hurricane Center to communicate its forecasts to the general public.

"After Hurricane Hugo in 1989," Mayfield recalled in his Oklahoma drawl, "I was talking to a behavioral scientist from Florida State. He said people don't respond to hurricane warnings. And I was insulted. Of course they do. But I have learned that he is absolutely right. People don't respond just to the phrase 'hurricane warning.' People respond to what they hear from local officials. You don't want the forecaster or the TV anchor making decisions on when to open shelters or when to reverse lanes."

Under Mayfield's guidance, the National Hurricane Center began to pay much more attention to how it presented its forecasts. It contrast to most government agencies, whose Web sites look as though they haven't been updated since the days when you got those free AOL CDs in the mail, the Hurricane Center takes great care in the design of its products, producing a series of colorful and attractive charts that convey information intuitively and accurately on everything from wind speed to storm surge.

The Hurricane Center also takes care in how it presents the uncertainty in its forecasts. "Uncertainty is the fundamental component of weather prediction," Mayfield said. "No forecast is complete without some description of that uncertainty." Instead of just showing a single track line for a hurricane's predicted path, for instance, their charts prominently feature a cone of uncertainty—"some people call it a cone of chaos," Mayfield said. This shows the range of places where the eye of the hurricane is most likely to make landfall.[51] Mayfield worries that even this isn't enough. Significant impacts like flash floods (which are often more deadly than the storm itself) can occur far from the center of the storm and long after peak wind speeds have died down. No people in New York City died from Hurricane Irene in 2011 despite massive media hype surrounding the storm, but three people did from flooding in landlocked Vermont[52] once the TV cameras were turned off.

What the Hurricane Center usually does not do is issue policy guidance to local officials, such as whether to evacuate a city. Instead, this function is outsourced to the National Weather Service's 122 local offices, who communicate with governors and mayors, sheriffs and police chiefs. The official reason for this is that the Hurricane Center figures the local offices will have better working knowledge of the cultures and the people they are dealing with on the ground. The unofficial reason, I came to recgonize after speaking with Mayfield, is that the Hurricane Center wants to keep its mission clear. The Hurricane Center and the Hurricane Center alone issues hurricane forecasts, and it needs those forecasts to be as accurate and honest as possible, avoiding any potential distractions.

But that aloof approach just wasn't going to work in New Orleans. Mayfield needed to pick up the phone.

Evacuation decisions are not easy, in part because evacuations themselves can be deadly; a bus carrying hospital evacuees from another 2005 storm, Hurricane Rita, burst into flames while leaving Houston, killing twenty-three elderly passengers.[53] "This is really tough with these local managers," Mayfield says. "They look at this probabilistic information and they've got to translate that into a decision. A go, no-go. A yes-or-no decision. They have to take a probabilistic decision and turn it into something deterministic."

In this case, however, the need for an evacuation was crystal clear, and the message wasn't getting through.

"We have a young man at the hurricane center named Matthew Green. Exceptional young man. Has a degree in meteorology. Coordinates warnings with the transit folks. His mother lived in New Orleans. For whatever reason, she was not leaving. Here's a guy who knows about hurricanes and emergency management and he couldn't get his own mother to evacuate."

So the Hurricane Center started calling local officials up and down the Gulf Coast. On Saturday, August 27—after the forecast had taken a turn for the worse but still two days before Katrina hit—Mayfield spoke with Governor Haley Barbour of Mississippi, who ordered a mandatory evacuation for its most vulnerable areas almost immediately,[54] and Governor Kathleen Blanco of Louisiana, who had already declared a state of emergency. Blanco told Mayfield that he needed to call Ray Nagin, the mayor of New Orleans, who had been much slower to respond.

Nagin missed Mayfield's call but phoned him back. "I don't remember exactly what I said," Mayfield told me. "We had tons of interviews over those two or three days. But I'm absolutely positive that I told him, You've got some tough decisions and some potential for a large loss of life." Mayfield told Nagin that he needed to issue a mandatory evacuation order, and to do so as soon as possible.

Nagin dallied, issuing a voluntary evacuation order instead. In the Big Easy, that was code for "take it easy"; only a mandatory evacuation order would convey the full force of the threat.[55] Most New Orleanians had not been alive when the last catastrophic storm, Hurricane Betsy, had hit the city in 1965. And those who had been, by definition, had survived it. "If I survived Hurricane Betsy, I can survive that one, too. We all ride the hurricanes, you know," an elderly resident who stayed in the city later told public officials.[56] Reponses like these were typical. Studies from Katrina and other storms have found that having survived a hurricane makes one *less* likely to evacuate the next time one comes.[57]

The reasons for Nagin's delay in issuing the evacuation order is a matter of some dispute—he may have been concerned that hotel owners might sue the city if their business was disrupted.[58] Either way, he did not call for a mandatory evacuation until Sunday at 11 A.M.[59]—and by that point the residents who had not gotten the message yet were thoroughly confused. One study found that about a third of residents who declined to evacuate the city had not heard the evacuation order at all. Another third heard it but said it did not give clear instructions.[60] Surveys of disaster victims are not always reliable—it is difficult for people to articulate why they behaved the way they did under significant emotional strain,[61] and a small percentage of the population will say they never heard an evacuation order even when it is issued early and often. But in this case, Nagin was responsible for much of the confusion.

There is, of course, plenty of blame to go around for Katrina—certainly to FEMA in addition to Nagin. There is also credit to apportion—most people did evacuate, in part because of the Hurricane Center's accurate forecast. Had Betsy topped the levees in 1965, before reliable hurricane forecasts were possible, the death toll would probably have been even greater than it was in Katrina.

One lesson from Katrina, however, is that accuracy is the best policy for a forecaster. It is forecasting's original sin to put politics, personal glory, or economic benefit before the truth of the forecast. Sometimes it is done with good intentions, but it always makes the forecast worse. The Hurricane Center works as hard as it can to avoid letting these things compromise its forecasts. It may not be a concidence that, in contrast to all the forecasting failures in this book, theirs have become 350 percent more accurate in the past twenty-five years alone.

"The role of a forecaster is to produce the best forecast possible," Mayfield says. It's so simple—and yet forecasters in so many fields routinely get it wrong.

5

DESPERATELY SEEKING SIGNAL

Just as the residents of L'Aquila, Italy, were preparing for bed on a chilly Sunday evening in April 2009, they felt a pair of tremors, each barely more perceptible than the rumbling of a distant freight train. The first earthquake, which occurred just before 11 P.M. local time, measured 3.9 on the magnitude scale,* a frequency strong enough to rattle nerves and loosen objects but little else. The second was even weaker, a magnitude 3.5; it would not have been powerful enough to awaken a sound sleeper.

But L'Aquila was on edge about earthquakes. The town, which sits in the foothills of the Apennine Mountains and is known for its ski resorts and medieval walls, had been experiencing an unusually large number of them—the two that Sunday were the seventh and eighth of at least magnitude 3 in the span

---

* News accounts often refer to the Richter scale, named after the Caltech seismologist Charles Richter. In fact, a different and more accurate scale—the moment magnitude scale, developed at Caltech in the late 1970s—is in more common use among seismologists today. The moment magnitude scale is designed to be comparable to the Richter scale—both scales are logarithmic, and a magnitude 8.0 earthquake is a very serious one under either definition. But they are not calculated in quite the same way. The earthquake magnitudes described in this chapter generally refer to the moment magnitude scale.

of about a week. Small earthquakes are not uncommon in this part of the world, but the rate is normally much less—about one such earthquake every two or three months. These were coming almost one hundred times as often.

Meanwhile, the citizens of a town a mountain pass away, Sulmona, had just survived an earthquake scare of their own. A technician named Giampaolo Giuliani, who worked at Italy's National Institute of Nuclear Physics, claimed to have detected unusually high levels of radon in the area. He theorized this might be a precursor to an earthquake and went so far as to tell Sulmona's mayor that an earthquake would strike the town on the afternoon of March 29. The mayor, impressed by the prediction, ordered vans carrying loudspeakers to drive about town, warning residents of the threat.[1]

No earthquake hit Sulmona that day. After the prediction failed, Giuliani was reported to local authorities for *procurato allarme* (bringing about alarm)—in essence, having yelled fire in a crowded theater. He was forced to remove his predictions from the Internet for fear of triggering further panic.

Authorities in L'Aquila told the residents the earthquake swarm* was nothing to worry about; the fault was helpfully discharging energy, explained Bernardo De Bernardinis, the deputy chief of Italy's Civil Protection Department,[2] reducing the threat of a major earthquake. He agreed with a reporter that they should sit back and enjoy a glass of wine;[3] De Bernardinis recommended a local specialty, a Montepulciano.

A major earthquake did hit L'Aquila, however. Measuring at magnitude 6.3, it came at 3:32 A.M. local time on Monday morning. Shaking houses from their foundations, caving in roofs, and turning furniture into projectiles, the quake killed more than 300 residents, left another 65,000 homeless, and caused more than $16 billion in damage.[4]

What We Do When Our Foundations Are Shaken

L'Aquila ought to have been better prepared. The city sits near a particularly violent type of fault known as a subduction zone, where the African Plate, one of

* Seismologists use the term "earthquake swarm" to refer to a series of small earthquakes.

the eight major tectonic plates that cover the earth's surface, slips slowly and inexorably beneath the Eurasian one. Its first significant earthquake was recorded in 1315, and earthquakes struck again in 1349, 1452, 1461, 1501, 1646, 1703, and 1706;[5] the most serious one, in 1786, had killed more than 5,000 people. Each time, often on direct order of the pope,[6] the town was rebuilt and repopulated.

Since then, L'Aquila had tempted fate for more than two centuries. An earthquake hit in 1958, but it was fairly minor—magnitude 5.0[7]—and only the town's oldest residents would have remembered it. The 2009 earthquake was much more powerful. The magnitude scale is logarithmic; a one-point increase in the scale indicates that the energy release has multiplied by thirty-two. Thus, the 2009 earthquake, magnitude 6.3, was about seventy-five times more powerful than the one that had hit L'Aquila in 1958. And it was about 3,000 times more powerful than the tremors—foreshocks to the major earthquake—that L'Aquila had experienced earlier that evening.

Still, while the 2009 earthquake was large by Italian standards, it was barely a hiccup on the global scale. The earthquake that devastated Japan in 2011 measured at magnitude 9.0 or 9.1—almost 11,000 times more powerful. And the largest earthquake recorded since reliable estimates were possible, which hit Chile in 1960 and measured magnitude 9.5, was about 60,000 times stronger than the L'Aquila quake.

Why, then, did L'Aquila—a fairly well-to-do town in a wealthy, industrialized nation—sustain such significant damage? One reason was the city's geology—L'Aquila sits on an ancient lake bed, which tends to amplify the earth's shaking. Mexico City was also built on an ancient lake bed,[8] and 10,000 were killed there in 1985 from an earthquake whose epicenter was more than two hundred miles away.

But the major reason was simply that the town had become complacent about the seismic danger that lay just fifteen kilometers underground. There was nothing resembling the proper level of earthquake readiness:[9] building codes, emergency supplies, community drills. Not only were centuries-old buildings leveled by the tremor, but so too were many modern ones, including a wing of a hospital that had been erected as recently as 2000. A little bit of warning would have saved untold lives there.

Had Giampaolo Giuliani provided that warning? In the Italian tabloids, he had become something of a savant and a martyr. Soft-spoken and disheveled, and often wearing the colors of the local soccer team, he played the role of the humble civil servant or absentminded professor whose insights had been ignored by the scientific establishment. He claimed that he had warned friends and family about the L'Aquila quake and was prevented from telling others only because of the police order against him. He demanded an apology from the authorities—not to him, he said, but to the people of L'Aquila.

Never mind that Giuliani had not actually predicted the earthquake. His prediction had been very specific: Sulmona, not L'Aquila, was at greater risk, and the earthquake would come in March rather than April. In fact, he had suggested to a local newspaper that the danger had passed. "To simplify the concepts," he said before launching into a rambling explanation about the lunar cycle, "the Earth-Moon system has come to visit at perihelion . . . the minimum distance from Earth, and aligned with the planet Venus. . . . I feel I can reassure my fellow citizens because the swarm will be diminishing with the end of March."[10]

Perihelion with the planet Venus? Radon gas? What did any of this have to do with earthquakes? And what about Giuliani's failed prediction in Sulmona? It didn't matter. When catastrophe strikes, we look for a signal in the noise—anything that might explain the chaos that we see all around us and bring order to the world again. Giuliani's rambling explanations were the closest thing available.

No type of catastrophe is more jarring to our sense of order than an earthquake. They quite literally shake our foundations. Whereas hurricanes descend upon us from the heavens and have sometimes been associated with metaphors for God's providence,* earthquakes come from deep underneath the surface and are more often taken to be signs of His wrath,[11] indifference,[12] or nonexistence. (The Lisbon earthquake of 1755 was a major spark for the development of secular philosophy.[13]) And whereas hurricanes—along with floods, tornadoes, and volcanoes—can often be forecasted in advance, earthquakes have defied centuries of efforts to predict them.

---

* The Japanese word *kamikaze* originally meant "divine wind," referring to typhoons in 1274 and 1281 that had helped to disperse a Mongol invasion.

Magic Toads and the Search for the Holy Grail

Pasadena, California, has long been the world's epicenter for earthquake research. It is home to the California Institute of Technology, where Charles Richter developed his famous logarithmic scale in 1935. The United States Geological Survey (USGS) also has a field office there, where most of its earthquake specialists reside. I traveled there in September 2009 to meet with Dr. Susan Hough, who is one of the USGS's top seismologists and who has written several books about earthquake prediction. She had watched Giuliani's television interviews with suspicion and had written a blistering editorial in the *New York Times*[14] that criticized both Giuliani and the attention paid to him.

Hough's editorial argued that Giuliani's success was merely coincidental. "The public heard about Mr. Giuliani's prediction because it appears to have been borne out," she wrote. "But there are scores of other [incorrect] predictions that the public never hears about."

If you have hundreds of people trying to make forecasts, and there are hundreds of earthquakes per year, inevitably someone is going to get one right. Giuliani's theories about radon gas and lunar cycles had been investigated many times over[15] by credentialed seismologists and had shown little or no ability to predict earthquakes. Giuliani had been lucky: the monkey who typed Shakespeare; the octopus who predicted the World Cup.

Hough's office at the USGS sits near a quiet corner of the Caltech campus where there are more eucalyptus trees than students. She seemed a little road weary when I met her, having just returned from a trip to Turkey where she'd been to study a system of earthquake faults. She has soft features and frizzy hair and her eyes are dark, tired—skeptical. "What's your day job?" she quizzed me a few moments after I greeted her.

At one point, she pulled a pocket-size globe off her desk, the sort that looks like it was bought at an airport gift shop. She took her index finger and drew a line across the surface of the globe, starting in the Sea of Japan and moving east–southeast.

"They are really concentrated in this belt—stretching from southern China through Greece," Hough explained, referring to the world's most destructive earthquakes. "It's a complicated earthquake zone, a lot of buildings with vulnerable construction. If you put a big earthquake under Tehran, you could kill a million people."

Indeed, almost all the deadliest earthquakes in modern history (figure 5-1) have occurred along the path that Hough outlined, one which passes through the Cradle of Civilization in the Middle East and through some of the most densely populated regions of the planet, including China and India. Often poor and crowded, these areas lack the luxury to prepare for a once-per-three-hundred-year catastrophe. But the death tolls can be catastrophic when earthquakes hit, stretching into the hundreds of thousands.*

FIGURE 5-1: DEADLIEST EARTHQUAKES SINCE 1900

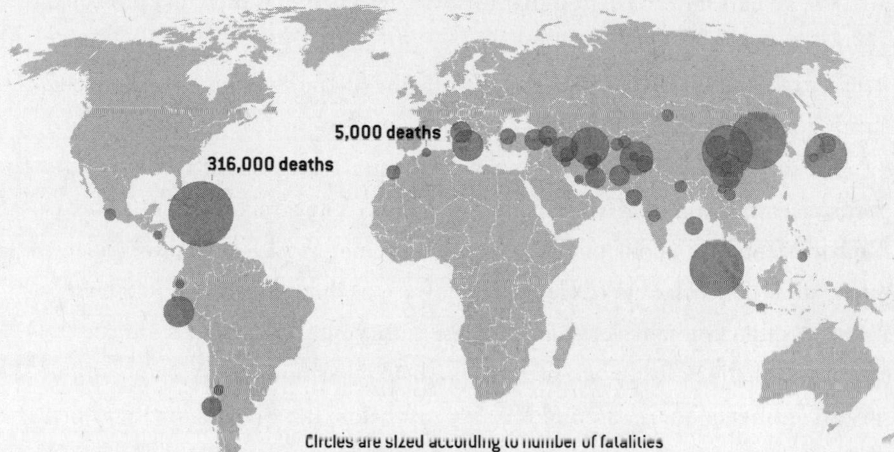

Earthquakes kill more people than hurricanes, in fact,[16] despite seeming like the rarer phenomenon.[17] Perhaps that is because they are so seldom predicted successfully. Whereas the landfall position of hurricanes can be forecasted at least three times more accurately now than they were even twenty-five years ago, the science of earthquake forecasting seems barely to have evolved since the ninth century A.D., when the Japanese first claimed to be able to an-

* The Haitian earthquake of 2010 was an exception to the pattern geographically, but not in how poverty and lax building standards contribute to immense death and destruction.

ticipate earthquakes by looking at the behavior of catfish.[18] (Cows, pigs, eels, rats, parakeets, seagulls, turtles, goldfish, and snakes have also been reported at various times to behave unusually in advance of an earthquake.)

Kooks like Giuliani are still taken seriously, and not just in the Italian tabloids.[19] The California Earthquake Prediction Council receives hundreds of unsolicited earthquake forecasts per year, most of which, the agency says, "discuss the strange behavior of household pets, intuition, Aunt Agatha's aching bunions, or other mysterious signs and portents that scientists simply don't understand."[20] Meanwhile, some of the stuff in academic journals is hard to distinguish from ancient Japanese folklore. A 2010 paper[21] in a relatively prestigious journal, *The Journal of Zoology*, observed that toads in a pond fifty miles from L'Aquila had stopped spawning five days before the major earthquake there.[22] Remarkably, it asserted that this was evidence that they had predicted the earthquake.

It's research like this that exhausts Hough. "If you look back in time, certainly going back to the 1970s, people would come up with some idea—they'd be optimistic—and then you wait ten years and that method would be debunked," she told me. "Ten years later, you have a new method and ten years later it's debunked. You just sort of sense a theme. Most top scientists at this point know better than to chase after a Holy Grail that probably doesn't exist."

But while Giuliani's close encounters with Venus or the toads are easy to dismiss, is there really no way at all to predict an earthquake? What about the swarm of smaller quakes around L'Aquila just before the Big One hit? Was that just a coincidence? The seismological community has a reputation for being very conservative. It was very slow to accept the theory of plate tectonics, for instance[23]—the now broadly accepted notion that the shifting of the earth's continental plates is the primary cause for earthquakes—not adopting it into their canon until the 1960s even though it was proposed in 1912. Had Hough's skepticism crossed the line into cynicism?

The official position of the USGS is even more emphatic: earthquakes cannot be predicted. "Neither the USGS nor Caltech nor any other scientists have ever predicted a major earthquake," the organization's Web site asserts.[24] "They do not know how, and they do not expect to know how any time in the foreseeable future."

Earthquakes cannot be predicted? This is a book about prediction, not a book that *makes* predictions, but I'm willing to stick my neck out: I predict that there will be more earthquakes in Japan next year than in New Jersey. And I predict that at some point in the next one hundred years, a major earthquake will hit somewhere in California.[25]

Both the USGS and I are playing some semantic games. The terms "prediction" and "forecast" are employed differently in different fields; in some cases, they are interchangeable, but other disciplines differentiate them. No field is more sensitive to the distinction than seismology. If you're speaking with a seismologist:

1. A **prediction** is a definitive and specific statement about when and where an earthquake will strike: *a major earthquake will hit Kyoto, Japan, on June 28.*
2. Whereas a **forecast** is a probabilistic statement, usually over a longer time scale: *there is a 60 percent chance of an earthquake in Southern California over the next thirty years.*

The USGS's official position is that earthquakes cannot be predicted. They can, however, be *forecasted*.

What We Know About How Earthquakes Behave

If you explore the USGS Web site, in fact, you'll find that it makes lots of tools available to help you *forecast* earthquakes. One particularly neat one is an application that lets you type in the longitude and latitude at any point in the United States; it will estimate the long-term probability of an earthquake there.[26] In figure 5-2, I've listed the probabilities for earthquakes in a variety of major U.S. cities as provided by the USGS Web site.

We all know that California is very seismically active; the USGS estimates that an earthquake of magnitude 6.8 or higher will hit San Francisco about once every thirty-five years. Many of you will also know that Alaska has many earthquakes—the second largest one in recorded history, magnitude 9.4, hit Anchorage in 1964.

FIGURE 5-2. FREQUENCY OF A MAJOR (>= MAGNITUDE 6.75) EARTHQUAKE WITHIN A 50-MILE RADIUS OF SELECT U.S. CITIES

Anchorage	1 per 30 years
San Francisco	1 per 35 years
Los Angeles	1 per 40 years
Seattle	1 per 150 years
Sacramento	1 per 180 years
San Diego	1 per 190 years
Salt Lake City	1 per 200 years
Portland, OR	1 per 500 years
Charleston, SC	1 per 600 years
Las Vegas	1 per 1,200 years
Memphis	1 per 2,500 years
Phoenix	1 per 7,500 years
New York	1 per 12,000 years
Boston	1 per 15,000 years
Philadelphia	1 per 17,000 years
St. Louis	1 per 23,000 years
Atlanta	1 per 30,000 years
Denver	1 per 40,000 years
Washington, DC	1 per 55,000 years
Chicago	1 per 75,000 years
Houston	1 per 100,000 years
Dallas	1 per 130,000 years
Miami	1 per 140,000 years

But did you know about Charleston, South Carolina? It is seismically active too; indeed, it experienced a magnitude 7.3 earthquake in 1886. The USGS estimates that there will be another big earthquake there about once per six hundred years. If you live in Seattle, you should probably have an earthquake plan ready; it is more earthquake-prone than many parts of California, the USGS says. But you don't need one if you live in Denver, which is a safe distance away from any continental boundaries.

This seems like an awful lot of very specific and user-friendly information for an organization whose party line is that it is impossible to predict earthquakes. But the USGS's forecasts employ a widely accepted seismological tool called the Gutenberg–Richter law. The theory, developed by Charles Richter and his Caltech colleague Beno Gutenberg in 1944, is derived from empirical statistics about earthquakes. It posits that there is a relatively simple relationship between the magnitude of an earthquake and how often one occurs.

If you compare the frequencies of earthquakes with their magnitudes, you'll find that the number drops off exponentially as the magnitude increases. While there are very few catastrophic earthquakes, there are literally millions of smaller ones—about 1.3 million earthquakes measuring between magnitude 2.0 and magnitude 2.9 around the world every year.[27] Most of these earthquakes go undetected—certainly by human beings and often by seismometers.[28] However, almost all earthquakes of magnitude 4.5 or greater are recorded today, however remote their location. Figure 5-3a shows the exponential decline in their frequencies, based on actual records of earthquakes from January 1964[29] through March 2012.[30]

It turns out that these earthquakes display a stunning regularity when you graph them in a slightly different way. In figure 5-3b, I've changed the vertical axis—which shows the frequency of earthquakes of different magnitudes—into a logarithmic scale.* Now the earthquakes form what is almost exactly a straight line on the graph. This pattern is characteristic of what is known as a power-law distribution, and it is the relationship that Richter and Gutenberg uncovered.

Something that obeys this distribution has a highly useful property: you can forecast the number of large-scale events from the number of small-scale ones, or vice versa. In the case of earthquakes, it turns out that for every increase of one point in magnitude, an earthquake becomes about ten times less frequent. So, for example, magnitude 6 earthquakes occur ten times more frequently than magnitude 7's, and one hundred times more often than magnitude 8's.

What's more, the Gutenberg–Richter law generally holds across regions of the globe as well as over the whole planet. Suppose, for instance, that we wanted

* Recall that the magnitude scale is already logarithmic, so this is what's technically known as a double-logarithmic plot.

FIGURE 5-3A: WORLDWIDE EARTHQUAKE FREQUENCIES, JANUARY 1964–MARCH 2012

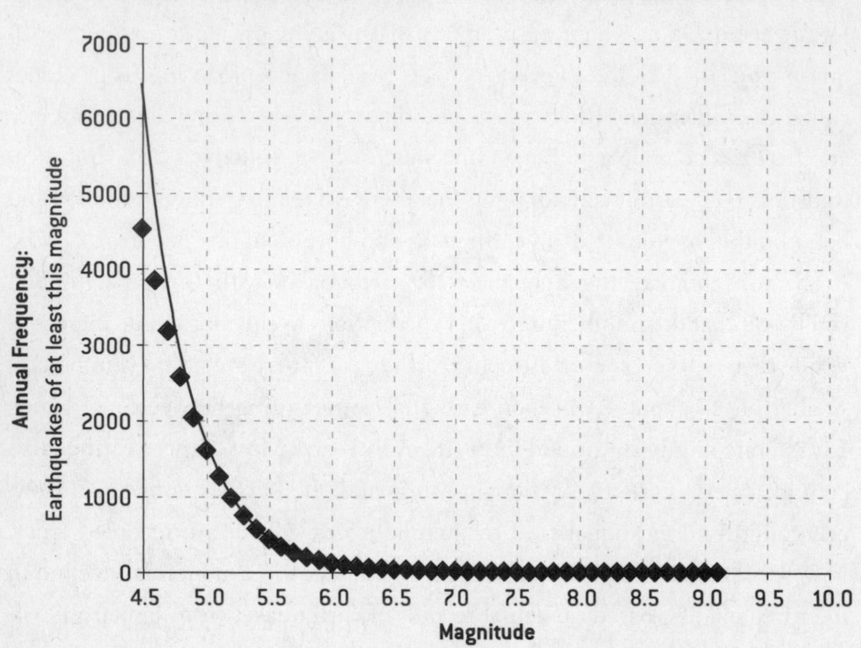

to make an earthquake forecast for Tehran, Iran. Fortunately, there hasn't been a catastrophic earthquake there since its seismicity began to be measured. But there have been a number of medium-size ones; between 1960 and 2009, there were about fifteen earthquakes that measured between 5.0 and 5.9 on the magnitude scale in the area surrounding the city.[31] That works out to about one for every three years. According to the power law that Gutenberg and Richter uncovered, that means that an earthquake measuring between 6.0 and 6.9 should occur about once every thirty years in Tehran.

Furthermore, it follows that an earthquake that measured 7.0 or greater would occur about once every three hundred years near Tehran. This is the earthquake that Susan Hough fears. The Haiti earthquake of 2010, which measured magnitude 7.0 and killed 316,000,[32] showed the apocalyptic consequences that earthquakes can produce in the developing world. Iran shares many of Haiti's problems—poverty, lax building codes, political corruption[33]—but it is much more densely populated. The USGS estimates, on the basis of high death

FIGURE 5-3B: WORLDWIDE EARTHQUAKE FREQUENCIES, JANUARY 1964–MARCH 2012, LOGARITHMIC SCALE

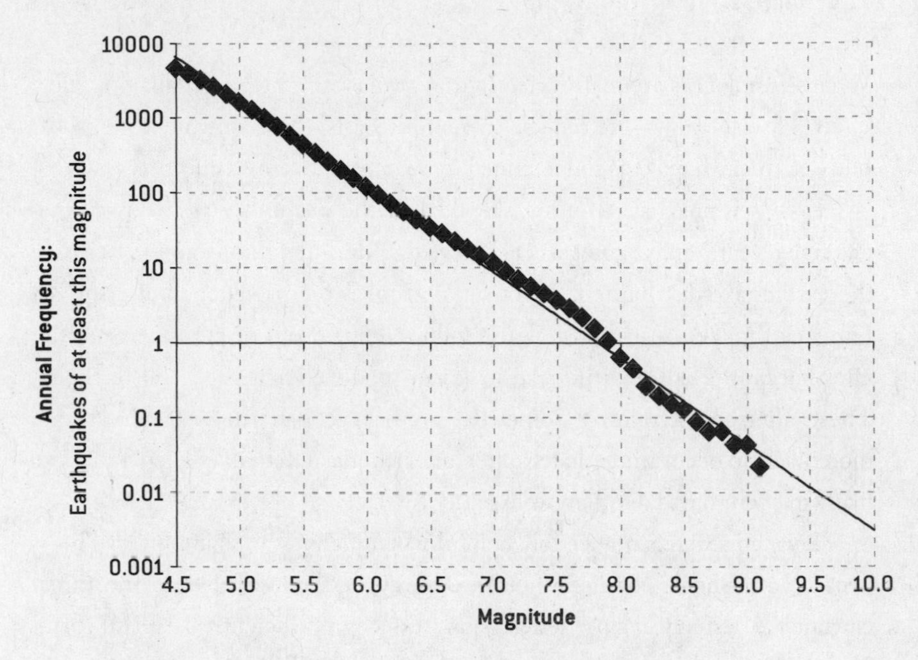

tolls from smaller earthquakes in Iran, that between 15 and 30 percent of Tehran's population could die in the event of a catastrophic tremor there.[34] Since there are about thirteen million people in Tehran's metro area, that would mean between two and four million fatalities.

What the Gutenberg–Richter law does not tell us anything about is *when* the earthquake would strike. (Nor does it suggest that Tehran is "due" for an earthquake if it hasn't experienced one recently.) Countries like Iran and Haiti do not have the luxury of making contingency plans for a once-every-three-hundred-year event. The earthquake forecasts produced using the Gutenberg–Richter law provide for a good general guide to the hazard in an area. But like weather forecasts determined from statistical records alone (it rains 35 percent of the time in London in March), they don't always translate into actionable intelligence (should I carry an umbrella?). Geological time scales occupy centuries or entire millennia; human life spans are measured in years.

The Temptation of the Signal

What seismologists are really interested in—what Susan Hough calls the "Holy Grail" of seismology—are *time-dependent* forecasts, those in which the probability of an earthquake is not assumed to be constant across time.

Even seismologists who are skeptical of the possibility of making time-dependent earthquake forecasts acknowledge that there are some patterns in the earthquake distribution. The most obvious is the presence of aftershocks. Large earthquakes are almost always followed by dozens or even thousands of aftershocks (the 2011 earthquake in Japan produced at least 1,200 of them). These aftershocks follow a somewhat predictable pattern.[35] Aftershocks are more likely to occur immediately after an earthquake than days later, and more likely to occur days later than weeks after the fact.

This, however, is not terribly helpful when it comes to saving lives. This is because aftershocks, by definition, are always less powerful than the initial earthquake. Usually, if a particular fault produces a sufficiently powerful earthquake, there will be a few aftershocks and then that'll be the end of the fireworks for a while. This isn't always the case, however. For example, the incredibly powerful earthquake that hit the New Madrid Fault on the Missouri-Tennessee border on December 16, 1811, evaluated by seismologists as magnitude 8.2, was followed just six hours later by another shock of about the same magnitude. And the fault was not yet quiesced: the December 16 quakes were succeeded by another magnitude 8.1 earthquake on January 23, and then yet another, even more powerful 8.3 earthquake on February 7. Which ones were the foreshocks? Which ones were the aftershocks? Any interpretation is about as useless as any other.

The question, of course, is whether we can predict earthquakes *before* the fact: can we tell the foreshocks and aftershocks apart in advance? When we look at data that shows the distribution of earthquakes across time and space, it tempts us with the possibility that there might be some signal in the noise.

Figure 5-4a, for instance, shows the distribution of earthquakes near L'Aquila[36] from 2006 until the magnitude 6.3 earthquake hit in 2009.[37] All the data in this chart, except the large black circle that indicates the main earthquake, shows earthquakes that occurred before the main shock. In the case of L'Aquila, there does seem to be a discernible pattern. A big cluster of earthquakes, measuring up to about magnitude 4, occurred just before the main shock in early 2009—much higher than the background rate of seismic activity in the area.

FIGURE 5-4A: EARTHQUAKES NEAR L'AQUILA, ITALY
JANUARY 1, 2006–APRIL 6, 2009

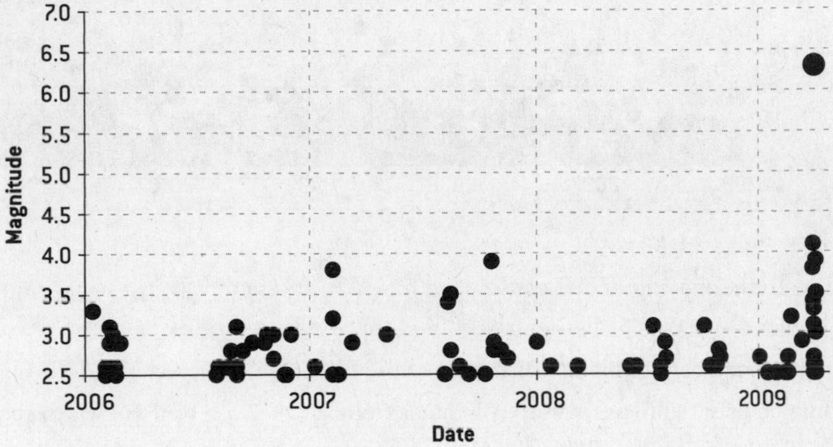

A more debatable case is the Japan earthquake of 2011. When we make one of these plots for the Tohoku region (figure 5-4b), we see, first of all, that it is much more seismically active than Italy. But are there patterns in the timing of the earthquakes there? There seem to be some; for instance, there is a cluster of earthquakes measuring between magnitude 5.5 and magnitude 7.0 in mid-2008. These, however, did not precipitate a larger earthquake. But we do see an especially large foreshock, magnitude 7.5, on March 9, 2011, preceding the magnitude 9.1 Tōhoku earthquake[38] by about fifty hours.

Only about half of major earthquakes are preceded by discernible foreshocks,[39] however. Haiti's was not (figure 5-4c). Instrumentation is not very good in most parts of the Caribbean, so we don't have records of magnitude 2 and 3 earthquakes, but seismometers in the United States and other areas should be able to pick up anything that registers at 4 or higher. The last time there had been even a magnitude 4 earthquake in the area was in 2005, five years

FIGURE 5-4B: EARTHQUAKES NEAR TŌHOKU, JAPAN
JANUARY 1, 2006–MARCH 11, 2011

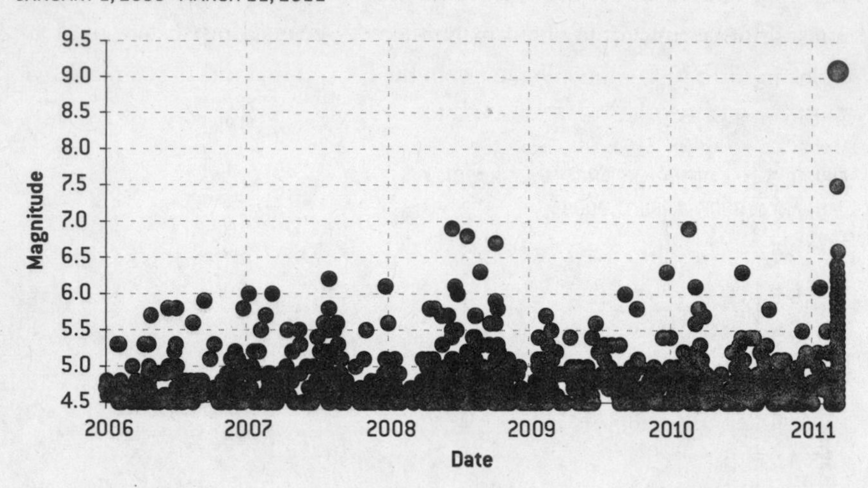

before the magnitude 7.0 earthquake hit in 2010. There was just no warning at all.

Complicating matters further are false alarms—periods of increased seismic activity that never result in a major tremor. One case well known to seismologists is a series of smaller earthquakes near Reno, Nevada, in early 2008.

FIGURE 5-4C: EARTHQUAKES NEAR LÉOGÂNE, HAITI
JANUARY 1, 2000–JANUARY 12, 2010

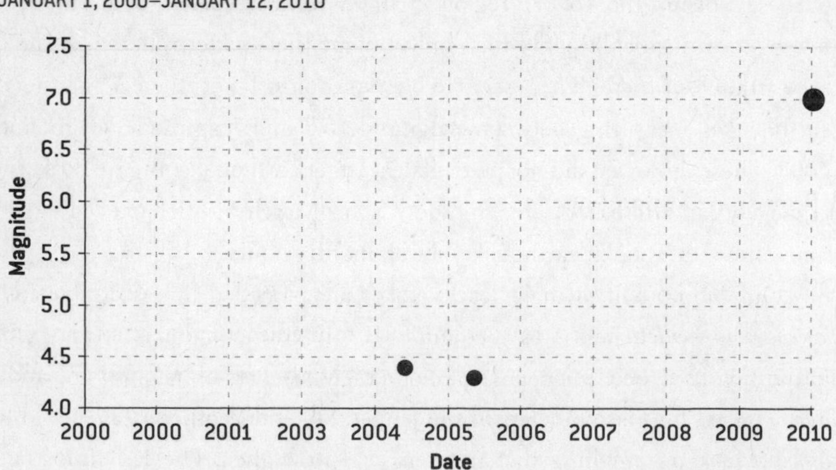

The Reno earthquake swarm looks a lot like the one we saw before L'Aquila in 2009. But it never amounted to anything much; the largest earthquake in the series was just magnitude 5.0 and no major earthquake followed.

FIGURE 5-4D: EARTHQUAKES NEAR RENO, NEVADA
JANUARY 1, 2006–DECEMBER 31, 2011

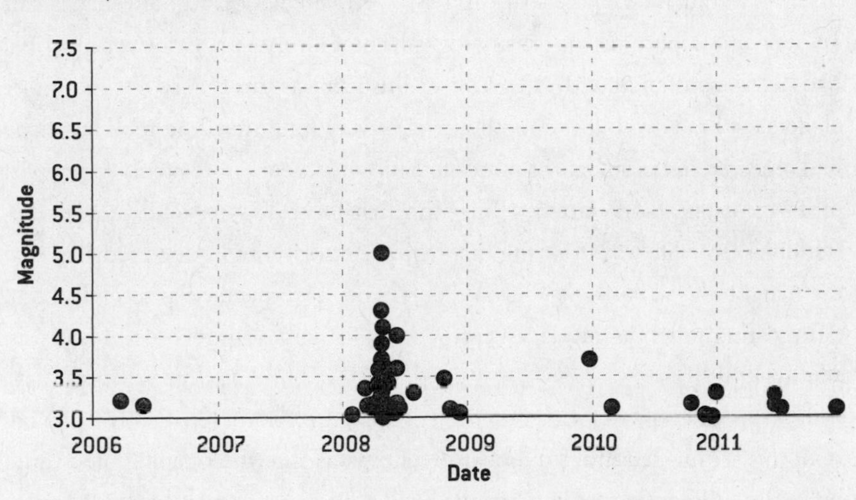

This is just a taste of the maddening array of data that seismologists observe. It seems to exist in a purgatory state—not quite random and not quite predictable. Perhaps that would imply that we could at least get halfway there and make some progress in forecasting earthquakes—even if we can never get to hard-and-fast predictions. But the historical record of attempts to predict earthquakes is one of almost complete failure.

A Parade of Failed Forecasts

Hough's 2009 book, *Predicting the Unpredictable: The Tumultuous Science of Earthquake Prediction,* is a history of efforts to predict earthquakes, and is as damning to that enterprise as Phil Tetlock's study was to political pundits. There just seems to have been no progress at all, and there have been many false alarms.

Lima, Peru

One of the more infamous cases involved a geophysicist named Brian Brady, who had a Ph.D. from MIT and worked at Colorado School of Mines. Brady asserted that a magnitude 9.2 earthquake—one of the largest in recorded history—would hit Lima, Peru, in 1981.[40] His prediction initially had a fair amount of support in the seismological community—an early version of it had been coauthored with a USGS scientist. But as the theory became more elaborate—Brady would eventually invoke everything from the rock bursts he had observed in his studies of mines to Einstein's theory of relativity in support of it—colleagues had started telling him that theory was beyond their understanding:[41] a polite way of saying that he was nuts. Eventually, he predicted that the magnitude 9.2 earthquake would be just one in a spectacular series in Peru, culminating in a magnitude 9.9 earthquake, the largest in recorded history, in August 1981.[42]

The prediction was leaked to the Peruvian media and terrified the population; this serious-seeming American scientist was sure their capital city would be in ruins. Their fear only intensified when it was reported that the Peruvian Red Cross had requested 100,000 body bags to prepare for the disaster. Tourism and property values declined,[43] and the U.S. government eventually dispatched a team of scientists and diplomats to Peru in an effort to calm nerves. It made front-page news when there was no Great Peruvian Earthquake in 1981 (or even a minor one).

Parkfield, California

If Lima had provided a warning that false alarms can extract a substantial psychological and economic cost on the population, it did not stop seismologists from seeking out the Holy Grail. While Brady had been something of a lone wolf, there were cases when earthquake prediction had much more explicit backing from the USGS and the rest of the seismological community. These efforts did not go so well either.

Among the most studied seismic zones in the world is Parkfield, California, which sits along the San Andreas Fault somewhere between Fresno, Bakersfield, and the next exit with an In-N-Out Burger. There had been earthquakes in Parkfield at what seemed to be regular intervals about twenty-two years apart: in 1857, 1881, 1901, 1922, 1934, and 1966. A USGS-sponsored paper[44] projected the trend forward and predicted with 95 percent confidence that there would be another such earthquake at some point between 1983 and 1993, most likely in 1988. The next significant earthquake to hit Parkfield did not occur until 2004, however, well outside of the prediction window.

Apart from being wrong, the Parkfield prediction also seemed to reinforce a popular misconception about earthquakes: that they come at regular intervals and that a region can be "due" for one if it hasn't experienced an earthquake in some time. Earthquakes result from a buildup of stress along fault lines. It might follow that the stress builds up until it is discharged, like a geyser erupting with boiling water, relieving the stress and resetting the process.

But the fault system is complex: regions like California are associated with multiple faults, and each fault has its own branches and tributaries. When an earthquake does strike, it may relieve the stress on one portion of a fault, but it can transfer it along to neighboring faults, or even to some faraway portion of the same fault.[45] Moreover, the stress on a fault is hard to observe directly—until an earthquake hits.

What this means is that if San Francisco is forecasted to have a major earthquake every thirty-five years, it does not imply that these will be spaced out evenly (as in 1900, 1935, 1970). It's safer to assume there is a 1 in 35 chance of an earthquake occurring every year, and that this rate does not change much over time regardless of how long it has been since the last one.

Mojave Desert, California

The Brady and Parkfield fiascoes seemed to suppress efforts at earthquake prediction for some time. But they came back with a vengeance in the 2000s, when newer and seemingly more statistically driven methods of earthquake prediction became the rage.

One such method was put forward by Vladimir Keilis-Borok, a Russian-born mathematical geophysicist who is now in his late eighties and teaches at UCLA. Keilis-Borok had done much to advance the theory of how earthquakes formed and first achieved notoriety in 1986 when, at a summit meeting in Reykjavík with Mikhail Gorbachev, President Reagan was handed a slip of paper predicting a major earthquake in the United States within the next five years, an event later interpreted to be the Loma Prieta quake that struck San Francisco in 1989.[46]

In 2004, Keilis-Borok and his team claimed to have made a "major breakthrough" in earthquake prediction.[47] By identifying patterns from smaller earthquakes in a given region, they said, they were able to predict large ones. The methods that Keilis-Borok applied to identify these patterns were elaborate and opaque,[48] representing past earthquakes with a series of eight equations, each of which was applied in combination with the others at all conceivable intervals of time and space. But, the team said, their method had correctly predicted 2003 earthquakes in San Simeon, California, and Hokkaido, Japan.

Whether the San Simeon and Hokkaido predictions were publicly communicated ahead of time remains unclear;[49] a search of the Lexis-Nexis database of newspapers reveals no mention of them in 2003.[50] When we are evaluating the success of a forecasting method, it is crucial to keep "retrodictions" and predictions separate; predicting the past is an oxymoron and obviously should not be counted among successes.[51]

By January 2004, however, Keilis-Borok had gone very public with another prediction:[52] an earthquake measuring at least magnitude 6.4 would hit an area of the Mojave Desert in Southern California at some point within the subsequent nine months. The prediction began to attract widespread attention: Keilis-Borok was featured in the pages of *Discover* magazine, the *Los Angeles Times*, and a dozen or so other mainstream publications. Someone from Governor Schwarzenegger's office called; an emergency panel was convened. Even the famously skeptical USGS was willing to give some credit; their Web site conceded that "the work of the Keilis-Borok team is a legitimate approach to earthquake prediction research."[53]

But no major earthquake hit the Mojave Desert that year, and indeed, almost a decade later, none has. The Keilis-Borok team has continued to make predictions about earthquakes in California, Italy, and Japan but with little success: a 2010 analysis found three hits but twenty-three misses among predictions that they had clearly enunciated ahead of time.[54]

Sumatra, Indonesia

There is another type of error, in which an earthquake of a given magnitude is deemed unlikely or impossible in a region—and then it happens. David Bowman, a former student of Keilis-Borok who is now the chair of the Department of Geological Sciences at Cal State Fullerton, had redoubled his efforts at earthquake prediction after the Great Sumatra Earthquake of 2004, the devastating magnitude 9.2 disaster that produced a tsunami and killed 230,000 people. Bowman's technique, like Keilis-Borok's, was highly mathematically driven and used medium-size earthquakes to predict major ones.[55] However, it was more elegant and ambitious, proposing a theory called accelerated moment release that attempted to quantify the amount of stress at different points in a fault system. In contrast to Keilis-Borok's approach, Bowman's system allowed him to forecast the likelihood of an earthquake along any portion of a fault; thus, he was not just predicting where earthquakes would hit, but also where they were unlikely to occur.

Bowman and his team did achieve some initial success; the massive aftershock in Sumatra in March 2005, measuring magnitude 8.6, had its epicenter in an area his method identified as high-risk. However, a paper that he published in 2006[56] also suggested that there was a particularly *low* risk of an earthquake on another portion of the fault, in the Indian Ocean adjacent to the Indonesian province of Bengkulu. Just a year later, in September 2007, a series of earthquakes hit exactly that area, culminating in a magnitude 8.5. Fortunately, the earthquakes occurred far enough offshore that fatalities were light, but it was devastating to Bowman's theory.

Between a Rock and a Hard Place

After the model's failure in 2007, Bowman did something that forecasters very rarely do. Rather than blame the failure on bad luck (his model had allowed for *some* possibility of an earthquake near Bengkulu, just not a high one), he reexamined his model and decided his approach to predicting earthquakes was fundamentally flawed—and gave up on it.

"I'm a failed predictor," Bowman told me in 2010. "I did a bold and stupid thing—I made a testable prediction. That's what we're supposed to do, but it can bite you when you're wrong."

Bowman's idea had been to identify the root causes of earthquakes—stress accumulating along a fault line—and formulate predictions from there. In fact, he wanted to understand how stress was changing and evolving throughout the entire system; his approach was motivated by chaos theory.

Chaos theory is a demon that can be tamed—weather forecasters did so, at least in part. But weather forecasters have a much better theoretical understanding of the earth's atmosphere than seismologists do of the earth's crust. They know, more or less, how weather works, right down to the molecular level. Seismologists don't have that advantage.

"It's easy for climate systems," Bowman reflected. "If they want to see what's happening in the atmosphere, they just have to look up. We're looking at rock. Most events occur at a depth of fifteen kilometers underground. We don't have a hope of drilling down there, realistically—sci-fi movies aside. That's the fundamental problem. There's no way to directly measure the stress."

Without that theoretical understanding, seismologists have to resort to purely statistical methods to predict earthquakes. You can create a statistical variable called "stress" in your model, as Bowman tried to do. But since there's no way to measure it directly, that variable is still just expressed as a mathematical function of past earthquakes. Bowman thinks that purely statistical approaches like these are unlikely to work. "The data set is incredibly noisy," he says. "There's not enough to do anything statistically significant in testing hypotheses."

What happens in systems with noisy data and underdeveloped theory—like earthquake prediction and parts of economics and political science—is a two-step process. First, people start to mistake the noise for a signal. Second, this noise pollutes journals, blogs, and news accounts with false alarms, undermining good science and setting back our ability to understand how the system really works.

Overfitting: The Most Important Scientific Problem You've Never Heard Of

In statistics, the name given to the act of mistaking noise for a signal is *overfitting.*

Suppose that you're some sort of petty criminal and I'm your boss. I deputize you to figure out a good method for picking combination locks of the sort you might find in a middle school—maybe we want to steal everybody's lunch money. I want an approach that will give us a high probability of picking a lock anywhere and anytime. I give you three locks to practice on—a red one, a black one, and a blue one.

After experimenting with the locks for a few days, you come back and tell me that you've discovered a foolproof solution. If the lock is red, you say, the combination is 27-12-31. If it's black, use the numbers 44-14-19. And if it's blue, it's 10-3-32.

I'd tell you that you've completely failed in your mission. You've clearly figured out how to open these three particular locks. But you haven't done anything to advance our theory of lock-picking—to give us some hope of picking them when we don't know the combination in advance. I'd have been interested in knowing, say, whether there was a good type of paper clip for picking these locks, or some sort of mechanical flaw we can exploit. Or failing that, if there's some trick to detect the combination: maybe certain types of numbers are used more often than others? You've given me an *overly specific* solution to a *general* problem. This is overfitting, and it leads to worse predictions.

The name overfitting comes from the way that statistical models are "fit" to match past observations. The fit can be too loose—this is called underfitting—in which case you will not be capturing as much of the signal as you could. Or it can be too tight—an overfit model—which means that you're fitting the noise in the data rather than discovering its underlying structure. The latter error is much more common in practice.

To see how this works, let's give ourselves an advantage that we'll almost never have in real life: we'll know *exactly* what the real data is supposed to look like. In figure 5-5, I've drawn a smooth parabolic curve, which peaks in the middle and trails off near the ends. This could represent any sort of real-world data that you might like: as we saw in chapter 3, for instance, it represents a pretty good description of how baseball players perform as they age, since they are better in the middle of their careers than at the end or the beginning.

FIGURE 5-5: TRUE DISTRIBUTION OF DATA

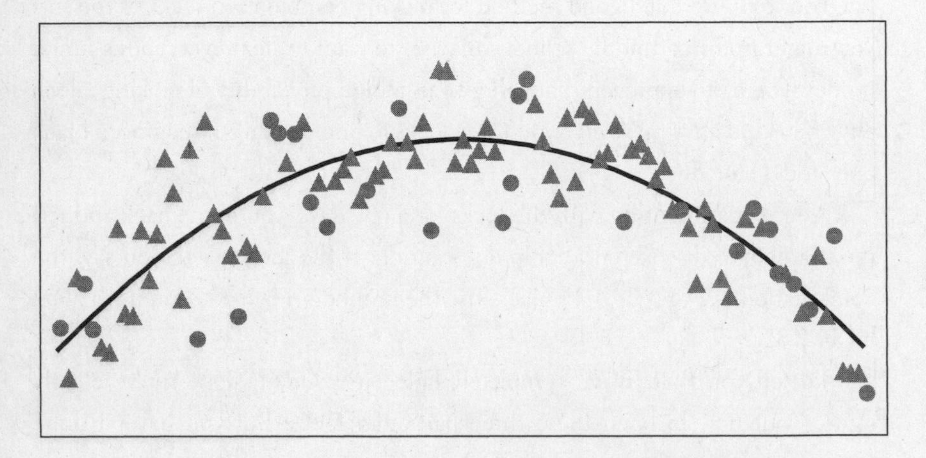

However, we do not get to observe this underlying relationship directly. Instead, it manifests itself through a series of individual data points and we have to infer the pattern from those. Moreover, these data points are affected by idiosyncratic circumstances—so there is some signal, but there is also some noise. In figure 5-5, I've plotted one hundred data points, represented by circles and triangles. This looks to be enough to detect the signal through the noise. Although there is some randomness in the data, it's pretty clear that they follow our curve.

What happens, however, when we have a more limited amount of data, as will usually be the case in real life? Then we have more potential to get ourselves in trouble. In figure 5-6a, I've limited us to about twenty-five of our one hundred observations. How would you connect these dots?

FIGURE 5-6A: LIMITED SAMPLE OF DATA

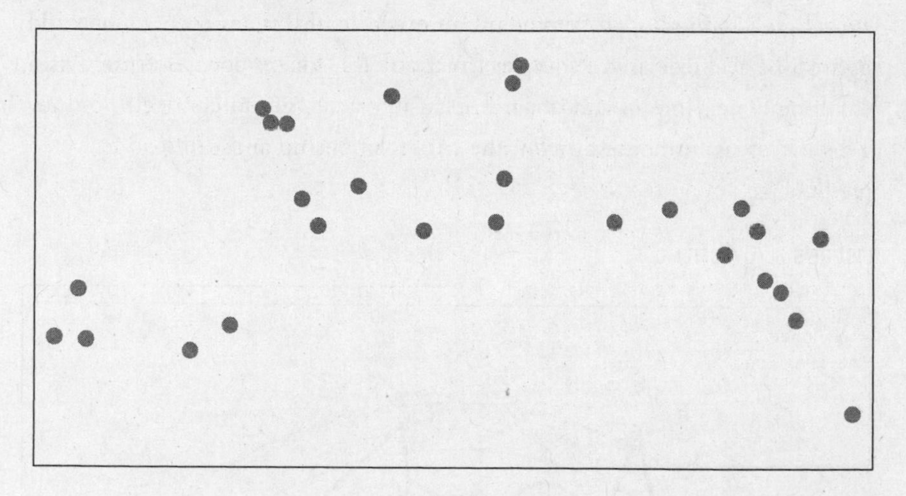

Knowing what the real pattern is supposed to be, of course, you'll still be inclined to fit the points with some kind of curve shape. Indeed, modeling this data with a simple mathematical expression called a quadratic equation does a very good job of re-creating the true relationship (figure 5-6b).

FIGURE 5-6B: WELL-FIT MODEL

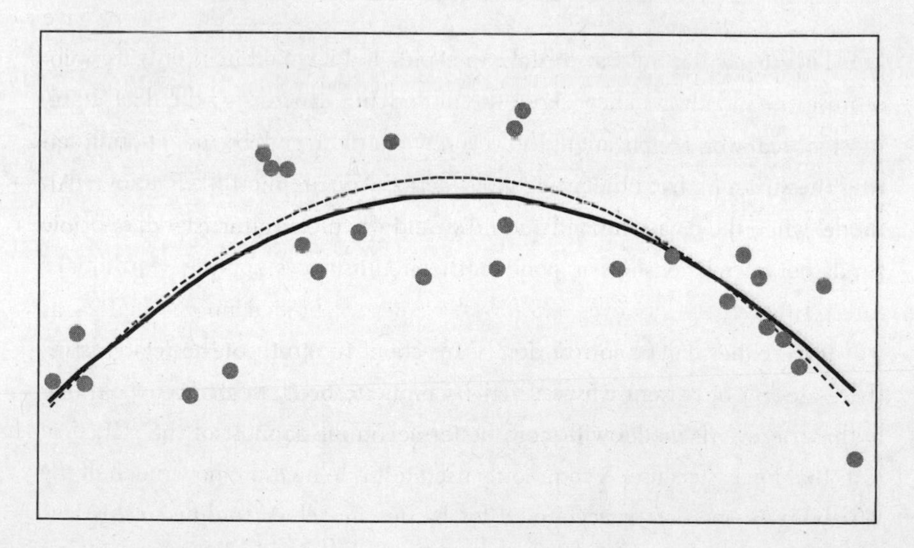

When we don't know the Platonic ideal for our data, however, sometimes we get greedy. Figure 5-6c represents an example of this: an overfit model. In figure 5-6c, we've devised a complex function[57] that chases down every outlying data point, weaving up and down implausibly as it tries to connect the dots. This moves us further away from the true relationship and will lead to worse predictions.

FIGURE 5-6C: OVERFIT MODEL

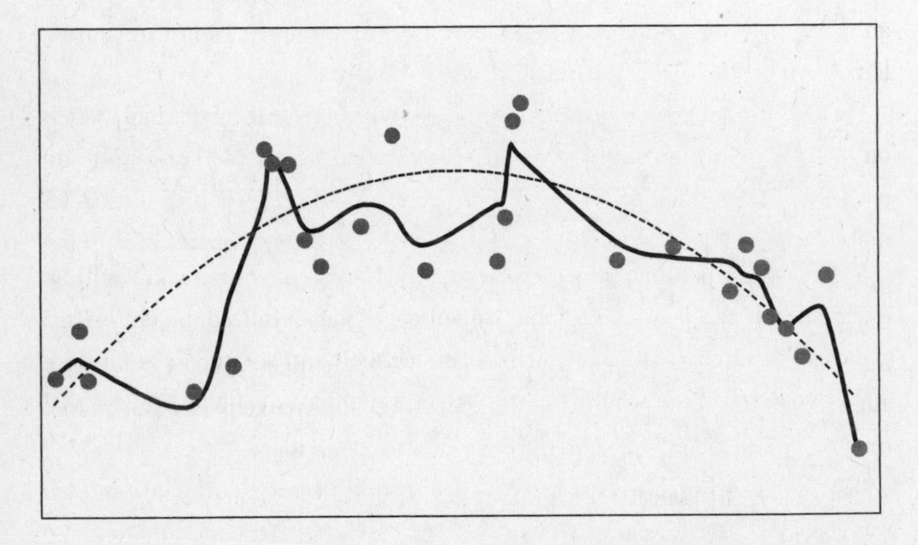

This seems like an easy mistake to avoid, and it would be if only we were omniscient and always knew about the underlying structure of the data. In almost all real-world applications, however, we have to work by induction, inferring the structure from the available evidence. You are most likely to overfit a model when the data is limited and noisy and when your understanding of the fundamental relationships is poor; both circumstances apply in earthquake forecasting.

If we either don't know or don't care about the truth of the relationship, there are lots of reasons why we may be prone to overfitting the model. One is that the overfit model will score better according to most of the statistical tests that forecasters use. A commonly used test is to measure how much of the variability in the data is accounted for by our model. According to this test, the overfit model (figure 5-6c) explains 85 percent of the variance, making it "better" than the properly fit one (figure 5-6b), which explains 56 percent. But the overfit model scores those extra points in essence by cheating—by fitting noise rather than signal. It actually does a much *worse* job of explaining the real world.[58]

As obvious as this might seem when explained in this way, many forecasters completely ignore this problem. The wide array of statistical methods available to researchers enables them to be no less fanciful—and no more scientific—than a child finding animal patterns in clouds.* "With four parameters I can fit an elephant," the mathematician John von Neumann once said of this problem.[59] "And with five I can make him wiggle his trunk."

Overfitting represents a double whammy: it makes our model look *better* on paper but perform *worse* in the real world. Because of the latter trait, an overfit model eventually will get its comeuppance if and when it is used to make real predictions. Because of the former, it may look superficially more impressive until then, claiming to make very accurate and newsworthy predictions and to represent an advance over previously applied techniques. This may make it easier to get the model published in an academic journal or to sell to a client, crowding out more honest models from the marketplace. But if the model is fitting noise, it has the potential to hurt the science.

As you may have guessed, something like Keilis-Borok's earthquake model was badly overfit. It applied an incredibly complicated array of equations to noisy data. And it paid the price—getting just three of its twenty-three predictions correct. David Bowman recognized that he had similar problems and pulled the plug on his model.

To be clear, these mistakes are usually honest ones. To borrow the title of another book, they play into our tendency to be fooled by randomness. We may even grow quite attached to the idiosyncrasies in our model. We may, without even realizing it, work backward to generate persuasive-sounding theories that rationalize them, and these will often fool our friends and colleagues as well as ourselves. Michael Babyak, who has written extensively on this problem,[60] puts

* If you feed a computer a string of coin tosses (a random mix of 1's and 0's representing heads and tails), and then test out statistical parameters to try to fit a pattern-matching model, eventually it will think it can call 60 percent or 70 percent or (if you include enough variables) 100 percent of coin flips correctly. All this is artificial, of course; over the long run, it will call exactly 50 percent of coin flips correctly, no more and no less.

the dilemma this way: "In science, we seek to balance curiosity with skepticism." This is a case of our curiosity getting the better of us.

An Overfit Model of Japan?

Our tendency to mistake noise for signal can occasionally produce some dire real-world consequences. Japan, despite being extremely seismically active, was largely unprepared for its devastating 2011 earthquake. The Fukushima nuclear reactor was built to withstand a magnitude 8.6 earthquake,[61] but not a 9.1. Archaeological evidence[62] is suggestive of historic tsunamis on the scale of the 130-foot waves that the 2011 earthquake produced, but these cases were apparently forgotten or ignored.

A magnitude 9.1 earthquake is an incredibly rare event in any part of the world: nobody should have been predicting it to the exact decade, let alone the exact date. In Japan, however, some scientists and central planners dismissed the possibility out of hand. This may reflect a case of overfitting.

In figure 5-7a, I've plotted the historical frequencies of earthquakes near the 2011 epicenter in Japan.[63] The data includes everything up through but not including the magnitude 9.1 earthquake on March 11. You'll see that the relationship almost follows the straight-line pattern that Gutenberg and Richter's method predicts. However, at about magnitude 7.5, there is a kink in the graph. There had been no earthquakes as large as a magnitude 8.0 in the region since 1964, and so the curve seems to bend down accordingly.

So how to connect the dots? If you go strictly by the Gutenberg–Richter law, ignoring the kink in the graph, you should still follow the straight line, as in figure 5-7b. Alternatively, you could go by what seismologists call a characteristic fit (figure 5-7c), which just means that it is descriptive of the historical frequencies of the earthquake in that area. In this case, that would mean that you took the kink in the historical data to be real—meaning, you thought there was some good reason why earthquakes larger than about magnitude 7.6 were unlikely to occur in the region.

FIGURE 5-7A: TŌHOKU, JAPAN EARTHQUAKE FREQUENCIES
JANUARY 1, 1964—MARCH 10, 2011

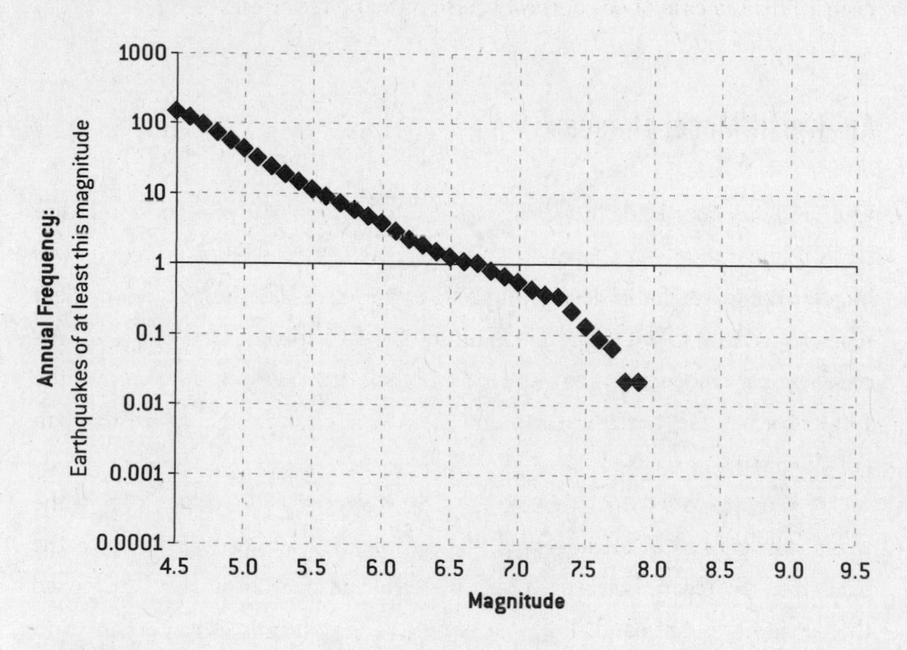

FIGURE 5-7B: TŌHOKU, JAPAN EARTHQUAKE FREQUENCIES
GUTENBERG-RICHTER FIT

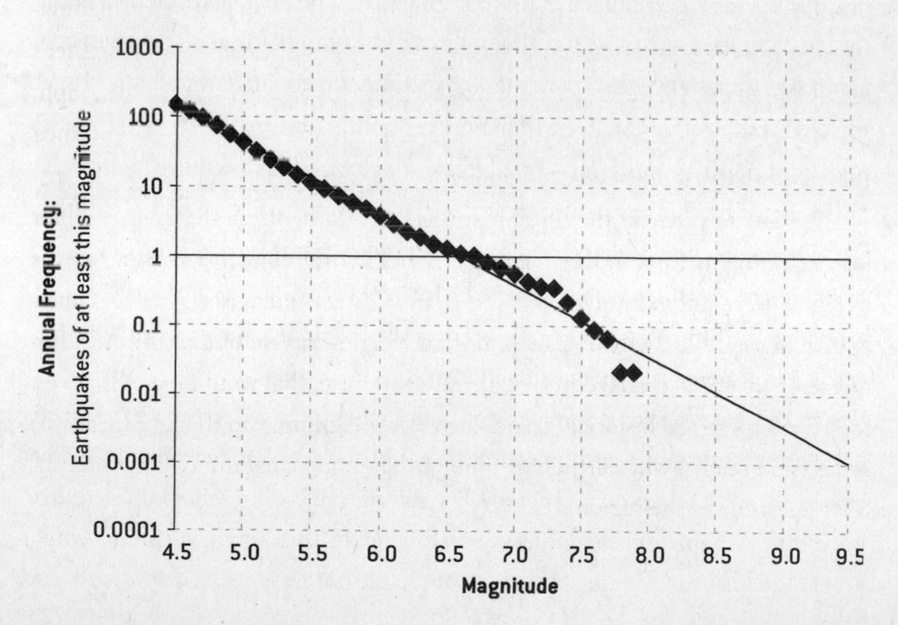

FIGURE 5-7C: TŌHOKU, JAPAN EARTHQUAKE FREQUENCIES
CHARACTERISTIC FIT

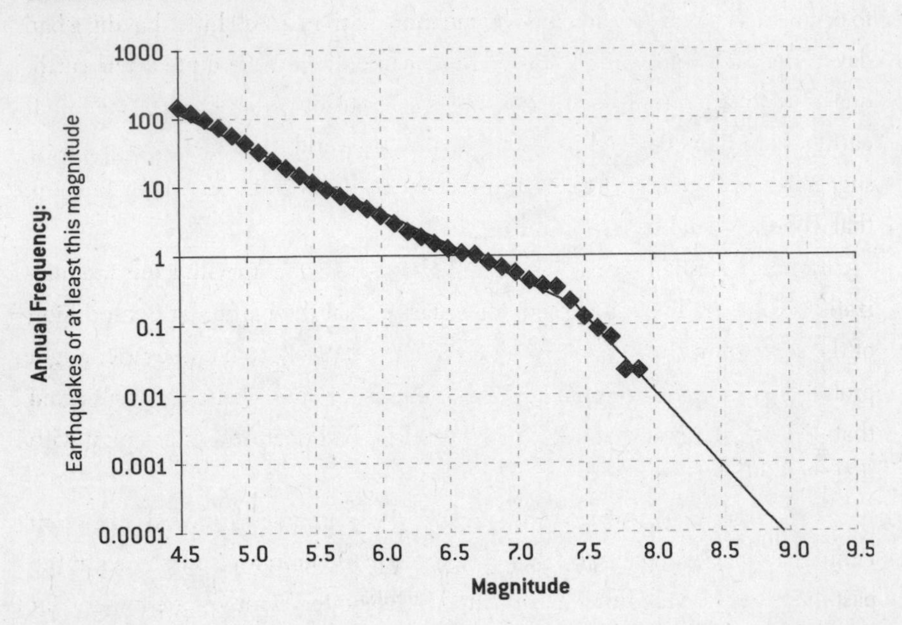

Here is another example where an innocuous-seeming choice of assumptions will yield radically distinct conclusions—in this case, about the probability of a magnitude 9 earthquake in this part of Japan. The characteristic fit suggests that such an earthquake was nearly impossible—it implies that one might occur about every 13,000 years. The Gutenberg–Richter estimate, on the other hand, was that you'd get one such earthquake every three hundred years. That's infrequent but hardly impossible—a tangible enough risk that a wealthy nation like Japan might be able to prepare for it.[64]

The characteristic fit matched the recent historical record from Tōhoku a bit more snugly. But as we've learned, this type of pattern-matching is *not* always a good thing—it could imply an overfit model, in which case it will do a worse job of matching the true relationship.

In this case, an overfit model would dramatically underestimate the likelihood of a catastrophic earthquake in the area. The problem with the characteristic fit is that it relied on an incredibly weak signal. As I mentioned, there had been no earthquake of magnitude 8 or higher in this region in the forty-five years or so prior to Tōhoku. However, these are rare events to begin with: the

Gutenberg–Richter law posits that they might occur only about once per thirty years in this area. It's not very hard at all for a once-per-thirty-year event to fail to occur in a forty-five-year window,[65] no more so than a .300 hitter having a bad day at the plate and going 0-for-5.[66] Meanwhile, there were quite a few earthquakes with magnitudes in the mid- to high 7's in this part of Japan. When such earthquakes had occurred in other parts of the world, they had almost always suggested the potential for larger ones. What justification was there to think that Tōhoku would be a special case?

Actually, seismologists in Japan and elsewhere came up with a few rationalizations for that. They suggested, for instance, that the particular composition of the seafloor in the region, which is old and relatively cool and dense, might prohibit the formation of such large earthquakes.[67] Some seismologists observed that, before 2004, no magnitude 9 earthquake had occurred in a region with that type of seafloor.

This was about like concluding that it was impossible for anyone from Pennsylvania to win the Powerball jackpot because no one had done so in the past three weeks. Magnitude 9 earthquakes, like lottery winners, are few and far between. Before 2004, in fact, only three of them had occurred in recorded history anywhere in the world. This wasn't nearly enough data to support such highly specific conclusions about the exact circumstances under which they might occur. Nor was Japan the first failure of such a theory; a similar one had been advanced about Sumatra[68] at a time when it had experienced lots of magnitude 7 earthquakes[69] but nothing stronger. Then the Great Sumatra Earthquake, magnitude 9.2,[70] hit in December 2004.

The Gutenberg–Richter law would not have predicted the exact timing of the Sumatra or Japan earthquakes, but it would have allowed for their possibility.[71] So far, it has held up remarkably well when a great many more elaborate attempts at earthquake prediction have failed.

The Limits of Earthquakes and Our Knowledge of Them

The very large earthquakes of recent years are causing seismologists to rethink what the upper bounds of earthquakes might be. If you look at figure 5-3b, which accounts for all earthquakes since 1964 (including Sumatra and Tōhoku) it now forms a nearly straight line though all the data points. A decade ago, you would have detected more of a kink in the graph (as in the Tōhoku chart in figure 5-7a). What this meant is that there were slightly fewer megaquakes than the Gutenberg–Richter law predicted. But recently we have been catching up.

Because they occur so rarely, it will take centuries to know what the true rate of magnitude 9 earthquakes is. It will take even longer to know whether earthquakes larger than magnitude 9.5 are possible. Hough told me that there may be some fundamental constraints on earthquake size from the geography of fault systems. If the largest continuous string of faults in the world ruptured together—everything from Tierra Del Fuego at the southern tip of South America all the way up through the Aleutians in Alaska—a magnitude 10 is about what you'd get, she said. But it is hard to know for sure.

Even if we had a thousand years of reliable seismological records, however, it might be that we would not get all that far. It may be that there are intrinsic limits on the predictability of earthquakes.

Earthquakes may be an inherently *complex* process. The theory of complexity that the late physicist Per Bak and others developed is different from chaos theory, although the two are often lumped together. Instead, the theory suggests that very simple things can behave in strange and mysterious ways when they interact with one another.

Bak's favorite example was that of a sandpile on a beach. If you drop another grain of sand onto the pile (what could be simpler than a grain of sand?), it can actually do one of three things. Depending on the shape and size of the pile, it might stay more or less where it lands, or it might cascade gently down the small hill toward the bottom of the pile. Or it might do something else: if the pile is too steep, it could destabilize the entire system and trigger a sand avalanche. Complex systems seem to have this property, with large periods of apparent stasis marked by sudden and catastrophic failures. These processes may not literally be random, but they are so irreducibly complex (right down to the last grain of sand) that it just won't be possible to predict them beyond a certain level.

The Beauty of the Noise

And yet complex processes produce order and beauty when you zoom out and look at them from enough distance. I use the terms signal and noise very loosely in this book, but they originally come from electrical engineering. There are different types of noise that engineers recognize—all of them are random, but they follow different underlying probability distributions. If you listen to true white noise, which is produced by random bursts of sound over a uniform distribution of frequencies, it is sibilant and somewhat abrasive. The type of noise associated with complex systems, called Brownian noise, is more soothing and sounds almost like rushing water.[72]

Meanwhile, the same tectonic forces that carve fault lines beneath the earth's surface also carve breathtaking mountains, fertile valleys, and handsome coastlines. What that means is that people will probably never stop living in them, despite the seismic danger.

Science on Trial

In a final irony of the L'Aquila earthquake, a group of seven scientists and public officials were quite literally put on trial for manslaughter in 2011.[73] Prosecutors from the city of L'Aquila alleged that they had failed to adequately notify the public about the risk of a Big One after the earthquake swarm there.

The trial was obviously ridiculous, but is there anything the scientists could have done better? Probably there was; there is fairly clear evidence that the risk of a major earthquake increases substantially—perhaps temporarily becoming one hundred to five hundred times higher than its baseline rate[74]—following an earthquake swarm. The risk was nevertheless extremely low—most earthquake swarms do not produce major quakes—but it was not quite right to imply that everything was normal and that people should sit down and have a glass of wine.

This book takes the view that the first duty of a forecaster is always fealty to the truth of the forecast. Politics, broadly defined, can get in the way of that. The seismological community is still scarred by the failed predictions in Lima and Parkfield, and by having to compete against the likes of Giuliani. This complicates their incentives and distracts them from their mission. Bad and irresponsible predictions can drive out good ones.

Hough is probably right that the Holy Grail of earthquake prediction will never be attained. Even if individual seismologists are behaving responsibly, we nevertheless have the collective output of the discipline to evaluate, which together constitutes thousands of hypotheses about earthquake predictability. The track record suggests that most of these hypotheses have failed and that magic-bullet approaches to earthquake prediction just aren't likely to work.

However, the track record of science as a whole is a remarkable one; that is also a clear signal. It is probably safe to conclude that the same method attempted over and over with little variation is unlikely to yield different results. But science often produces "unpredictable" breakthroughs.

One area in which seismologists have made some progress is in the case of very short term earthquake forecasts, as might have been relevant in L'Aquila. Next to the Gutenberg–Richter law, the knowledge that major earthquakes essentially always produce aftershocks is the most widely accepted finding in the discipline. Some seismologists I spoke with, like John Rundle of UC Davis and Tom Jordan of the University of Southern California, are concentrating more on these near-term forecasts and increasingly take the view that they should be communicated clearly and completely to the public.

Jordan's research, for instance, suggests that aftershocks sometimes move in a predictable geographic direction along a fault line. If they are moving in the direction of a population center, they can potentially be more threatening to life and property even if they are becoming less powerful. For instance, the magnitude 5.8 earthquake in Christchurch, New Zealand, in 2011, which killed 185, was an aftershock of a 7.0 earthquake that occurred in September 2010 in a remote part of the country.[75] When it comes to aftershocks, there is clearly a lot of signal, so this may be the more natural place to focus.

Finally, technology is always pushing forward. Recent efforts by NASA and by Rundle to measure fault stress through remote sensing systems like GPS satellites have shown some promise.[76] Although the efforts are crude for the time being, there is potential to increase the amount of data at seismologists' disposal and get them closer to understanding the root causes of earthquakes.

These methods may eventually produce some forward progress. If success in earthquake prediction has been almost nonexistent for millennia, the same was true for weather forecasting until about forty years ago. Or it may be that as we develop our understanding of complexity theory—itself a very new branch of science—we may come to a more emphatic conclusion that earthquakes are not really predictable at all.

Either way, there will probably be some failed predictions first. As the memory of our mistakes fades, the signal will again seem to shimmer over the horizon. Parched for prediction we will pursue it, even if it is a mirage.

6

HOW TO DROWN IN THREE FEET OF WATER

Political polls are dutifully reported with a margin of error, which gives us a clue that they contain some uncertainty. Most of the time when an economic prediction is presented, however, only a single number is mentioned. The economy will create 150,000 jobs next month. GDP will grow by 3 percent next year. Oil will rise to $120 per barrel.

This creates the perception that these forecasts are amazingly accurate. Headlines expressing surprise at any minor deviation from the prediction are common in coverage of the economy:

> **Unexpected Jump in Unemployment**
> **Rate to 9.2% Stings Markets**
> **—*Denver Post*, July 9, 2011**[1]

If you read the fine print of that article, you'd discover that the "unexpected" result was that the unemployment rate had come in at 9.2 percent—rather than 9.1 percent[2] as economists had forecasted. If a one-tenth of a percentage point error is enough to make headlines, it seems like these forecasts must ordinarily be very reliable.

Instead, economic forecasts are blunt instruments at best, rarely being able to anticipate economic turning points more than a few months in advance. Fairly often, in fact, these forecasts have failed to "predict" recessions *even once they were already under way*: a majority of economists did not think we were in one when the three most recent recessions, in 1990, 2001, and 2007, were later determined to have begun.[3]

Forecasting something as large and complex as the American economy is a very challenging task. The gap between how well these forecasts actually do and how well they are perceived to do is substantial.

Some economic forecasters wouldn't want you to know that. Like forecasters in most other disciplines, they see uncertainty as the enemy—something that threatens their reputation. They don't estimate it accurately, making assumptions that lower the amount of uncertainty in their forecast models but that don't improve their predictions in the real world. This tends to leave us less prepared when a deluge hits.

The Importance of Communicating Uncertainty

In April 1997, the Red River of the North flooded Grand Forks, North Dakota, overtopping the town's levees and spilling more than two miles into the city.*[4] Although there was no loss of life, nearly all of the city's 50,000 residents had to be evacuated, cleanup costs ran into the billions of dollars,[5] and 75 percent of the city's homes were damaged or destroyed.[6]

Unlike a hurricane or an earthquake, the Grand Forks flood may have been a preventable disaster. The city's floodwalls could have been reinforced using sandbags.[7] It might also have been possible to divert the overflow into depopulated areas—into farmland instead of schools, churches, and homes.

Residents of Grand Forks had been aware of the flood threat for months. Snowfall had been especially heavy in the Great Plains that winter, and the

* The political scientist Roger Pielke Jr., who was brought in to consult on the disaster, alerted me to this story.

National Weather Service, anticipating runoff as the snow melted, had predicted the waters of the Red River would crest to forty-nine feet, close to the all-time record.

There was just one small problem. The levees in Grand Forks had been built to handle a flood of fifty-one feet. Even a small miss in the forty-nine-foot prediction could prove catastrophic.

In fact, the river crested to fifty-four feet. The Weather Service's forecast hadn't been perfect by any means, but a five-foot miss, two months in advance of a flood, is pretty reasonable—about as well as these predictions had done on average historically. The margin of error on the Weather Service's forecast—based on how well their flood forecasts had done in the past—was about plus or minus nine feet. That implied about a 35 percent chance of the levees being overtopped.[8]

FIGURE 6-1: FLOOD PREDICTION WITH MARGIN OF ERROR[9]

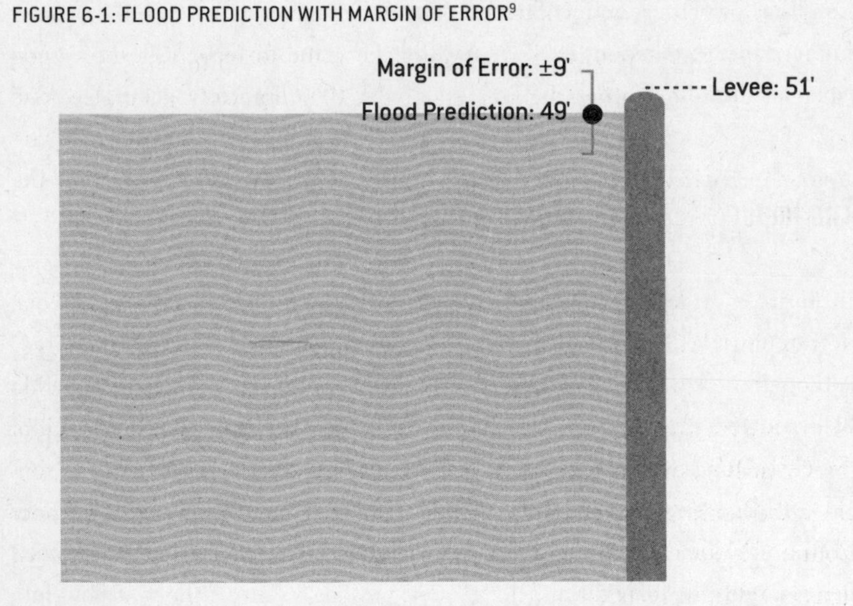

The problem is that the Weather Service had explicitly avoided communicating the uncertainty in their forecast to the public, emphasizing only the forty-nine-foot prediction. The forecasters later told researchers that they were afraid the public might lose confidence in the forecast if they had conveyed any uncertainty in the outlook.

Instead, of course, it would have made the public much better prepared—and possibly able to prevent the flooding by reinforcing the levees or diverting the river flow. Left to their own devices, many residents became convinced they didn't have anything to worry about. (Very few of them bought flood insurance.[10]) A prediction of a forty-nine-foot crest in the river, expressed without any reservation, seemed to imply that the flood would hit forty-nine feet exactly; the fifty-one-foot levees would be just enough to keep them safe. Some residents even interpreted the forecast of forty-nine feet as representing the *maximum* possible extent of the flood.[11]

An oft-told joke: a statistician drowned crossing a river that was only three feet deep *on average*. On average, the flood might be forty-nine feet in the Weather Service's forecast model, but just a little bit higher and the town would be inundated.

The National Weather Service has since come to recognize the importance of communicating the uncertainty in their forecasts accurately and honestly to the public, as we saw in chapter 4. But this sort of attitude is rare among other kinds of forecasters, especially when they predict the course of the economy.

Are Economists Rational?

Now consider what happened in November 2007. It was just one month before the Great Recession officially began. There were already clear signs of trouble in the housing market: foreclosures had doubled,[12] and the mortgage lender Countrywide was on the verge of bankruptcy.[13] There were equally ominous signs in credit markets.[14]

Economists in the Survey of Professional Forecasters, a quarterly poll put out by the Federal Reserve Bank of Philadelphia, nevertheless foresaw a recession as relatively unlikely. Instead, they expected the economy to grow at a just slightly below average rate of 2.4 percent in 2008. And they thought there was almost no chance of a recession as severe as the one that actually unfolded.

The Survey of Professional Forecasters is unique in that it asks economists to explicitly indicate a range of outcomes for where they see the economy headed. As I have emphasized throughout this book, a probabilistic consideration of outcomes is an essential part of a scientific forecast. If I asked you to forecast the total that will be produced when you roll a pair of six-sided dice, the correct answer is not any single number but an enumeration of possible outcomes and their respective probabilities, as in figure 6-2. Although you will roll 7 more often than any other number, it is not intrinsically any more or any less consistent with your forecast than a roll of 2 or 12, provided that each number comes up in accordance with the probability you assign it over the long run.

FIGURE 6-2: FORECASTED PROBABILITY DISTRIBUTION: SUM OF PAIR OF DICE

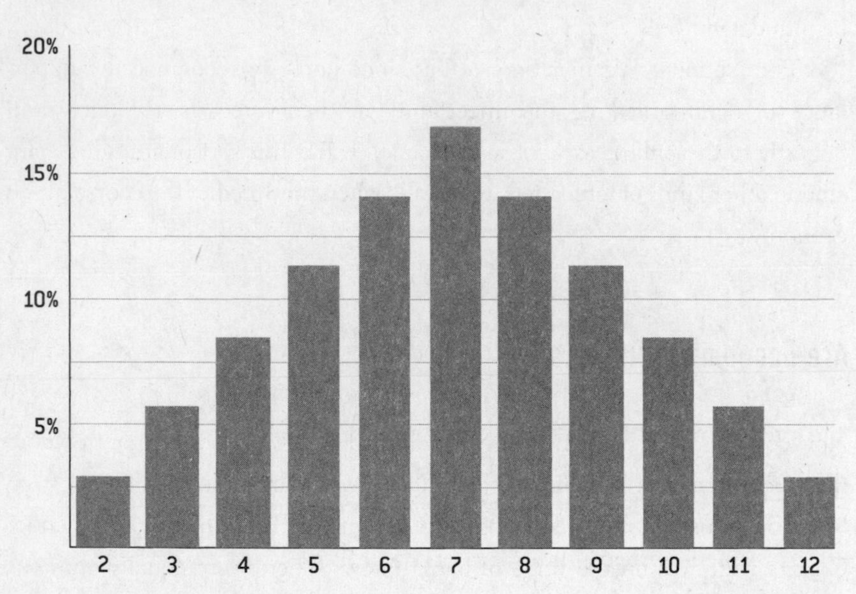

The economists in the Survey of Professional Forecasters are asked to do something similar when they forecast GDP and other variables—estimating, for instance, the probability that GDP might come in at between 2 percent and 3 percent, or between 3 percent and 4 percent. This is what their forecast for GDP looked like in November 2007 (figure 6-3):

FIGURE 6-3: FORECASTED PROBABILITY DISTRIBUTION: REAL U.S. GDP GROWTH (2008)
Survey of Professional Forecasters, November 2007

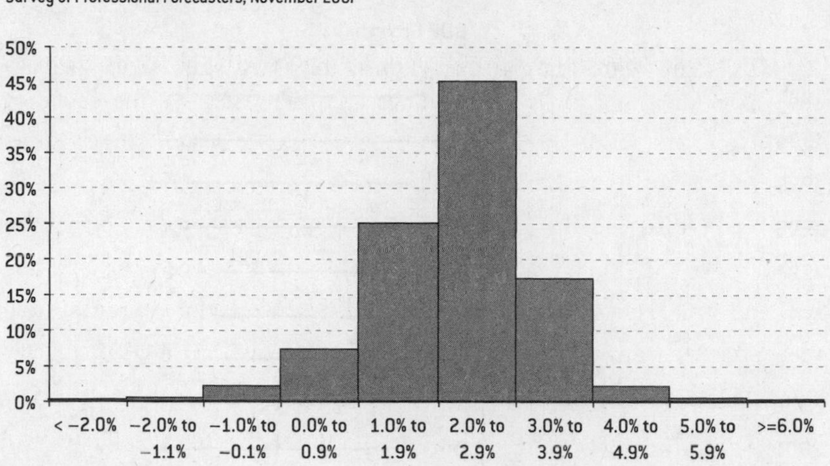

As I mentioned, the economists in this survey thought that GDP would end up at about 2.4 percent in 2008, slightly below its long-term trend. This was a very bad forecast: GDP actually *shrank* by 3.3 percent once the financial crisis hit. What may be worse is that the economists were extremely confident in their bad prediction. They assigned only a 3 percent chance to the economy's shrinking by any margin over the whole of 2008.[15] And they gave it only about a 1-in-500 chance of shrinking by at least 2 percent, as it did.[16]

Indeed, economists have for a long time been much too confident in their ability to predict the direction of the economy. In figure 6-4, I've plotted the forecasts of GDP growth from the Survey of Professional Forecasters for the eighteen years between 1993 and 2010.[17] The bars in the chart represent the 90 percent prediction intervals as stated by the economists.

A prediction interval is a range of the most likely outcomes that a forecast provides for, much like the margin of error in a poll. A 90 percent prediction interval, for instance, is supposed to cover 90 percent of the possible real-world outcomes, leaving only the 10 percent of outlying cases at the tail ends of the distribution. If the economists' forecasts were as accurate as they claimed, we'd expect the actual value for GDP to fall within their prediction interval nine times out of ten, or all but about twice in eighteen years.

FIGURE 6-4: GDP FORECASTS: 90 PERCENT PREDICTION INTERVALS AGAINST ACTUAL RESULTS

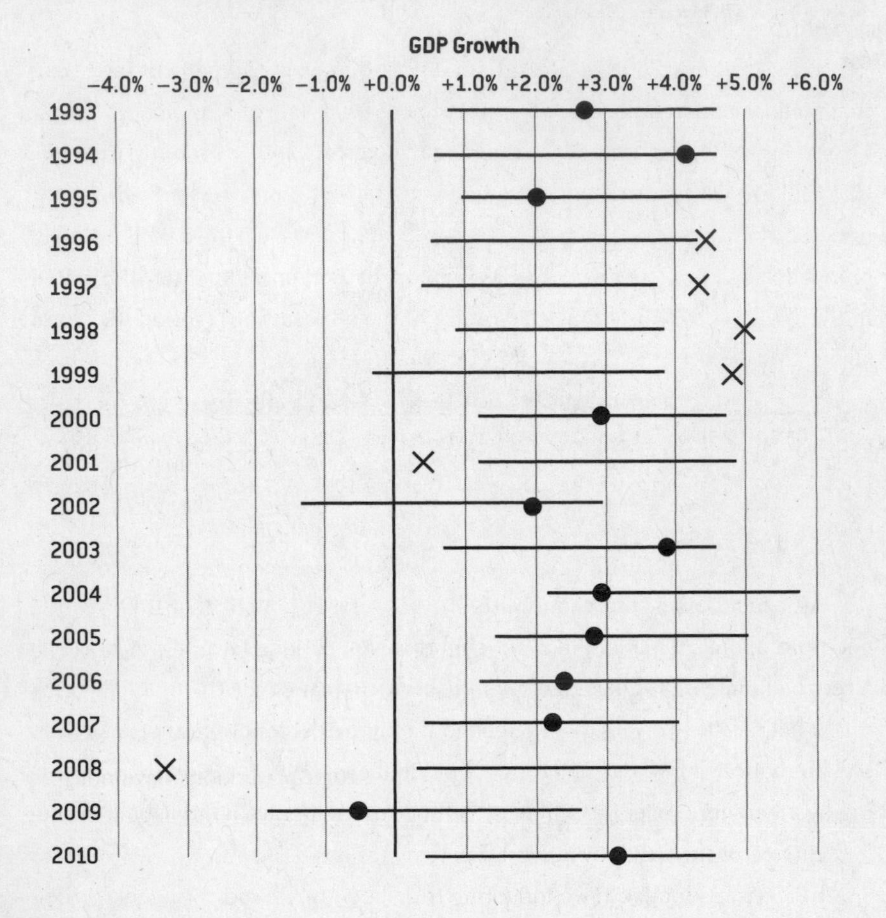

In fact, the actual value for GDP fell outside the economists' prediction interval six times in eighteen years, or fully one-third of the time. Another study,[18] which ran these numbers back to the beginnings of the Survey of Professional Forecasters in 1968, found even worse results: the actual figure for GDP fell outside the prediction interval almost *half* the time. There is almost no chance[19] that the economists have simply been unlucky; they fundamentally overstate the reliability of their predictions.

In reality, when a group of economists give you their GDP forecast, the true 90 percent prediction interval—based on how these forecasts have actually performed[20] and not on how accurate the economists claim them to be—spans about 6.4 points of GDP (equivalent to a margin of error of plus or minus 3.2 percent).*

When you hear on the news that GDP will grow by 2.5 percent next year, that means it could quite easily grow at a spectacular rate of 5.7 percent instead. Or it could fall by 0.7 percent—a fairly serious recession. Economists haven't been able to do any better than that, and there isn't much evidence that their forecasts are improving. The old joke about economists' having called nine out of the last six recessions correctly has some truth to it; one actual statistic is that in the 1990s, economists predicted only 2 of the 60 recessions around the world a year ahead of time.[21]

Economists aren't unique in this regard. Results like these are the rule; experts either aren't very good at providing an honest description of the uncertainty in their forecasts, or they aren't very interested in doing so. This property of overconfident predictions has been identified in many other fields, including medical research, political science, finance, and psychology. It seems to apply both when we use our judgment to make a forecast (as Phil Tetlock's political scientists did) and when we use a statistical model to do so (as in the case of the failed earthquake forecasts that I described in chapter 5).

But economists, perhaps, have fewer excuses than those in other professions for making these mistakes. For one thing, their predictions have not just been overconfident but also quite poor in a real-world sense, often missing the actual GDP figure by a very large and economically meaningful margin. For another, organized efforts to predict variables like GDP have been around for many years, dating back to the Livingston Survey in 1946, and these results are well-documented and freely available. Getting feedback about how well our predictions have done is one way—perhaps the essential way—to improve them. Economic forecasters get more feedback than people in most other professions, but they haven't chosen to correct for their bias toward overconfidence.

Isn't economics supposed to be the field that studies the rationality of human behavior? Sure, you might expect someone in another field—an anthropologist, say—to show bias when he makes a forecast. But not an economist.

---

* The 95 percent prediction interval—the standard that political polls use—is even larger: 9.1 percentage points, equivalent to a margin of error of plus or minus 4.6 points.

Actually, however, that may be part of the problem. Economists understand a lot about rationality—which means they also understand a lot about how our incentives work. If they're making biased forecasts, perhaps this is a sign that they don't have much incentive to make good ones.

"Nobody Has a Clue"

Given the track record of their forecasts, there was one type of economist I was most inclined to seek out—an economist who would be honest about how difficult his job is and how easily his forecast might turn out to be wrong. I was able to find one: Jan Hatzius, the chief economist at Goldman Sachs.

Hatzius can at least claim to have been more reliable than his competitors in recent years. In November 2007, a time when most economists still thought a recession of any kind to be unlikely, Hatzius penned a provocative memo entitled "Leveraged Losses: Why Mortgage Defaults Matter." It warned of a scenario in which millions of homeowners could default on their mortgages and trigger a domino effect on credit and financial markets, producing trillions of dollars in losses and a potentially very severe recession—pretty much exactly the scenario that unfolded. Hatzius and his team were also quick to discount the possibility of a miraculous postcrisis recovery. In February 2009, a month after the stimulus package had been passed and the White House had claimed it would reduce unemployment to 7.8 percent by the end of 2009, Hatzius projected unemployment to rise to 9.5 percent[22] (quite close to the actual figure of 9.9 percent).

Hatzius, a mellow to the point of melancholy German who became Goldman Sachs's chief economist in 2005,[23] eight years after starting at the firm, draws respect even from those who take a skeptical view of the big banks. "[Jan is] very good," Paul Krugman told me. "I hope that Lloyd Blankfein's malevolence won't spill over to Jan and his people." Hatzius also has a refreshingly humble attitude about his ability to forecast the direction of the U.S. economy.

"Nobody has a clue," he told me when I met him at Goldman's glassy office on West Street in New York. "It's hugely difficult to forecast the business cycle. Understanding an organism as complex as the economy is very hard."

As Hatzius sees it, economic forecasters face three fundamental challenges. First, it is very hard to determine cause and effect from economic statistics alone. Second, the economy is always changing, so explanations of economic behavior that hold in one business cycle may not apply to future ones. And third, as bad as their forecasts have been, the data that economists have to work with isn't much good either.

Correlations Without Causation

The government produces data on literally 45,000 economic indicators each year.[24] Private data providers track as many as *four million* statistics.[25] The temptation that some economists succumb to is to put all this data into a blender and claim that the resulting gruel is haute cuisine. There have been only eleven recessions since the end of World War II.[26] If you have a statistical model that seeks to explain eleven outputs but has to choose from among four million inputs to do so, many of the relationships it identifies are going to be spurious. (This is another classic case of overfitting—mistaking noise for a signal—the problem that befell earthquake forecasters in chapter 5.)

Consider how creative you might be when you have a stack of economic variables as thick as a phone book. A once-famous "leading indicator" of economic performance, for instance, was the winner of the Super Bowl. From Super Bowl I in 1967 through Super Bowl XXXI in 1997, the stock market[27] gained an average of 14 percent for the rest of the year when a team from the original National Football League (NFL) won the game.[28] But it fell by almost 10 percent when a team from the original American Football League (AFL) won instead.

Through 1997, this indicator had correctly "predicted" the direction of the stock market in twenty-eight of thirty-one years. A standard test of statistical significance,[29] if taken literally, would have implied that there was only about a 1-in-4,700,000 possibility that the relationship had emerged from chance alone.

It *was* just a coincidence, of course. And eventually, the indicator began to perform badly. In 1998, the Denver Broncos, an original AFL team, won the Super Bowl—supposedly a bad omen. But rather than falling, the stock market gained 28 percent amid the dot-com boom. In 2008, the NFL's New York Giants came from behind to upset the AFL's New England Patriots on David Tyree's spectacular catch—but Tyree couldn't prevent the collapse of the housing bubble, which caused the market to crash by 35 percent. Since 1998, in fact, the stock market has done about 10 percent *better* when the AFL team won the Super Bowl, exactly the opposite of what the indicator was fabled to predict.

How does an indicator that supposedly had just a 1-in-4,700,000 chance of failing flop so badly? For the same reason that, even though the odds of winning the Powerball lottery are only 1 chance in 195 million,[30] somebody wins it every few weeks. The odds are hugely against any one person winning the lottery—but millions of tickets are bought, so somebody is going to get lucky. Likewise, of the millions of statistical indicators in the world, a few will have happened to correlate especially well with stock prices or GDP or the unemployment rate. If not the winner of the Super Bowl, it might be chicken production in Uganda. But the relationship is merely coincidental.

Although economists might not take the Super Bowl indicator seriously, they can talk themselves into believing that other types of variables—anything that has any semblance of economic meaning—are critical "leading indicators" foretelling a recession or recovery months in advance. One forecasting firm brags about how it looks at four hundred such variables,[31] far more than the two or three dozen major ones that Hatzius says contain most of the economic substance.* Other forecasters have touted the predictive power of such relatively obscure indicators as the ratio of bookings-to-billings at semiconductor companies.[32] With so many economic variables to pick from, you're sure to find something that fits the noise in the past data well.

It's much harder to find something that identifies the signal; variables that

* The substantive variables fall into about a dozen major categories: growth (as measured by GDP and its components), jobs, inflation, interest rates, wages and income, consumer confidence, industrial production, sales and consumer spending, asset prices (like stocks and homes), commodity prices (like oil futures), and measures of fiscal policy and government spending. As you can see, this already gives you plenty to work with, so there is little need to resort to four hundred variables.

are leading indicators in one economic cycle often turn out to be lagging ones in the next. Of the seven so-called leading indicators in a 2003 *Inc.* magazine article,[33] all of which had been good predictors of the 1990 and 2001 recessions, only two—housing prices and temporary hiring—led the recession that began in 2007 to any appreciable degree. Others, like commercial lending, did not begin to turn downward until a year *after* the recession began.

Even the well-regarded Leading Economic Index, a composite of ten economic indicators published by the Conference Board, has had its share of problems. The Leading Economic Index has generally declined a couple of months in advance of recessions. But it has given roughly as many false alarms—including most infamously in 1984, when it sharply declined for three straight months,[34] signaling a recession, but the economy continued to zoom upward at a 6 percent rate of growth. Some studies have even claimed that the Leading Economic Index has no predictive power at all when applied in real time.[35]

"There's very little that's really predictive," Hatzius told me. "Figuring out what's truly causal and what's correlation is very difficult to do."

Most of you will have heard the maxim "correlation does not imply causation." Just because two variables have a statistical relationship with each other does not mean that one is responsible for the other. For instance, ice cream sales and forest fires are correlated because both occur more often in the summer heat. But there is no causation; you don't light a patch of the Montana brush on fire when you buy a pint of Häagen Dazs.

If this concept is easily expressed, however, it can be hard to apply in practice, particularly when it comes to understanding the causal relationships in the economy. Hatzius noted, for instance, that the unemployment rate is usually taken to be a lagging indicator. And sometimes it is. After a recession, businesses may not hire new employees until they are confident about the prospects for recovery, and it can take a long time to get all the unemployed back to work again. But the unemployment rate can also be a leading indicator for consumer demand, since unemployed people don't have much ability to purchase new goods and services. During recessions, the economy can fall into a vicious cycle: businesses won't hire until they see more consumer demand, but consumer demand is low because businesses aren't hiring and consumers can't afford their products.

Consumer confidence is another notoriously tricky variable. Sometimes consumers are among the first to pick up warning signs in the economy. But they can also be among the last to detect recoveries, with the public often perceiving the economy to be in recession long after a recession is technically over. Thus, economists debate whether consumer confidence is a leading or lagging indicator,[36] and the answer may be contingent on the point in the business cycle the economy finds itself at. Moreover, since consumer confidence affects consumer behavior, there may be all kinds of feedback loops between expectations about the economy and the reality of it.

An Economic Uncertainty Principle

Perhaps an even more problematic set of feedback loops are those between economic forecasts and economic policy. If, for instance, the economy is forecasted to go into recession, the government and the Federal Reserve will presumably take steps to ameliorate the risk or at least soften the blow. Part of the problem, then, is that forecasters like Hatzius have to predict political decisions as well as economic ones, which can be a challenge in a country where the Congress has a 10 percent approval rating.

But this issue also runs a little deeper. As pointed out by the Nobel Prize–winning economist Robert Lucas[37] in 1976, the past data that an economic model is premised on resulted in part from policy decisions in place at the time. Thus, it may not be enough to know what current policy makers will do; you also need to know what fiscal and monetary policy looked like during the Nixon administration. A related doctrine known as Goodhart's law, after the London School of Economics professor who proposed it,[38] holds that once policy makers begin to target a particular variable, it may begin to lose its value as an economic indicator. For instance, if the government artificially takes steps to inflate housing prices, they might well increase, but they will no longer be good measures of overall economic health.

At its logical extreme, this is a bit like the observer effect (often mistaken for a related concept, the Heisenberg uncertainty principle): once we begin to measure something, its behavior starts to change. Most statistical models are built on the notion that there are independent variables and dependent variables, inputs and outputs, and they can be kept pretty much separate from one another.[39] When it comes to the economy, they are all lumped together in one hot mess.

An Ever-Changing Economy

Even if they could resolve all these problems, economists would still have to contend with a moving target. The American and global economies are always evolving, and the relationships between different economic variables can change over the course of time.

Historically, for instance, there has been a reasonably strong correlation between GDP growth and job growth. Economists refer to this as Okun's law. During the Long Boom of 1947 through 1999, the rate of job growth[40] had normally been about half the rate of GDP growth, so if GDP increased by 4 percent during a year, the number of jobs would increase by about 2 percent.

The relationship still exists—more growth is certainly better for job seekers. But its dynamics seem to have changed. After each of the last couple of recessions, considerably fewer jobs were created than would have been expected during the Long Boom years. In the year after the stimulus package was passed in 2009, for instance, GDP was growing fast enough to create about two million jobs according to Okun's law.[41] Instead, an additional 3.5 million jobs were *lost* during the period.

Economists often debate about what the change means. The most pessimistic interpretation, advanced by economists including Jeffrey Sachs of Columbia University, is that the pattern reflects profound structural problems in the American economy: among them, increasing competition from other countries, an imbalance between the service and manufacturing sectors, an aging population, a declining middle class, and a rising national debt. Under this theory, we have entered a new and unhealthy normal, and the problems may get worse unless fundamental changes are made. "We were underestimating the role of global change in causing U.S. change," Sachs told me. "The loss of jobs internationally to China and emerging markets have really jolted the American economy."

The bigger question is whether the volatility of the 2000s is more representative of the long-run condition of the economy—perhaps the long boom years had been the outlier. During the Long Boom, the economy was in recession only 15 percent of the time. But the rate was more than twice that—36 percent—from 1900 through 1945.[42]

Although most economists believe that some progress has been made in stabilizing the business cycle, we may have been lucky to avoid more problems. This particularly holds in the period between 1983 and 2006—a subset of the Long Boom that is sometimes called the Great Moderation—when the economy was in recession just 3 percent of the time. But much of the growth was fueled by large increases in government and consumer debt, as well as by various asset-price bubbles. Advanced economies have no divine right to grow at Great Moderation rates: Japan's, which grew at 5 percent annually during the 1980s, has grown by barely one percent per year since then.[43]

This may be one reason why forecasters and policy makers were taken so much by surprise by the depth of the 2007 recession. Not only were they failing to account for events like the Great Depression*—they were sometimes calibrating their forecasts according to the Great Moderation years, which were an outlier, historically speaking.

Don't Throw Out Data

The Federal Open Market Committee, which is charged with setting interest rates, is required by law to release macroeconomic forecasts to Congress at least twice per year. The Fed was in some ways ahead of the curve by late 2007: their forecasts of GDP growth were slightly more bearish than those issued by private-sector forecasters, prompting them to lower interest rates four times toward the end of the year.

Still, in the Fed's extensive minutes from a late October 2007 meeting, the term "recession" was not used even once in its discussion of the economy.[44] The Fed is careful with its language, and the possibility of a recession may nevertheless have been implied through the use of phrases like *downside risks*. But they

* This may be in part because the economic data from before World War II is quite incomplete.

were not betting on a recession (their forecast still projected growth), and there was little indication that they were entertaining the possibility of as severe a recession as actually unfolded.

Part of the reason may have been that the Fed was looking at data from the Great Moderation years to set their expectations for the accuracy of their forecasts. In particular, they relied heavily upon a paper that looked at how economic forecasts had performed from 1986 through 2006.[45] The problem with looking at only these years is that they contained very little economic volatility: just two relatively mild recessions in 1990–1991 and in 2001. "By gauging current uncertainty with data from the mid-1980s on," the authors warned, "we are implicitly assuming that the calm conditions since the Great Moderation will persist into the future." This was an awfully big assumption to make. The Fed may have concluded that a severe recession was unlikely in 2007 in part because they had chosen to ignore years in which there were severe recessions.

A forecaster should almost never ignore data, especially when she is studying rare events like recessions or presidential elections, about which there isn't very much data to begin with. Ignoring data is often a tip-off that the forecaster is overconfident, or is overfitting her model—that she is interested in showing off rather than trying to be accurate.

In this particular case, it was not obvious that economists had improved much at forecasting the business cycle. In figure 6-5a, I've compared predicted levels of GDP growth from the Survey of Professional Forecasters against the actual figures for the years 1968 through 1985—these are the years the Fed could have looked at but chose to throw out. You'll see there's quite a lot of economic volatility in this period, such as during the inflation-driven recessions of the mid-1970s and early 1980s. Still, the results are not completely discouraging for forecasters, in that the forecasted and actual outcomes have a reasonably strong correlation with one another.

If you make the same plot for the years 1986 through 2006 (as in figure 6-5b), you'll find just the reverse. Most of the data points—both the forecasted values for GDP and the actual ones—are bunched closely together in a narrow range between about 2 percent and 5 percent annual growth. Because there was so little volatility during this time, the average error in the forecast was less

FIGURE 6-5A: GDP FORECASTS VERSUS ACTUAL GDP, 1968–1985

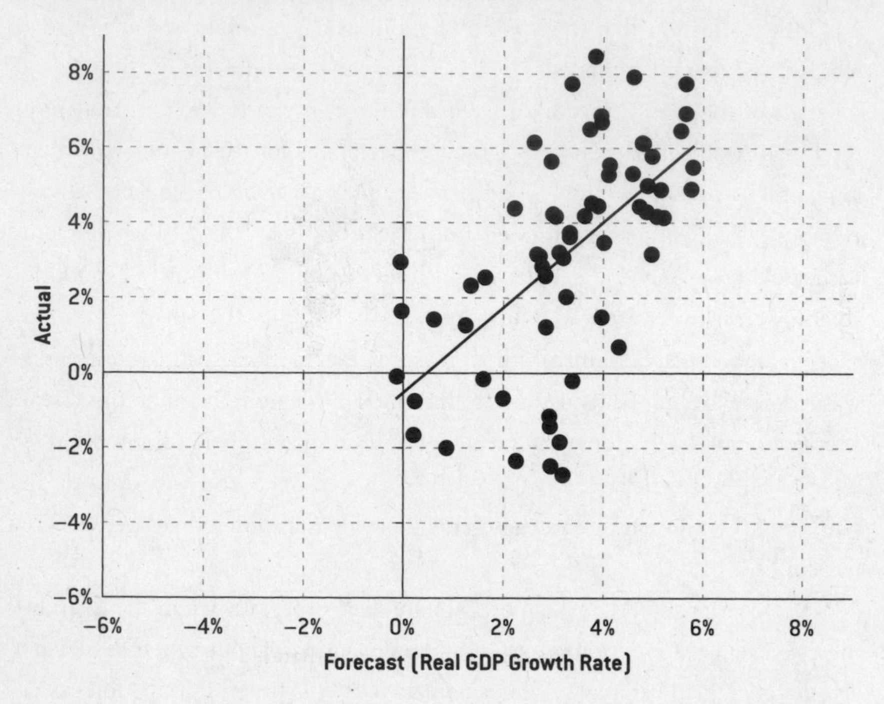

than in the previous period.* However, to the extent there was any variability in the economy, like the mild recessions of 1990–91 or in 2001, the forecasts weren't doing a very good job of capturing it—in fact, there was almost no correlation between the predicted and actual results. There was little indication that economists had become more skilled at forecasting the course of the economy. Instead, their jobs had become temporarily easier because of the calm economic winds, as a weather forecaster in Honolulu faces an easier task than one in Buffalo.

The other rationale you'll sometimes hear for throwing out data is that there has been some sort of fundamental shift in the problem you are trying to solve. Sometimes these arguments are valid to a certain extent: the American economy is a constantly evolving thing and periodically undergoes structural shifts (recently, for instance, from an economy dominated by manufacturing

* The root-mean squared error for the forecasts in these years was 1.1 points of GDP, versus 2.3 points for the years 1968–85.

FIGURE 6-5B: GDP FORECASTS VERSUS ACTUAL GDP, 1986–2006

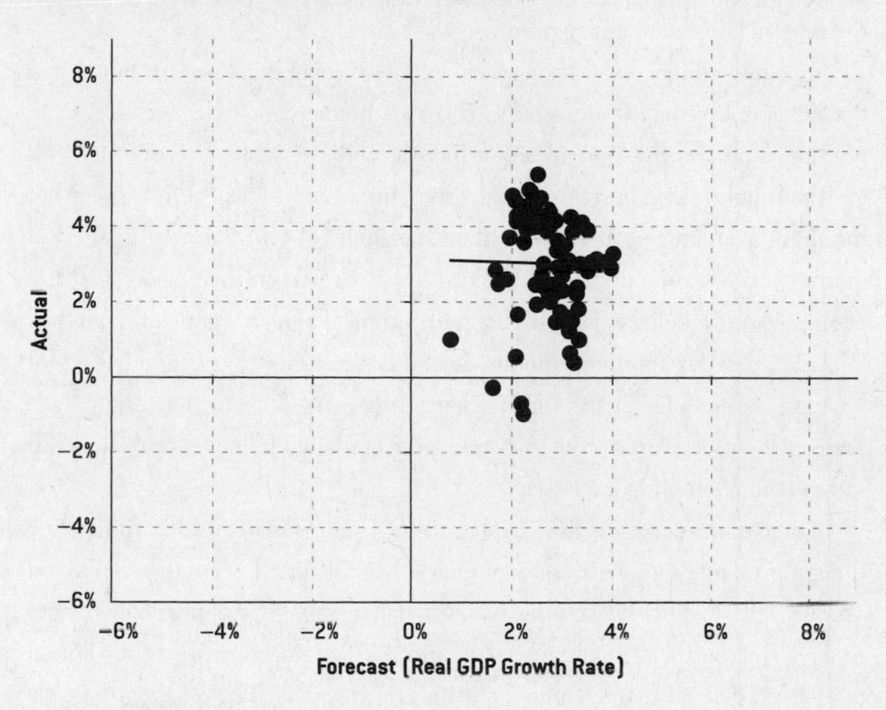

to one dominated by its service sector). This isn't baseball, where the game is always played by the same rules.

The problem with this is that you never know when the *next* paradigm shift will occur, and whether it will tend to make the economy more volatile or less so, stronger or weaker. An economic model conditioned on the notion that nothing major will change is a useless one. But anticipating these turning points is not easy.

Economic Data Is Very Noisy

The third major challenge for economic forecasters is that their raw data isn't much good. I mentioned earlier that economic forecasters rarely provide their prediction intervals when they produce their forecasts—probably because doing so would undermine the public's confidence in their expertise. "Why do people not give intervals? Because they're embarrassed," Hatzius says. "I think that's the reason. People are embarrassed."

The uncertainty, however, applies not just to economic forecasts but also to the economic variables themselves. Most economic data series are subject to revision, a process that can go on for months and even years after the statistics are first published. The revisions are sometimes enormous.[46] One somewhat infamous example was the government's estimate of GDP growth in the last quarter of 2008. Initially reported as "only" a 3.8 percent rate of decline, the economy is now believed to have been declining at almost 9 percent. Had they known the real size of the economic hole, the White House's economists might have pushed for a larger stimulus package in January 2009, or they might have realized how deep the problems were and promoted a longer-term solution rather than attempting a quick fix.

Large errors like these have been fairly common. Between 1965 and 2009,[47] the government's initial estimates of quarterly GDP were eventually revised, on average, by 1.7 points. That is the average change; the range of possible changes in each quarterly GDP is higher still, and the margin of error[48] on an initial quarterly GDP estimate is plus or minus 4.3 percent. That means there's a chance that the economy will turn out to have been in recession even if the government had initially reported above-average growth, or vice versa. The government first reported that the economy had grown by 4.2 percent in the fourth quarter of 1977, for instance, but that figure was later revised to negative 0.1 percent.[49]

So we should have some sympathy for economic forecasters.[50] It's hard enough to know where the economy is going. But it's much, much harder if you don't know where it is to begin with.

A Butterfly Flaps Its Wings in Brazil and Someone Loses a Job in Texas

The challenge to economists might be compared to the one faced by weather forecasters. They face two of the same fundamental problems.

First, the economy, like the atmosphere, is a dynamic system: everything affects everything else and the systems are perpetually in motion. In meteorology, this problem is quite literal, since the weather is subject to chaos theory—a butterfly flapping its wings in Brazil can theoretically cause a tornado in Texas. But in loosely the same way, a tsunami in Japan or a longshoreman's strike in Long Beach can affect whether someone in Texas finds a job.

Second, weather forecasts are subject to uncertain initial conditions. The probabilistic expression of weather forecasts ("there's a 70 percent chance of rain") arises not because there is any inherent randomness in the weather. Rather, the problem is that meteorologists assume they have imprecise measurements of what the initial conditions were like, and weather patterns (because they are subject to chaos theory) are extremely sensitive to changes in the initial conditions. In economic forecasting, likewise, the quality of the initial data is frequently quite poor.

Weather prediction, however, is one of the real success stories in this book. Forecasts of everything from hurricane trajectories to daytime high temperatures have gotten much better than they were even ten or twenty years ago, thanks to a combination of improved computer power, better data-collection methods, and old-fashioned hard work.

The same cannot be said for economic forecasting. Any illusion that economic forecasts were getting better ought to have been shattered by the terrible mistakes economists made in advance of the recent financial crisis.[51]

If the meteorologist shares some of the economist's problems of a dynamic system with uncertain initial conditions, she has a wealth of hard science to make up for it. The physics and chemistry of something like a tornado are not all that complicated. That does not mean that tornadoes are easy to predict. But meteorologists have a strong fundamental understanding of what causes tornadoes to form and what causes them to dissipate.

Economics is a much softer science. Although economists have a reasonably sound understanding of the basic systems that govern the economy, the cause and effect are all blurred together, especially during bubbles and panics when the system is flushed with feedback loops contingent on human behavior.

Nevertheless, if discerning cause and effect is difficult for economists, it is probably better to try than just give up. Consider again, for instance, what Hatzius wrote on November 15, 2007:

> The likely mortgage credit losses pose a significantly bigger macroeconomic risk than generally recognized. . . . The macroeconomic consequences could be quite dramatic. If leveraged investors see $200 [billion in] aggregate credit loss, they might need to scale back their lending by $2 trillion. This is a large shock. . . . It is easy to see how such a shock could produce a substantial recession or a long period of very sluggish growth.

Consumers had been extended too much credit, Hatzius wrote, to pay for homes that the housing bubble had made unaffordable. Many of them had stopped making their mortgage payments, and there were likely to be substantial losses from this. The degree of leverage in the system would compound the problem, paralyzing the credit market and the financial industry more broadly. The shock might be large enough to trigger a severe recession.

And this is exactly how the financial crisis played out. Not only was Hatzius's forecast correct, but it was also right for the right reasons, explaining the causes of the collapse and anticipating the effects. Hatzius refers to this chain of cause and effect as a "story." It is a story about the economy—and although it might be a data-driven story, it is one grounded in the real world.

In contrast, if you just look at the economy as a series of variables and equations without any underlying structure, you are almost certain to mistake noise for a signal and may delude yourself (and gullible investors) into thinking you are making good forecasts when you are not. Consider what happened to one of Hatzius's competitors, the forecasting firm ECRI.

In September 2011, ECRI predicted a near certainty of a "double dip" recession. "There's nothing that policy makers can do to head it off," it advised.[52] "If you think this is a bad economy, you haven't seen anything yet." In interviews, the managing director of the firm, Lakshman Achuthan, suggested the recession would begin almost immediately if it hadn't started already.[53] The firm described the reasons for its prediction in this way:

> ECRI's recession call isn't based on just one or two leading indexes, but on dozens of specialized leading indexes, including the U.S. Long Leading Index . . . to be followed by downturns in the Weekly Leading Index and other shorter-leading indexes. In fact, the most reliable forward-looking in-

dicators are now collectively behaving as they did on the cusp of full-blown recessions.[54]

There's plenty of jargon, but what is lacking in this description is any actual economic substance. Theirs was a story about data—as though data itself caused recessions—and not a story about the economy. ECRI actually seems quite proud of this approach. "Just as you do not need to know exactly how a car engine works in order to drive safely," it advised its clients in a 2004 book, "You do not need to understand all the intricacies of the economy to accurately read those gauges."[55]

This kind of statement is becoming more common in the age of Big Data.[56] Who needs theory when you have so much information? But this is categorically the wrong attitude to take toward forecasting, especially in a field like economics where the data is so noisy. Statistical inferences are much stronger when backed up by theory or at least some deeper thinking about their root causes. There were certainly reasons for economic pessimism in September 2011[57]—for instance, the unfolding debt crisis in Europe—but ECRI wasn't looking at those. Instead, it had a random soup of variables that mistook correlation for causation.[58]

Indeed, the ECRI forecast seemed to demark an economic turning point—but it was a *positive* one. The S&P 500 gained 21 percent in the five months after ECRI announced its recession call,[59] while GDP growth registered at a fairly healthy clip of 3.0 percent in the last quarter of 2011 instead of going into recession. ECRI kicked the can down the road, "clarifying" the call to say that it extended all the way into 2012 even though this is not what they had implied originally.[60]

When Biased Forecasts Are Rational

If you're looking for an economic forecast, the best place to turn is the average or aggregate prediction rather than that of any one economist. My research into the Survey of Professional Forecasters suggests that these aggregate forecasts are about 20 percent more accurate[61] than the typical individual's forecast at predicting GDP, 10 percent better at predicting unemployment, and 30 percent better at predicting inflation. This property—group forecasts beat individual ones—has been found to be true in almost every field in which it has been studied.

And yet while the notion that aggregate forecasts beat individual ones is an important empirical regularity, it is sometimes used as a cop-out when forecasts might be improved. The aggregate forecast is made up of individual forecasts; if those improve, so will the group's performance. Moreover, even the aggregate economic forecasts have been quite poor in any real-world sense, so there is plenty of room for progress.

Most economists rely on their judgment to some degree when they make a forecast, rather than just take the output of a statistical model as is. Given how noisy the data is, this is probably helpful. A study[62] by Stephen K. McNess, the former vice president of the Federal Reserve Bank of Boston, found that judgmental adjustments to statistical forecasting methods resulted in forecasts that were about 15 percent more accurate. The idea that a statistical model would be able to "solve" the problem of economic forecasting was somewhat in vogue during the 1970s and 1980s when computers came into wider use. But as was the case in other fields, like earthquake forecasting during that time period, improved technology did not cover for the lack of theoretical understanding about the economy; it only gave economists faster and more elaborate ways to mistake noise for a signal. Promising-seeming models failed badly at some point or another and were consigned to the dustbin.[63]

Invoking one's judgment, however, also introduces the potential for bias. You may make the forecast that happens to fit your economic incentives or your political beliefs. Or you may be too proud to change your story even when the facts and circumstances demand it. "I do think that people have the tendency, which needs to be actively fought," Hatzius told me, "to see the information flow the way you want to see it."

Are some economists better at managing this trade-off than others? Is the economist who called the last recession right more likely to get the next one too? This question has an interesting answer.

Statistical tests designed to identify predictive skill have generally come up with negative results when applied to the Survey of Professional Forecasters.[64] That is, if you look at that survey, there doesn't seem to be much evidence that some economists are consistently better than others. Studies of another panel, the Blue Chip Economic Survey, have more often come up with positive findings, however.[65] There is clearly a lot of luck involved in economic forecasting—economists who are permanently bearish or bullish are guaranteed to be right every now and then. But the studies of the Blue Chip panel seem to find that some economists do a little bit better than others over the long run.

What is the difference between the two surveys? The Survey of Professional Forecasters is conducted anonymously: each economist is assigned a random ID number that remains constant from survey to survey, but nothing is revealed about just who he is or what he does. In the Blue Chip panel, on the other hand, everybody's forecast has his name and reputation attached to it.

When you have your name attached to a prediction, your incentives may change. For instance, if you work for a poorly known firm, it may be quite rational for you to make some wild forecasts that will draw big attention when they happen to be right, even if they aren't going to be right very often. Firms like Goldman Sachs, on the other hand, might be more conservative in order to stay within the consensus.

Indeed, this exact property has been identified in the Blue Chip forecasts:[66] one study terms the phenomenon "rational bias."[67] The less reputation you have, the less you have to lose by taking a big risk when you make a prediction. Even if you know that the forecast is dodgy, it might be rational for you to go after the big score. Conversely, if you have already established a good reputation, you might be reluctant to step too far out of line even when you think the data demands it.

Either of these reputational concerns potentially distracts you from the goal of making the most honest and accurate forecasts—and they probably worsen forecasts on balance. Although the differences are modest, historically the anonymous participants in the Survey of Professional Forecasters have done slightly better at predicting GDP and unemployment than the reputation-minded Blue Chip panelists.[68]

Overcoming Bias

If it can be rational to produce bad forecasts, that implies there are consumers of these forecasts who aid and abet them. Just as there are political pundits who make careers out of making implausible claims to partisan audiences, there are bears, bulls, and contrarians who will always have a constituency in the marketplace for economic ideas. (Sometimes economic forecasts have expressly political purposes too. It turns out that the economic forecasts produced by the White House, for instance, have historically been among the least accurate of all,[69] regardless of whether it's a Democrat or a Republican in charge.)

When it comes to economic forecasting, however, the stakes are higher than for political punditry. As Robert Lucas pointed out, the line between economic forecasting and economic policy is very blurry; a bad forecast can make the real economy worse.

There may be some hope at the margin for economic forecasting to benefit from further technological improvements. Things like Google search traffic patterns, for instance, can serve as leading indicators for economic data series like unemployment.

"The way we think about it is if you take something like initial claims on unemployment insurance, that's a very good predictor for unemployment rates, which is a good predictor for economic activity," I was told by Google's chief economist, Hal Varian, at Google's headquarters in Mountain View, California. "We can predict unemployment initial claims earlier because if you're in a company and a rumor goes around that there are going to be layoffs, then people start searching 'where's the unemployment office,' 'how am I going to apply for unemployment,' and so on. It's a slightly leading indicator."

Still, the history of forecasting in economics and other fields suggests that technological improvements may not help much if they are offset by human biases, and there is little indication that economic forecasters have overcome these. For instance, they do not seem to have been chastened much by their experience with the Great Recession. If you look at the forecasts for GDP growth that the Survey of Professional Forecasters made in November 2011 (figure 6-6), it still exhibited the same tendency toward overconfidence that we

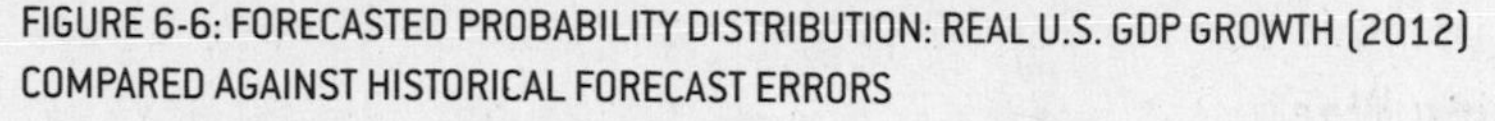

FIGURE 6-6: FORECASTED PROBABILITY DISTRIBUTION: REAL U.S. GDP GROWTH (2012) COMPARED AGAINST HISTORICAL FORECAST ERRORS

Survey of Professional Forecasters, November 2011

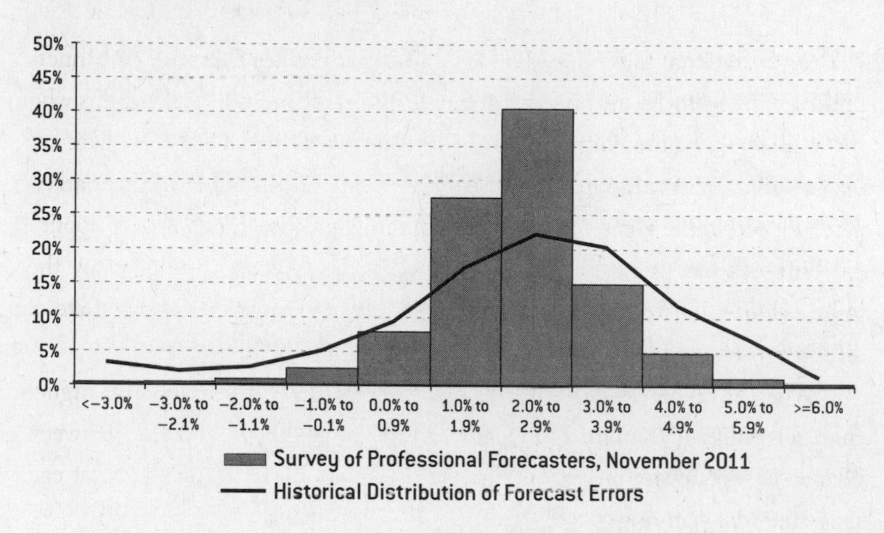

saw in 2007, with forecasters discounting both upside and downside economic scenarios far more than is justified by the historical accuracy of their forecasts.[70]

If we want to reduce these biases—we will never be rid of them entirely—we have two fundamental alternatives. One might be thought of as a supply-side approach—creating a market for accurate economic forecasts. The other might be a demand-side alternative: reducing demand for inaccurate and overconfident ones.

Robin Hanson, an economist at George Mason University, is an advocate of the supply-side alternative. I met him for lunch at one of his favorite Moroccan places in northern Virginia. He's in his early fifties but looks much younger (despite being quite bald), and is a bit of an eccentric. He plans to have his head cryogenically frozen when he dies.[71] He is also an advocate of a system he calls "futarchy" in which decisions on policy issues are made by prediction markets[72] rather than politicians. He is clearly not a man afraid to challenge the conventional wisdom. Instead, Hanson writes a blog called Overcoming Bias, in which he presses his readers to consider which cultural taboos, ideological beliefs, or misaligned incentives might constrain them from making optimal decisions.

"I think the most interesting question is how little effort we actually put into forecasting, even on the things we say are important to us," Hanson told me as the food arrived.

"In an MBA school you present this image of a manager as a great decision maker—the scientific decision maker. He's got his spreadsheet and he's got his statistical tests and he's going to weigh the various options. But in fact real management is mostly about managing coalitions, maintaining support for a project so it doesn't evaporate. If they put together a coalition to do a project, and then at the last minute the forecasts fluctuate, you can't dump the project at the last minute, right?

"Even academics aren't very interested in collecting a track record of forecasts—they're not very interested in making clear enough forecasts to score," he says later. "What's in it for them? The more fundamental problem is that we have a demand for experts in our society but we don't actually have that much of a demand for accurate forecasts."

Hanson, in order to address this deficiency, is an advocate of prediction markets—systems where you can place bets on a particular economic or policy outcome, like whether Israel will go to war with Iran, or how much global temperatures will rise because of climate change. His argument for these is pretty simple: they ensure that we have a financial stake in being accurate when we make forecasts, rather than just trying to look good to our peers.

We will revisit the idea of prediction markets in chapter 11; they are not a panacea, particularly if we make the mistake of assuming that they can never go wrong. But as Hansen says, they can yield some improvement by at least getting everyone's incentives in order.

One of the most basic applications might simply be markets for predicting macroeconomic variables like GDP and unemployment. There are already a variety of direct and indirect ways to bet on things like inflation, interest rates, and commodities prices, but no high-volume market for GDP exists.

There could be a captive audience for these markets: common stocks have become more highly correlated with macroeconomic risks in recent years,[73] so they could provide a means of hedging against them. These markets would also provide real-time information to policy makers, essentially serving as continuously updated forecasts of GDP. Adding options to the markets—bets on, say, whether GDP might grow by 5 percent, or decline by 2 percent—would punish overconfident forecasters and yield more reliable estimates of the uncertainties inherent in forecasting the economy.

The other solution, the "demand-side" approach, is slower and more incremental. It simply means that we have to be better consumers of forecasts. In the context of the economic forecasting, that might mean turning the spotlight away from charlatans with "black box" models full of random assortments of leading indicators and toward people like Jan Hatzius who are actually talking economic substance. It might also mean placing more emphasis on the noisiness of economic indicators and economic forecasts. Perhaps initial estimates of GDP should be reported with margins of error, just as political polls are.

More broadly, it means recognizing that the amount of confidence someone expresses in a prediction is not a good indication of its accuracy—to the contrary, these qualities are often inversely correlated. Danger lurks, in the economy and elsewhere, when we discourage forecasters from making a full and explicit account of the risks inherent in the world around us.

7

ROLE MODELS

The flu hit Fort Dix like clockwork every January; it had almost become a rite of passage. Most of the soldiers would go home for Christmas each year, fanning out to all corners of the United States for their winter break. They would then return to the base, well-fed and well-rested, but also carrying whichever viruses might have been going around their hometowns. If the flu was anywhere in the country, it was probably coming back with them. Life in the cramped setting of the barracks, meanwhile, offered few opportunities for privacy or withdrawal. If someone—anyone—had caught the flu back home, he was more likely than not to spread it to the rest of the platoon. You could scarcely conjure a scenario more favorable to transmission of the disease.

Usually this was no cause for concern; tens of millions of Americans catch the flu in January and February every year. Few of them die from it, and young, healthy men like David Lewis, a nineteen-year-old private from West Ashley, Massachusetts, who had returned to Fort Dix that January, are rarely among the exceptions. So Lewis, even though he'd been sicker than most of the recruits and ordered to stay in the barracks, decided to join his fellow privates on a fifty-mile march through the snow-blanketed marshlands of central New Jersey. He was in no mood to let a little fever bother him—it was 1976, the year of the nation's bicentennial, and the country needed order and discipline in the uncertain days following Watergate and Vietnam.[1]

But Lewis never made it back to the barracks: thirteen miles into the march, he collapsed and was later pronounced dead. An autopsy revealed that Lewis's lungs were flush with blood: he had died of pneumonia, a common complication of flu, but not usually one to kill a healthy young adult like Lewis.

The medics at Fort Dix had already been nervous about that year's flu bug. Although some of the several hundred soldiers who had gotten ill that winter had tested positive for the A/Victoria flu strain—the name for the common and fairly benign virus that was going around the world that year[2]—there were others like Lewis who had suffered from an unidentified and apparently much more severe type of flu. Samples of their blood were sent to the Center for Disease Control (CDC) in Atlanta for further testing.

Two weeks later the CDC revealed the identity of the mysterious virus. It was not a new type of flu after all but instead something altogether more disturbing, a ghost from epidemics past: influenza virus type H1N1, more commonly known as the swine flu. H1N1 had been responsible for the worst pandemic in modern history: the Spanish flu of 1918–20, which afflicted a third of humanity and killed 50 million,[3] including 675,000 in the United States. For reasons of both science and superstition, the disclosure sent a chill though the nation's epidemiological community. The 1918 outbreak's earliest manifestations had also come at a military base, Fort Riley in Kansas, where soldiers were busy preparing to enter World War I.[4] Moreover, there was a belief at that time—based on somewhat flimsy scientific evidence—that a major flu epidemic manifested itself roughly once every ten years.[5] The flu had been severe in 1938, 1947, 1957, and 1968;[6] in 1976, the world seemed due for the next major pandemic.

A series of dire predictions soon followed. The concern was not an immediate outbreak—by the time the CDC had positively identified the H1N1 strain, flu season had already run its course. But scientists feared that it foreshadowed something much worse the following winter. There had never been a case, a prominent doctor noted to the *New York Times*,[7] in which a newly identified strain of the flu had failed to outcompete its rivals and become the global hegemon: wimpy A/Victoria stood no chance against its more virulent and ingenious rival. And if H1N1 were anywhere near as deadly as the 1918 version had been, the consequences might be very bad indeed. Gerald Ford's secretary of health, F. David Mathews, predicted that one million Americans would die, eclipsing the 1918 total.[8]

President Ford found himself in a predicament. The vaccine industry, somewhat like the fashion industry, needs at least six months of lead time to know what the hip vaccine is for the new season; the formula changes a little bit every year. If they suddenly had to produce a vaccine that guarded against H1N1—and particularly if they were going to produce enough of it for the entire nation—they would need to get started immediately. Meanwhile, Ford was struggling to overcome a public perception that he was slow-witted and unsure of himself—an impression that grew more entrenched every weekend with Chevy Chase's bumbling-and-stumbling caricature of him on NBC's new hit show, *Saturday Night Live*. So Ford took the resolute step of asking Congress to authorize some 200 million doses of vaccine, and ordered a mass vaccination program, the first the country had seen since Jonas Salk had developed the polio vaccine in the 1950s.

The press portrayed the mass vaccination program as a gamble.[9] But Ford thought of it as a gamble between money and lives, and one that he was on the right side of. Overwhelming majorities in both houses of Congress approved his plans at a cost of $180 million.[10]

By summer, however, there were serious doubts about the government's plans. Although summer is the natural low season for the flu in the United States,[11] it was winter in the Southern Hemisphere, when flu is normally at its peak. And nowhere, from Auckland to Argentina, were there any signs of H1N1; instead, the mild and common A/Victoria was the dominant strain again. Indeed, the roughly two hundred cases at Fort Dix remained the only confirmed cases of H1N1 anywhere in the world, and Private Lewis's the only death. Criticism started to pour in from all quarters: from the assistant director of the CDC,[12] the World Health Organization,[13] the prestigious British medical journal *The Lancet*,[14] and the editorial pages of the *New York Times*, which was already characterizing the H1N1 threat a "false alarm."[15] No other Western country had called for such drastic measures.

Instead of admitting that they had overestimated the threat, the Ford administration doubled down, preparing a series of frightening public service announcements that ran in regular rotation on the nation's television screens that fall.[16] One mocked the naïveté of those who refused flu shots—*"I'm the healthiest fifty-five-year-old you've ever seen—I play golf every weekend!"* the balding everyman says, only to be shown on his deathbed moments later. Another featured a female narrator tracing the spread of the virus from one person to the next, dishing about it in breathy tones as though it were an STD—"Betty's mother gave it to the cabdriver . . . and to one of the *charming* stewardesses . . . and then she gave it to her friend Dottie, who had a heart condition and died."

The campy commercials were intended to send a very serious message: *Be afraid, be very afraid*. Americans took the hint. Their fear, however, manifested itself as much toward the vaccine as toward the disease itself. Throughout American history, the notion of the government poking needles into everyone's arm has always provoked more than its fair share of anxiety. But this time there was a more tangible basis for public doubt. In August of that year, under pressure from the drug companies, Congress and the White House had agreed to indemnify them from legal liability in the event of manufacturing defects. This was widely read as a vote of no-confidence; the vaccine looked as though it was being rushed out without adequate time for testing. Polls that summer showed that only about 50 percent of Americans planned to get vaccinated, far short of the government's 80 percent goal.[17]

The uproar did not hit a fever pitch until October, when the vaccination program began. On October 11, a report surfaced from Pittsburgh that three senior citizens had died shortly after receiving their flu shots; so had two elderly persons in Oklahoma City; so had another in Fort Lauderdale.[18] There was no evidence that any of the deaths were linked to the vaccinations—elderly people die every day, after all.[19] But between the anxiety about the government's vaccination program and the media's dubious understanding of statistics,[20] every death of someone who'd gotten a flu shot became a cause for alarm. Even Walter Cronkite, the most trusted man in America—who had broken from his trademark austerity to admonish the media for its sensational handling of the story—could not calm the public down. Pittsburgh and many other cities shuttered their clinics.[21]

By late fall, another problem had emerged, this one far more serious. About five hundred patients, after receiving their shots, had begun to exhibit the symptoms of a rare neurological condition known as Guillain–Barré syndrome, an autoimmune disorder that can cause paralysis. This time, the statistical evidence was far more convincing: the usual incidence of Guillain–Barré in the general population is only about one case per million persons.[22] In contrast, the rate in the vaccinated population had been ten times that—five hundred cases out of the roughly fifty million people who had been administered the vaccine. Although scientists weren't positive why the vaccines were causing Guillain–Barré, manufacturing defects triggered by the rush production schedule were a plausible culprit,[23] and the consensus of the medical community[24] was that the vaccine program should be shut down for good, which the government finally did on December 16.

In the end, the outbreak of H1N1 at Fort Dix had been completely isolated; there was never another confirmed case anywhere in the country.[25] Meanwhile, flu deaths from the ordinary A/Victoria strain were slightly below average in the winter of 1976–77.[26] It had been much ado about nothing.

The swine flu fiasco—as it was soon dubbed—was a disaster on every level for President Ford, who lost his bid for another term to the Democrat Jimmy Carter that November.[27] The drug makers had been absolved of any legal responsibility, leaving more than $2.6 billion in liability claims[28] against the United States government. It seemed like every local paper had run a story about the poor waitress or schoolteacher who had done her duty and gotten the vaccine, only to have contracted Guillain–Barré. Within a couple of years, the number of Americans willing to take flu shots dwindled to only about one million,[29] potentially putting the nation in grave danger had a severe strain hit in 1978 or 1979.[30]

Ford's handling of H1N1 was irresponsible on a number of levels. By invoking the likelihood of a 1918-type pandemic, he had gone against the advice of medical experts, who believed at the time that the chance of such a worst-case outcome was no higher than 35 percent and perhaps as low as 2 percent.[31]

Still, it was not clear what had caused H1N1 to disappear just as suddenly as it emerged. And predictions about H1N1 would fare little better when it came back some thirty-three years later. Scientists at first missed H1N1 when it reappeared in 2009. Then they substantially overestimated the threat it might pose once they detected it.

A Sequel to the Swine Flu Fiasco?

The influenza virus is perpetuated by birds—particularly wild seafaring birds like albatrosses, seagulls, ducks, swans, and geese, which carry its genes from one continent to another but rarely become sick from the disease. They pass it along to other species, especially pigs and domesticated fowl like chickens,[32] which live in closer proximity to humans. Chickens can become ill from the flu, but they can usually cope with it well enough to survive and pass it along to their human keepers. Pigs are even better at this, because they are receptive to both human and avian viruses as well as their own, providing a vessel for different strains of the virus to mix and mutate together.[33]

The perfect incubator for the swine flu, then, would be a region in which each of three conditions held:

1. It would be a place where humans and pigs lived in close proximity—that is, somewhere where pork was a staple of the diet.
2. It would be a place near the ocean where pigs and seafaring birds might intermingle.
3. And it would probably be somewhere in the developing world, where poverty produced lower levels of hygiene and sanitation, allowing animal viruses to be transmitted to humans more easily.

This mix almost perfectly describes the conditions found in Southeast Asian countries like China, Indonesia, Thailand, and Vietnam (China alone is home to about half the world's pigs[34]). These countries are very often the source for the flu, both the annual strains and the more unusual varieties that can potentially become global pandemics.* So they have been the subject of most of the medical community's attention, especially in recent years because of the fear over another strain of the virus. H5N1, better known as bird flu or avian flu, has been simmering for some years in East Asia and could be extremely deadly if it mutated in the wrong way.

These circumstances are not exclusive to Asia, however. The Mexican state of Veracruz, for instance, provides similarly fertile conditions for the flu. Veracruz has a coastline on the Gulf of Mexico, and Mexico is a developing country with a culinary tradition that heavily features pork.[35] It was in Veracruz—where very few scientists were looking for the flu[36]—that the 2009 outbreak of H1N1 began.[37]

By the end of April 2009, scientists were bombarded with alarming statistics about the swine flu in Veracruz and other parts of Mexico. There were reports of about 1,900 cases of H1N1 in Mexico and some 150 deaths. The ratio of these two quantities is known as the case fatality rate and it was seemingly very high—about 8 percent of the people who had acquired the flu had apparently died from it, which exceeded the rate during the Spanish flu epidemic.[38] Many of the dead, moreover, were relatively young and healthy adults, another characteristic of severe outbreaks. And the virus was clearly quite good at reproducing itself; cases had already been detected in Canada, Spain, Peru, the United Kingdom, Israel, New Zealand, Germany, the Netherlands, Switzerland, and Ireland, in addition to Mexico and the United States.[39]

It suddenly appeared that H1N1—not H5N1—was the superbug that scientists had feared all along. Mexico City was essentially shut down; European countries warned their citizens against travel to either Mexico or the United States. Hong Kong and Singapore, notoriously jittery about flu pandemics, saw their stock markets plunge.[40]

The initial worries soon subsided. Swine flu had indeed spread extremely rapidly in the United States—from twenty confirmed cases on April 26 to 2,618

* In contrast, consider India, where much of the population is either vegetarian or Muslim—two groups that do not consume any pork. (Even in the United States or Great Britain, you're unlikely to find pork on the menu at an authentic Indian restaurant.) Although India meets the other two conditions, it has very rarely been the source for the flu.

some fifteen days later.[41] But most cases were surprisingly mild, with just three deaths confirmed in the United States, a fatality rate comparable to the seasonal flu. Just a week after the swine flu had seemed to have boundless destructive potential, the CDC recommended that closed schools be reopened.

The disease had continued to spread across the globe, however, and by June 2009 the WHO had declared it a level 6 pandemic, its highest classification. Scientists feared the disease might follow the progression of the Spanish flu of 1918, which had initially been fairly mild, but which came back in much deadlier second and third waves (figure 7-1). By August, the mood had again grown more pessimistic, with U.S. authorities describing a "plausible scenario" in which as much as half the population might be infected by swine flu and as many as 90,000 Americans might die.[42]

FIGURE 7-1: DEATH RATE FROM 1918–19 H1N1 OUTBREAK

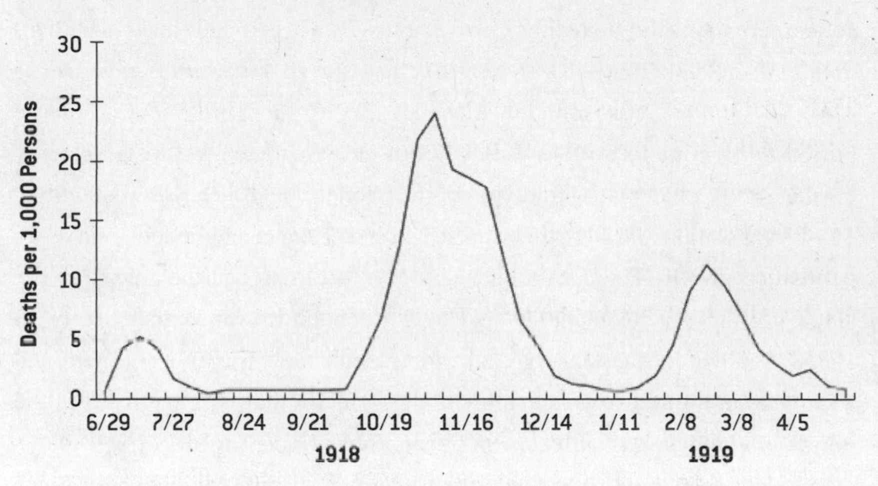

Those predictions also proved to be unwarranted, however. Eventually, the government reported that a total of about fifty-five million Americans had become infected with H1N1 in 2009—about one sixth of the U.S. population rather than one half—and 11,000 had died from it.[43] Rather than being an unusually severe strain of the virus, H1N1 had in fact been exceptionally mild, with a fatality rate of just 0.02 percent. Indeed, there were slightly *fewer* deaths from the flu in 2009–10 than in a typical season.[44] It hadn't quite been the epic embarrassment of 1976, but there had been failures of prediction from start to finish.

There are no guarantees that flu predictions will do better the next time around. In fact, the flu and other infectious diseases have several properties that make them intrinsically very challenging to predict.

The Dangers of Extrapolation

Extrapolation is a very basic method of prediction—usually, much too basic. It simply involves the assumption that the current trend will continue indefinitely, into the future. Some of the best-known failures of prediction have resulted from applying this assumption too liberally.

At the turn of the twentieth century, for instance, many city planners were concerned about the increasing use of horse-drawn carriages and their main pollutant: horse manure. Knee-deep in the issue in 1894, one writer in the *Times* of London predicted that by the 1940s, every street in London would be buried under nine feet of the stuff.[45] About ten years later, fortunately, Henry Ford began producing his prototypes of the Model T and the crisis was averted.

Extrapolation was also the culprit in several failed predictions related to population growth. Perhaps the first serious effort to predict the growth of the global population was made by an English economist, Sir William Petty, in 1682.[46] Population statistics were not widely available at the time and Petty did a lot of rather innovative work to infer, quite correctly, that the growth rate in the human population was fairly slow in the seventeenth century. Incorrectly, however, he assumed that things would always remain that way, and his predictions implied that global population might be just over 700 million people in 2012.[47] A century later, the Industrial Revolution began, and the population began to increase at a much faster rate. The actual world population, which surpassed seven billion in late 2011,[48] is about ten times higher than Petty's prediction.

The controversial 1968 book *The Population Bomb*, by the Stanford biologist Paul R. Ehrlich and his wife, Anne Ehrlich, made the opposite mistake, quite wrongly predicting that hundreds of millions of people would die from starvation in the 1970s.[49] The reasons for this failure of prediction were myriad, including the Ehrlichs' tendency to focus on doomsday scenarios to draw attention to their cause. But one major problem was that they had assumed the record-high fertility rates in the free-love era of the 1960s would continue on indefinitely, meaning that there would be more and more hungry mouths to feed.* "When I wrote *The Population Bomb* I thought our interests in sex and children were so strong that it would be hard to change family size," Paul Ehrlich told me in a brief interview. "We found out that if you treat women decently and give them job opportunities, the fertility rate goes down." Other scholars who had not made such simplistic assumptions realized this at the time; population projections issued by the United Nations in the 1960s and 1970s generally did a good job of predicting what the population would look like thirty or forty years later.[50]

Extrapolation tends to cause its greatest problems in fields—including population growth and disease—where the quantity that you want to study is growing exponentially. In the early 1980s, the cumulative number of AIDS cases diagnosed in the United States was increasing in this exponential fashion:[51] there were 99 cases through 1980, then 434 through 1981, and eventually 11,148 through 1984. You can put these figures into a chart, as some scholars did at the time,[52] and seek to extrapolate the pattern forward. Doing so would have yielded a prediction that the number of AIDS cases diagnosed in the United States would rise to about 270,000 by 1995. This would not have been a very good prediction; unfortunately it was too low. The actual number of AIDS cases was about 560,000 by 1995, more than twice as high.

Perhaps the bigger problem from a statistical standpoint, however, is that precise predictions aren't really possible to begin with when you are extrapolating on an exponential scale. A properly applied version[53] of this method, which accounted for its margin of error, would have implied that there could be as few as 35,000 AIDS cases through 1995 or as many as 1.8 million. That's much too broad a range to provide for much in the way of predictive insight.

---

* If you had assumed that the global population would continue to grow at a rate of 2.1 percent per year, the rate that it did in 1968 when Ehrlich published his book, global population would project to almost nine billion by 2012, considerably above its actual figure of seven billion.

FIGURE 7-2: CUMULATIVE AIDS CASES DIAGNOSED IN THE UNITED STATES: ACTUAL THROUGH 1984 AND EXTRAPOLATED TO 1995

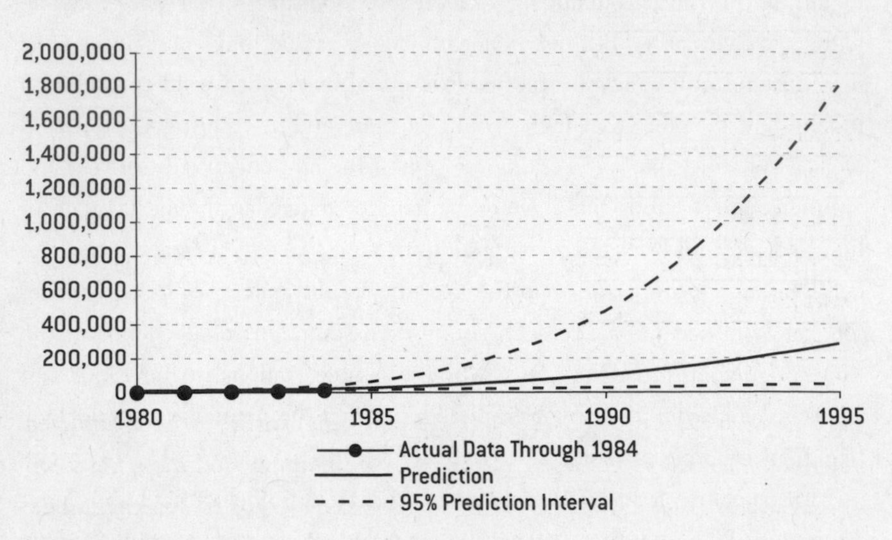

Why Flu Predictions Failed in 2009

Although the statistical methods that epidemiologists use when a flu outbreak is first detected are not quite as simple as the preceding examples, they still face the challenge of making extrapolations from a small number of potentially dubious data points.

One of the most useful quantities for predicting disease spread is a variable called the basic reproduction number. Usually designated as R_0, it measures the number of uninfected people that can expect to catch a disease from a single infected individual. An R_0 of 4, for instance, means that—in the absence of vaccines or other preventative measures—someone who gets a disease can be expected to pass it along to four other individuals before recovering (or dying) from it.

In theory, any disease with an R_0 greater than 1 will eventually spread to the entire population in the absence of vaccines or quarantines. But the numbers are sometimes much higher than this: R_0 was about 3 for the Spanish flu, 6 for smallpox, and 15 for measles. It is perhaps well into the triple digits for malaria, one of the deadliest diseases in the history of civilization, which still accounts for about 10 percent of all deaths in some parts of the world today.[54]

FIGURE 7-3: MEDIAN ESTIMATES OF R_0 FOR VARIOUS DISEASES[55]

Disease	R_0
Malaria	150
Measles	15
Smallpox	6
HIV/AIDS	3.5
SARS	3.5
H1N1 (1918)	3
Ebola (1995)	1.8
H1N1 (2009)	1.5
Seasonal flu	1.3

The problem is that reliable estimates of R_0 can usually not be formulated until well after a disease has swept through a community and there has been sufficient time to scrutinize the statistics. So epidemiologists are forced to make extrapolations about it from a few early data points. The other key statistical measure of a disease, the fatality rate, can similarly be difficult to measure accurately in the early going. It is a catch-22; a disease cannot be predicted very accurately without this information, but reliable estimates of these quantities are usually not available until the disease has begun to run its course.

Instead, the data when an infectious disease first strikes is often misreported. For instance, the figures that I gave you for early AIDS diagnoses in the United States were those that were available only years after the fact. Even these updated statistics did a rather poor job at prediction. However, if you were relying on the data that was actually available to scientists at the time,[56] you would have done even worse. This is because AIDS, in its early years, was poorly understood and was highly stigmatized, both among patients and sometimes also among doctors.[57] Many strange syndromes with AIDS-like symptoms went undiagnosed or misdiagnosed—or the opportunistic infections that AIDS can cause were mistaken for the principal cause of death. Only years later when doctors began to reopen old case histories did they come close to developing good estimates of the prevalence of AIDS in its early years.

Inaccurate data was also responsible for some of the poor predictions about swine flu in 2009. The fatality rate for H1N1 was apparently extremely high in Mexico in 2009 but turned out to be extremely low in the United States. Although some of that has to do with differences in how effective the medical care was in each country, much of the discrepancy was a statistical illusion.

The case fatality rate is a simple ratio: the number of deaths caused by a disease divided by the number of cases attributed to it. But there are uncertainties on both sides of the equation. On the one hand, there was some tendency in Mexico to attribute deaths related to other forms of the flu, or other diseases entirely, to H1N1. Laboratory tests revealed that as few as one-quarter of the deaths supposedly linked to H1N1 in fact showed distinct signs of it. On the other hand, there was surely underreporting, probably by orders of magnitude, in the number of cases of H1N1. Developing countries like Mexico have neither the sophisticated reporting systems of the United States nor a culture of going to the doctor at the first sign of illness;[58] that the disease spread so rapidly once the United States imported it suggests that there might have been thousands, perhaps even tens of thousands, of mild cases of the flu in Mexico that had never become known to authorities.

In fact, H1N1 may have been circulating in southern and central Mexico for months before it came to the medical community's attention (especially since they were busy looking for bird flu in Asia). Reports of an attack of respiratory illness had surfaced in the small town of La Gloria, Veracruz, in early March 2009, where the majority of the town had taken ill, but Mexican authorities initially thought it was caused by a more common strain of the virus called H3N2.[59]

In contrast, swine flu was the subject of obsessive media speculation from the very moment that it entered the United States. Few cases would have gone unnoticed. With these higher standards of reporting, the case fatality rate in the United States was reasonably reliable and took some of the worst-case scenarios off the table—but not until it was too late to undo some of the alarming predictions that had been made on the public record.

Self-Fulfilling and Self-Canceling Predictions

In many cases involving predictions about human activity, the very act of prediction can alter the way that people behave. Sometimes, as in economics, these changes in behavior can affect the outcome of the prediction itself, either nullifying it or making it more accurate. Predictions about the flu and other infectious diseases are affected by both sides of this problem.

A case where a prediction can bring itself about is called a self-fulfilling prediction or a self-fulfilling prophecy. This can happen with the release of a political poll in a race with multiple candidates, such as a presidential primary. Voters in these cases may behave tactically, wanting to back a candidate who could potentially win the state rather than waste their vote, and a well-publicized poll is often the best indication of whether a candidate is fit to do that. In the late stages of the Iowa Republican caucus race in 2012, for example, CNN released a poll that showed Rick Santorum surging to 16 percent of the vote when he had been at about 10 percent before.[60] The poll may have been an outlier—other surveys did not show Santorum gaining ground until after the CNN poll had been released.[61] Nevertheless, the poll earned Santorum tons of favorable media coverage and some voters switched to him from ideologically similar candidates like Michele Bachmann and Rick Perry. Before long, the poll had fulfilled its own destiny, with Santorum eventually winning Iowa while Bachmann and Perry finished far out of the running.

More subtle examples of this involve fields like design and entertainment, where businesses are essentially competing with one another to predict the consumer's taste—but also have some ability to influence it through clever marketing plans. In fashion, there is something of a cottage industry to predict which colors will be popular in the next season[62]—this must be done a year or so in advance because of the planning time required to turn around a clothing line. If a group of influential designers decide that brown will be the hot color next year and start manufacturing lots of brown clothes, and they get models and celebrities to wear brown, and stores begin to display lots of brown in their windows and their catalogs, the public may well begin to comply with the trend. But they're responding more to the marketing of brown than expressing some deep underlying preference for it. The designer may look like a savant for having "anticipated" the in color, but if he had picked white or black or lavender instead, the same process might have unfolded.[63]

Diseases and other medical conditions can also have this self-fulfilling property. When medical conditions are widely discussed in the media, people are more likely to identify their symptoms, and doctors are more likely to diagnose (or misdiagnose) them. The best-known case of this in recent years is autism. If you compare the number of children who are diagnosed as autistic[64] to the frequency with which the term *autism* has been used in American newspapers,[65] you'll find that there is an almost perfect one-to-one correspondence (figure 7-4), with both having increased markedly in recent years. Autism, while not properly thought of as a disease, presents parallels to something like the flu.

"It's a fascinating phenomenon that we've seen. In diseases that have no causal mechanism, news events precipitate increased reporting," I was told by Dr. Alex Ozonoff of the Harvard School for Public Health. Ozonoff received his training in pure mathematics and is conversant in many data-driven fields, but now concentrates on applying rigorous statistical analysis to the flu and other infectious diseases. "What we find again and again and again is that the more a particular condition is on people's minds and the more it's a current topic of discussion, the closer the reporting gets to 100 percent."

FIGURE 7-4: AUTISM—MEDIA COVERAGE AND DIAGNOSIS 1992–2008

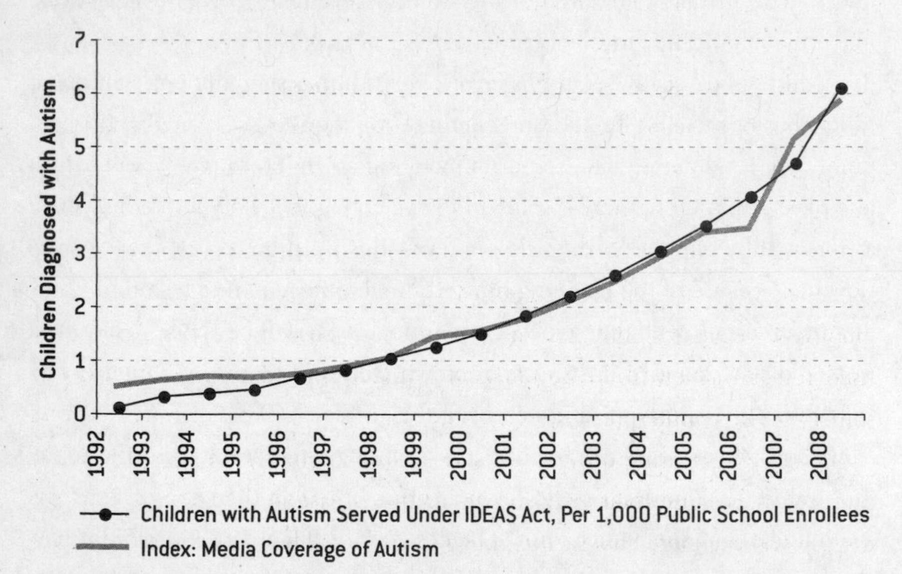

Ozonoff thinks this phenomenon may have been responsible for some of the velocity with which swine flu seemed to have spread throughout the United States in 2009. The disease was assuredly spreading rapidly, but some of the sharp statistical increase may have come from people reporting symptoms to their doctors which they might otherwise have ignored.

If doctors are looking to make estimates of the rate at which the incidence of a disease is expanding in the population, the number of *publicly reported* cases may provide misleading estimates of it. The situation can be likened to crime reporting: if the police report an increased number of burglaries in a neighborhood, is that because they are being more vigilant and are catching crimes that they had missed before, or have made it easier to report them?* Or is it because the neighborhood is becoming more dangerous? These problems are extremely vexing for anyone looking to make predictions about the flu in its early stages.

Self-Canceling Predictions

A self-canceling prediction is just the opposite: a case where a prediction tends to undermine itself. One interesting case is the GPS navigation systems that are coming into more and more common use. There are two major north-to-south routes through Manhattan: the West Side Highway, which borders the Hudson River, and the FDR Drive, which is on Manhattan's east side. Depending on her destination, a driver may not strongly prefer either thoroughfare. However, her GPS system will tell her which one to take, depending on which has less traffic—it is predicting which route will make for the shorter commute. The problem comes when a lot of other drivers are using the same navigation systems—all of a sudden, the route will be flooded with traffic and the "faster" route will turn out to be the slower one. There is already some theoretical[66] and empirical[67] evidence that this has become a problem on certain commonly used routes in New York, Boston, and London, and that these systems can sometimes be counterproductive.

This self-defeating quality can also be a problem for the accuracy of flu predictions because their goal, in part, is to increase public awareness of the disease and therefore change the public's behavior. The most effective flu pre-

* New York, for instance, does not allow you to file a police report online, while San Francisco does, as I found out when my rental car was broken into there in a reporting trip for this book. San Francisco is doing a better job of helping citizens and visitors to report and prevent crimes. But perversely, this makes its reported crime rate higher.

diction might be one that *fails* to come to fruition because it motivates people toward more healthful choices.

Simplicity Without Sophistication

The Finnish scientist Hanna Kokko likens building a statistical or predictive model to drawing a map.[68] It needs to contain enough detail to be helpful and do an honest job of representing the underlying landscape—you don't want to leave out large cities, prominent rivers and mountain ranges, or major highways. Too much detail, however, can be overwhelming to the traveler, causing him to lose his way. As we saw in chapter 5 these problems are not purely aesthetic. Needlessly complicated models may fit the noise in a problem rather than the signal, doing a poor job of replicating its underlying structure and causing predictions to be worse.

But how much detail is too much—or too little? Cartography takes a lifetime to master and combines elements of both art and science. It probably goes too far to describe model building as an art form, but it does require a lot of judgment.

Ideally, however, questions like Kokko's can be answered empirically. Is the model working? If not, it might be time for a different level of resolution. In epidemiology, the traditional models that doctors use are quite simple—and they are not working that well.

The most basic mathematical treatment of infectious disease is called the SIR model (figure 7-5). The model, which was formulated in 1927,[69] posits that there are three "compartments" in which any given person might reside at any given time: S stands for being susceptible to a disease, I for being infected by it, and R for being recovered from it. For simple diseases like the flu, the movement from compartment to compartment is entirely in one direction: from S to I to R. In this model, a vaccination essentially serves as a shortcut,* allowing a person to progress from S to R without getting ill. The mathematics behind the

---

* A vaccine typically contains a small and very weak microorganism that is derived from the bug that it is hoping to prevent—just enough for you to develop some immunity to it without becoming sick.

model is relatively straightforward, boiling down to a handful of differential equations that can be solved in a few seconds on a laptop.

FIGURE 7-5: SCHEMATIC OF SIR MODEL

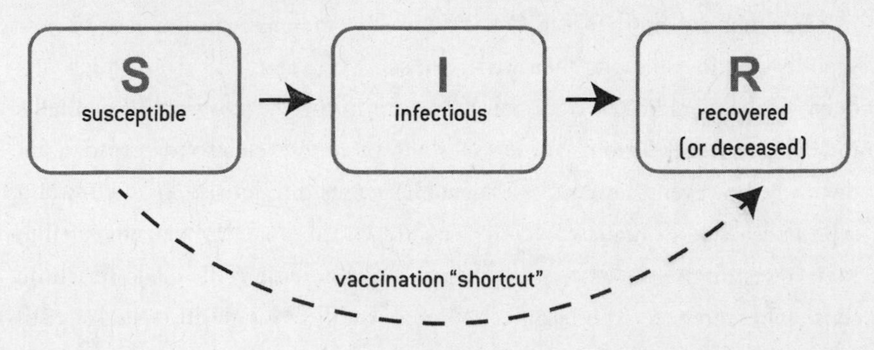

The problem is that the model requires a lot of assumptions to work properly, some of which are not very realistic in practice. In particular, the model assumes that everybody in a given population behaves the same way—that they are equally susceptible to a disease, equally likely to be vaccinated for it, and that they intermingle with one another at random. There are no dividing lines by race, gender, age, religion, sexual orientation, or creed; everybody behaves in more or less the same way.

An HIV Paradox in San Francisco

It is easiest to see why these assumptions are flawed in the case of something like a sexually transmitted disease.

The late 1990s and early 2000s were accompanied by a marked rise in unprotected sex in San Francisco's gay community,[70] which had been devastated by the HIV/AIDS pandemic two decades earlier. Some researchers blamed this on increasing rates of drug use, particularly crystal methamphetamine, which is often associated with riskier sexual behavior. Others cited the increasing effectiveness of antiretroviral therapy—cocktails of medicine that can extend the lives of HIV-positive patients for years or decades: gay men no longer saw an HIV diagnosis as a death sentence. Yet other theories focused on generational patterns—the San Francisco of the 1980s, when the AIDS epidemic was at its peak, was starting to feel like ancient history to a younger generation of gay men.[71]

The one thing the experts agreed on was that as unprotected sex increased, HIV infection rates were liable to do so as well.[72]

But that did not happen. Other STDs did increase: the number of new syphilis diagnoses among men who have sex with men (MSM)[73]—which had been virtually eradicated from San Francisco in the 1990s—rose substantially, to 502 cases in 2004 from 9 in 1998.[74] Rates of gonorrhea also increased. Paradoxically, however, the number of new HIV cases did not rise. In 2004, when syphilis reached its highest level in years, the number of HIV diagnoses fell to their lowest figure since the start of the AIDS epidemic. This made very little sense to researchers; syphilis and HIV are normally strongly correlated statistically, and they also have a causal relationship, since having one disease can make you more vulnerable to acquiring the other one.[75]

The solution to the paradox, it now appears, is that gay men had become increasingly effective at "serosorting"—that is, they were choosing sex partners with the same HIV status that they had. How they were able to accomplish this is a subject of some debate, but it has been documented by detailed behavioral studies in San Francisco,[76] Sydney,[77] London, and other cities with large gay populations. It may be that public health campaigns—some of which, wary of

FIGURE 7-6: NEW DIAGNOSES OF HIV AND SYPHILIS, GAY MEN (MSM), SAN FRANCISCO, 1998–2004

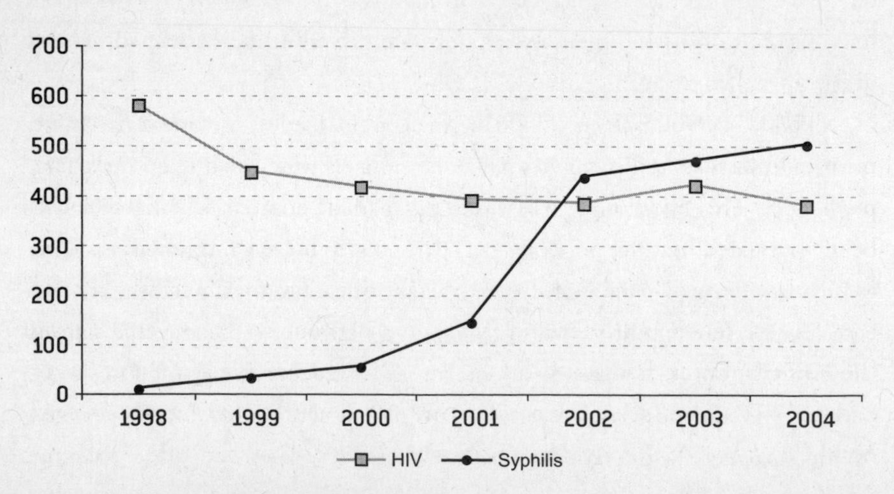

"condom fatigue," instead focused on the notion of "negotiated safety"—were having some positive effect. It may be that the Internet, which to some extent has displaced the gay bar as the preferred place to pick up a sex partner, has different norms for disclosure: many men list their HIV status in their profiles, and it may be easier to ask tough questions (and to get honest responses) from the privacy of one's home than in the din of the dance hall.[78]

Whatever the reason, it was clear that this type of specific, localized behavior was confounding the simpler disease models—and fortunately in this case it meant that they were overpredicting HIV. Compartmental models like SIR assume that every individual has the rate of susceptibility to a disease. That won't work as well for diseases that require more intimate types of contact, or where risk levels are asymmetric throughout different subpopulations. You don't just randomly walk into the grocery store and come home with HIV.

How the Models Failed at Fort Dix

Even in the case of simpler diseases, however, the compartmental models have sometimes failed because of their lax assumptions. Take measles, for instance. Measles is the first disease that most budding epidemiologists learn about in their Ph.D. programs because it is the easiest one to study. "Measles is the model system for infectious disease," says Marc Lipsitch, a colleague of Ozonoff's at Harvard. "It's unambiguous. You can do a blood test, there's only one strain, and everyone who has it is symptomatic. Once you have it you don't have it again." If there's any disease that the SIR models should be able to handle, it should be measles.

But in the 1980s and early 1990s, there were a series of unusually severe measles outbreaks in Chicago that epidemiologists were having a tough time predicting. The traditional models suggested that enough Chicagoans had been vaccinated that the population should have achieved a condition known as "herd immunity"—the biological equivalent of a firewall in which the disease has too few opportunities to spread and dies out. In some years during the 1980s, however, as many as a thousand Chicagoans—most of them young children—caught measles; the problem was so frightening that the city ordered nurses to go door-to-door to administer shots.[79]

Dr. Robert Daum, a pediatrician and infectious disease specialist who works at University of Chicago hospitals, has studied these measles outbreaks in great depth. Daum is a doctor's doctor, with a dignified voice, a beard, and a detached sense of humor. He had just returned from Haiti, where he had assisted with relief efforts for the 2010 earthquake, when I visited him and two of his colleagues in Chicago.

Chicago, where I lived for thirteen years, is a city of neighborhoods. Those neighborhoods are often highly segregated, with little mixing across racial or socioeconomic lines. Daum discovered that the neighborhoods also differed in their propensity toward vaccination: inner-city residents in Chicago's mostly poor, mostly black South Side were less likely to have had their children get their MMR (measles, mumps, and rubella) shots. Those unvaccinated children were going to school together, playing together, coughing and sneezing on one another. They were violating one of the assumptions of the SIR model called random mixing, which assumes that any two members of the population are equally likely to come into contact with each other. And they were spreading measles.

This phenomenon of nonrandom mixing may also have been the culprit in the swine flu fiasco of 1976, when scientists were challenged to extrapolate the national H1N1 threat from the cases they had seen at Fort Dix. The swine flu strain—now known as A/New Jersey/76—appeared so threatening in part because it had spread very quickly throughout the base: 230 confirmed cases were eventually diagnosed in a period of two or three weeks.[80] Thus, scientists inferred that it must have had a very high basic reproduction ratio (R_0)—perhaps as high as the 1918 pandemic, which had an R_0 of about 3.

A military platoon, however, is an usually disease-prone environment. Soldiers are in atypically close contact with one another, in cramped settings in which they may be sharing essential items like food and bedding materials, and in which there is little opportunity for privacy. Moreover, they are often undergoing strenuous physical exercise—temporarily depleting their immune systems—and the social norm of the military is to continue to report to work even when you are sick. Infectious disease has numerous opportunities to be passed along, and so transmission will usually occur very quickly.

Subsequent study[81] of the outbreak at Fort Dix has revealed that the rapid spread of the disease was caused by these circumstantial factors, rather than by the disease's virulence. Fort Dix just wasn't anything like a random neighborhood or workplace somewhere in America. In fact, A/New Jersey/76 was nothing much to worry about at all—its R_0 was a rather wimpy 1.2, no higher than that of the seasonal flu. Outside a military base, or a roughly analogous setting like a college dormitory or a prison, it wasn't going to spread very far. The disease had essentially lived out its life span at Fort Dix, running out of new individuals to infect.

The fiasco over A/New Jersey/76—like the HIV/syphilis paradox in San Francisco, or the Chicago measles outbreaks of the 1980s—speaks to the limitations of models that make overly simplistic assumptions. I certainly do not mean to suggest that you should always prefer complex models to simple ones; as we have seen in other chapters in this book, complex models can also lead people astray. And because complex models often give more precise (but *not* necessarily more accurate) answers, they can trip a forecaster's sense of overconfidence and fool him into thinking he is better at prediction than he really is.

Still, while simplicity can be a virtue for a model, a model should at least be *sophisticatedly* simple.[82] Models like SIR, although they are useful for understanding disease, are probably too blunt to help predict its course.

SimFlu

Weather forecasts provide one of the relatively few examples of more complex models that have substantially improved prediction. It has taken decades of work, but by creating what is essentially a physical simulation of the atmosphere, meteorologists are able to do much better than purely statistical approaches to weather prediction.

An increasing number of groups are looking to apply a similar approach to disease prediction using a technique known as agent-based modeling. I visited some researchers at the University of Pittsburgh who are at the forefront of developing these techniques. The Pittsburgh team calls their model FRED, which stands for "framework for the reconstruction of epidemic dynamics." The name is also a tip of the hat to Fred Rogers, the Pittsburgher who was the host of *Mister Rogers' Neighborhood*.

Pittsburgh, like Chicago, is a city of neighborhoods. The Pittsburgh researchers think about neighborhoods when they think about disease, and so FRED is essentially a sort of SimPittsburgh—a very detailed simulation in which every person is represented by an "agent" who has a family, a social network, a place of residence, and a set of beliefs and behaviors that are consistent with her socioeconomic status.

Dr. John Grefenstette, one of the scientists on the Pittsburgh team, has spent most of his life in the city and still has traces of its distinct accent. He explained how FRED is organized: "They have schools and workplaces and hospitals all placed according to the right geographical distribution. They have a quite complicated setup where they assign children to schools; they don't all go to the closest school—and some of schools are small and some of them are real large. And so you get this synthetic sort of a SimCity population."

Dr. Grefenstette and his amiable colleague Dr. Shawn Brown showed me the results of some of FRED's simulations, with waves of disease colorfully rippling zip code by zip code through SimPittsburgh or SimWashington or Sim-Philadelphia. But FRED is also very serious business. These models take few shortcuts: literally everybody in a city, county, or state might be represented. Some agent-based models even seek to simulate the entire country or the entire world. Like weather models, they require an exponential number of calculations to be made and therefore require supercomputers to run.

These models also require a lot of data. It's one thing to get the demographics right, which can be estimated fairly accurately through the census. But the models also need to account for human behavior, which can be much less predictable. Exactly how likely is a twenty-six-year-old Latina single mother to get vaccinated, for instance? You could draw up a survey and ask her—agent-based models rely fairly heavily on survey data. But people notoriously lie about (or misremember) their health choices: people claim they wash their hands more often than they really do, for instance,[83] or that they use condoms more often than they really do.[84]

One fairly well-established principle, Dr. Grefenstette told me, is that people's willingness to engage in inconvenient but healthful measures like vaccination is tied to the risk they perceive of acquiring a disease. Our SimPittsburgher will get a flu shot if she concludes that the risk from swine flu is serious, but not if she doesn't. But how might her perception change if her neighbor gets sick, or her child does? What if there are a bunch of stories about the flu on the local news? The self-fulfilling and self-canceling properties of disease prediction are therefore still highly pertinent to these agent-based models. Because they are dynamic and allow an agent's behavior to change over time, they may be more capable of handling these questions.

Or consider Dr. Daum and his team at the University of Chicago, who are building an agent-based model to study the spread of a dangerous disease called MRSA, an antibiotic-resistant staph infection that can cause ordinary abrasions like cuts, scrapes, and bruises to develop into life-threatening and sometimes untreatable infections. MRSA is a complicated disease with many pathways for transmission: it can be spread through fairly casual contact like hugging, or through open wounds, or through an exchange of bodily fluids like sweat or blood. It can also sometimes linger on different types of surfaces like countertops or towels. One fairly common setting for MRSA is locker rooms, where athletes may share equipment; MRSA outbreaks have been reported among football teams ranging from high school to the NFL. Making matters more confusing still, many people carry the MRSA bacteria without ever becoming sick from it or showing any symptoms at all.

In their attempt to model MRSA, Daum and his colleagues must ask themselves questions such as these: Which types of people use a Band-Aid when they have a cut? How common is hugging in different types of cultures? How many people in a neighborhood have been to prison, where staph infections are common?

These are the sorts of questions that a traditional model can't even hope to address, and where agent-based models can at least offer the chance of more accurate predictions. But the variables that the Pittsburgh and Chicago teams must account for are vast and wide-ranging—as will necessarily be the case when you're trying to simulate the behavior of every individual in an entire population. Their work often takes detours into cognitive psychology, behavioral economics, ethnography, and even anthropology: agent-based models are used to study HIV infection in communities as diverse as the jungles of Papua New Guinea[85] and the gay bars of Amsterdam.[86] They require extensive knowledge of local customs and surroundings.

Agent-based modeling is therefore an exceptionally ambitious undertaking, and the groups working in the field are often multidisciplinary All-Star teams composed of some of the best and brightest individuals in their respective professions. But for all that brainpower, their efforts are often undermined by a lack of data. "Even for H1N1, it's been difficult to get detailed geographical data on who got sick, when, and where," Grefenstette laments. "And it's amazing how difficult it is to get data on past outbreaks."

When speaking to the Pittsburgh and Chicago teams, I was sometimes reminded of the stories you read about the beautiful new shopping malls in China, which come with every imaginable frill—Roman columns, indoor roller coasters, Venetian canals—but don't yet have any stores or people in them. Both the Chicago and Pittsburgh teams have come to a few highly useful and actionable conclusions—Dr. Grefenstette figured out, for instance, that school closures can backfire if they are too brief or occur too soon, and the U. of C. team surmised that the unusually large number of MRSA cases in inner-city Chicago was caused by the flow of people into and out of the Cook County Jail. But mostly, the models are at least a few years ahead of themselves, waiting to feed off data that does not yet exist.

The agent-based models—unlike weather forecast models that can be refined on a daily basis—are also hard to test. Major diseases come around only every so often. And even if the models are right, they might be victims of their own success because of the self-canceling property of a successful disease prediction. Suppose that the model suggests that a particular intervention—say, closing the schools in one county—might be highly effective. And the intervention works! The progress of the disease in the real world will then be slowed. But it might make the model look, in retrospect, as though it had been too pessimistic.

The Pittsburgh and Chicago teams have therefore been hesitant to employ their models to make specific predictions. Other teams were less cautious in advance of the 2009 swine flu outbreak and some issued very poor predictions about it,[87] sometimes substantially underestimating how far the flu would spread.

For the time being the teams are mostly limited to what Dr. Daum's colleague Chip Macal calls "modeling for insights." That is, the agent-based models might help us to perform experiments that can teach us about infectious disease, but they are unlikely to help us predict an outbreak—for now.

How to Proceed When Prediction Is Hard

The last two major flu scares in the United States proved not to live up to the hype. In 1976, there was literally no outbreak of H1N1 beyond the cases at Fort Dix; Ford's mass vaccination program had been a gross overreaction. In 2009, the swine flu infected quite a number of people but killed very few of them. In both instances, government predictions about the magnitude of the outbreak had missed to the high side.

But there are no guarantees the error will be in the same direction the next time the flu comes along. A human-adapted strain of avian flu, H5N1 could kill hundreds of millions of people. A flu strain that was spread as easily as the 2009 version of H1N1, but had the fatality ratio of the 1918 version, would have killed 1.4 million Americans. There are also potential threats from non-influenza viruses like SARS, and even from smallpox, which was eradicated from the world in 1977 but which could potentially be reintroduced into society as a biological weapon by terrorists, potentially killing millions. The most serious epidemics, almost by definition, can progress very rapidly: in 2009, it took H1N1 only about a week to go from a disease almost completely undetected by the medical community to one that appeared to have the potential to kill tens of millions of people.

The epidemiologists I spoke with for this chapter—in a refreshing contrast to their counterparts in some other fields—were strongly aware of the limitations of their models. "It's stupid to predict based on three data points," Marc Lipsitch told me, referring to the flu pandemics in 1918, 1957, and 1968. "All you can do is plan for different scenarios."

If you can't make a good prediction, it is very often harmful to pretend that you can. I suspect that epidemiologists, and others in the medical community, understand this because of their adherence to the Hippocratic oath. *Primum non nocere*: First, do no harm.

Much of the most thoughtful work on the use and abuse of statistical models and the proper role of prediction comes from people in the medical profession.[88] That is not to say there is nothing on the line when an economist makes a prediction, or a seismologist does. But because of medicine's intimate connection with life and death, doctors tend to be appropriately cautious. In their field, stupid models kill people. It has a sobering effect.

There is something more to be said, however, about Chip Macal's idea of "modeling for insights." The philosophy of this book is that prediction is as much a means as an end. Prediction serves a very central role in hypothesis testing, for instance, and therefore in all of science.[89]

As the statistician George E. P. Box wrote, "All models are wrong, but some models are useful."[90] What he meant by that is that all models are simplifications of the universe, as they must necessarily be. As another mathematician said, "The best model of a cat is a cat."[91] Everything else is leaving out some sort of detail. How pertinent that detail might be will depend on exactly what problem we're trying to solve and on how precise an answer we require.

Nor are statistical models the only tools we use that require us to make approximations about the universe. Language, for instance, is a type of model, an approximation that we use to communicate with one another. All languages contain words that have no direct cognate in other languages, even though they are both trying to explain the same universe. Technical subfields have their own specialized language. To you or me, the color of some of the text on the front cover of this book is yellow. To a graphic designer, that term is too approximate—instead, it's Pantone 109.

But, Box wrote, some models are useful. It seems to me that the work the Chicago or Pittsburgh teams are doing with their agent-based models is extremely useful. Figuring out how different ethnic groups think about vaccination, how disease is transmitted throughout different neighborhoods in a city, or how people react to news reports about the flu are each important problems in their own right.

A good model can be useful even when it fails. "It should be a given that whatever forecast we make on average will be wrong," Ozonoff told me. "So usually it's about understanding how it's wrong, and what to do when it's wrong, and minimizing the cost to us when it's wrong."

The key is in remembering that a model is a tool to help us understand the complexities of the universe, and never a substitute for the universe itself. This is important not just when we make predictions. Some neuroscientists, like MIT's Tomasso Poggio, think of the entire way our brains process information as being through a series of approximations.

This is why it is so crucial to develop a better understanding of ourselves, and the way we distort and interpret the signals we receive, if we want to make better predictions. The first half of this book has largely been concerned with where these approximations have been serving us well and where they've been failing us. The rest of the book is about how to make them better, a little bit at a time.

8

LESS AND LESS AND LESS WRONG*

The sports bettor Haralabos "Bob" Voulgaris lives in a gleaming, modernist house in the Hollywood Hills of Los Angeles—all metal and glass, with a pool in the back, like something out of a David Hockney painting. He spends every night from November through June watching the NBA, five games at a time, on five Samsung flat screens (the DirecTV guys had never seen anything like it). He escapes to his condo at Palms Place in Las Vegas whenever he needs a short break, and safaris in Africa when he needs a longer one. In a bad year, Voulgaris makes a million dollars, give or take. In a good year, he might make three or four times that.

So Bob enjoys some trappings of the high life. But he doesn't fit the stereotype of the cigar-chomping gambler in a leisure suit. He does not depend on insider tips, crooked referees, or other sorts of hustles to make his bets. Nor does he have a "system" of any kind. He uses computer simulations, but does not rely upon them exclusively.

What makes him successful is the way that he analyzes information. He is

* The title of this chapter is inspired by a line from the poem "The Road to Wisdom," by the Danish mathematician Piet Hein: "to err and err and err again, but less and less and less."

not just hunting for patterns. Instead, Bob combines his knowledge of statistics with his knowledge of basketball in order to identify meaningful *relationships* in the data.

This requires a lot of hard work—and sometimes a lot of guts. It required a big, calculated gamble to get him to where he is today.

Voulgaris grew up in Winnipeg, Manitoba, a hardworking but frostbitten city located ninety miles north of the Minnesota border. His father had once been quite wealthy—worth about $3 million dollars at his peak—but he blew it all gambling. By the time Voulgaris was twelve, his dad was broke. By the time he was sixteen, he realized that if he was going to get the hell out of Winnipeg, he needed a good education and would have to pay for it himself. So while attending the University of Manitoba, he looked for income wherever he could find it. In the summers, he'd go to the far northern reaches of British Columbia to work as a tree climber; the going rate was seven cents per tree. During the school year, he worked as an airport skycap, shuttling luggage back and forth for Winnipeggers bound for Toronto or Minneapolis or beyond.

Voulgaris eventually saved up to buy out a stake in the skycap company that he worked for and, before long, owned much of the business. By the time he was a college senior, in 1999, he had saved up about $80,000.

But $80,000 still wasn't a *lot* of money, Voulgaris thought—he'd seen his dad win and lose several times that amount many times over. And the job prospects for a philosophy major from the University of Manitoba weren't all that promising. He was looking for a way to accelerate his life when he came across a bet that he couldn't resist.

That year, the Los Angeles Lakers had hired the iconoclastic coach Phil Jackson, who had won six championships with the Chicago Bulls. The Lakers had plenty of talent: their superstar center, the seven-foot-one behemoth Shaquille O'Neal, was at the peak of his abilities, and their twenty-one-year-old guard Kobe Bryant, just four years out of high school, was turning into a superstar in his own right. Two great players—a big man like O'Neal and a scorer like Bryant—has long been a formula for success in the NBA, especially when they are paired with a great coach like Jackson who could manage their outsize egos.

And yet conventional wisdom was skeptical about the Lakers. They had never gotten into a rhythm the previous year, the strike-shortened season of 1998–99, when they churned through three coaches and finished 31-19, eliminated in four straight games by the San Antonio Spurs in the second round of the playoffs. Bryant and O'Neal were in a perpetual feud, with O'Neal apparently jealous that Bryant—still not old enough to drink legally—was on the verge of eclipsing him in popularity, his jersey outselling O'Neal's in Los Angeles sporting goods stores.[1] The Western Conference was strong back then, with cohesive and experienced teams like San Antonio and Portland, and the rap was that the Lakers were too immature to handle them.

When the Lakers were blown out by Portland in the third game of the regular season, with O'Neal losing his cool and getting ejected midway through the game, it seemed to confirm all the worst fears of the pundits and the shock jocks. Even the hometown *Los Angeles Times* rated the Lakers as just the seventh-best team in the NBA[2] and scolded Vegas handicappers for having given them relatively optimistic odds, 4-to-1 against, of winning the NBA title before the season had begun.

Just a couple of weeks into the 1999–2000 regular season, the Vegas bookmakers had begun to buy into the skepticism and had lengthened the Lakers' odds to 6½ to 1, making for a much better payout for anyone who dared to buck the conventional wisdom. Voulgaris was never a big believer in conventional wisdom—it's in large part its shortcomings that make his lifestyle possible—and he thought this was patently insane. The newspaper columnists and the bookies were placing too much emphasis on a small sample of data, ignoring the bigger picture and the context that surrounded it.

The Lakers weren't even playing that badly, Voulgaris thought. They had won five of their first seven games despite playing a tough schedule, adjusting to a new coach, and working around an injury to Bryant, who had hurt his wrist in the preseason and hadn't played yet. The media was focused on their patchy 1998–99 season, which had been interrupted by the strike and the coaching changes, while largely ignoring their 61-21 record under more normal circumstances in 1997–98. Voulgaris had watched a lot of Lakers games: he liked what Jackson was doing with the club. So he placed $80,000—his entire life savings less a little he'd left over for food and tuition—on the Lakers to win the NBA championship. If he won his bet, he'd make half a million dollars. If he lost it, it would be back to working double shifts at the airport.

Initially, Voulgaris's instincts were looking very good. From that point in the season onward, the Lakers won 62 of their remaining 71 contests, including three separate winning streaks of 19, 16, and 11 games. They finished at 67-15, one of the best regular-season records in NBA history. But the playoffs were another matter: the Western Conference was brutally tough in those years, and even with home-court advantage throughout the playoffs—their reward for their outstanding regular season—winning four series in a row would be difficult for the Lakers.

Los Angeles survived a scare against a plucky Sacramento Kings team in the first round of the playoffs, the series going to a decisive fifth game, and then waltzed past Phoenix in the Western Conference Semifinals. But in the next round they drew the Portland Trail Blazers, who had a well-rounded and mature roster led by Michael Jordan's former sidekick—and Jackson's former pupil—Scottie Pippen. Portland would be a rough matchup for the Lakers: although they lacked the Lakers' talent, their plodding, physical style of play often knocked teams out of their rhythm.[3]

The Lakers won the first game of the best-of-seven series fairly easily, but then the roller-coaster ride began. They played inexplicably poorly in the second game in Los Angeles, conceding twenty consecutive points to Portland in the third quarter[4] and losing 106-77, their most lopsided defeat of the season.[5]

The next two games were played at the Rose Garden in Portland, but in Game 3, the Lakers gathered themselves after falling down by as many as thirteen points in the first half, with Bryant swatting away a shot in the final seconds to preserve a two-point victory.[6] They defied gravity again in Game 4, overcoming an eleven-point deficit as O'Neal, a notoriously poor free-throw shooter, made all nine of his attempts.[7] Trailing three games to one in the series, the Trail Blazers were "on death's door," as Jackson somewhat injudiciously put it.[8]

But in the fifth game, at the Staples Center in Los Angeles, the Lakers couldn't shoot the ball straight, making just thirty of their seventy-nine shots in a 96-88 defeat. And in the sixth, back in Portland, they fell out of rhythm early and never caught the tune, as the Blazers marched to a 103-93 win. Suddenly the series was even again, with the deciding Game 7 to be played in Los Angeles.

The prudent thing for a gambler would have been to hedge his bet. For instance, Voulgaris could have put $200,000 on Portland, who were 3-to-2 underdogs, to win Game 7. That would have locked in a profit. If the Blazers won, he would make more than enough from his hedge to cover the loss of his original $80,000 bet, still earning a net profit of $220,000.[9] If the Lakers won instead, his original bet would still pay out—he'd lose his hedge, but net $320,000 from both bets combined.* That would be no half-million-dollar score, but still pretty good.

But there was a slight problem: Voulgaris didn't have $200,000. Nor did he know anybody else who did, at least not anybody he could trust. He was a twenty-three-year-old airport skycap living in his brother's basement in Winnipeg. It was literally Los Angeles or bust.

Early on in the game his chances didn't look good. The Blazers went after O'Neal at every opportunity, figuring they'd either force him to the free-throw line, where every shot was an adventure, or get him into foul trouble instead as he retaliated. Halfway through the second quarter, the strategy was working to a tee, as O'Neal had picked up three fouls and hadn't yet scored from the field. Then Portland went on a ferocious run midway through the third quarter, capped off by a Pippen three-pointer that gave them a sixteen-point lead as boos echoed throughout the Staples Center.[10]

Voulgaris's odds at that point were very long. Rarely did a team[11] that found itself in the Lakers' predicament—down sixteen points with two minutes left to play in the third quarter—come back to win the game; it can be calculated that the odds were about 15-to-1 against their doing so.[12] His bet—his ticket out of Winnipeg—looked all but lost.[13]

But early in the fourth quarter, the downside to Portland's brutally physical style of play suddenly became clear. Their players were beaten-up and fatigued,

---

* This assumes that the Lakers would beat the Indiana Pacers, the Eastern Conference champions, in the NBA Finals, against whom they'd be heavily favored. Voulgaris could have hedged his bet again if he wanted to mitigate that slim risk.

running on fumes and adrenaline. The Lakers were playing before their home crowd, which physiologists have shown provides athletes with an extra burst of testosterone when they need it most.[14] And the Lakers were the younger team, with a more resilient supply of energy.

Portland, suddenly, couldn't hit a shot, going more than six minutes without scoring early in the fourth quarter, right as the Lakers were quickening their pace. L.A. brought their deficit down to single digits, then five points, then three, until Brian Shaw hit a three-pointer to even the score with four minutes left, and Bryant knotted two free-throws a couple of possessions later to give them the lead. Although Portland's shooting improved in the last few minutes, it was too late, as the Lakers made clear with a thunderous alley-oop between their two superstars, Bryant and O'Neal, to clinch the game.

Two weeks later, the Lakers disposed of the Indiana Pacers in efficient fashion to win their first NBA title since the Magic Johnson era. And Bob the skycap was halfway to becoming a millionaire.

How Good Gamblers Think

How did Voulgaris know that his Lakers bet would come through? He didn't. Successful gamblers—and successful forecasters of any kind—do not think of the future in terms of no-lose bets, unimpeachable theories, and infinitely precise measurements. These are the illusions of the sucker, the sirens of his overconfidence. Successful gamblers, instead, think of the future as speckles of probability, flickering upward and downward like a stock market ticker to every new jolt of information. When their estimates of these probabilities diverge by a sufficient margin from the odds on offer, they may place a bet.

The Vegas line on the Lakers at the time that Voulgaris placed his bet, for instance, implied that they had a 13 percent chance of winning the NBA title. Voulgaris did not think the Lakers' chances were 100 percent or even 50 percent—but he was confident they were quite a bit higher than 13 percent. Perhaps more like 25 percent, he thought. If Voulgaris's calculation was right, the bet had a theoretical profit of $70,000.

FIGURE 8-1: HOW VOULGARIS SAW HIS LAKERS BET

Outcome	Probability	Net Profit
Lakers win championship	25%	+$520,000
Lakers do not win championship	75%	–$80,000
Expected profit		+$70,000

If the future exists in shades of probabilistic gray to the forecaster, however, the present arrives in black and white. Bob's theoretical profit of $70,000 consisted of a 25 percent chance of winning $520,000 and a 75 percent chance of losing $80,000 averaged together. Over the long term, the wins and losses will average out: the past and the future, to a good forecaster, can resemble one another more than either does the present since both can be expressed in terms of long-run probabilities. But this was a one-shot bet. Voulgaris needed to have a pretty big edge (the half dozen different reasons he thought the bookies undervalued the Lakers), and a pretty big head on his shoulders, in order to make it.

FIGURE 8-2: THE WORLD THROUGH THE EYES OF A SUCCESSFUL GAMBLER

Now that Voulgaris has built up a bankroll for himself, he can afford to push smaller edges. He might place three or four bets on a typical night of NBA action. While the bets are enormous by any normal standard they are small compared with his net worth, small enough that he can seem glumly indifferent about them. On the night that I visited, he barely blinked an eye when, on one of the flat screens, the Utah Jazz inserted a seven-foot-two Ukrainian stiff named Kyrylo Fesenko into the lineup, a sure sign that they were conceding the game and that Voulgaris would lose his $30,000 bet on it.

Voulgaris's big secret is that he doesn't have a big secret. Instead, he has a thousand *little* secrets, quanta of information that he puts together one vector at a time. He has a program to simulate the outcome of each game, for instance. But he relies on it only if it suggests he has a very clear edge or it is supplemented by other information. He watches almost every NBA game—some live, some on tape—and develops his own opinions about which teams are playing up to their talent and which aren't. He runs what is essentially his own scouting service, hiring assistants to chart every player's defensive positioning on every play, giving him an advantage that even many NBA teams don't have. He follows the Twitter feeds of dozens of NBA players, scrutinizing every 140-character nugget for relevance: a player who tweets about the club he's going out to later that night might not have his head in the game. He pays a lot of attention to what the coaches say in a press conference and the code that they use: if the coach says he wants his team to "learn the offense" or "play good fundamental basketball," for instance, that might suggest he wants to slow down the pace of the game.

To most people, the sort of things that Voulgaris observes might seem trivial. And in a sense, they are: the big and obvious edges will have been noticed by other gamblers, and will be reflected in the betting line. So he needs to dig a little deeper.

Late in the 2002 season, for instance, Voulgaris noticed that games involving the Cleveland Cavaliers were particularly likely to go "over" the total for the game. (There are two major types of sports bets, one being the point spread and the other being the over-under line or *total*—how many points both teams will score together.) After watching a couple of games closely, he quickly detected the reason: Ricky Davis, the team's point guard and a notoriously selfish player, would be a free agent at the end of the year and was doing everything he could to improve his statistics and make himself a more marketable commodity. This meant running the Cavaliers' offense at a breakneck clip in an effort to create as many opportunities as possible to accumulate points and assists. Whether or not this was good basketball didn't much matter: the Cavaliers were far out of playoff contention.[15] As often as not, the Cavaliers' opponents would be out of contention as well and would be happy to return the favor, engaging them in an unspoken pact to play loose defense and trade baskets in an attempt to improve one another's stats.[16] Games featuring the Cavaliers suddenly went from 192 points per game to 207 in the last three weeks of the season.[17] A bet on the over was not quite a sure thing—there are no sure things—but it was going to be highly profitable.

Patterns like these can sometimes seem obvious in retrospect: *of course* Cavaliers games were going to be higher-scoring if they had nothing left to play for but to improve their offensive statistics. But they can escape bettors who take too narrow-minded a view of the statistics without considering the context that produce them. If a team has a couple of high-scoring games in a row, or even three or four, it usually doesn't mean anything. Indeed, because the NBA has a long season—thirty teams playing eighty-two games each—little streaks like these will occur all the time.[18] Most of them are suckers' bets: they will have occurred for reasons having purely to do with chance. In fact, because the bookmakers will usually have noticed these trends as well, and may have overcompensated for them when setting the line, it will sometimes be smart to bet the other way.

So Voulgaris is *not* just looking for patterns. Finding patterns is easy in any kind of data-rich environment; that's what mediocre gamblers do. The key is in determining whether the patterns represent noise or signal.

But although there isn't any one particular key to why Voulgaris might or might not bet on a given game, there is a particular type of thought process that helps govern his decisions. It is called Bayesian reasoning.

The Improbable Legacy of Thomas Bayes

Thomas Bayes was an English minister who was probably born in 1701—although it may have been 1702. Very little is certain about Bayes's life, even though he lent his name to an entire branch of statistics and perhaps its most famous theorem. It is not even clear that anybody knows what Bayes looked like; the portrait of him that is commonly used in encyclopedia articles may have been misattributed.[19]

What is in relatively little dispute is that Bayes was born into a wealthy family, possibly in the southeastern English county of Hertfordshire. He traveled far away to the University of Edinburgh to go to school, because Bayes was a member of a Nonconformist church rather than the Church of England, and was banned from institutions like Oxford and Cambridge.[20]

Bayes was nevertheless elected as a Fellow of the Royal Society despite a relatively paltry record of publication, where he may have served as a sort of in-house critic or mediator of intellectual debates. One work that most scholars attribute to Bayes—although it was published under the pseudonym John Noon[21]—is a tract entitled "Divine Benevolence."[22] In the essay, Bayes considered the age-old theological question of how there could be suffering and evil in the world if God was truly benevolent. Bayes's answer, in essence, was that we should not mistake our human imperfections for imperfections on the part of God, whose designs for the universe we might not fully understand. "Strange therefore . . . because he only sees the lowest part of this scale, [he] should from hence infer a defeat of happiness in the whole," Bayes wrote in response to another theologian.[23]

Bayes's much more famous work, "An Essay toward Solving a Problem in the Doctrine of Chances,"[24] was not published until after his death, when it was brought to the Royal Society's attention in 1763 by a friend of his named Richard Price. It concerned how we formulate probabilistic beliefs about the world when we encounter new data.

Price, in framing Bayes's essay, gives the example of a person who emerges into the world (perhaps he is Adam, or perhaps he came from Plato's cave) and sees the sun rise for the first time. At first, he does not know whether this is typical or some sort of freak occurrence. However, each day that he survives and the sun rises again, his confidence increases that it is a permanent feature of nature. Gradually, through this purely statistical form of inference, the probability he assigns to his prediction that the sun will rise again tomorrow approaches (although never exactly reaches) 100 percent.

The argument made by Bayes and Price is *not* that the world is intrinsically probabilistic or uncertain. Bayes was a believer in divine perfection; he was also an advocate of Isaac Newton's work, which had seemed to suggest that nature follows regular and predictable laws. It is, rather, a statement—expressed both mathematically and philosophically—about how we learn about the universe: that we learn about it through approximation, getting *closer and closer to the truth* as we gather more evidence.

This contrasted[25] with the more skeptical viewpoint of the Scottish philosopher David Hume, who argued that since we could not be *certain* that the sun would rise again, a prediction that it would was inherently no more rational than one that it wouldn't.[26] The Bayesian viewpoint, instead, regards rationality as a *probabilistic* matter. In essence, Bayes and Price are telling Hume, don't blame nature because you are too daft to understand it: if you step out of your skeptical shell and make some predictions about its behavior, perhaps you will get a little closer to the truth.

Probability and Progress

We might notice how similar this claim is to the one that Bayes made in "Divine Benevolence," in which he argued that we should not confuse our own fallibility for the failures of God. Admitting to our own imperfections is a necessary step on the way to redemption.

However, there is nothing intrinsically religious about Bayes's philosophy.[27] Instead, the most common mathematical expression of what is today recognized as Bayes's theorem was developed by a man who was very likely an atheist,[28] the French mathematician and astronomer Pierre-Simon Laplace.

Laplace, as you may remember from chapter 4, was the poster boy for scientific determinism. He argued that we could predict the universe perfectly—given, of course, that we knew the position of every particle within it and were quick enough to compute their movement. So why is Laplace involved with a theory based on probabilism instead?

The reason has to do with the disconnect between the perfection of nature and our very human imperfections in measuring and understanding it. Laplace was frustrated at the time by astronomical observations that appeared to show anomalies in the orbits of Jupiter and Saturn—they seemed to predict that Jupiter would crash into the sun while Saturn would drift off into outer space.[29] These predictions were, of course, quite wrong, and Laplace devoted much of his life to developing much more accurate measurements of these planets' orbits.[30] The improvements that Laplace made relied on probabilistic inferences[31] in lieu of exacting measurements, since instruments like the telescope were still very crude at the time. Laplace came to view probability as a waypoint between ignorance and knowledge. It seemed obvious to him that a more thorough understanding of probability was essential to scientific progress.[32]

The intimate connection between probability, prediction, and scientific progress was thus well understood by Bayes and Laplace in the eighteenth century—the period when human societies were beginning to take the explosion of information that had become available with the invention of the printing press several centuries earlier, and finally translate it into sustained scientific, technological, and economic progress. The connection is essential—equally to predicting the orbits of the planets and the winner of the Lakers' game. As we will see, science may have stumbled later when a different statistical paradigm, which deemphasized the role of prediction and tried to recast uncertainty as resulting from the errors of our measurements rather than the imperfections in our judgments, came to dominate in the twentieth century.

The Simple Mathematics of Bayes's Theorem

If the philosophical underpinnings of Bayes's theorem are surprisingly rich, its mathematics are stunningly simple. In its most basic form, it is just an algebraic expression with three known variables and one unknown one. But this simple formula can lead to vast predictive insights.

Bayes's theorem is concerned with conditional probability. That is, it tells us the probability that a theory or hypothesis is true *if* some event has happened.

Suppose you are living with a partner and come home from a business trip to discover a strange pair of underwear in your dresser drawer. You will probably ask yourself: what is the probability that your partner is cheating on you? The *condition* is that you have found the underwear; the *hypothesis* you are interested in evaluating is the probability that you are being cheated on. Bayes's theorem, believe it or not, can give you an answer to this sort of question—provided that you know (or are willing to estimate) three quantities:

- First, you need to estimate the probability of the underwear's appearing *as a condition of the hypothesis being true*—that is, you are being cheated upon. Let's assume for the sake of this problem that you are a woman and your partner is a man, and the underwear in question is a pair of panties. If he's cheating on you, it's certainly easy enough to imagine how the panties got there. Then again, even (and perhaps especially) if he is cheating on you, you might expect him to be more careful. Let's say that the probability of the panties' appearing, conditional on his cheating on you, is 50 percent.
- Second, you need to estimate the probability of the underwear's appearing *conditional on the hypothesis being false*. If he isn't cheating, are there some innocent explanations for how they got there? Sure, although not all of them are pleasant (they could be *his* panties). It could be that his luggage got mixed up. It could be that a platonic female friend of his, whom you trust, stayed over one night. The panties could be a gift to you that he forgot to wrap up. None of these theories is inherently untenable, although some verge on dog-ate-my-homework excuses. Collectively you put their probability at 5 percent.
- Third and most important, you need what Bayesians call a *prior probability* (or simply a *prior*). What is the probability you would have assigned to him cheating on you *before* you found the underwear? Of course, it might be hard to be entirely objective about this now that the panties have made themselves known. (Ideally, you establish your priors before you start to examine the evidence.) But sometimes, it is possible to estimate a number like this empirically. Studies have found, for instance, that about 4 percent of married partners cheat on their spouses in any given year,[33] so we'll set that as our prior.

If we've estimated these values, Bayes's theorem can then be applied to establish a *posterior possibility*. This is the number that we're interested in: how likely is it that we're being cheated on, given that we've found the underwear? The calculation (and the simple algebraic expression that yields it) is in figure 8-3.

FIGURE 8-3: BAYES'S THEOREM—UNDERWEAR EXAMPLE

PRIOR PROBABILITY		
Initial estimate of how likely it is that he is cheating on you.	x	4%
A NEW EVENT OCCURS: MYSTERIOUS UNDERWEAR ARE FOUND		
Probability of underwear appearing conditional on his cheating on you.	y	50%
Probability of underwear appearing if he is *not* cheating on you.	z	5%
POSTERIOR PROBABILITY		
Revised estimate of how likely it is that he is cheating on you, given that you've found the underwear.	$\frac{xy}{xy + z(1-x)}$	29%

As it turns out, this probability is still fairly low: 29 percent. This may still seem counterintuitive—aren't those panties pretty incriminating? But it stems mostly from the fact that you had assigned a low prior probability to him cheating. Although an innocent man has fewer plausible explanations for the appearance of the panties than a guilty one, you had started out thinking he was an innocent man, so that weighs heavily into the equation.

When our priors are strong, they can be surprisingly resilient in the face of new evidence. One classic example of this is the presence of breast cancer among women in their forties. The chance that a woman will develop breast cancer in her forties is fortunately quite low—about 1.4 percent.[34] But what is the probability if she has a positive mammogram?

Studies show that if a woman does *not* have cancer, a mammogram will incorrectly claim that she does only about 10 percent of the time.[35] If she does have cancer, on the other hand, they will detect it about 75 percent of the time.[36] When you see those statistics, a positive mammogram seems like very bad news indeed. But if you apply Bayes's theorem to these numbers, you'll come to a different conclusion: the chance that a woman in her forties has breast cancer *given that she's had a positive mammogram* is still only about 10 percent. These false positives dominate the equation because very few young women have breast cancer to begin with. For this reason, many doctors recommend that women do not begin getting regular mammograms until they are in their fifties and the prior probability of having breast cancer is higher.[37]

Problems like these are no doubt challenging. A recent study that polled the statistical literacy of Americans presented this breast cancer example to them—and found that just 3 percent of them came up with the right probability estimate.[38] Sometimes, slowing down to look at the problem visually (as in figure 8-4) can provide a reality check against our inaccurate approximations. The visualization makes it easier to see the bigger picture—because breast cancer is so rare in young women, the fact of a positive mammogram is not all that telling.

FIGURE 8-4: BAYES'S THEOREM—MAMMOGRAM EXAMPLE

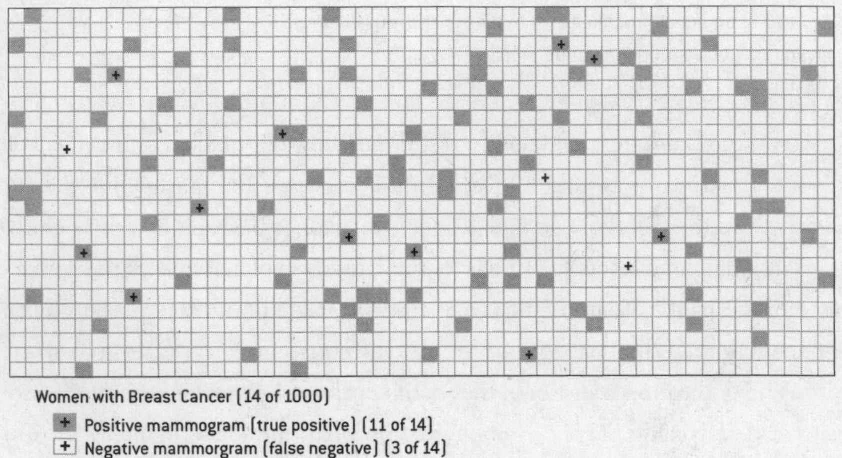

Women with Breast Cancer (14 of 1000)
- Positive mammogram (true positive) (11 of 14)
- Negative mammorgram (false negative) (3 of 14)

Women Without Breast Cancer (986 of 1000)
- Positive mammogram (false positive) (99 of 986)
- Negative mammorgram (true negative) (887 of 986)

Usually, however, we focus on the newest or most immediately available information, and the bigger picture gets lost. Smart gamblers like Bob Voulgaris have learned to take advantage of this flaw in our thinking. He made a profitable bet on the Lakers in part because the bookmakers placed much too much emphasis on the Lakers' first several games, lengthening their odds of winning the title from 4 to 1 to 6½ to 1, even though their performance was about what you might expect from a good team that had one of its star players injured. Bayes's theorem requires us to think through these problems more carefully and can be very useful for detecting when our gut-level approximations are much too crude.

This is not to suggest that our priors always dominate the new evidence, however, or that Bayes's theorem inherently produces counterintuitive results. Sometimes, the new evidence is so powerful that it overwhelms everything else, and we can go from assigning a near-zero probability of something to a near-certainty of it almost instantly.

Consider a somber example: the September 11 attacks. Most of us would have assigned almost no probability to terrorists crashing planes into buildings in Manhattan when we woke up that morning. But we recognized that a terror attack was an obvious possibility once the first plane hit the World Trade Center. And we had no doubt we were being attacked once the second tower was hit. Bayes's theorem can replicate this result.

For instance, say that before the first plane hit, our estimate of the possibility of a terror attack on tall buildings in Manhattan was just 1 chance in 20,000, or 0.005 percent. However, we would also have assigned a very low probability to a plane hitting the World Trade Center by accident. This figure can actually be estimated empirically: in the previous 25,000 days of aviation over Manhattan[39] prior to September 11, there had been two such accidents: one involving the Empire State Building in 1945 and another at 40 Wall Street in 1946. That would make the possibility of such an accident about 1 chance in 12,500 on any given day. If you use Bayes's theorem to run these numbers (figure 8-5a), the probability we'd assign to a terror attack increased from 0.005 percent to 38 percent the moment that the first plane hit.

FIGURE 8-5A: BAYES'S THEOREM—TERROR ATTACK EXAMPLE

PRIOR PROBABILITY		
Initial estimate of how likely it is that terrorists would crash planes into Manhattan skyscrapers.	x	0.005%
A NEW EVENT OCCURS: FIRST PLANE HITS WORLD TRADE CENTER		
Probability of plane hitting if terrorists are attacking Manhattan skyscrapers.	y	100%
Probability of plane hitting if terrorists are *not* attacking Manhattan skyscrapers (i.e. an accident).	z	0.008%
POSTERIOR PROBABILITY		
Revised estimate of probability of terror attack, given first plane hitting World Trade Center.	$\frac{xy}{xy + z(1-x)}$	38%

The idea behind Bayes's theorem, however, is not that we update our probability estimates just once. Instead, we do so continuously as new evidence presents itself to us. Thus, our posterior probability of a terror attack after the first plane hit, 38 percent, becomes our *prior* possibility before the second one did. And if you go through the calculation again, to reflect the second plane hitting the World Trade Center, the probability that we were under attack becomes a near-certainty—99.99 percent. One accident on a bright sunny day in New York was unlikely enough, but a second one was almost a literal impossibility, as we all horribly deduced.

FIGURE 8-5B: BAYES'S THEOREM—TERROR ATTACK EXAMPLE

PRIOR PROBABILITY		
Revised estimate of probability of terror attack, given first plane hitting World Trade Center.	x	38%
A NEW EVENT OCCURS: SECOND PLANE HITS WORLD TRADE CENTER		
Probability of plane hitting if terrorists are attacking Manhattan skyscrapers.	y	100%
Probability of plane hitting if terrorists are *not* attacking Manhattan skyscrapers (i.e. an accident).	z	0.008%
POSTERIOR PROBABILITY		
Revised estimate of probability of terror attack, given second plane hitting World Trade Center.	$\frac{xy}{xy + z(1-x)}$	99.99%

I have deliberately picked some challenging examples—terror attacks, cancer, being cheated on—because I want to demonstrate the breadth of problems to which Bayesian reasoning can be applied. Bayes's theorem is not any kind of magic formula—in the simple form that we have used here, it consists of nothing more than addition, subtraction, multiplication, and division. We have to provide it with information, particularly our estimates of the prior probabilities, for it to yield useful results.

However, Bayes's theorem does require us to think probabilistically about the world, even when it comes to issues that we don't like to think of as being matters of chance. This does not require us to have taken the position that the world is intrinsically, *metaphysically* uncertain—Laplace thought everything from the orbits of the planets to the behavior of the smallest molecules was governed by orderly Newtonian rules, and yet he was instrumental in the development of Bayes's theorem. Rather, Bayes's theorem deals with *epistemological* uncertainty—the limits of our knowledge.

The Problem of False Positives

When we fail to think like Bayesians, false positives are a problem not just for mammograms but for all of science. In the introduction to this book, I noted the work of the medical researcher John P. A. Ioannidis. In 2005, Ioannidis published an influential paper, "Why Most Published Research Findings Are False,"[40] in which he cited a variety of statistical and theoretical arguments to claim that (as his title implies) the *majority* of hypotheses deemed to be true in journals in medicine and most other academic and scientific professions are, in fact, false.

Ioannidis's hypothesis, as we mentioned, looks to be one of the true ones; Bayer Laboratories found that they could not replicate about *two-thirds* of the positive findings claimed in medical journals when they attempted the experiments themselves.[41] Another way to check the veracity of a research finding is to see whether it makes accurate predictions in the real world—and as we have seen throughout this book, it very often does not. The failure rate for predictions made in entire fields ranging from seismology to political science appears to be extremely high.

"In the last twenty years, with the exponential growth in the availability of information, genomics, and other technologies, we can measure millions and millions of potentially interesting variables," Ioannidis told me. "The expectation is that we can use that information to make predictions work for us. I'm not saying that we haven't made any progress. Taking into account that there are a couple of million papers, it would be a shame if there wasn't. But there are obviously not a couple of million discoveries. Most are not really contributing much to generating knowledge."

This is why our predictions may be *more* prone to failure in the era of Big Data. As there is an exponential increase in the amount of available information, there is likewise an exponential increase in the number of hypotheses to investigate. For instance, the U.S. government now publishes data on about 45,000 economic statistics. If you want to test for relationships between all combinations of two pairs of these statistics—is there a causal relationship between the bank prime loan rate and the unemployment rate in Alabama?—that gives you literally one billion hypotheses to test.*

But the number of *meaningful* relationships in the data—those that speak to causality rather than correlation and testify to how the world really works—is orders of magnitude smaller. Nor is it likely to be increasing at nearly so fast a rate as the information itself; there isn't any more truth in the world than there was before the Internet or the printing press. Most of the data is just noise, as most of the universe is filled with empty space.

Meanwhile, as we know from Bayes's theorem, when the underlying incidence of something in a population is low (breast cancer in young women; truth in the sea of data), false positives can dominate the results if we are not careful. Figure 8-6 represents this graphically. In the figure, 80 percent of true scientific hypotheses are correctly deemed to be true, and about 90 percent of false hypotheses are correctly rejected. And yet, because true findings are so rare, about two-thirds of the findings deemed to be true are actually false!

Unfortunately, as Ioannidis figured out, the state of published research in *most* fields that conduct statistical testing is probably very much like what you see in figure 8-6.† Why is the error rate so high? To some extent, this entire book represents an answer to that question. There are many reasons for it—some having to do with our psychological biases, some having to do with common methodological errors, and some having to do with misaligned incentives. Close to the root of the problem, however, is a flawed type of statistical thinking that these researchers are applying.

---

* The number of possible combinations is calculated as 45,000 times 44,999 divided by two, which is 1,012,477,500.

† One difference is that the negative findings are probably kept in a file drawer rather than being published (about 90 percent of the papers published in academic journals today document positive findings rather than negative ones). However, that does not mask the problem of false positives in the findings that do make it to publication.

FIGURE 8-6: A GRAPHICAL REPRESENTATION OF FALSE POSITIVES

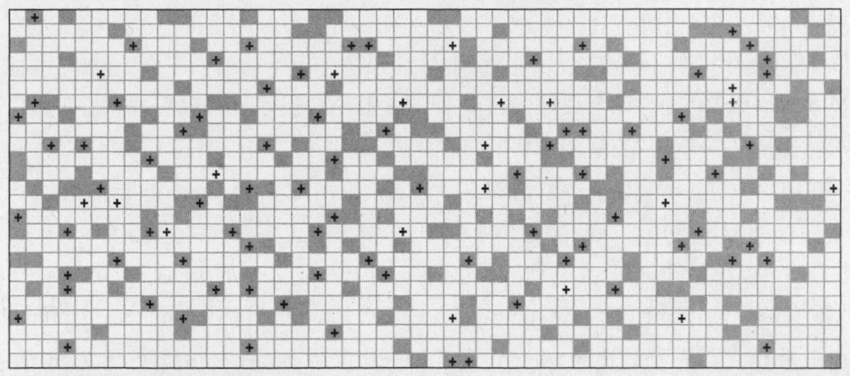

True Hypotheses (100 of 1000)
- Positive statistical test (true positive) (80 of 100)
- Negative statistical test (false negative) (20 of 100)

False Hypotheses (900 of 1000)
- Positive statistical test (false positive) (180 of 900)
- Negative statistical test (true negative) (720 of 900)

When Statistics Backtracked from Bayes

Perhaps the chief intellectual rival to Thomas Bayes—although he was born in 1890, almost 120 years after Bayes's death—was an English statistician and biologist named Ronald Aylmer (R. A.) Fisher. Fisher was a much more colorful character than Bayes, almost in the English intellectual tradition of Christopher Hitchens. He was handsome but a slovenly dresser,[42] always smoking his pipe or his cigarettes, constantly picking fights with his real and imagined rivals. He was a mediocre lecturer but an incisive writer with a flair for drama, and an engaging and much-sought-after dinner companion. Fisher's interests were wide-ranging: he was one of the best biologists of his day and one of its better geneticists, but was an unabashed elitist who bemoaned the fact that the poorer classes were having more offspring than the intellectuals.[43] (Fisher dutifully had eight children of his own.)

Fisher is probably more responsible than any other individual for the statistical methods that remain in wide use today. He developed the terminology of the statistical significance test and much of the methodology behind it. He was also no fan of Bayes and Laplace—Fisher was the first person to use the term

"Bayesian" in a published article, and he used it in a derogatory way,[44] at another point asserting that the theory "must be wholly rejected."[45]

Fisher and his contemporaries had no problem with the formula called Bayes's theorem *per se*, which is just a simple mathematical identity. Instead, they were worried about how it might be applied. In particular, they took issue with the notion of the Bayesian prior.[46] It all seemed too subjective: we have to stipulate, in advance, how likely we think something is before embarking on an experiment about it? Doesn't that cut against the notion of objective science?

So Fisher and his contemporaries instead sought to develop a set of statistical methods that they hoped would free us from any possible contamination from bias. This brand of statistics is usually called "frequentism" today, although the term "Fisherian" (as opposed to Bayesian) is sometimes applied to it.[47]

The idea behind frequentism is that uncertainty in a statistical problem results exclusively from collecting data among just a sample of the population rather than the whole population. This makes the most sense in the context of something like a political poll. A survey in California might sample eight hundred people rather than the eight million that will turn out to vote in an upcoming election there, producing what's known as sampling error. The margin of error that you see reported alongside political polls is a measure of this: exactly how much error is introduced because you survey eight hundred people in a population of eight million? The frequentist methods are designed to quantify this.

Even in the context of political polling, however, sampling error does not always tell the whole story. In the brief interval between the Iowa Democratic caucus and New Hampshire Democratic Primary in 2008, about 15,000 people were surveyed[48] in New Hampshire—an enormous number in a small state, enough that the margin of error on the polls was theoretically just plus-or-minus 0.8 percent. The actual error in the polls was about ten times that, however: Hillary Clinton won the state by three points when the polls had her losing to Barack Obama by eight. Sampling error—the *only* type of error that frequentist statistics directly account for—was the least of the problem in the case of the New Hampshire polls.

Likewise, some polling firms consistently show a bias toward one or another party:[49] they could survey all 200 million American adults and they still wouldn't get the numbers right. Bayes had these problems figured out 250 years ago. If you're using a biased instrument, it doesn't matter how many measurements you take—you're aiming at the wrong target.

Essentially, the frequentist approach toward statistics seeks to wash its hands of the reason that predictions most often go wrong: human error. It views uncertainty as something intrinsic to the experiment rather than something intrinsic to our ability to understand the real world. The frequentist method also implies that, as you collect more data, your error will eventually approach zero: this will be both necessary and sufficient to solve any problems. Many of the more problematic areas of prediction in this book come from fields in which useful data is sparse, and it is indeed usually valuable to collect more of it. However, it is hardly a golden road to statistical perfection if you are not using it in a sensible way. As Ioannidis noted, the era of Big Data only seems to be worsening the problems of false positive findings in the research literature.

Nor is the frequentist method particularly objective, either in theory or in practice. Instead, it relies on a whole host of assumptions. It usually presumes that the underlying uncertainty in a measurement follows a bell-curve or normal distribution. This is often a good assumption, but not in the case of something like the variation in the stock market. The frequentist approach requires defining a sample population, something that is straightforward in the case of a political poll but which is largely arbitrary in many other practical applications. What "sample population" was the September 11 attack drawn from?

The bigger problem, however, is that the frequentist methods—in striving for immaculate statistical procedures that can't be contaminated by the researcher's bias—keep him hermetically sealed off from the real world. These methods discourage the researcher from considering the underlying context or plausibility of his hypothesis, something that the Bayesian method demands in the form of a prior probability. Thus, you will see apparently serious papers published on how toads can predict earthquakes,[50] or how big-box stores like Target beget racial hate groups,[51] which apply frequentist tests to produce "statistically significant" (but manifestly ridiculous) findings.

Data Is Useless Without Context

Fisher mellowed out some toward the end of his career, occasionally even praising Bayes.[52] And some of the methods he developed over his long career (although not the ones that are in the widest use today) were really compromises between Bayesian and frequentist approaches. In the last years of his life, however, Fisher made a grievous error of judgment that helps to demonstrate the limitations of his approach.

The issue concerned cigarette smoking and lung cancer. In the 1950s, a large volume of research—some of it using standard statistical methods and some using Bayesian ones[53]—claimed there was a connection between the two, a connection that is of course widely accepted today.

Fisher spent much of his late life fighting against these conclusions, publishing letters in prestigious publications including *The British Medical Journal* and *Nature*.[54] He did not deny that the statistical relationship between cigarettes and lung cancer was fairly strong in these studies, but he claimed it was a case of correlation mistaken for causation, comparing it to a historical correlation between apple imports and marriage rates in England.[55] At one point, he argued that lung cancer caused cigarette smoking and not the other way around[56]—the idea, apparently, was that people might take up smoking for relief from their lung pain.

Many scientific findings that are commonly accepted today would have been dismissed as hooey at one point. This was sometimes because of the cultural taboos of the day (such as in Galileo's claim that the earth revolves around the sun) but at least as often because the data required to analyze the problem did not yet exist. We might let Fisher off the hook if, it turned out, there was not compelling evidence to suggest a linkage between cigarettes and lung cancer by the 1950s. Scholars who have gone back and looked at the evidence that existed at the time have concluded, however, that there was plenty of it—a wide variety of statistical and clinical tests conducted by a wide variety of researchers in a wide variety of contexts demonstrated the causal relationship between them.[57] The idea was quickly becoming the scientific consensus.

So why did Fisher dismiss the theory? One reason may have been that he was a paid consultant of the tobacco companies.[58] Another may have been that he was a lifelong smoker himself. And Fisher liked to be contrarian and controversial, and disliked anything that smacked of puritanism. In short, he was biased, in a variety of ways.

But perhaps the bigger problem is the way that Fisher's statistical philosophy tends to conceive of the world. It emphasizes the objective purity of the experiment—every hypothesis could be tested to a perfect conclusion if only enough data were collected. However, in order to achieve that purity, it denies the need for Bayesian priors or any other sort of messy real-world context. These methods neither require nor encourage us to think about the plausibility of our hypothesis: the idea that cigarettes cause lung cancer competes on a level playing field with the idea that toads predict earthquakes. It is, I suppose, to Fisher's credit that he recognized that correlation does not always imply causation. However, the Fisherian statistical methods do not encourage us to think about *which* correlations imply causations and which ones do not. It is perhaps no surprise that after a lifetime of thinking this way, Fisher lost the ability to tell the difference.

Bob the Bayesian

In the Bayesian worldview, prediction is the yardstick by which we measure progress. We can perhaps never know the truth with 100 percent certainty, but making correct predictions is the way to tell if we're getting closer.

Bayesians hold the gambler in particularly high esteem.[59] Bayes and Laplace, as well as other early probability theorists, very often used examples from games of chance to explicate their work. (Although Bayes probably did not gamble much himself,[60] he traveled in circles in which games like cards and billiards were common and were often played for money.) The gambler makes predictions (good), and he makes predictions that involve estimating probabilities (great), and when he is willing to put his money down on his predictions (even better), he discloses his beliefs about the world to everyone else. The most practical definition of a Bayesian prior might simply be the odds at which you are willing to place a bet.*

And Bob Voulgaris is a particularly Bayesian type of gambler. He likes betting on basketball precisely because it is a way to test himself and the accuracy of his theories. "You could be a general manager in sports and you could be like, Okay, I'll get this player and I'll get that player," he told me toward the end of our interview. "At the end of the day you don't really know if you're right or wrong. But at the end of the day, the end of the season, I know if I'm right or wrong because I know if I'm winning money or I'm losing it. That's a pretty good validation."

Voulgaris soaks up as much basketball information as possible because everything could potentially shift his probability estimates. A professional sports bettor like Voulgaris might place a bet only when he thinks he has at least a 54 percent chance of winning it. This is just enough to cover the "vigorish" (the cut a sportsbook takes on a winning wager), plus the risk associated with putting one's money into play. And for all his skill and hard work—Voulgaris is among the best sports bettors in the world today—he still gets only about 57 percent of his bets right. It is just exceptionally difficult to do much better than that.

A small piece of information that improves Voulgaris's estimate of his odds from 53 percent to 56 percent can therefore make all the difference. This is the sort of narrow margin that gamblers, whether at the poker table or in the stock market, make their living on. Fisher's notion of statistical significance, which uses arbitrary cutoffs devoid of context† to determine what is a "significant" finding and what isn't,[61] is much too clumsy for gambling.

But this is not to suggest that Voulgaris avoids developing *hypotheses* around what he's seeing in the statistics. (The problem with Fisher's notion of

---

* Or more properly, the odds you would set as a betting line so as to be indifferent between either side of the bet. Most Bayesians do require that priors avoid what is called a Dutch book, where the odds are incoherent. If you establish a set of prior probabilities on each of the thirty teams winning the NBA championship, they have to add up to 100 percent exactly since this represents an exhaustive set of possibilities.

† It has been found that because 95 percent confidence in a statistical test is Fisher's traditional dividing line between "signficiant" and "insignificant," researchers are much more likely to report findings that statistical tests classify as 95.1 percent certain than those they classify as 94.9 percent certain—a practice that seems more superstitious than scientific.

hypothesis testing is not with having hypotheses but with the way Fisher recommends that we test them.)[62] In fact, this is critical to what Voulgaris does. Everyone can see the statistical patterns, and they are soon reflected in the betting line. The question is whether they represent signal or noise. Voulgaris forms hypotheses from his basketball knowledge so that he might tell the difference more quickly and more accurately.

Voulgaris's approach to betting basketball is one of the purer distillations of the scientific method that you're likely to find (figure 8-7). He observes the world and asks questions: why are the Cleveland Cavaliers so frequently going over on the total? He then gathers information on the problem, and formulates a hypothesis: the Cavaliers are going over because Ricky Davis is in a contract year and is trying to play at a fast pace to improve his statistics. The difference between what Voulgaris does and what a physicist or biologist might do is that he demarcates his predictions by placing bets on them, whereas a scientist would hope to validate her prediction by conducting an experiment.

FIGURE 8-7: SCIENTIFIC METHOD

Step in Scientific Method[63]	**Sports Betting Example**
Observe a phenomenon	Cavaliers games are frequently going over the game total.
Develop a hypothesis to explain the phenomenon	Cavaliers games are going over because Ricky Davis is playing for a new contract and trying to score as many points as possible.
Formulate a prediction from the hypothesis	Davis's incentives won't change until the end of the season. Therefore: (i) he'll continue to play at a fast pace, and, (ii) future Cavaliers games will continue to be high-scoring as a result.
Test the prediction	Place your bet.

If Voulgaris can develop a strong hypothesis about what he is seeing in the data, it can enable him to make more aggressive bets. Suppose, for instance, that Voulgaris reads some offhand remark from the coach of the Denver Nuggets about wanting to "put on a good show" for the fans. This is probably just idle chatter, but it *might* imply that the team will start to play at a faster pace in order to increase ticket sales. If this hypothesis is right, Voulgaris might expect that an over bet on Nuggets games will win 70 percent of the time as opposed to the customary 50 percent. As a consequence of Bayes's theorem, the stronger Voulgaris's belief in his hypothesis, the more quickly he can begin to make profitable bets on Nuggets games. He might be able to do so after watching just a game or two, observing whether his theory holds in practice—quickly enough that Vegas will have yet to catch on. Conversely, he can avoid being distracted by statistical patterns, like the Lakers' slow start in 1999, that have little underlying meaning but which other handicappers might mistake for a signal.

The Bayesian Path to Less Wrongness

But are Bob's probability estimates subjective or objective? That is a tricky question.

As an empirical matter, we all have beliefs and biases, forged from some combination of our experiences, our values, our knowledge, and perhaps our political or professional agenda. One of the nice characteristics of the Bayesian perspective is that, in explicitly acknowledging that we have prior beliefs that affect how we interpret new evidence, it provides for a very good *description* of how we react to the changes in our world. For instance, if Fisher's prior belief was that there was just a 0.00001 percent chance that cigarettes cause lung cancer, that helps explain why all the evidence to the contrary couldn't convince him otherwise. In fact, there is nothing prohibiting you under Bayes's theorem from holding beliefs that you believe to be *absolutely* true. If you hold there is a 100 percent probability that God exists, or a 0 percent probability, then under Bayes's theorem, *no* amount of evidence could persuade you otherwise.

I'm not here to tell you whether there are things you should believe with *absolute* and *unequivocal* certainty or not.* But perhaps we should be more

* Although bear in mind that one of the conclusions of this book is that people are overconfident; we probably have too many beliefs that tend toward the 0 percent or 100 percent end of the spectrum.

honest about declaiming these. Absolutely nothing useful is realized when one person who holds that there is a 0 percent probability of something argues against another person who holds that the probability is 100 percent. Many wars—like the sectarian wars in Europe in the early days of the printing press—probably result from something like this premise.

This does not imply that all prior beliefs are equally correct or equally valid. But I'm of the view that we can never achieve perfect objectivity, rationality, or accuracy in our beliefs. Instead, we can strive to be *less* subjective, *less* irrational, and *less* wrong. Making predictions based on our beliefs is the best (and perhaps even the only) way to test ourselves. If objectivity is the concern for a greater truth beyond our personal circumstances, and prediction is the best way to examine how closely aligned our personal perceptions are with that greater truth, the most objective among us are those who make the most accurate predictions. Fisher's statistical method, which saw objectivity as residing within the confines of a laboratory experiment, is less suitable to this task than Bayesian reasoning.

One property of Bayes's theorem, in fact, is that our beliefs should converge toward one another—and toward the truth—as we are presented with more evidence over time. In figure 8-8, I've worked out an example wherein three inves-

FIGURE 8-8: BAYESIAN CONVERGENCE

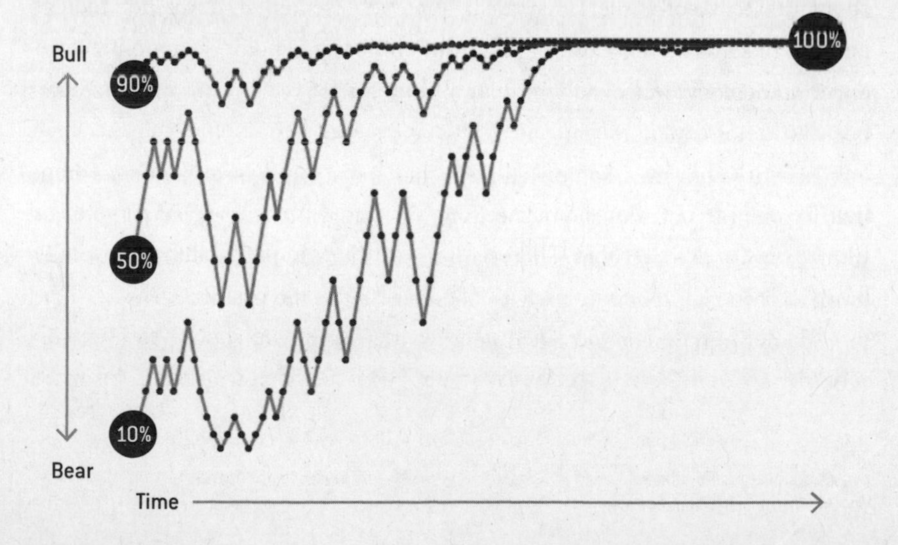

tors are trying to determine whether they are in a bull market or a bear market. They start out with very different beliefs about this—one of them is optimistic, and believes there's a 90 percent chance of a bull market from the outset, while another one is bearish and says there's just a 10 percent chance. Every time the market goes up, the investors become a little more bullish relative to their prior, while every time it goes down the reverse occurs. However, I set the simulation up such that, although the fluctuations are random on a day-to-day basis, the market increases 60 percent of the time over the long run. Although it is a bumpy road, eventually all the investors correctly determine that they are in a bull market with almost (although not *exactly*, of course) 100 percent certainty.

In theory, science should work this way. The notion of scientific consensus is tricky, but the idea is that the opinion of the scientific community converges *toward* the truth as ideas are debated and new evidence is uncovered. Just as in the stock market, the steps are not always forward or smooth. The scientific community is often too conservative about adapting its paradigms to new evidence,[64] although there have certainly also been times when it was too quick to jump on the bandwagon. Still, provided that everyone is on the Bayesian train,* even incorrect beliefs and quite wrong priors are revised toward the truth in the end.

Right now, for instance, we may be undergoing a paradigm shift in the statistical methods that scientists are using. The critique I have made here about the flaws of Fisher's statistical approach is neither novel nor radical: prominent scholars in fields ranging from clinical psychology[65] to political science[66] to ecology[67] have made similar arguments for years. But so far there has been little fundamental change.

Recently, however, some well-respected statisticians have begun to argue that frequentist statistics should no longer be taught to undergraduates.[68] And some professions have considered banning Fisher's hypothesis test from their journals.[69] In fact, if you read what's been written in the past ten years, it's hard to find anything that *doesn't* advocate a Bayesian approach.

---

* And that they don't hold priors that they believe to be *exactly* 100 percent true or *exactly* 0 percent true; these will not and *cannot* change under Bayes's theorem.

Bob's money is on Bayes, too. He does not literally apply Bayes's theorem every time he makes a prediction. But his practice of testing statistical data in the context of hypotheses and beliefs derived from his basketball knowledge is very Bayesian, as is his comfort with accepting probabilistic answers to his questions.

It will take some time for textbooks and traditions to change. But Bayes's theorem holds that we will converge toward the better approach. Bayes's theorem predicts that the Bayesians will win.

9

RAGE AGAINST THE MACHINES

The twenty-seven-year-old Edgar Allan Poe, like many others before him, was fascinated by the Mechanical Turk, a contraption that had once beaten Napoleon Bonaparte and Benjamin Franklin at chess. The machine, constructed in Hungary in 1770 before Poe or the United States of America were born, had come to tour Baltimore and Richmond in the 1830s after having wowed audiences around Europe for decades. Poe deduced that it was an elaborate hoax, its cogs and gears concealing a chess master who sat in its cupboards and manipulated its levers to move the pieces about the board and nod its turban-covered head every time that it put its opponent into check.

Poe is regarded as the inventor of the detective story,[1] and some of his work in sleuthing the hoax was uncanny. He was rightly suspicious, for instance, that a man (later determined to be the German chess master William Schlumberger) could always be found packing and unpacking the machine but was nowhere to be seen while the game was being played (aha! he was in the box).

FIGURE 9-1: THE MECHANICAL TURK

What was truly visionary about Poe's essay on the Mechanical Turk, however, was his grasp of its implications for what we now call artificial intelligence (a term that would not be coined until 120 years later). His essay expressed a very deep and very modern ambivalence about the prospect that computers might be able to imitate, or improve on, the higher functions of man.

Poe recognized just how impressive it might be for a machine to play chess at all. The first mechanical computer, what Charles Babbage called the difference engine, had barely been conceived of at the time that Poe wrote his exposé. Babbage's proposed computer, which was never fully built during his lifetime, might at best hope to approximate some elementary functions like logarithms in addition to carrying out addition, subtraction, multiplication, and division. Poe thought of Babbage's work as impressive enough—but still, all it did was take predictable inputs, turn a few gears, and spit out predictable outputs. There was no intelligence there—it was purely mechanistic. A computer that could play chess, on the other hand, verged on being miraculous because of the judgment required to play the game well.

Poe claimed that if this chess-playing machine were real, it must by definition play chess flawlessly; machines do not make computational errors. He took the fact that the Turk did not play perfect chess—it won most of its games but lost a few—as further proof that it was not a machine but a human-controlled apparatus, full of human imperfections.

Although Poe's logic was flawed, this reverence for machines is still with us today. We regard computers as astonishing inventions, among the foremost expressions of human ingenuity. Bill Gates often polls as among the most admired men in America,[2] and Apple and Google among our most admired companies.[3] And we expect computers to behave flawlessly and effortlessly, somehow overcoming the imperfections of their creators.

Moreover, we view the calculations of computer programs as unimpeachably precise and perhaps even prophetic. In 2012, a pair of British teenagers were accused of defrauding investors out of more than a million dollars by promoting their stock-picking "robot" named MARL,[4] which they claimed could process "1,986,832 mathematical calculations per second" while avoiding "human 'gut feelings,'" allowing investors to double their money every few hours by following MARL's recommendations for penny stocks.[5]

Even when computer predictions do not inspire our credulity, they can spark our fears; computers that run programs to forecast the chance of survival for hospital patients, for instance, are sometimes written about in news accounts[6] as though they are cousins of the HAL 9000, the computer from *2001: A Space Odyssey*, which decided it had no more use for the astronauts and tried to suffocate them.

As we enter the era of Big Data, with information and processing power increasing at exponential rates, it may be time to develop a healthier attitude toward computers and what they might accomplish for us. Technology is beneficial as a labor-saving device, but we should not expect machines to do our thinking for us.

The Birth of the Chess Computer

The Spanish engineer Leonardo Torres y Quevedo built a version of the Mechanical Turk in 1912, which he called *El Ajedrecista* (the chess player). Although *El Ajedrecista* is sometimes regarded as the first computer game,[7] it was extremely limited in its functionality, restricted to determining positions in an endgame in which there are just three pieces left on the board. (*El Ajedrecista* also did not have any stereotypical Turkish headgear.)

The father of the modern chess computer was MIT's Claude Shannon, a mathematician regarded as the founder of information theory, who in 1950 published a paper called "Programming a Computer for Playing Chess."[8] Shannon identified some of the algorithms and techniques that form the backbone of chess programs today. He also recognized why chess is such an interesting problem for testing the powers of information-processing machines.

Chess, Shannon realized, has an exceptionally clear and distinct goal—achieving checkmate. Moreover, it follows a relatively simple set of rules and has no element of chance or randomness. And yet, as anybody who has played chess has realized (I am not such a good player myself), using those simple rules to achieve that simple goal is not at all easy. It requires deep concentration to survive more than a couple of dozen moves into a chess game, let alone to actually win one. Shannon saw chess as a litmus test for the power of computers and the sort of abilities they might someday possess.

But Shannon, in contrast to some who came after him, did not hold the romanticized notion that computers might play chess in the same way that humans do. Nor did he see their victory over humans at chess as being inevitable. Instead, he saw four potential advantages for computers:

1. They are very fast at making calculations.
2. They won't make errors, unless the errors are encoded in the program.
3. They won't get lazy and fail to fully analyze a position or all the possible moves.
4. They won't play emotionally and become overconfident in an apparent

winning position that might be squandered or grow despondent in a difficult one that might be salvaged.

These were to be weighed, Shannon thought, against four distinctly human advantages:

1. Our minds are flexible, able to shift gears to solve a problem rather than follow a set of code.
2. We have the capacity for imagination.
3. We have the ability to reason.
4. We have the ability to learn.

It seemed like a fair fight to Shannon. But that was only the case for a few fleeting moments in the mid-1990s, when the Russian grandmaster Garry Kasparov—the best chess player of all time—went up against what was then one of the most advanced computers ever built, IBM's Deep Blue.

Before their match, humans were winning the fight—it wasn't even close. Yet computers have prevailed ever since, and will continue to do so for as long as we live.

Chess, Prediction, and Heuristics

In accordance with Bayes's theorem, prediction is fundamentally a type of information-processing activity—a matter of using new data to test our hypotheses about the objective world, with the goal of coming to truer and more accurate conceptions about it.

Chess might be thought of as analogous to prediction. The players must process information—the position of the thirty-two pieces on the board and their possible moves. They use this information to devise strategies to place their opponent in checkmate. These strategies in essence represent different hypotheses about how to win the game. Whoever succeeds in that task had the better hypothesis.

Chess is deterministic—there is no real element of luck involved. But the same is theoretically true of the weather, as we saw in chapter 4. Our knowledge of both systems is subject to considerable imperfections. In weather, much of the problem is that our knowledge of the initial conditions is incomplete. Even though we have a very good idea of the rules by which the weather system behaves, we have incomplete information about the position of all the molecules that form clouds and rainstorms and hurricanes. Hence, the best we can do is to make probabilistic forecasts.

In chess, we have both complete knowledge of the governing rules *and* perfect information—there are a finite number of chess pieces, and they're right there in plain sight. But the game is still very difficult for us. Chess speaks to the constraints on our information-processing capabilities—and it might tell us something about the best strategies for making decisions despite them. The need for prediction arises not necessarily because the world itself is uncertain, but because understanding it fully is beyond our capacity.[9]

Both computer programs and human chess masters therefore rely on making simplifications to forecast the outcome of the game. We can think of these simplifications as "models," but *heuristics* is the preferred term in the study of computer programming and human decision making. It comes from the same Greek root word from which we derive *eureka*.[10] A heuristic approach to problem solving consists of employing rules of thumb when a deterministic solution to a problem is beyond our practical capacities.

Heuristics are very useful things, but they necessarily produce biases and blind spots.[11] For instance, the heuristic "When you encounter a dangerous animal, run away!" is often a useful guide but not when you meet a grizzly bear; she may be startled by your sudden movement and she can easily outrun you. (Instead, the National Park Service advises you to remain as quiet and as still as possible when you encounter a grizzly bear and even to play dead if necessary.[12]) Humans and computers apply different heuristics when they play chess. When they play against each other, the game usually comes down to who can find his opponent's blind spots first.

Kasparov's Failed Prediction

In January 1988, Garry Kasparov, the top-rated chess player in the world from 1986 until his retirement in 2005,[13] predicted that no computer program would be able to defeat a human grandmaster at chess until at least the year 2000.[14] "If any grandmaster has difficulties playing computers," he quipped at a press conference in Paris, "I would be happy to provide my advice."[15] Later that same year, however, the Danish grandmaster Bent Larsen was defeated by a program named Deep Thought, a graduate-school project by several students at Carnegie Mellon University.

The garden-variety grandmaster, however, was no Kasparov, and when Deep Thought squared off against Kasparov in 1989 it was resoundingly defeated. Kasparov has always respected the role of computing technology in chess, and had long studied with computers to improve his game, but he offered Deep Thought only the faintest praise, suggesting that one day a computer could come along that might require him to exert his "100 percent capabilities" in order to defeat it.[16]

The programmers behind Deep Thought, led by Feng-hsiung Hsu and Murray Campbell, were eventually hired by IBM, where their system evolved into Deep Blue. Deep Blue did defeat Kasparov in the first game of a match in Philadelphia in 1996, but Kasparov rebounded to claim the rest of the series fairly easily. It was the next year, in a rematch in New York, when the unthinkable happened. Garry Kasparov, the best and most intimidating chess player in history, was intimidated by a computer.

In the Beginning . . .

A chess game, like everything else, has three parts: the beginning, the middle and the end. What's a little different about chess is that each of these phases tests different intellectual and emotional skills, making the game a mental triathlon of speed, strength, and stamina.

In the beginning of a chess game the center of the board is void, with pawns, rooks, and bishops neatly aligned in the first two rows awaiting instructions from their masters. The possibilities are almost infinite. White can open the game in any of twenty different ways, and black can respond with twenty of its own moves, creating 4,000 possible sequences after the first full turn. After the second full turn, there are 71,852 possibilities; after the third, there are 9,132,484. The number of possibilities in an entire chess game, played to completion, is so large that it is a significant problem even to estimate it, but some mathematicians put the number as high as $10^{10^{50}}$. These are astronomical numbers: as Diego Rasskin-Gutman has written, "There are more possible chess games than the number of atoms in the universe."[17]

It might seem that at the beginning of the game, when all the pieces are still on the board and the number of possibilities is most limitless, computers are at their greatest strength. As IBM's Web site bragged before the match against Kasparov, their computer could calculate 200 million positions per second. "Incidentally, Garry Kasparov can evaluate approximately three positions per second," it noted snidely.[18] How did Kasparov have a chance?

But chess computers had long been rather poor at the opening phase of the game. Although the number of possibilities was the most limitless, the objectives were also the least clear. When there are $10^{10^{50}}$ branches on the tree, calculating 3 moves per second or 200 million is about equally fruitless unless you are harnessing that power in a directed way.

Both computer and human players need to break a chess game down into intermediate goals: for instance, capturing an opponent's pawn or putting their king into check. In the middle of the match, once the pieces are locked in combat and threaten one another, there are many such strategic objectives available. It is a matter of devising tactics to accomplish them, and forecasting which might have the most beneficial effects on the remainder of the game. The goals of the opening moves, however, are more abstract. Computers struggle with abstract and open-ended problems, whereas humans understand heuristics like "control the center of the board" and "keep your pawns organized" and can devise any number of creative ways to achieve them.

Moreover, because the opening moves are more routine to players than positions they may encounter later on, humans can rely on centuries' worth of experience to pick the best moves. Although there are theoretically twenty moves that white might play to open the game, more than 98 percent of competitive chess games begin with one of the best four.[19]

The problem for humans is that computer programs can systematize this knowledge by studying statistics. Chess databases contain the results of literally hundreds of thousands of games and like any other database can be mined for predictive insight. IBM's programmers studied things like how often each sequence of opening moves had been played and how strong the players were who played them, as well as how often each series of moves resulted in wins, losses, and draws for their respective sides.[20] The computer's heuristics for analyzing these statistics could potentially put it on a level playing field with human intuition and experience, if not well ahead of it. "Kasparov isn't playing a computer, he's playing the ghosts of grandmasters past," IBM's Web site said in reference to Deep Blue's deep databases.[21]

Kasparov's goal, therefore, in his first game of his six-game match against Deep Blue in 1997, was to take the program out of database-land and make it fly blind again. The opening move he played was fairly common; he moved his knight to the square of the board that players know as f3. Deep Blue responded on its second move by advancing its bishop to threaten Kasparov's knight—undoubtedly because its databases showed that such a move had historically reduced white's winning percentage* from 56 percent to 51 percent.

Those databases relied on the assumption, however, that Kasparov would respond as almost all other players had when faced with the position,[22] by moving his knight back out of the way. Instead, he ignored the threat, figuring that Deep Blue was bluffing,[23] and chose instead to move one of his pawns to pave the way for his bishop to control the center of the board.

Kasparov's move, while sound strategically, also accomplished another objective. He had made just three moves and Deep Blue had made just two, and yet the position they had now achieved (illustrated in figure 9-2) had literally occurred just once before in master-level competition[24] out of the hundreds of thousands of games in Deep Blue's database.

Even when very common chess moves are played, there are so many pos-

* Winning percentage, as I use the term here, refers to the number of points that each side scores out of the possible total, where one point is awarded for winning a game and half a point is awarded for a draw. If you play ten games, winning five, drawing three, and losing two, you have a winning percentage of 65 percent.

sible branches on the tree that databases are useless after perhaps ten or fifteen moves. In any long game of chess, it is quite likely that you and your opponent will eventually reach some positions that literally no two players in the history of humanity have encountered before. But Kasparov had taken the database out after just three moves. As we have learned throughout this book, purely statistical approaches toward forecasting are ineffective at best when there is not a sufficient sample of data to work with.

Deep Blue would need to "think" for itself.

FIGURE 9-2: POSITION AFTER KASPAROV'S 3RD MOVE IN GAME 1

The Chess Player's Dilemma: Breadth Versus Depth

The middle of a chess game (simply called the *midgame*) potentially favors the strengths of the computer. With the pieces free to move in the center of the board, there are an average of around forty possible moves rather than twenty for every turn.[25] That might not seem like a huge difference, but because the tree of possibilities explodes exponentially, it quickly adds up. Suppose, for instance, that you want to calculate just three full turns ahead (that is, three moves each for you and your opponent, or six total). At the beginning of the game, this function is approximated by the number twenty taken to the sixth power, which is sixty-four million positions, already a gargantuan number. In the midgame, however, you have to calculate forty to the sixth power, or 4.1 billion possibilities. Deep Blue could calculate all those positions in just twenty seconds. Kasparov would require literally forty-three years to do so, even without pausing to eat, sleep, or go to the bathroom.

Great players like Kasparov do not delude themselves into thinking they can calculate all these possibilities. This is what separates elite players from amateurs. In his famous study of chess players, the Dutch psychologist Adriaan de Groot found that amateur players, when presented with a chess problem, often frustrated themselves by looking for the perfect move, rendering themselves incapable of making any move at all.[26]

Chess masters, by contrast, are looking for a *good* move—and certainly if at all possible the *best* move in a given position—but they are more *forecasting* how the move might favorably dispose their position than trying to enumerate every possibility. It is "pure fantasy," the American grandmaster Reuben Fine wrote,[27] to assume that human chess players have calculated every position to completion twenty or thirty moves in advance.

It is not quite as simple as saying that "the perfect is the enemy of the good." If you want to really master a game like chess, you may need to get beyond such simple heuristics. Nevertheless, we are not capable of making *perfect* decisions when presented with more information than we can process in a limited amount of time. Acknowledging those imperfections may free us to make the best decisions that we can in chess and in other contexts that involve forecasting.

This is not to say that grandmasters like Kasparov don't have any calculations to perform. At the very least, Kasparov might need to devise a tactic, a precise sequence of perhaps three to five moves, to capture an opponent's piece or accomplish some other near-term objective. For each of those moves, he will need to think about how his opponent might respond—all the possible variations in the play—and whether any of them could nullify his tactic. He will also need to consider whether the opponent has laid any traps for him; a strong-looking position can be turned into a checkmate in just a few moves if a player's king is not protected.

Chess players learn through memory and experience where to concentrate their thinking. Sometimes this involves probing many branches of the tree but just a couple of moves down the line; at other times, they focus on just one branch but carry out the calculation to a much greater depth. This type of trade-off between breadth and depth is common anytime that we face a complicated problem. The Defense Department and the CIA, for instance, must decide whether to follow up on a broader array of signals in predicting and preventing potential terrorist attacks, or instead to focus on those consistent with what they view as the most likely threats. Elite chess players tend to be good at metacognition—thinking about the way they think—and correcting themselves if they don't seem to be striking the right balance.

Strategy Versus Tactics

Computer chess machines, to some extent, get to have it both ways. They use heuristics to prune their search trees, focusing more of their processing power on more promising branches rather than calculating every one to the same degree of depth. But because they are so much faster at calculation, they don't have to compromise as much, evaluating all the possibilities a little bit *and* the most important-seeming ones in greater detail.

But computer chess programs can't always see the bigger picture and think strategically. They are very good at calculating the tactics to achieve some near-term objective but not very strong at determining which of these objectives are most important in the grander scheme of the game.

Kasparov tried to exploit the blind spots in Deep Blue's heuristics by baiting it into mindlessly pursuing plans that did not improve its strategic position. Computer chess programs often prefer short-term objectives that can be broken down and quantized and that don't require them to evaluate the chessboard as a holistic organism. A classic example of the computer's biases is its willingness to accept *sacrifices*; it is often very agreeable when a strong player offers to trade a better piece for a weaker one.

The heuristic "Accept a trade when your opponent gives up the more powerful piece" is *usually* a good one—but not necessarily when you are up against a player like Kasparov and he is willing to take the seemingly weaker side of the deal; he knows the tactical loss is outweighed by strategic gain. Kasparov offered Deep Blue such a trade thirty moves into his first game, sacrificing a rook for a bishop,* and to his delight Deep Blue accepted. The position that resulted, as shown in figure 9-3a, helps to illustrate some of the blind spots that result from the computer's lack of strategic thinking.

FIGURE 9-3A: POSITION AFTER KASPAROV'S 32ND MOVE IN GAME 1

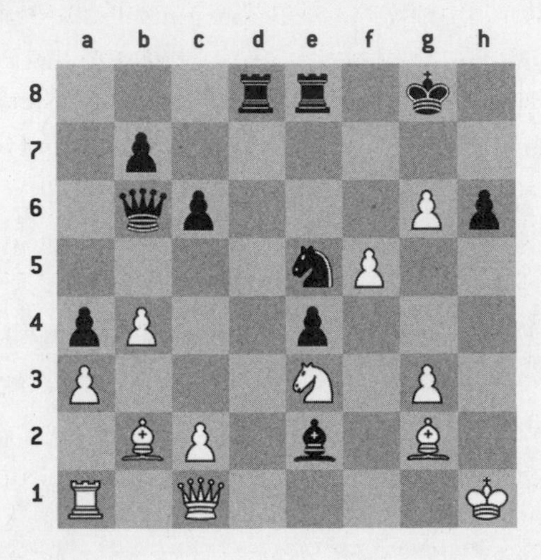

Kasparov and Deep Blue each had their own ways of simplifying the position shown in figure 9-3a. Computers break complicated problems down into discrete elements. To Deep Blue, for instance, the position might look more like what you see in figure 9-3b, with each piece assigned a different point value. If you add up the numbers in this way, Deep Blue had the equivalent of a one-pawn advantage over Kasparov, which converts to a win or draw the vast majority of the time.[28]

FIGURE 9-3B: DISCRETE EVALUATION OF POSITION

* Rooks are typically evaluated by computers (and humans) as being about 60 percent more valuable than bishops, other things being equal.

Humans, instead, are more capable of focusing on the most important elements and seeing the strategic whole, which sometimes adds up to more than the sum of its parts. To Kasparov, the position looked more like what you see in figure 9-3c, and it was a very good position indeed. What Kasparov sees is that he has three pawns advancing on Deep Blue's king, which has little protection. The king will either need to move out of the way—in which case, Kasparov can move his pawns all the way to Deep Blue's home row and promote* them to queens—or it could potentially be mated. Meanwhile, Kasparov's queen and his bishop, although they are in the lower left-hand corner of the board for now, are capable of moving diagonally across it with relatively little obstruction; they are thus adding to the pressure the vulnerable king already faces from the pawns. Kasparov does not yet know exactly how Deep Blue's king will be mated, but he knows that faced with such pressure the odds are heavily in his favor. Indeed, the strength of Kasparov's position would soon become apparent to Deep Blue, which would resign the game thirteen moves later.

FIGURE 9-3C: HOLISTIC EVALUATION OF POSITION

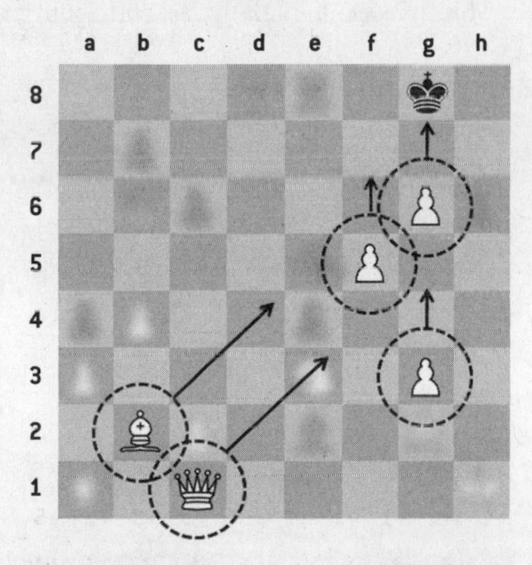

* Promotion in chess refers to what happens when a pawn reaches the opponent's eighth rank. It can be exchanged for a bishop, knight, rook or (as is almost always most advantageous) a second queen.

"Typical computer weakness," Kasparov later said. "I'm sure it was very pleased with the position, but the consequences were too deep for it to judge the position correctly."[29]

"Human Outcalculates a Calculator," trumpeted the headline in the *New York Times*,[30] which published no fewer than four articles about the game the next day.

But the game had not been without a final twist. It was barely noticed by commentators at the time, but it may have altered chess history.

The Beginning of the End

In the final stage of a chess game, the *endgame*, the number of pieces on the board are fewer, and winning combinations are sometimes more explicitly calculable. Still, this phase of the game necessitates a lot of precision, since closing out a narrowly winning position often requires dozens of moves to be executed properly without any mistakes. To take an extreme case, the position illustrated in figure 9-4 has been shown to be a winning one for white no matter what black does, but it requires white to execute literally 262 consecutive moves correctly.*

FIGURE 9-4: A WIN FOR WHITE . . . IN 262 MOVES

* In practice, this position is not winning, because chess employs a rule that automatically calls the game a draw if no pieces have been captured and no pawns have been advanced after fifty consecutive moves.

A human player would almost certainly not solve the position in figure 9-4. However, humans have a lot of practice in completing endgames that might take ten, fifteen, twenty, or twenty-five moves to finish.

The endgame can be a mixed blessing for computers. There are few intermediate tactical goals left, and unless a computer can literally solve the position to the bitter end, it may lose the forest for the trees. However, just as chess computers have databases to cover the opening moves, they also have databases of these endgame scenarios. Literally all positions in which there are six or fewer pieces on the board have been solved to completion. Work on seven-piece positions is mostly complete—some of the solutions are intricate enough to require as many as 517 moves—but computers have memorized exactly which are the winning, losing, and drawing ones.

Thus, something analogous to a black hole has emerged by this stage of the game: a point beyond which the gravity of the game tree becomes inescapable, when the computer will draw all positions that should be drawn and win all of them that should be won. The abstract goals of this autumnal phase of a chess game are replaced by a set of concrete ones: get your queenside pawn to *here*, and you *will* win; induce black to move his rook *there*, and you *will* draw.

Deep Blue, then, had some incentive to play on against Kasparov in Game 1. Its circuits told it that its position was a losing one, but even great players like Kasparov make serious blunders about once per seventy-five moves.[31] One false step by Kasparov might have been enough to trigger Deep Blue's sensors and allow it to find a drawing position. Its situation was desperate, but not quite hopeless.

Instead, Deep Blue did something very strange, at least to Kasparov's eyes. On its forty-fourth turn, Deep Blue moved one of its rooks into white's first row rather than into a more conventional position that would have placed Kasparov's king into check. The computer's move seemed completely pointless. At a moment when it was under assault from every direction, it had essentially passed its turn, allowing Kasparov to advance one of his pawns into black's second row, where it threatened to be promoted to a queen. Even more strangely, Deep Blue resigned the game just one turn later.

What had the computer been thinking? Kasparov wondered. He was used to seeing Deep Blue commit *strategic* blunders—for example, accepting the bishop-rook exchange—in complex positions where it simply couldn't think deeply enough to recognize the implications. But this had been something different: a *tactical* error in a relatively simple position—exactly the sort of mistake that computers *don't* make.

FIGURE 9-5: DEEP BLUE'S PREPLEXING MOVE

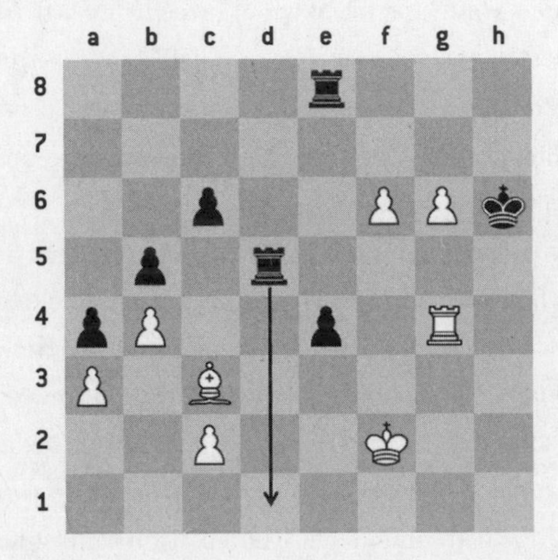

"How can a computer commit suicide like that?" Kasparov asked Frederic Friedel, a German chess journalist who doubled as his friend and computer expert, when they studied the match back at the Plaza Hotel that night.[32] There were some plausible explanations, none of which especially pleased Kasparov. Perhaps Deep Blue had indeed committed "suicide," figuring that since it was bound to lose anyway, it would rather not reveal any more to Kasparov about how it played. Or perhaps, Kasparov wondered, it was part of some kind of elaborate hustle? Maybe the programmers were sandbagging, hoping to make the hubristic Kasparov overconfident by throwing the first game?

Kasparov did what came most naturally to him when he got anxious and began to pore through the data. With the assistance of Friedel and the computer program Fritz, he found that the conventional play—black moving its rook into the sixth column and checking white's king—wasn't such a good move for Deep Blue after all: it would ultimately lead to a checkmate for

Kasparov, although it would still take more than twenty moves for him to complete it.

But what this implied was downright frightening. The only way the computer would pass on a line that would have required Kasparov to spend twenty moves to complete his checkmate, he reasoned, is if it had found another one that would take him longer. As Friedel recalled:

> Deep Blue had actually worked it all out, down to the very end and simply chosen the least obnoxious losing line. "It probably saw mates in 20 and more," said Garry, thankful that he had been on the right side of these awesome calculations.[33]

To see twenty moves ahead in a game as complex as chess was once thought to be impossible for both human beings and computers. Kasparov's proudest moment, he once claimed, had come in a match in the Netherlands in 1999, when he had visualized a winning position some *fifteen* moves in advance.[34] Deep Blue was thought to be limited to a range of six to eight moves ahead in most cases. Kasparov and Friedel were not *exactly* sure what was going on, but what had seemed to casual observers like a random and inexplicable blunder instead seemed to them to reveal great wisdom.

Kasparov would never defeat Deep Blue again.

Edgar Allan Kasparov

In the second game, the computer played more aggressively, never allowing Kasparov into a comfortable position. The critical sequence came about thirty-five turns in. The material was relatively even: each player had their queen, one of their bishops, both of their rooks and seven of their pawns. But Deep Blue, playing white, had slightly the better of it by having the next move and a queen that had ample room to maneuver. The position (figure 9-6) wasn't quite threatening to Kasparov, but there was the *threat of a threat:* the next few moves would open up the board and determine whether Deep Blue had a chance to win or whether the game was inevitably headed for a draw.

FIGURE 9-6: DEEP BLUE'S OPTIONS IN 36TH MOVE OF GAME 2

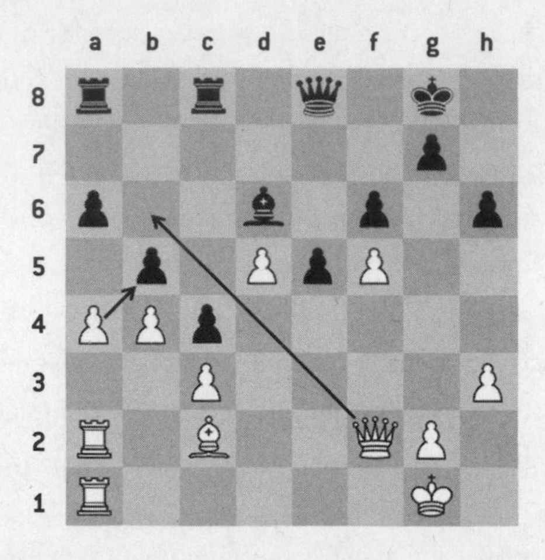

Deep Blue had a couple of moves to consider. It could move its queen into a more hostile position; this would have been the more tactical play. Or it could exchange pawns with white, opening up the left-hand side of the board. This would create a more open, elegant, and strategic position.

The grandmasters commenting on the match uniformly expected Deep Blue to take the first option and advance its queen.[35] It was the somewhat more obvious move, and it would be more in character for Deep Blue: computers prefer busy, complex, computationally demanding positions. But after "thinking" for an unusually long time, Deep Blue instead chose the pawn exchange.[36]

Kasparov looked momentarily relieved, since the pawn swap put less immediate pressure on him. But the more he evaluated the position, the less comfortable he became, biting his knuckles, burying his head in his hands—one audience member thought he caught Kasparov crying.[37] Why had Deep Blue not elected to press forward with its queen? Its actual move was not manifestly inferior—indeed, it's a move he could have imagined one of his flesh-and-blood rivals, like his longtime nemesis Anatoly Karpov, trying under the right conditions. But a computer would need a good tactical reason for it—and he simply couldn't figure out what that reason was. *Unless* his suspicion was correct—Deep Blue was capable of foreseeing twenty or more moves down the road.

Kasparov and Deep Blue played out about eight more turns. To the reporters and experts watching the game, it had become obvious that Kasparov, who had played defensively from the start, had no chance to win. But he might still be able to bring the game to a draw. Then to the surprise of the audience, Kasparov resigned the game after the forty-fifth move. The computer can't have miscalculated, he thought, not when it could think twenty moves ahead. He knew that Deep Blue was going to win, so why deplete his energy when there were four more games left to play?

The crowd in the auditorium burst into robust applause:[38] it had been a well-played game, much more so than the first one, and if Deep Blue's checkmate did not seem quite as inevitable to them as it had to Kasparov, it was surely because they hadn't thought about the position as deeply as he had. But they saved their real admiration for Deep Blue: it had played like a human being. "Nice style!" exclaimed Susan Polgar, the women's world champion, to the *New York Times*.[39] "The computer played a champion's style, like Karpov." Joel Benjamin, a grandmaster who had been assisting the Deep Blue team, agreed: "This was not a computer-style game. This was real chess!"

Kasparov hurried out of the Equitable Center that night without speaking to the media, but he had taken his fellow grandmasters' comments to heart. Perhaps Deep Blue was *literally* human, and not in any existential sense. Perhaps like the Mechanical Turk two centuries earlier, a grandmaster was working surreptitiously to pull its gears behind the scenes. Perhaps Benjamin, a strong player who had once drawn with Kasparov, had not just been coaching Deep Blue but actually intervening on its behalf during the games.

With their minds so powerfully wired to detect patterns, chess champions have a reputation for being slightly paranoid. At a press conference the next day, Kasparov accused IBM of cheating. "Maradona called it the hand of God," he said of the computer's play.[40] It was a coy allusion to the goal that the great Argentinean soccer player Diego Maradona had scored in an infamous 1986 World Cup match against England. Replays later revealed that the ball had been put into the net not off his head, but instead, illegally, off his left hand. Maradona claimed that he had scored the goal "*un poco con la cabeza de Maradona y otro poco con la mano de Dios*"—a little with the head of Maradona and a little with the hand of God. Kasparov, likewise, seemed to think Deep Blue's circuitry had been supplemented with a superior intelligence.

Kasparov's two theories about Deep Blue's behavior were, of course, mutually contradictory—as Edgar Allan Poe's conceptions of the Mechanical Turk had been. The machine was playing much too well to *possibly* be a computer—or the machine had an intelligence so vast that no human had any hope of comprehending it.

Still, quitting the second game had been a mistake: Deep Blue had not in fact clinched its victory, as Friedel and Yuri Dokhoian, Kasparov's most trusted assistant, sheepishly revealed to him over lunch the next day. After playing out the position on Fritz overnight, they had found a line which in just seven more turns would have forced Deep Blue into perpetual check and given Kasparov his tie.* "That was all?" Kasparov said, staring blankly at the traffic on Fifth Avenue. "I was so impressed by the deep positional play of the computer that I didn't think there was any escape."[41]

Although at 1-1 the match was still even, Kasparov's confidence was deeply shaken. He had never lost a competitive chess match in his life; now, he was on the ropes. And making matters worse, he had committed a chess sin of the highest order: resigning a game that he could have played to a draw. It was an embarrassing, unprecedented mistake. The journalists and grandmasters covering the match couldn't recall the last time a champion made such an error.

Kasparov resolved that he wouldn't be able to beat Deep Blue by playing the forceful, intimidating style of chess that had made him World Champion. Instead, he would have to try to trick the computer with a cautious and unconventional style, in essence playing the role of the hacker who prods a program for vulnerabilities. But Kasparov's opening move in the third game, while unusual enough to knock Deep Blue out of its databases, was too inferior to yield anything better than a draw. Kasparov played better in the fourth and fifth games, seeming to have the advantage at points in both of them, but couldn't overcome the gravity of Deep Blue's endgame databases and drew both of them as well. The match was square at one win for each player and three ties, with one final game to play.

On the day of the final game, Kasparov showed up at the Equitable Center looking tired and forlorn; Friedel later recalled that he had never seen him in

---

* A more recent analysis of the second game, conducted in 2007 with a computer that surpasses both Fritz and Deep Blue in skill, suggests that Kasparov may not have been able to draw the position if Deep Blue had played the exact right sequence of moves. Still, Kasparov may well have drawn the game had he played it out in 1997.

such a dark mood. Playing the black pieces, Kasparov opted for something called the Caro-Kann Defense. The Caro-Kann is considered somewhat weak—black's winning percentage with it is 44.7 percent historically—although far from irredeemable for a player like Karpov who knows it well. But Kasparov did not know the Caro-Kann; he had rarely played it in tournament competition. After just a few moves, he was straining, taking a long time to make decisions that were considered fairly routine. And on his seventh move, he committed a grievous blunder, offering a knight sacrifice one move too early. Kasparov recognized his mistake almost immediately, slumping down in his chair and doing nothing to conceal his displeasure. Just twelve moves later—barely an hour into the game—he resigned, storming away from the table.

Deep Blue had won. Only, it had done so less with a bang than an anticlimactic whimper. Was Kasparov simply exhausted, exacerbating his problems by playing an opening line with which he had little familiarity? Or, as the grandmaster Patrick Wolff concluded, had Kasparov thrown the game,[42] to delegitimize Deep Blue's accomplishment? Was there any significance to the fact that the line he had selected, the Caro-Kann, was a signature of Karpov, the rival whom he had so often vanquished?

But these subtleties were soon lost to the popular imagination. Machine had triumphed over man! It was like when HAL 9000 took over the spaceship. Like the moment when, exactly thirteen seconds into "Love Will Tear Us Apart," the synthesizer overpowers the guitar riff, leaving rock and roll in its dust.[43]

Except it wasn't true. Kasparov had been the victim of a large amount of human frailty—and a tiny software bug.

How to Make a Chess Master Blink

Deep Blue was born at IBM's Thomas J. Watson Center—a beautiful, crescent-shaped, retro-modern building overlooking the Westchester County foothills. In its lobby are replicas of early computers, like the ones designed by Charles Babbage. While the building shows a few signs of rust—too much wood paneling and too many interior offices—many great scientists have called it home, including the mathematician Benoit Mandelbrot, and Nobel Prize winners in economics and physics.

I visited the Watson Center in the spring of 2010 to see Murray Campbell, a mild-mannered and still boyish-looking Canadian who was one of the chief engineers on the project since its days as Deep Thought at Carnegie Mellon. (Today, Campbell oversees IBM's statistical modeling department.) In Campbell's office is a large poster of Kasparov glaring menacingly at a chessboard with the caption:

How Do You Make a Computer Blink?
Kasparov vs. Deep Blue
May 3–11, 1997

In the end, Kasparov, not Deep Blue, blinked, although not quite for the reasons that Campbell and his team were expecting.

Deep Blue was designed with the goal of beating Kasparov and Kasparov specifically. The team tried to predict which opening sequences Kasparov was most likely to use and develop strong counterattacks to them. (Kasparov, indeed, averted the trap by playing opening moves that he had rarely used before in tournament competition.) Because of its mediocre performance against Kasparov in 1996 and its problems against like-minded players in training matches, meanwhile, Deep Blue's processing power was doubled and its heuristics were refined.[44] Campbell knew that Deep Blue needed to probe more deeply (but perhaps more selectively) into the search tree to match wits with Kasparov's deep strategic thinking. At the same time, the system was designed to be slightly biased toward complicated positions, which played more to its strengths.

"Positions that are good for computers are complex positions with lots of pieces on the board so there's lots of legal moves available," Campbell told me. "We want the positions where tactics are more important than strategy. So you can do some minor things to encourage that."

In this sense, Deep Blue was more "human" than any chess computer before or since. Although game theory does not come into play in chess to the same degree it does in games of incomplete information like poker, the opening sequences are one potential exception. Making a slightly inferior move to throw your opponent off-balance can undermine months of his preparation time—or months of yours if he knows the right response to it. But most computers try to play "perfect" chess rather than varying their game to match up well against their opponent. Deep Blue instead did what most human players do and leaned into positions where Campbell thought it might have a comparative advantage.

Feature or Bug?

Still, Kasparov's skills were so superior in 1997 that it was really just a matter of programming Deep Blue to play winning chess.

In theory, programming a computer to play chess is easy: if you let a chess program's search algorithms run for an indefinite amount of time, then all $10^{10^{50}}$ positions can be solved by brute force. "There is a well-understood algorithm to solve chess," Campbell told me. "I could probably write the program in half a day that could solve the game if you just let it run long enough." In practice, however, "it takes the lifetime of the universe to do that," he lamented.

Teaching a chess computer how to beat a World Champion, instead, often comes down to a banal process of trial and error. Does allotting the program more time in the endgame and less in the midgame improve performance on balance? Is there a better way to evaluate the value of a knight vis-à-vis a bishop in the early going? How quickly should the program prune dead-looking branches on its search tree even if it knows there is some residual chance that a checkmate or a trap might be lurking there?

By tweaking these parameters and seeing how it played with the changes, Campbell put Deep Blue through many trials. But sometimes it still seemed to make errors, playing strange and unexpected moves. When this happened, Campbell had to ask the age-old programmer's question: was the new move a feature of the program—a eureka moment that indicated it was growing yet more skilled? Or was it a bug?

My general advice, in the broader context of forecasting, is to lean heavily toward the "bug" interpretation when your model produces an unexpected or hard-to-explain result. It is too easy to mistake noise for a signal. Bugs can undermine the hard work of even the strongest forecasters.

Bob Voulgaris, the millionaire basketball bettor I introduced to you in chapter 8, one year decided that he wanted to bet baseball. The simulator he designed consistently recommended "under" bets on the Philadelphia Phillies and the bets weren't doing very well. It turned out that the error came down to a single misplaced character in 10,000 lines of code: his assistant had mistakenly coded the Phillies' home ballpark—Citizens Bank Park, a compact field that boosts offense and home runs—as P-H-l rather than P-H-I. That one line of code had been enough to swamp the signal in his program and tie up Voulgaris's capital in betting on the noise. Voulgaris was so dismayed by the bug that he stopped using his baseball-betting program entirely.

The challenge for Campbell is that Deep Blue long ago became better at chess than its creators. It might make a move that they wouldn't have played, but they wouldn't necessarily know if it was a bug.

"In the early stages of debugging Deep Blue, when it would make a move that was unusual, I would say, 'Oh, there's something wrong,'" Campbell told me. "We'd dig in and look at the code and eventually figure out what the problem was. But that happened less and less as time went on. As it continued to make these unusual moves, we'd look in and see that it had figured out something that is difficult for humans to see."

Perhaps the most famous moves in chess history were made by the chess prodigy Bobby Fischer in the so-called "Game of the Century" in 1956 (figure 9-7). Fischer, just thirteen years old at the time, made two dramatic sacrifices in his game against the grandmaster Donald Byrne—at one point offering up a knight for no apparent gain, then a few moves later, deliberately leaving his queen unguarded to advance one of his bishops instead. Both moves were entirely right; the destruction that Fischer realized on Byrne from the strategic gain in his position became obvious just a few moves later. However, few grandmasters then or today would have considered Fischer's moves. Heuristics like "Never give up your queen except for another queen or an immediate checkmate" are too powerful, probably because they serve a player well 99 percent of the time.

FIGURE 9-7: BOBBY FISCHER'S FAMOUS SACRIFICES (1956)

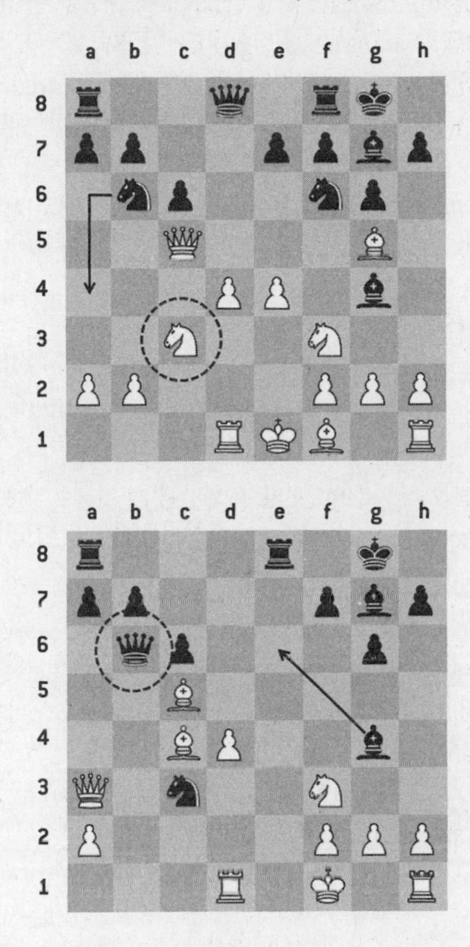

When I put the positions into my midrange laptop and ran them on the computer program Fritz, however, it identified Fischer's plays after just a few seconds. In fact, the program considers any moves *other* than the ones that Fischer made to be grievous errors. In searching through all possible moves, the program identified the situations where the heuristic should be discarded.

We should probably not describe the computer as "creative" for finding the moves; instead, it did so more through the brute force of its calculation speed. But it also had another advantage: it did not let its hang-ups about the right way to play chess get in the way of identifying the right move in those particular circumstances. For a human player, this would have required the creativity and confidence to see beyond the conventional thinking. People marveled at Fischer's skill because he was so young, but perhaps it was for exactly that reason that he found the moves: he had the full breadth of his imagination at his disposal. The blind spots in our thinking are usually of our own making and they can grow worse as we age. Computers have their blind spots as well, but they can avoid these failures of the imagination by at least considering all possible moves.

Nevertheless, there were some bugs in Deep Blue's inventory: not many, but a few. Toward the end of my interview with him, Campbell somewhat mischievously referred to an incident that had occurred toward the end of the first game in their 1997 match with Kasparov.

"A bug occurred in the game and it may have made Kasparov misunderstand the capabilities of Deep Blue," Campbell told me. "He didn't come up with the theory that the move that it played was a bug."

The bug had arisen on the forty-fourth move of their first game against Kasparov; unable to select a move, the program had defaulted to a last-resort fail-safe in which it picked a play *completely at random*. The bug had been inconsequential, coming late in the game in a position that had already been lost; Campbell and team repaired it the next day. "We had seen it once before, in a test game played earlier in 1997, and thought that it was fixed," he told me. "Unfortunately there was one case that we had missed."

In fact, the bug was anything but *unfortunate* for Deep Blue: it was likely what allowed the computer to beat Kasparov. In the popular recounting of Kasparov's match against Deep Blue, it was the second game in which his problems originated—when he had made the almost unprecedented error of forfeiting a position that he could probably have drawn. But what had inspired Kasparov to commit this mistake? His anxiety over Deep Blue's forty-fourth move in the first game—the move in which the computer had moved its rook for no apparent purpose. Kasparov had concluded that the counterintuitive play must be a sign of superior intelligence. He had never considered that it was simply a bug.

For as much as we rely on twenty-first-century technology, we still have

Edgar Allan Poe's blind spots about the role that these machines play in our lives. The computer had made Kasparov blink, but only because of a design flaw.

What Computers Do Well

Computers are very, very fast at making calculations. Moreover, they can be counted on to calculate faithfully—without getting tired or emotional or changing their mode of analysis in midstream.

But this does not mean that computers produce perfect forecasts, or even necessarily good ones. The acronym GIGO ("garbage in, garbage out") sums up this problem. If you give a computer bad data, or devise a foolish set of instructions for it to analyze, it won't spin straw into gold. Meanwhile, computers are not very good at tasks that require creativity and imagination, like devising strategies or developing theories about the way the world works.

Computers are most useful to forecasters, therefore, in fields like weather forecasting and chess where the system abides by relatively simple and well-understood laws, but where the equations that govern the system must be solved many times over in order to produce a good forecast. They seem to have helped very little in fields like economic or earthquake forecasting where our understanding of root causes is blurrier and the data is noisier. In each of those fields, there were high hopes for computer-driven forecasting in the 1970s and 1980s when computers became more accessible to everyday academics and scientists, but little progress has been made since then.

Many fields lie somewhere in between these two poles. The data is often good but not great, and we have some understanding of the systems and processes that generate the numbers, but not a perfect one. In cases like these, it may be possible to improve predictions through the process that Deep Blue's programmers used: trial and error. This is at the core of business strategy for the company we most commonly associate with Big Data today.

When Trial and Error Works

Visit the Googleplex in Mountain View, California, as I did in late 2009, and it isn't always clear when somebody is being serious and when they're joking around. It's a culture that fosters creativity, with primary colors, volleyball courts, and every conceivable form of two-wheeled vehicle. Google people, even its engineers and economists, can be whimsical and offbeat.

"There are these experiments running all the time," said Hal Varian, the chief economist at Google, when I met him there. "You should think of it as more of an organism, a living thing. I have said that we should be concerned about what happens when it comes alive, like Skynet.* But we made a deal with the governor of California"—at the time, Arnold Schwarzenegger—"to come and aid us."

Google performs extensive testing on search and its other products. "We ran six thousand experiments on search last year and probably another six thousand or so on the ad monetization side," he said. "So Google is doing on a rough order of ten thousand experiments a year."

Some of these experiments are highly visible—occasionally involving rolling out a whole new product line. But most are barely noticeable: moving the placement of a logo by a few pixels, or slightly permuting the background color on an advertisement, and then seeing what effect that has on click-throughs or monetization. Many of the experiments are applied to as few as 0.5 percent of Google's users, depending on how promising the idea seems to be.

When you search for a term on Google, you probably don't think of yourself as participating in an experiment. But from Google's standpoint, things are a little different. The search results that Google returns, and the order in which they appear on the page, represent their *prediction* about which results you will find most useful.

How is a subjective-seeming quality like "usefulness" measured and pre-

* Skynet was the evil computer in the *Terminator* series. It gets very mad when you confuse it with HAL 9000.

dicted? If you search for a term like *best new mexican restaurant*, does that mean you are planning a trip to Albuquerque? That you are looking for a Mexican restaurant that opened recently? That you want a Mexican restaurant that serves *Nuevo Latino* cuisine? You probably should have formed a better search query, but since you didn't, Google can convene a panel of 1,000 people who made the same request, show them a wide variety of Web pages, and have them rate the utility of each one on a scale of 0 to 10. Then Google would display the pages to you in order of the highest to lowest average rating.

Google cannot do this for every search request, of course—not when they receive hundreds of millions of search requests per day. But, Varian told me, they do use human evaluators on a series of representative search queries. Then they see which statistical measurements are best correlated with these human judgments about relevance and usefulness. Google's best-known statistical measurement of a Web site is PageRank,[45] a score based on how many other Web pages link to the one you might be seeking out. But PageRank is just one of two hundred signals that Google uses[46] to approximate the human evaluators' judgment.

Of course, this is not such an easy task—two hundred signals applied to an almost infinite array of potential search queries. This is why Google places so much emphasis on experimentation and testing. The product you know as Google search, as good as it is, will very probably be a little bit different tomorrow.

What makes the company successful is the way it combines this rigorous commitment to testing with its freewheeling creative culture. Google's people are given every inducement to do what people do much better than computers: come up with ideas, a *lot* of ideas. Google then harnesses its immense data to put these ideas to the test. The majority of them are discarded very quickly, but the best ones survive.

Computer programs play chess in this way, exploring almost all possible options in at least some depth, but focusing their resources on the more promising lines of attack. It is a very Bayesian process: Google is always at a running start, refining its search algorithms, never quite seeing them as finished.

In most cases, we cannot test our ideas as quickly as Google, which gets feedback more or less instantaneously from hundreds of millions of users around the world. Nor do we have access to a supercomputer, as Deep Blue's engineers did. Progress will occur at a much slower rate.

Nevertheless, a commitment to testing ourselves—actually seeing how well our predictions work in the real world rather than in the comfort of a statistical model—is probably the best way to accelerate the learning process.

Overcoming Our Technological Blind Spot

In many ways, we are our greatest technological constraint. The slow and steady march of human evolution has fallen out of step with technological progress: evolution occurs on millennial time scales, whereas processing power doubles roughly every other year.

Our ancestors who lived in caves would have found it advantageous to have very strong, perhaps almost hyperactive pattern-recognition skills—to be able to identify in a split-second whether that rustling in the leaves over yonder was caused by the wind or by an encroaching grizzly bear. Nowadays, in a fast-paced world awash in numbers and statistics, those same tendencies can get us into trouble: when presented with a series of random numbers, we see patterns where there aren't any. (Advertisers and politicians, possessed of modern guile, often prey on the primordial parts of our brain.)

Chess, however, makes for a happy ending. Kasparov and Deep Blue's programmers saw each other as antagonists, but they each taught us something about the complementary roles that computer processing speed and human ingenuity can play in prediction.

In fact, the best game of chess in the world right now might be played neither by man nor machine.[47] In 2005, the Web site ChessBase.com, hosted a "freestyle" chess tournament: players were free to supplement their own insight with any computer program or programs that they liked, and to solicit advice over the Internet. Although several grandmasters entered the tournament, it was won neither by the strongest human players nor by those using the most highly regarded software, but by a pair of twentysomething amateurs from New Hampshire, Steven Cramton and Zackary "ZakS" Stephen, who surveyed a combination of three computer programs to determine their moves.[48] Cramton and Stephen won because they were neither awed nor intimidated by technology. They knew the strengths and weakness of each program and acted less as players than as coaches.

Be wary, however, when you come across phrases like "the computer thinks the Yankees will win the World Series." If these are used as shorthand for a more precise phrase ("the output of the computer program is that the Yankees will win the World Series"), they may be totally benign. With all the information in the world today, it's certainly helpful to have machines that can make calculations much faster than we can.

But if you get the sense that the forecaster means this more literally—that he thinks of the computer as a sentient being, or the model as having a mind of its own—it may be a sign that there isn't much thinking going on at all. Whatever biases and blind spots the forecaster has are sure to be replicated in his computer program.

We have to view technology as what it always has been—a tool for the betterment of the human condition. We should neither worship at the altar of technology nor be frightened by it. Nobody has yet designed, and perhaps no one ever will, a computer that thinks like a human being.[49] But computers are themselves a reflection of human progress and human ingenuity: it is not really "artificial" intelligence if a human designed the artifice.

10

THE POKER BUBBLE

The year 2003 was the start of the "poker boom," a sort of bubble economy in which the number of new and inexperienced players was growing exponentially and even a modicum of poker skill could be parlayed into large profits. The phenomenon had two immediate and related causes. One was the 2003 World Series of Poker in Las Vegas, which was won by a twenty-seven-year-old amateur, a Nashville accountant with the auspicious name of Chris Moneymaker. Moneymaker was the literal embodiment of the poker everyman: a slightly pudgy office drone who, through a never-ending series of daring bluffs and lucky draws, had turned the $39 he'd paid to enter an online qualifying tournament into a $2.5 million purse.

ESPN turned Moneymaker's achievement into a six-part miniseries, played on nearly continuous repeat on weekday evenings until baseball season finally came along to fill the void. It was terrific advertising for the "sport" of poker, which until that time had a reputation for being seedy, archaic, and intimidating. Suddenly, every balding, five-foot-eight accountant who had long ago given up on his dream of being the next Michael Jordan or Derek Jeter could see in Moneymaker someone who looked just like him, who had a job just like his, and who in a matter of weeks had gone from rank amateur to the winner of the biggest poker tournament in the world.

But the ESPN broadcasts presented a highly sanitized version of what reality actually looks like at the poker table. For one thing, out of the necessity of compressing more than forty hours of play involving more than eight hundred players into six hours of broadcasts, they showed only a small fraction of the hands as they were actually played. What's more, because of the ingenious invention of the "hole cam"—pinhole-size cameras installed around the edge of the table beside each player—the cards of not just Moneymaker but those of each of his opponents were revealed to the home audience as the hand was being played out, giving the audience the feeling of being clairvoyant. Poker is a pretty easy game if you know what cards your opponent holds.

Moneymaker was cast as the protagonist who could do no wrong. Hands that a sober analysis might have concluded he'd played poorly were invariably praised by the announcers—rash bluffs became *gutsy* ones, premature folds became *perceptive* ones. Moneymaker was not some slightly-above-average schmoe getting the cards of his life*[1] but a poker savant who was cunning enough to have developed into a world-class player almost overnight.

The viewer was led to believe that poker is easy to learn, easy to profit from, and incredibly action-packed—none of which is true. But that didn't stop many of them from concluding that only a ticket to Las Vegas separated them from life as the next Chris Moneymaker. The number of participants in the World Series of Poker's $10,000 main event exploded, from 839 the year that Moneymaker won it to 8,773 just three years later.

I was one of those people.[2] I lived the poker dream for a while, and then it died. I learned that poker sits at the muddy confluence of the signal and the noise. My years in the game taught me a great deal about the role that chance plays in our lives and the delusions it can produce when we seek to understand the world and predict its course.

---

* Moneymaker has made "only" about $110,000 per year from poker tournaments since his World Series win, before accounting for his substantial entry fees into tournaments.

FIGURE 10-1: WORLD SERIES OF POKER MAIN EVENT PARTICIPANTS, 1970–2006

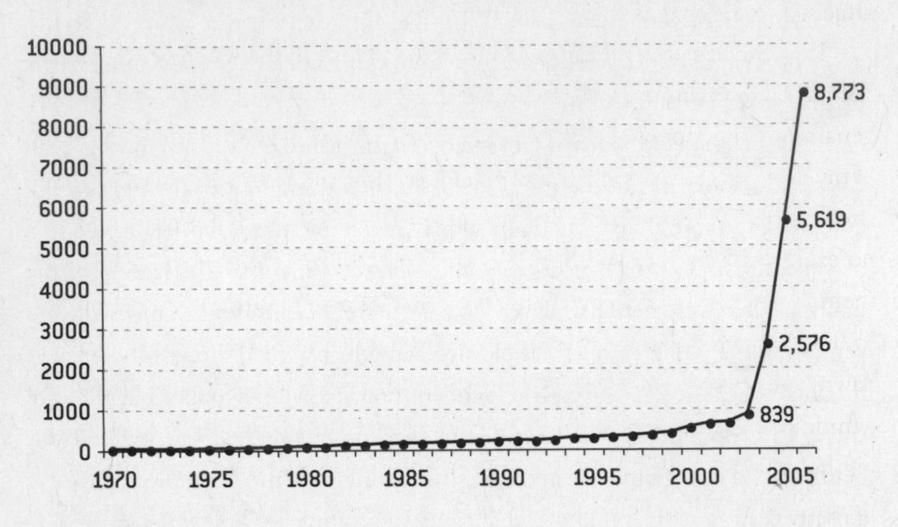

The Start of a Poker Dream

The other catalyst of the poker boom was the Internet. Internet poker had existed in some form since 1998, but it began to go mainstream in 2003 as companies like Party Poker and PokerStars became more aggressive about marketing their way through the legal quagmire of Internet gambling. Players from all around the world flocked to the online games, overcoming their reservations about the security and legality of the virtual cardrooms. They offered a whole host of conveniences: 24/7 access to every form of poker at stakes ranging from pennies a hand to hundreds of dollars; a faster-paced experience (a computer chip can shuffle much faster than a human dealer can—and there's no need to tip it); the comfort of playing from home rather than at a smoky, decrepit casino.

I had a job not unlike Moneymaker's, as I described earlier, working as an economic consultant for the accounting firm KPMG. One of my buddies at work suggested that we start a regular game, gambling just enough to make ourselves nervous. I had a smidgen of poker experience, having made a few four-in-the-morning sojourns to the Soaring Eagle Indian Casino in Mount Pleasant, Michigan. But I was rusty and needed practice and began to look online. A clunky, buggy site named Pacific Poker was making players an offer that was hard to refuse: $25 in real-money poker chips, with almost no strings attached.[3]

I lost the initial $25 fairly quickly, but the players in the Pacific Poker games did not seem much more sophisticated than the mix of ex-convicts and septuagenarians who populated the games at the Soaring Eagle. So I deposited $100 of my own. Almost all professional poker players begin their careers on winning streaks—the ones that lose at first are usually sensible enough to quit—and I was no exception. My bankroll began to grow, by $50 or $100 a night at first and them sometimes by $500 or $1,000. After about three months, my winnings hit $5,000; I began staying up all night to play, taking a cab to work at the crack of dawn and faking my way through the workday. After six months and $15,000 in winnings, I quit my job, leaving the exciting world of international tax consulting behind to split my time between playing cards and working for Baseball Prospectus. It was liberating; I felt as though I'd hacked the system somehow.

I have no idea whether I was really a good player at the very outset. But the bar set by the competition was low, and my statistical background gave me an advantage. Poker is sometimes perceived to be a highly psychological game, a battle of wills in which opponents seek to make perfect reads on one another by staring into one another's souls, looking for "tells" that reliably betray the contents of the other hands. There is a little bit of this in poker, especially at the higher limits, but not nearly as much as you'd think. (The psychological factors in poker come mostly in the form of self-discipline.) Instead, poker is an incredibly mathematical game that depends on making probabilistic judgments amid uncertainty, the same skills that are important in any type of prediction.

How Poker Players Predict Hands

Good poker players are not distinguished by their ability to predict which cards will come next, as though they had ESP. Only the most superstitious or paranoid players believe that the order of the shuffle is anything other than random. Only the very worst ones will have failed to commit the most basic odds calculations to memory: that a flush has about a 1-in-3 chance of coming in with two cards to come, or that a pair of aces will beat a pair of kings about

80 percent of the time. The core analytic skill, rather, is what players call "hand reading": in figuring which cards your opponent might hold, and how they might affect her decisions throughout the rest of the hand.

This is an extremely challenging problem, especially in Texas hold 'em, the most commonly played variant of the game. In hold 'em, the cards are kept facedown and nothing is certain about an opponent's hand until all the bets are in and the pot is pushed someone's way. Each player begins the hand with one of 1,326 possible hand combinations. Everything from a majestic pair of aces to the lowly seven-deuce are among them, and nothing but the player's love of money prevents her from playing the one hand as though it is the other one.

However, players can use their hand-reading skills to develop what amounts to a forecast of their opponent's likely range of hands. Poker players often talk about "putting opponents on a hand," and sometimes they proceed as though they know exactly what two cards the opponent holds. But the best players always entertain numerous hypotheses, which they weigh and balance against the opponent's actions. A good poker forecast is probabilistic. It should become more precise as the hand is played out, but it is usually not possible to predict *exactly* which of the 1,326 possible hands your opponent holds until the cards are revealed, particularly if your opponent is skilled and is deliberately behaving unpredictably.[4]

Indeed, information is so hard to come by in Texas hold 'em that players begin to make estimates about their opponents' range of hands even before any of the cards are dealt. In online games, this is often done through data mining: you'll have statistics on how loose or tight, how passive or aggressive, each opponent's play has been in previous games. In brick-and-mortar casinos, it is done through players' past histories with one another—or, failing that, through what amounts to ethnic profiling. Players from Sweden, Lebanon, and China, for instance, have a reputation for being more aggressive than those from France, England, or India. Younger players are presumed to be looser and more aggressive than older ones. Men are assumed to be more likely to bluff than women. These stereotypes, like any others, are not always true: at the hold 'em games I used to play in at the Bellagio in Las Vegas, the best players were very often women, and they were good in part because they were much more aggressive than their opponents assumed. But poker players don't have the time for political correctness. Even if the stereotype that women play more conservatively than men is false 45 percent of the time, the fact that it might be true 55 percent of the time gives them something to work with.

Once the game begins, these crude assumptions are supplanted by more reliable information: how the player has played previous hands at the table that day and how she is playing this one. The process is fundamentally and profoundly Bayesian, with each player updating their probabilistic assessments after each bet, check, and call. If you doubt the practical uses of Bayes's theorem, you have probably never witnessed a poker game.

A QUICK PRIMER ON TEXAS HOLD 'EM

It is easy to access the rules of Texas hold 'em online or in other books, but I'll point out a few basics for the uninitiated, to introduce you to the terminology used in the next few pages. These rules are simple compared with those of other card games. But much as is the case in chess, the relatively simple rules create a game with exceptional tactical and strategic depth.

The game begins when two personal cards (called down cards or **hole cards**) are dealt facedown to each player. A round of betting ensues at this stage. These personal cards then start to be combined with a series of **community cards** (also called **the board**) that are dealt faceup and shared between all players at the table. Each player seeks to formulate his best five-card poker hand between his down cards and the community cards. The community cards are revealed sequentially, with a round of betting separating each stage. The first three cards are revealed simultaneously and are called the **flop** (one of the many pieces of colorful vocabulary that poker players use). The fourth community card, called the **turn**, is revealed next. Finally the last card, the **river**, is exposed, and a final round of betting takes place. More often than not, all players but one will have folded by this point. If not, the players' down cards are finally flipped faceup and the best hand at the **showdown** wins the pot.

The ranking of poker hands goes thusly:

A straight flush (K♥ Q♥ J♥ T♥ 9♥)

Four-of-a-kind (7♣ 7♥ 7♠ 7♦ 2♥)

A full house (Q♣ Q♥ Q♠ 5♦ 5♣)

A flush (A♠ J♠ 9♠ 4♠ 2♠)

A straight (8♣ 7♦ 6♠ 5♦ 4♥)

Three-of-a-kind or a **set** (9♦ 9♠ 9♥ A♣ 2♥)

Two pair (A♥ A♠ 3♣ 3♦ 7♠)

One pair (K♠ K♥ 9♣ 8♣ 6♦)

An unpaired high card (A♣ Q♠ 8♦ 5♦ 3♠).

Otherwise-tied hands are awarded to the player with the highest-ranking cards: for instance, an ace-high flush beats a 9-high flush. When pairs are of the same rank, the tie is broken by the third-highest card (or **kicker**). For example, the hand (8♥ 8♠ K♣ 7♥ 5♥) beats the hand (8♥ 8♠ Q♦ 7♥ 6♠) since the kicker is a king rather than a queen.

A Not So Simple Poker Hand

Say that you're playing in a $5/$10 no-limit hold 'em game at the Bellagio.* The first few players fold and you find yourself looking at some decent cards, a pair of eights (8♠ 8♣). So you raise to $25 and are called by just one player, a sixty-something man whom we will call the Lawyer.

The Lawyer is a nice guy—a little chatty in between hands, but quiet enough once the cards are dealt. We have learned that he has done fairly well for himself as a partner at an intellectual property law firm on the East Coast. You might imagine him wearing a polo shirt and periodically texting his friend to check on tee times. He had one beer when the cocktail waitress first came around before switching to coffee. He is not likely to be intimidated by these medium-size stakes and isn't acting like it.

---

* The $5 and $10 increments refer to the obligatory bets, called **blind bets,** that rotate around the table and force the action. **No-limit** refers to the fact that players may bet any amount up to the amount sitting in front of them in chips, on any betting round. However, the betting in a no-limit hold 'em game is constrained by the amount that the lesser of the two players has before the hand begins. If you start the hand with $500 in chips and are playing against Bill Gates who has $500,000,000, the most you are putting at risk is $500; any bets that Gates makes above that amount are considered void. So he can't force you to round up an additional $499,999,500 in chips just so that you might call his bluff.

When the Lawyer first sat down at the table with us, we had been torn between two hypotheses about him: first, that he might be playing to show off a little, and therefore might make a few wild moves and bluffs, and second that he might take a more legalistic and by-the-book approach. Our subsequent observation has confirmed that the latter profile is more likely to be correct: he seems to be a mediocre player, avoiding catastrophic errors but without applying much finesse. The Lawyer is by no means the worst player at the table, but he's probably not a long-term winner. Still, we haven't played against him for all that long and aren't completely sure about any of this.

So what do we know about the Lawyer's hand so far? The only thing we know absolutely without doubt is that he cannot hold any hands that contain the 8♠ or the 8♣, since those cards are in our hand. Unfortunately, that reduces the number of starting hand combinations only to 1,225 from 1,326, each of which was equally likely to be dealt when the hand began.

However, the Lawyer has given us some other information: he has chosen to call our bet. That means his hand is likely to be at least decent: most players, including the Lawyer, fold a clear majority of their hands against a raise before the flop. It also means he is unlikely to have an *extremely* strong hand like a pair of aces, since he called not just in lieu of folding but also in lieu of reraising, although this possibility cannot be discounted completely.*

We can start to form a probabilistic, Bayesian assessment of what hands the Lawyer might hold. From past experience with players like this, we know that his range of hands is likely to include pairs like 9♦ 9♠. It might include some hands with aces, especially if both cards are of the same suit (like the hand A♥ 5♥), meaning that it can more easily make flushes. Likewise, it might include what are called "suited connectors"—hands like 6♠ 5♠ that have consecutive cards of the same suit and which can make both flushes and straights. Finally, he might call with hands that contain two face cards like the king of clubs and the jack of diamonds (K♣ J♦).

If we had enough time, we could enumerate the Lawyer's hand range completely, with all 1,326 possibilities assigned a probability from 0 percent to 100 percent given his action so far, as in figure 10-2a. This is how a com-

---

* Occasionally a player might call rather than raise with a hand like a pair of aces for purposes of deception.

puter, which can sort through the possibilities very quickly, might think about his play.

FIGURE 10-2A: PROBABALISTIC REPRESENTATION OF OPPONENT'S HAND RANGE

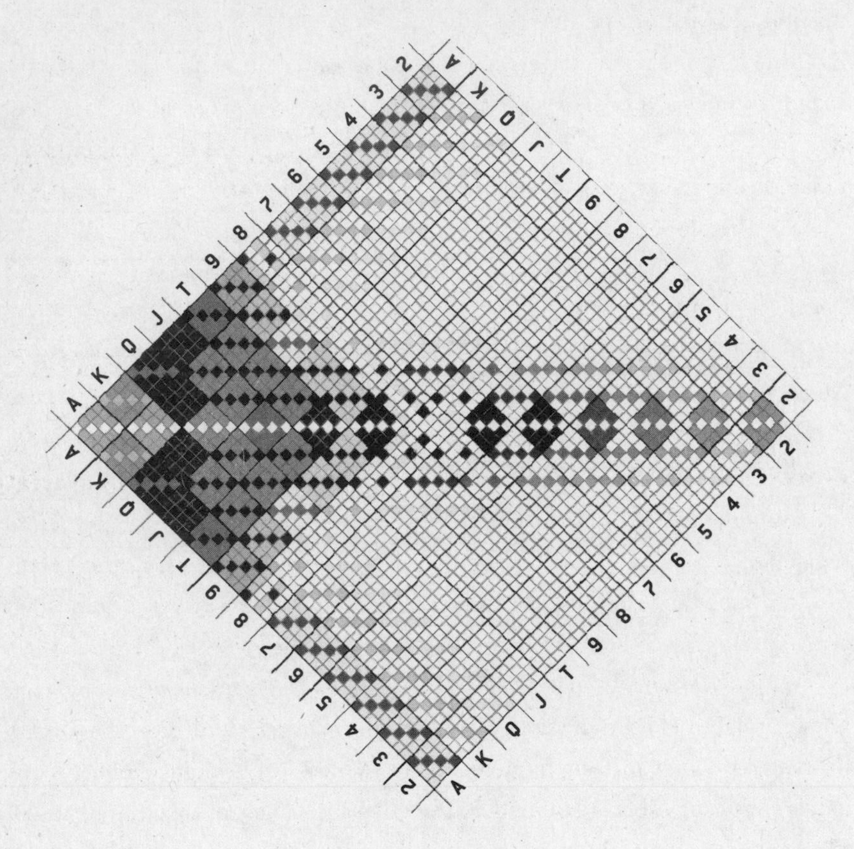

A matrix like this is far too complicated to keep track of under real-world conditions, however, so what players seek to do instead is break their opponent's range into groups of hands that might function in about the same ways as one another (figure 10-2b). In this case, the group of hands that would most concern us is if the Lawyer began the hand with a pair higher than our eights.

Fortunately, the probability of this is low: a player is blessed with a pair only rarely in hold 'em. Indeed, when the cards are first dealt in hold 'em, the chance of starting out with a pair of nines or better is only about 3 percent. However, we need to update our estimate of this probability because of his call: the Lawyer is throwing out many of his worthless hands, increasing the chance that he has a strong one. According to our estimate, the chance he holds a pair higher than our eights has already risen to about 6 percent given how he has played the hand thus far.

FIGURE 10-2B: OPPONENT'S HAND GROUPINGS BEFORE FLOP

Hand Type	Example Hand	Prior Probability Before Call	Posterior Probability After Call
Pair, 8s through As	J♥ J♣	3%	6%
Pair, 2s through 7s	6♠ 6♦	3%	12%
Suited aces	A♥ 9♥	4%	15%
Unsuited aces	A♠ Q♥	11%	16%
Suited face cards	Q♠ J♠	2%	10%
Unsuited face cards	K♣ T♠	5%	16%
Suited connectors	7♦ 6♦	2%	8%
Misc. hands	J♣ 8♣	70%	17%

The other 94 percent of the time, the Lawyer starts out with a worse hand than ours. The problem is that there are five cards still left to come, and while it is relatively hard for them to improve our pair (we would need to catch one of the two remaining eights in the deck), he could more easily make a higher pair, straight, or flush.

The Lawyer takes a swig of coffee as the dealer arranges the flop cards on the center of the table. They consist of two clubs—a king and a three—along with the nine of hearts.

K♣ 9♥ 3♣

These cards have not improved our hand. Our hope is that they have not improved the Lawyer's either, in which case our pair of eights will still be best. So we make a relatively modest bet of $35 into the $65 pot. The Lawyer pauses for a moment and calls.

This call is not great news for us, as we can see by further refining our estimate of his hand range. The key, following Bayes's theorem, is to think in terms of conditional probabilities. For instance, if the Lawyer started with a hand like K♣ J♦, now giving him a pair of kings, how likely is it that he'd call us again? (He'd almost certainly at least call having made top pair, but might he have raised instead?) What about if he began the hand with a smaller pair than ours, like 7♥ 7♠—how likely is it that he'd call rather than fold? If we had the luxury of time, we could go through all 1,326 hand combinations individually and revise our estimates accordingly (figure 10-3).

Our actual estimates at the table will not be quite as precise as this. Still, we can come to some broad probabilistic characterizations of his hand range given his call. About 30 percent of the time, the Lawyer's hand connected strongly with the flop and he has a pair of kings or better—a good hand that is unlikely to fold without a lot of pressure. There is also about a 20 percent chance that he has a pair worse than kings but better than our eights. These hands beat ours as well, but the Lawyer might be more likely to fold them later if we keep betting aggressively.

Meanwhile, there is about a 25 percent chance that the Lawyer has a draw to a straight or a flush, which is behind us for now but has many ways to improve. Finally, there is a 25 percent chance that he has a worse pair than ours or has almost nothing and is continuing in the hand solely in the hope of bluffing us later; these are the most favorable cases.

You can see how complicated poker decisions become. Some of these possibilities imply that we should continue to bet our hand as aggressively as possible. Others imply we should take a more cautious approach, while still others would mean we should be preparing to fold.

Just as we are contemplating this thorny decision, the dealer puts out the ideal card and makes our life easy. It is one of the two remaining eights in the deck, the 8♦, giving us three of a kind. The only way we are beaten is if the Lawyer started out with a pair of nines or a pair of kings and made a higher set on the flop, and has played them passively in order to trap us. (Poker players term this "slowplaying.") Still, we should not be thinking so defensively. In sorting through his possible hands, we should have the better one perhaps 98 percent of the time. So this time we make a relatively large bet: $100 into the $135 pot.

FIGURE 10-3: OPPONENT'S HAND GROUPINGS ON FLOP

Hand Type	Example Hand	Prior Probability Before Call	Posterior Probability After Call
Set (Ks, 9s, 3s)	9♦ 9♠	2%	2%
Two pair	K♥ 9♥	2%	2%
One pair, kings or better	K♠ J♦	15%	28%
One pair, 8s through Qs	9♦ 8♦	13%	20%
One pair, 7s or worse	A♠ 3♠	15%	20%
Flush draw, no pair	8♣ 6♣	6%	12%
Straight draw (no flush draw)	Q♦ T♠	12%	12%
Other unpaired hand	A♥ J♣	35%	4%

The Lawyer calls us once more. Because he is likely to have folded out his weaker pairs and weaker draws, we can now narrow his hand range even further. In fact, of the 1,326 hands that he might have started out with, not more than about seventy-five are all that likely at this stage. Often, he will have a pair of kings, a hand that we were previously worried about but now beat. To the extent we are concerned, it's mostly about another club coming, which could still give him a flush.

FIGURE 10-4: OPPONENT'S HAND GROUPINGS ON TURN

Hand Type	Example Hand	Prior Probability Before Call	Posterior Probability After Call
Set of Ks or 9s	9♦ 9♠	2%	1%
Set of 3s	3♥ 3♣	1%	1%
Two pair	K♣ 9♥	2%	2%
One pair, kings or better	K♥ J♠	28%	45%
One pair, 3s through Qs	9♦ 7♦	40%	33%
Flush draw, no pair	A♣ 2♣	12%	14%
Straight draw (no flush draw)	J♦ T♠	13%	3%
Other unpaired hand	A♠ Q♥	3%	1%

Instead, the final card is the harmless-looking five of spades, which does not complete the flush:

K♣ 9♥ 3♣ 8♦ 5♠

We bet $250 into the $335 pot, hoping that the Lawyer will call us with a worse hand. Suddenly, however, he springs to life. "I'm all-in," he says to the dealer in a voice just barely louder than a whisper. He neatly pushes his remaining chips—about $1,200—into the pot.

What the hell just happened? We now need to put our Bayesian thinking skills to the test. If our forecast of his hand range is off, we could easily make a $1,200 mistake.

We look at the board and realize there is one *exact* hand, one from among the 1,326 random combinations, that seems *most* consistent with his play. The specific hand is a seven and a six of clubs (7♣ 6♣). It is a suited connector, so we think he would have called with it before the flop. On the flop, this hand made a flush draw with four clubs, and we didn't bet enough to push him off it. On the turn, the hand missed its flush but nevertheless became stronger: the 8♦ that made our hand three-of a-kind gave the Lawyer the possibility of making a straight with any ten or five. If that was indeed his hand, the 5♠ on the river made his straight, which beats our three-of-a-kind and would explain why he is now betting so boldly.

So should we fold? Even if you have never played poker, it is worth pausing for a moment to consider what you'd do.

The answer is that you should very probably *not* fold. In fact, against many players, you should be pleased that more money is going into the pot.

The solution comes from Bayes's theorem. It's true that the all-in bet is an extremely powerful move—it conveys much more information than his calls before. But before the Lawyer went all-in, we would have assigned a very low probability—perhaps 1 percent—to his holding *exactly* the seven and six of clubs, just one possible hand out of the myriad combinations. Unless we are extremely confident that the 7♣ 6♣ is about the *only* hand that he'd make this play with, folding could be a huge mistake. Our hand only needs to be good about 35 percent of the time to make the call correct mathematically.

In fact, there are some alternate possibilities for his hand. The Lawyer could have a set of threes or possibly a set of fives, which still lose to our set of eights. He could plausibly have made two-pair with a hand like K♥ 5♥. Some players would play a pair of aces in this way. In *his* Bayesian model of *our* hand range, the Lawyer might reasonably figure that hands like this are better than ours even though they are not—good enough to go all-in—and he might be willing to get a lot of money in with them.

There are also a couple of hands apart from the straight which would beat us. If the Lawyer was slowplaying a set of nines the whole way, or a set of kings, he'll now get our money. This is counterbalanced by the possibility of a complete bluff. If the Lawyer had a flush draw that missed, the only way he can win the pot is by bluffing at it.

As Arthur Conan Doyle once said, "Once you eliminate the impossible, whatever remains, no matter how improbable, must be the truth." This is sound logic, but we have a lot of trouble distinguishing the impossible from the highly improbable and sometimes get in trouble when we try to make too fine a distinction. *All* of the opponent's hands are in some way highly improbable at this stage; this has been an unusual hand. It is a matter of weighing improbabilities against other improbabilities, and the calculation weighs against the hypothesis that he has 7♣ 6♣ exactly. If we ran the possibilities through a computer, it might think there's something like a two-thirds probability that we still have the best hand (figure 10-5).

FIGURE 10-5. OPPONENT'S HAND GROUPINGS ON RIVER

Hand Type	Example Hand	Prior Probability Before All-In Raise	Posterior Probability After All-In Raise
Straight	7♣ 6♣	1%	16%
Set of Ks or 9s	9♦ 9♠	2%	17%
Set of 5s or 3s	5♥ 5♣	2%	19%
Two pair	K♥ 5♥	3%	20%
One pair, Ks or better	A♠ A♦	44%	15%
One pair, 8s through Qs[5]	8♥ 7♥	35%	4%
No pair (pure bluff)	7♠ 2♥	13%	9%

In practice, poker players might differ a lot in how they assess the probabilities for his hand. Skilled poker players are probably better than 99.9 percent of the population at making reasonably good probabilistic judgments under uncertainty. In fact, I don't know of a single game or intellectual exercise that better refines these skills. However, when I posted this hand on Two Plus Two, an online forum for professional poker players, assessments ranged from that we were nearly certain to have the best hand to that we were nearly certain to be beat.[6] My view is that both these assessments are overconfident. We should not proceed as though we don't know *anything* about the opponent's hand, but in general our predictive errors come in thinking that there is more certainty in the world than there really is. In this case, seeking to put the opponent on an *exact* hand would imply a fold, while a fuller assessment of the probabilities—coupled with the favorable adds from the pot—means that we should call instead.

Schrödinger's Poker Hand

If this hand came up in a televised tournament on ESPN that showed us each player's hole cards, the analysis from the commentators might be different. They might assert that the fold was obvious if they knew that the opponent held 7♣ 6♣. In a parallel universe where the hand played out exactly the same way but the opponent had 3♥ 3♠ instead, they'd tell us how thrilled we should be to get more money into the pot.

In a televised game in 2009, two world-class players, Tom Dwan and Phil Ivey, played a hand in which the pot size eventually reached more than a million dollars.[7] In the hand, Ivey caught a miracle card on the turn to make him a 5-high straight. Unfortunately, the same card also gave Dwan a 7-high straight,* the only possible better hand. "If anybody can get away from this, it's Phil Ivey," one of the announcers said, implying that it would be a sign of superior poker talent if he folded. In fact, throwing away the hand would have been a terrible play. Given what Ivey knew at the time, and how aggressively he and

* Ivey's hand was A♣ 2♦ while Dwan's was 7♥ 6♥. The flop was J♣ 3♦ 5♣ and the turn was the 4♥, making Ivey a 5-high straight and Dwan a 7-high straight.

Dwan play, he should have expected to have the best hand at least 90 percent of the time. If Ivey *hadn't* lost all his chips on the hand, he would have been playing badly.

While television coverage has been a great boon to poker, it leaves many casual players with misleading impressions about the right way to play it, focusing too much on the results and not enough on the correct decision-making process.

"It's not very common that you can narrow someone's holdings down to one hand," Dwan told me. "Definitely much less common than most pros and TV shows would have you believe."

Making Ourselves Unpredictable

Dwan was once better known by his online screen name "durrrr," which he selected because he figured it would put the other players on tilt if they lost to him. Dwan deposited $50 at the online site Full Tilt Poker at age seventeen, later dropping out of Boston College to play poker full-time.[8] He rose through the ranks to become the apex predator in the online-poker food chain.[9] Millions of dollars flowed through him each month; sometimes he lost it but more often he won.[10]

By the time I spoke with him in 2012, Dwan was widely considered among the best no-limit hold 'em players in the world.[11] He has a reputation for being creative, aggressive, and most of all, fearless. In 2009, he challenged any player in the world, except for his close friend Phil Galfond, to play him head-to-head at very favorable odds. Three strong players eventually took him on and Dwan won two of these matches.

Yet for all his apparent bravado—Dwan is fairly low-key in person—[12] his approach to thinking about poker and the world in general is highly probabilistic. He profits because his opponents are too sure of themselves. "It's important in most areas of life to come up with a probability instead of a yes or no," he told me. "It's a huge flaw that people make in a lot of areas that they analyze, whether they're trying to form a fiscal union, pay for groceries, or hoping that they don't get fired."

Dwan seeks to exploit these tendencies by deliberately obfuscating his play. If the most important technical skill in poker is learning how to forecast your opponent's hand range, the next-most-important one is making your own play unpredictable. "The better people are, the less certain you're going to be about what they have or what they're doing or what their range is," Dwan says. "And they'll be more apt to manipulate that to take advantage of your judgments."

While I'll never be the player that Dwan is, I took advantage of this in my own way during my days as a poker professional. In the soft online games of the mid-2000s, I could make money by playing conservative, tight poker, but I soon discovered that a more aggressive style could make me even more. The idea was to find the blind spots that my opponents might have in estimating my hand range.

When you raise before the flop, for instance, the opponent will typically put you on big cards like those containing aces, kings, and queens. You will have those hands sometimes, of course. But I would also raise with hands like the ones we were worried about the Lawyer having, hands with small cards like 7♣ 6♣. What I found is that when big cards came on the board, like an ace or king, the opponent would often give me credit for catching those cards and fold. If smaller cards came instead, meanwhile, I'd often have made a pair or some kind of good draw. Sometimes, I'd even make an unlikely-seeming hand like a straight with these cards, which could send my opponents into tilt. One interesting thing about poker is that the very best players and the very worst ones both play quite randomly, although for different reasons.* Thus, you can sometimes fool opponents into thinking you are a weak player even if you are likely to take their money.

Eventually, some of my opponents caught on to my more aggressive style, but this wasn't all bad. It meant that they were more likely to call down when I did have a "predictable" hand like a pair of kings, making these hands more profitable for me.

In fact, bluffing and aggressive play is not just a luxury in poker but a necessity—otherwise your play is just too predictable. Poker games have be-

---

* Whereas the very best players deliberately insert some randomness into their play to make it harder to read their hands, the worst ones may play randomly because they don't know what the right play is to begin with.

come *extremely* aggressive since I stopped playing regularly five years ago, and game theory[13] as well as computer simulations[14] strongly suggest this is the optimal approach. Blitzing your opponent with a deluge of possibilities is the best way to complicate his probability calculations.

Sometimes you may also be able to identify situations where your opponents' intuitive estimates of the probabilities are too crude. Whenever a poker player thinks that his opponent might *never* play a certain hand in a certain way—*never* bluff in a certain situation, for instance—that's when you have the opportunity to exploit him by confusing his sense of the improbable and the impossible.

"There were a bunch of things I did that I knew were extremely suboptimal but made me a really large amount of money for a long portion of time," Dwan told me. "It's only in the last few years people finally started realizing and getting better."

Dwan's main game, no-limit hold 'em, is especially fertile for such a strategy because you potentially control, through the size of your bets, the amount of money at stake on each decision. Some choices that Dwan makes involve no more than $100, while others might be for stakes of $10,000, $100,000 or even more. Make a few extra decisions right in million-dollar pots, and the collective sum of what you do for $100 at a time hardly matters at all.

I mostly played limit hold 'em instead, where the betting increment is fixed on each round. (Until very recently, this was the most popular game outside of tournaments; ten years ago, there were often no more than two or three no-limit games running anywhere in the United States.[15]) Limit poker offers fewer opportunities for creativity. Still, until practice caught up with theory, I had a couple of very successful years by exploiting an aggressive approach. In both 2004 and 2005, I made an income from poker in the six figures, with my cumulative profits from the game peaking at about $400,000 overall.

The Prediction Learning Curve

The difference between Dwan and me is that, while he is willing to take on almost literally any other player for any stakes at any time, I was merely in the upper middle class of poker players and needed to be in a game with some

FIGURE 10-6: THE PARETO PRINCIPLE OF PREDICTION

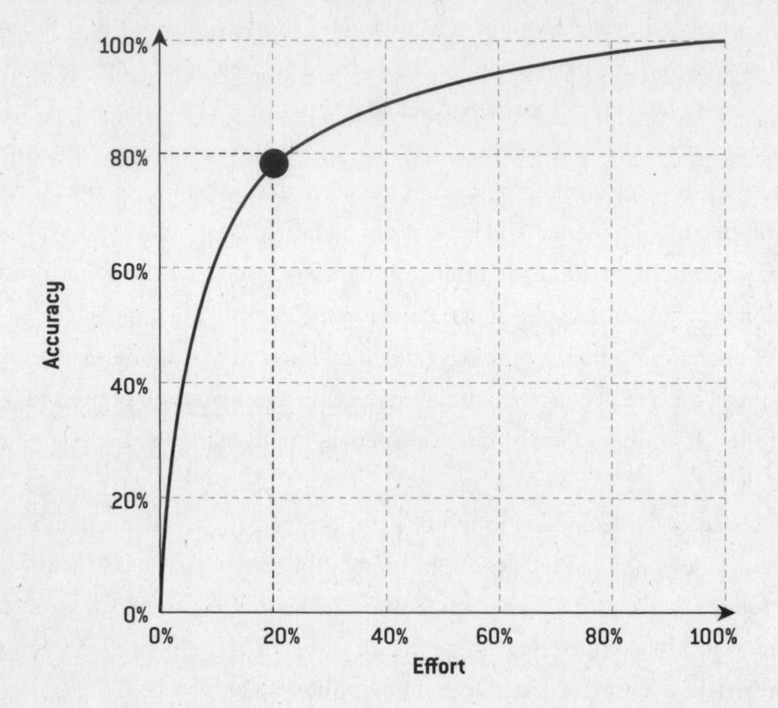

bad ones to be a favorite to make money. Fortunately, there were plenty of these bad players—what poker players call fish—during the poker boom years.

There is a learning curve that applies to poker and to most other tasks that involve some type of prediction. The key thing about a learning curve is that it really is a curve: the progress we make at performing the task is not linear. Instead, it usually looks something like this (figure 10-6)—what I call the Pareto Principle of Prediction.

What you see is a graph that consists of *effort* on one axis and *accuracy* on the other. You could label the axes differently—for instance, *experience* on the one hand and *skill* on the other. But the same general idea holds. By effort or experience I mean the amount of money, time, or critical thinking that you are willing to devote toward a predictive problem. By accuracy or skill I mean how reliable the predictions will prove to be in the real world.

The name for the curve comes from the well-known business maxim called the Pareto principle or 80-20 rule (as in: 80 percent of your profits come from 20 percent of your customers[16]). As I apply it here, it posits that getting a few basic things right can go a long way. In poker, for instance, simply learning to fold your worst hands, bet your best ones, and make some effort to consider what your opponent holds will substantially mitigate your losses. If you are willing to do this, then perhaps 80 percent of the time you will be making the same decision as one of the best poker players like Dwan—even if you have spent only 20 percent as much time studying the game.

This relationship also holds in many other disciplines in which prediction is vital. The first 20 percent often begins with having the right data, the right technology, and the right incentives. You need to have some information—more of it rather than less, ideally—and you need to make sure that it is quality-controlled. You need to have some familiarity with the tools of your trade—having top-shelf technology is nice, but it's more important that you know how to use what you have. You need to care about accuracy—about getting at the objective truth—rather than about making the most pleasing or convenient prediction, or the one that might get you on television.

Then you might progress to a few intermediate steps, developing some rules of thumb (heuristics) that are grounded in experience and common sense and some systematic process to make a forecast rather than doing so on an ad hoc basis.

These things aren't exactly *easy*—many people get them wrong. But they aren't hard either, and by doing them you may be able to make predictions 80 percent as reliable as those of the world's foremost expert.

Sometimes, however, it is not so much how good your predictions are in an absolute sense that matters but *how good they are relative to the competition.* In poker, you can make 95 percent of your decisions correctly and still lose your shirt at a table full of players who are making the right move 99 percent of the time. Likewise, beating the stock market requires outpredicting teams of investors in fancy suits with MBAs from Ivy League schools who are paid seven-figure salaries and who have state-of-the-art computer systems at their disposal.

In cases like these, it can require a *lot* of extra effort to beat the competition. You will find that you soon encounter diminishing returns. The extra experience that you gain, the further wrinkles that you add to your strategy, and the additional variables that you put into your forecasting model—these will only make a marginal difference. Meanwhile, the helpful rules of thumb that you developed—now you will need to learn the exceptions to them.

However, when a field is highly competitive, it is only through this painstaking effort around the margin that you can make any money. There is a "water level" established by the competition and your profit will be like the tip of an iceberg: a small sliver of competitive advantage floating just above the surface, but concealing a vast bulwark of effort that went in to support it.

I've tried to avoid these sorts of areas. Instead, I've been fortunate enough to take advantage of fields where the water level was set pretty low, and getting the basics right counted for a lot. Baseball, in the pre-*Moneyball* era, used to be one of these. Billy Beane got an awful lot of mileage by recognizing a few simple things, like the fact that on-base percentage is a better measure of a player's offensive performance than his batting average. Nowadays pretty much everyone realizes that. In politics, I'd expect that I'd have a small edge at best if there were a dozen clones of FiveThirtyEight. But often I'm effectively "competing" against political pundits, like those on *The McLaughlin Group*,

FIGURE 10-7: THE PARETO PRINCIPLE OF PREDICTION IN COMPETITIVE ENVIRONMENTS

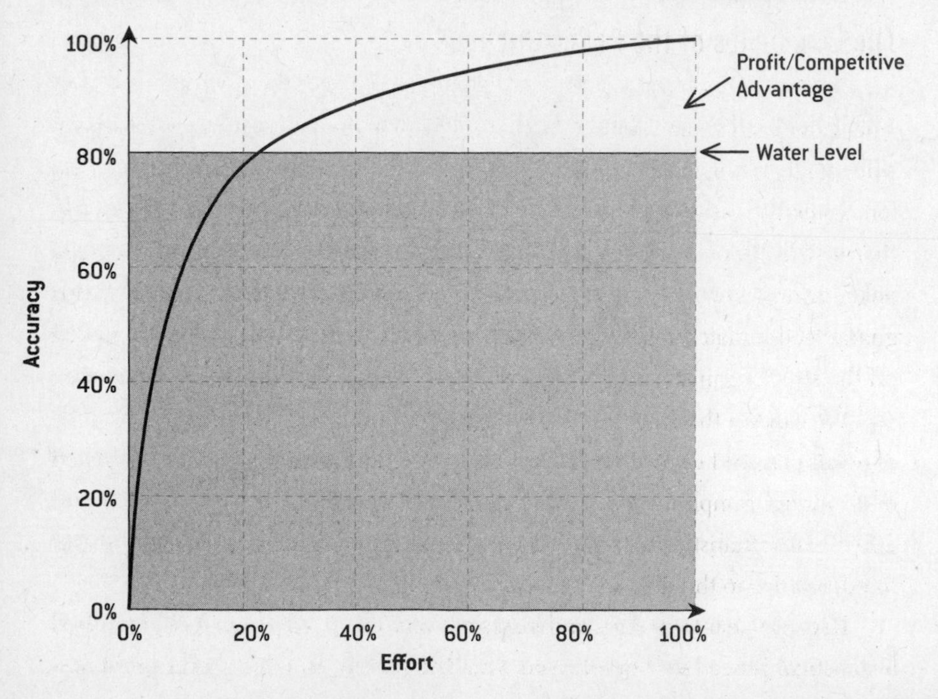

who aren't really even trying to make accurate predictions. Poker was also this way in the mid-2000s. The steady influx of new and inexperienced players who thought they had learned how to play the game by watching TV kept the water level low.

If you have strong analytical skills that might be applicable in a number of disciplines, it is very much worth considering the strength of the competition. It is often possible to make a profit by being pretty good at prediction in fields where the competition succumbs to poor incentives, bad habits, or blind adherence to tradition—or because you have better data or technology than they do. It is *much* harder to be *very* good in fields where everyone else is getting the basics right—and you may be fooling yourself if you think you have much of an edge.

In general, society does need to make the extra effort at prediction, even though it may entail a lot of hard work with little immediate reward—or we need to be more aware that the approximations we make come with trade-offs. But if you're approaching prediction as more of a business proposition, you're usually better off finding someplace where you can be the big fish in a small pond.

The Economics of the Poker Bubble

The Pareto Principle of Prediction implies that the worst forecasters—those who aren't getting even the first 20 percent right—are much worse than the best forecasters are good. Put another way, average forecasters are closer to the top than to the bottom of the pool. I'm sure that I'd lose a ton of money if I played poker against Dwan. But I'd gladly play him if, as part of the deal, I were also guaranteed a match for the same stakes against some random person I picked off the street, against whom I'd expect to make back my losses and then some.

We can test this hypothesis empirically by examining the statistical records of poker players. I evaluated the data from an online poker site, which consisted of a random sampling of no-limit hold 'em players over a period in 2008 and 2009. These statistics told me how much money the players won or lost per hand, relative to the stakes they were playing.[17]

Because near-term wins and losses are very much subject to luck, I applied a statistical procedure[18] to estimate what the players' true long-term profitability was. I then ordered the players by their skill level and broke them down into ten equal-size quadrants. The top quadrant—consisting of the top 10 percent of the player pool*—corresponds to the best player at a typical ten-person table.[19] The bottom 10 percent, meanwhile, are the biggest fish.

Figure 10-8a represents my estimate of how skilled the players in each quadrant really are, measured as money won or lost per one hundred hands in a no-limit hold 'em game with $5/$10 blinds. The figures include both money won and lost to the other players and that lost to the casino, which either takes a small percentage of each pot (known as the rake) or charges an hourly fee for dealing the game.[20]

FIGURE 10-8A: ESTIMATED MONEY WON OR LOST PER 100 HANDS IN A $5/$10 NO-LIMIT HOLD 'EM GAME

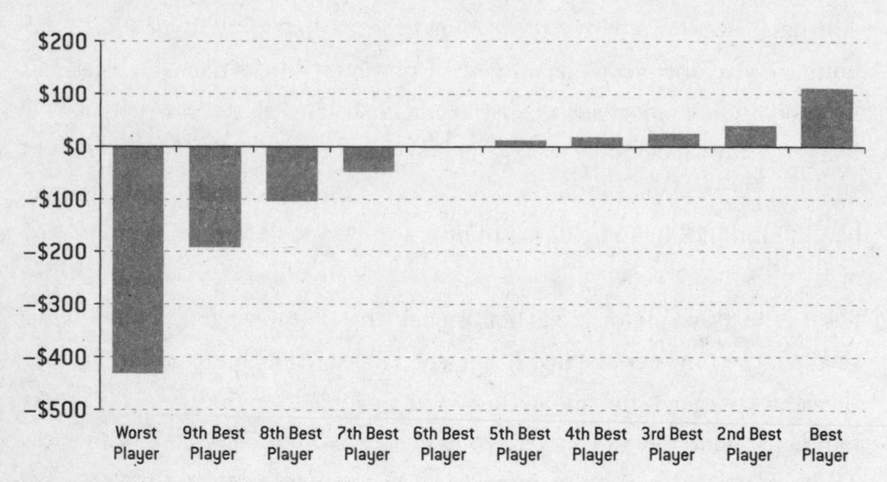

I estimate that the very best player at the table in one of these games is averaging a profit of about $110 per one hundred hands played over the long run.

* My analysis of the players was weighted by the number of hands that each player was dealt. In poker, a very large proportion of the total hands are dealt to a very small percentage of the total players, those who play the game every day rather than once every month or every year. In fact, the online poker environment is a hyperbolized version of the 80-20 rule: about 80 percent of the hands in the database were dealt to just 2 percent of the total players. Since you are much more likely to encounter a player from the 2 percent of prolific players than if you selected at random from all players who had played even one hand, the analysis would create an unrealistic impression of the economy of a poker table if not conducted in this way.

That's a nice wage in an online casino, where hands are dealt very quickly and you could get almost that many hands during an hour or two.* It's less attractive in a traditional casino, where it might take four hours to play the same number of hands, and translates to wage of $25 or $30 per hour.

The key insight, however, is that the worst players at the table are losing money *much* faster than even the best ones are making it. For instance, I estimate that the worst player in the game—the biggest fish—was losing at a rate of more than $400 per one hundred hands. This player is so poor that he would literally be better off folding every hand, which would cost him only $150 per one hundred hands instead.

Here you see the statistical echo of the 80/20 rule: there's a much larger difference between the very worst players and the average ones than between the average ones and the best. The better players are doing just a few things differently from one another, while those at the lower end of the curve are getting even the basics wrong, diverging wildly from optimal strategy.

In the classic poker movie *Rounders*,[21] Matt Damon's character advises us that if you can't spot the sucker in your first half hour at the table, then you must be the sucker. I don't think this is quite true: it may be that the game doesn't have any suckers. It is emphatically the case, however, that if you can't spot one or two bad players in the game, you probably shouldn't be playing in it. In poker, the line between success and failure is very thin and the presence of a single fish can make the difference.

In the game I just described, the one fish was feeding a lot of hungry mouths. His presence was worth about $40 per 100 hands to the other players. That subsidy was enough that about half of them were making money, even after the house's cut. Poker abides by a "trickle up" theory of wealth: the bottom 10 percent of players are losing money quickly enough to support a relatively large middle class of break-even players.

But what happens when the fish—the sucker—busts out, as someone losing money at this rate is bound to do? Several of the marginally winning players

* Online players often also "multitable," meaning play in more than one game at once, which further multiplies their win rate. This is physically impossible (or at least disallowed) in a brick-and-mortar casino.

FIGURE 10-8B: ESTIMATED MONEY WON OR LOST PER 100 HANDS IN A $5/$10 NO-LIMIT HOLD 'EM GAME AFTER FISH BUSTS OUT

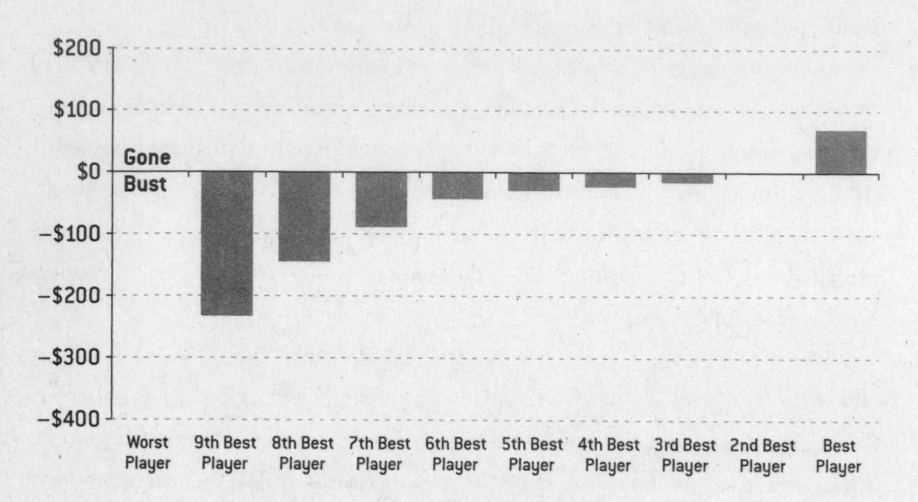

turn into marginally losing ones (figure 10-8b). In fact, we now estimate that *only* the very best player at the table is still making money over the long run, and then less than he did before.

What's more, the subtraction of the fish from the table can have a cascading effect on the other players. The one who was formerly the next-to-worst player is now the sucker, and will be losing money at an even faster rate than before. So he may bust out too, in turn making the remaining players' task yet more challenging. The entire equilibrium of the poker ecosystem can be thrown out of balance.

How, in fact, do poker games sustain themselves if the worst players are a constant threat to go broke? Sometimes there are fishy players with bottomless pockets: PokerKingBlog.com has alleged that Guy Laliberté, the CEO of Cirque du Soleil, lost as much as $17 million in online poker games in 2008,[22] where he sought to compete in the toughest high-stakes games against opponents like Dwan. Whatever the number, Laliberté is a billionaire who was playing the game for the intellectual challenge and to him this was almost nothing, the equivalent of the average American losing a few hundred bucks at blackjack.

Much more commonly, the answer is that there is not just one fishy player who loses money in perpetuity but a steady stream of them who take their turn in the barrel, losing a few hundred or a few thousand dollars and then quitting.

At a brick-and-mortar casino like the Bellagio, these players might wander in from the craps table, or from one of its nightclubs, or after going on a winning streak in a tournament or a smaller-stakes game.

In the online poker environment of my experience, the fish population was more irregular and depended on the regulatory environment in different countries, the amount of advertising that the poker sites were doing, and perhaps even the time of year.[23] During the poker boom years, however, the player pool was expanding so rapidly that there was always a wealth of fishes.

That was about to change.

The Poker Bubble Bursts

In October 2006 the outgoing Republican Congress, hoping to make headway with "values voters" before the midterm elections[24] but stymied on more pressing issues, passed a somewhat ambiguous law known as the Unlawful Internet Gambling Enforcement Act (UIGEA). The UIGEA, strictly speaking, didn't make online poker illegal. What it did, rather, was to target the third-party companies that facilitated the movement of money into and out of the poker sites. Sure, you could play poker, the law said, in effect—but you couldn't have any chips. Meanwhile, the Department of Justice began targeting companies that were offering online gambling to Americans. David Carruthers, the CEO of an offshore site known as BetOnSports PLC, was arrested on a layover in Dallas while changing planes on a trip from the United Kingdom to Costa Rica. Other prosecutions soon followed.

All this scared the hell out of many online poker players—as well as many of the proprietors of the games. Party Poker, then the largest online poker site, locked Americans out of the games two weeks after the UIGEA passed; its stock crashed by 65 percent in twenty-four hours.[25] Other companies stayed in business, developing workarounds to the new law, but it had become harder to get your money in and riskier to take it back out.

I had made most of my money from Party Poker, which advertised aggressively and was known for having the fishiest players. During the two-week grace period after Party Poker made its announcement but kept the games open to

Americans, the games there were fishier than ever, sometimes taking on a *Lord of the Flies* mentality. I had some of my winningest poker days during this period.

Once Party Poker shut Americans out, however, and I shifted my play to tougher sites like PokerStars, I found that I wasn't winning anymore. In fact, I was losing—a lot: about $75,000 during the last few months of 2006, most of it in one horrible evening. I played through the first several months of 2007 and continued to lose—another $60,000 or so. At that point, no longer confident that I could beat the games, I cashed out the rest of my money and quit.

My conclusion at the time was that the composition of the player pool had changed dramatically. Many of the professional players, reliant on the game for income, had soldiered on and kept playing, but most of the amateurs withdrew their funds or went broke. The fragile ecology of the poker economy was turned upside down—without those weak players to prop the game up, the water level had risen, and some of the sharks turned into suckers.[26]

Meanwhile, even before the new law passed, my play had begun to deteriorate, or at least cease to improve. I had hit a wall, playing uncreative and uninspired poker. When I did play, I combined the most dangerous trait of the professional player—the sense that I was entitled to win money—with the bad habits of the amateur, playing late into the evening, sometimes after having been out with friends.

In retrospect, things worked out pretty fortunately for me. The extra time I had on my hands—and my increased interest in the political process following the passage of the UIGEA—eventually led to the development of FiveThirtyEight. And while it wasn't fun to lose a third of my winnings, it was better than losing all of them. Some players who continued in the game were not so lucky. In 2011, the "Black Friday" indictments filed by the Department of Justice shut down many of the online poker sites for good,[27] some of which proved to be insolvent and did not let players cash out their bankrolls.

I've sometimes wondered what would have happened if I'd played on. Poker is so volatile that it's possible for a theoretically winning player to have a losing streak that persists for months, or even for a full year. The flip side of this is that it's possible for a losing player to go on a long winning streak before he realizes that he isn't much good.

Luck Versus Skill in Poker

Luck and skill are often portrayed as polar opposites. But the relationship is a little more complicated than that.

Few of us would doubt, for instance, that major-league baseball players are highly skilled professionals. It just isn't easy to hit a baseball thrown at ninety-eight miles per hour with a piece of ash, and some human beings are a wee bit more refined in their talent for this than others. But there is *also* a lot of luck in baseball—you can hit the ball as hard as hell and still line out to the second baseman. It takes a lot of time for these skill differences to become obvious; even a couple of months' worth of data is not really enough. In figure 10-9, I've plotted the batting averages achieved by American League players in April 2011 on one axis, and the batting averages for the same players in May 2011 on the other one.[28] There seems to be no correlation between the two. (A player named Brendan Ryan, for instance, hit .184 in April but .384 in May.) And yet, we know from looking at statistics over the longer term—what baseball players do over whole seasons or over the course of their careers—that hitting ability differs substantially from player to player.[29]

FIGURE 10-9: BATTING AVERAGE FOR AMERICAN LEAGUE PLAYERS, APRIL AND MAY 2011

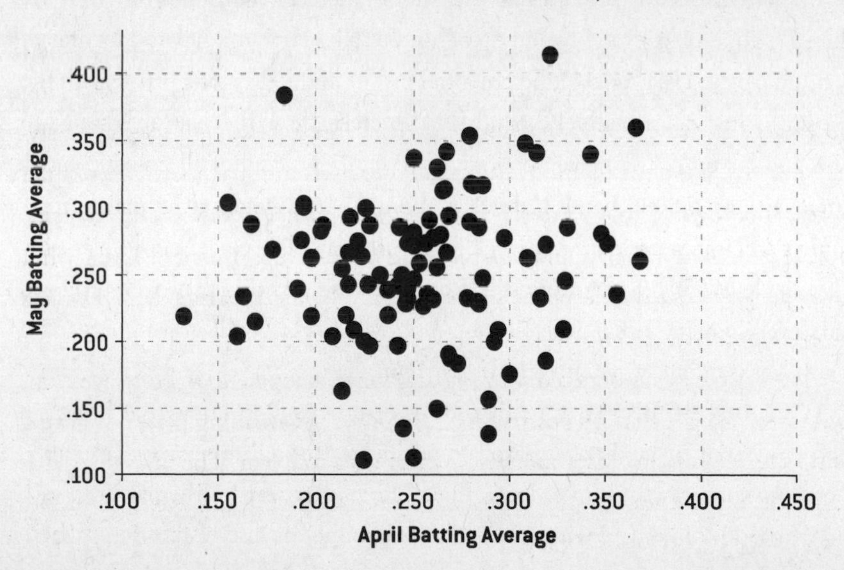

Poker is very much like baseball in this respect. It involves tremendous luck *and* tremendous skill. The opposite of poker would be something like tic-tac-toe (figure 10-10). There is no element of luck in the game, but there isn't much skill either. A precocious second grader could do just as well as Bill Gates at it.

FIGURE 10-10: SKILL VERSUS LUCK MATRIX

	Low luck	**High luck**
Low skill	Tic-Tac-Toe	Roulette
High skill	Chess	Poker

Still, it can take a long time for poker players to figure out how good they really are. The luck component is especially strong in limit hold 'em, the game that I specialized in. Correct strategy in this game implies that you will scrap and fight for many pots, and staying in so many hands to the end means that a lot depends on the luck of the deal. A very good limit hold 'em player, in a game where the betting increments are $100 and $200, might make $200 for every one hundred hands played. However, the volatility in his results—as measured by a statistic called standard deviation—is likely to be about sixteen times that, or about $3,200 for every one hundred hands.[30]

What this means is that even after literally tens of thousands of hands are played, a good player might wind up behind or a bad one might wind up ahead. In figure 10-11, I've modeled the potential profits and losses for a player with the statistics I just described. The bands in the chart show the plausible range of wins and losses for the player, enough to cover 95 percent of all possible cases. After he plays 60,000 hands—about as many as he'd get in if he played forty hours a week in a casino every week for a full year—the player could plausibly have made $275,000 or have lost $35,000. In essence, this player could go to work every day for a year and still lose money. This is why it is sometimes said that poker is a hard way to make an easy living.

Of course, if this player really did have some way to know that he was a long-term winner, he'd have reason to persevere through his losses. In reality, there's no sure way for him to know that. The proper way for the player to estimate his odds of being a winner, instead, is to apply Bayesian statistics,[31] where

FIGURE 10-11: PLAUSIBLE WIN RATES FOR SKILLED LIMIT HOLD 'EM PLAYER, $100/$200 GAME

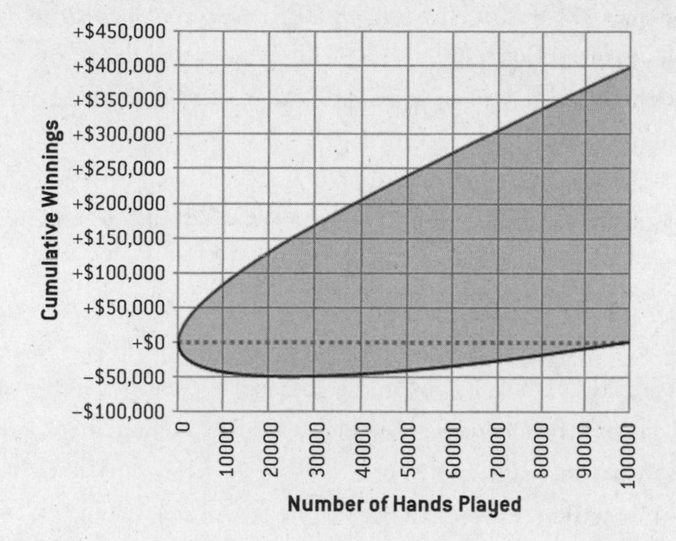

he revises his belief about how good he really is, on the basis of both his results and his prior expectations.

If the player is being honest with himself, he should take quite a skeptical attitude toward his own success, even if he is winning at first. The player's prior belief should be informed by the fact that the average poker player by definition loses money, since the house takes some money out of the game in the form of the rake while the rest is passed around between the players.[32] The Bayesian method described in the book *The Mathematics of Poker*, for instance, would suggest that a player who had made $30,000 in his first 10,000 hands at a $100/$200 limit hold 'em game was nevertheless *more likely than not* to be a long-term loser.

Our Poker Delusions

Most players, as you might gather, are not quite this honest with themselves. I certainly wasn't when I was living in the poker bubble. Instead, they start out with the assumption that they are winning players—until they have the truth beaten into them.

"Poker is all about people who think they're favorites when they're not," Dwan told me. "People can have some pretty deluded views on poker."

Another player, Darse Billings, who developed a computer program that competed successfully[33] against some of the world's best limit hold 'em players,* put it even more bluntly.

"There is no other game that I know of where humans are so smug, and think that they just play like wizards, and then play so badly," he told me. "Basically it's because they don't know anything, and they think they must be God-like, and the truth is that they aren't. If computer programs feed on human hubris, then in poker they will eat like kings."

This quality, of course, is not unique to poker. As we will see in chapter 11, much of the same critique can be applied to traders on Wall Street, who often think they can beat market benchmarks like the S&P 500 when they usually cannot. More broadly, overconfidence is a huge problem in any field in which prediction is involved.

Poker is not a game like roulette, where results are determined purely by luck and nobody would make money if they took an infinite number of spins on the wheel. Nor are poker players very much like roulette players; they are probably much more like investors, in fact. According to one study of online poker players, 52 percent have at least a bachelor's degree[34]—about twice the rate in the U.S. population as a whole, and four times the rate among those who purchase lottery tickets.[35] Most poker players are smart enough to know that some players *really do* make money over the long term—and this is what can get them in trouble.

Why We Tilt

Tommy Angelo pursued a poker dream before it was cool. In 1990, at the age of thirty-two, he quit his job as a drummer and pianist for a country rock band to play poker full-time.[36]

---

* While computer programs like Billings's have become quite good at limit hold 'em, they are not yet very good at the more strategically challenging game of no-limit hold 'em.

"I was hooked on it," Angelo told me when I spoke with him in 2012. "I loved the idea of being a professional poker player when I first heard the words. The whole idea was so glorious, of not having a job. It's like you're beating society, making all your money on your wits alone. I couldn't imagine anything more appealing."

But Angelo, like most poker players, had his ups and downs—not just in his results but also in the quality of his play. When he was playing his best, he was very good. But he wasn't always playing his best—very often, he was on tilt.

"I was a great tilter," Angelo reflected in his book, *Elements of Poker*, referring to a state of overaggressive play brought on by a loss of perspective.[37] "I knew all the different kinds. I could do steaming tilt, simmering tilt, too loose tilt, too tight tilt, too aggressive tilt, too passive tilt, playing too high tilt, playing too long tilt, playing too tired tilt, entitlement tilt, annoyed tilt, injustice tilt, frustration tilt, sloppy tilt, revenge tilt, underfunded tilt, overfunded tilt, shame tilt, distracted tilt, scared tilt, envy tilt, this-is-the-worst-pizza-I've-ever-had tilt, I-just-got-showed-a-bluff tilt, and of course, the classics: I-gotta-get-even tilt, and I-only-have-so-much-time-to-lose-this-money tilt, also known as demolition tilt."

What Angelo eventually came to realize is that, for all his skill, his periodic bouts of tilt prevented him from doing much more than scrape by. As we have seen, it is considerably easier to lose money at poker when you play badly than to make money when you are playing well. Meanwhile, the edge even for a long-term winning player is quite small. It is quite plausible for someone who plays at a world-class level 90 percent of the time to lose money from the game overall if he tilts the other 10 percent of the time.

Angelo realized the full extent of his tilt problems once he was in his forties, after he had started writing about the game and coaching other players. He is naturally a perceptive person, and what started out as strategy sessions often turned into therapy sessions.

"I was coaching so many different types of people with so many different types of problems related to poker," he told me. "The problems were very easy to see in another person. I'd say here's a guy that's just as smart as me. And I know for a *fact* he's delusional about his skill. And I know that if everyone else is delusional, I have to be too."

Every poker player tilts at least to some degree, Angelo thinks. "If someone came up to me and said 'I don't tilt,' my mind registers, 'There's another delusional statement from a delusional human.' It happens all the time."

I had my own types of tilt when I played poker actively. I wasn't one of those throw-stuff-around-the-room type of guys. Nor would I turn into a crazy maniac when I tilted, trying to play every hand (although my "A-game" was wild enough). Sometimes I'd even tighten up my game a bit. However, I'd play mechanically, unthinkingly, for long stretches of time and often late into the night—I'd just call a lot and hope that the pot got pushed my way. I had given up, deep down, on really trying to beat the game.

I realize now (I'm not sure that I did when I was playing) what my tilt triggers were. The biggest one was a sense of *entitlement*. I didn't mind so much when I wasn't getting many cards and had to fold for long stretches of time—that, I realized, was part of the statistical variance in the game. But when I thought I had played particularly well—let's say I correctly detected an opponent's bluff, for instance—but then he caught a miracle card on the river and beat my hand anyway, *that's* what could really tilt me. I thought I had earned the pot, but he made the money.

By tilting, I could bring things back into a perverse equilibrium: I'd start playing badly enough that I *deserved* to lose. The fundamental reason that poker players tilt is that this balance is so often out of whack: over the short term, and often over the medium term, a player's results aren't very highly correlated with his skill. It certainly doesn't help matters when players have an unrealistic notion about their skill level, as they very often will. "We tend to latch onto data that supports our theory," Angelo told me. "And the theory is usually 'I'm better than they are.'"

Beyond Results-Oriented Thinking

In the United States, we live in a very results-oriented society. If someone is rich or famous or beautiful, we tend to think they *deserve* to be those things. Often, in fact, these factors are self-reinforcing: making money begets more opportunities to make money; being famous provides someone with more ways to leverage their celebrity; standards of beauty may change with the look of a Hollywood starlet.

This is not intended as a political statement, an argument for (or against) greater redistribution of wealth or anything like that. As an empirical matter, however, success is determined by some combination of hard work, natural talent, and a person's opportunities and environment—in other words, some combination of noise and signal. In the U.S., we tend to emphasize the signal component most of the time—except perhaps when it comes to our own shortcomings, which we tend to attribute to bad luck. We can account for our neighbors' success by the size of their home, but we don't know as much about the hurdles they had to overcome to get there.

When it comes to prediction, we're *really* results-oriented. The investor who calls the stock market bottom is heralded as a genius, even if he had some buggy statistical model that just happened to get it right. The general manager who builds a team that wins the World Series is assumed to be better than his peers, even if, when you examine his track record, the team succeeded despite the moves he made rather than because of them. And this is certainly the case when it comes to poker. Chris Moneymaker wouldn't have been much of a story if the marketing pitch were "Here's some slob gambler who caught a bunch of lucky cards."

Sometimes we take consideration of luck too far in the other direction, when we excuse predictions that really were bad by claiming they were unlucky. The credit-rating agencies used a version of this excuse when their incompetence helped to usher in the financial collapse. But as a default, just as we perceive more signal than there really is when we make predictions, we also tend to attribute more skill than is warranted to successful predictions when we assess them later.

Part of the solution is to apply more rigor in how we evaluate predictions. The question of how skillful a forecast is can often be addressed through empirical methods; the long run is achieved more quickly in some fields than in others. But another part of the solution—and sometimes the *only* solution when the data is very noisy—is to focus more on process than on results. If the sample of predictions is too noisy to determine whether a forecaster is much good, we can instead ask whether he is applying the attitudes and aptitudes that we know are correlated with forecasting success over the long run. (In a sense, we'll be predicting how good his predictions will be.)

Poker players tend to understand this more than most other people, if only because they tend to experience the ups and downs in such a visceral way. A high-stakes player like Dwan might experience as much volatility in a single session of poker as a stock market investor would in his lifetime. Play well and win; play well and lose; play badly and lose; play badly and win: every poker player has experienced each of these conditions so many times over that they know there is a difference between process and results.

If you talk with the very best players, they don't take any of their success for granted; they focus as much as they can on self-improvement. "Anyone who thinks they've gotten good enough, good enough that they've solved poker, they're getting ready for a big downswing," Dwan told me.

Angelo tries to speed the process along with his clients. "We're walking around in this cloud of noise all the time," he said. "Very often we don't see what's going on accurately." Angelo's methods for achieving this are varied and sometimes unconventional: he is an advocate of meditation, for instance. Not all his clients meditate, but the broader idea is to increase their level of self-awareness, encouraging them to develop a better sense for which things are and are not within their control.*

When we play poker, we control our decision-making process but not how the cards come down. If you correctly detect an opponent's bluff, but he gets a lucky card and wins the hand anyway, you should be pleased rather than angry, because you played the hand as well as you could. The irony is that by being *less* focused on your results, you may achieve better ones.

Still, we are imperfect creatures living in an uncertain world. If we make a prediction and it goes badly, we can never really be certain whether it was our fault or not, whether our model was flawed or we just got unlucky. The closest approximation to a solution is to achieve a state of equanimity with the noise and the signal, recognizing that both are an irreducible part of our universe, and devote ourselves to appreciating each for what it is.

---

* Meditation helps people achieve this, in part, by encouraging focus on posture and breathing—things that are within our control but which we normally take for granted.

11

IF YOU CAN'T BEAT 'EM . . .

In 2009, a year after a financial crisis had wrecked the global economy, American investors traded $8 million in stocks every *second* that the New York Stock Exchange was open for business. Over the course of the typical trading day, the volume grew to $185 billion, roughly as much as the economies of Nigeria, the Philippines or Ireland produce in an entire year. Over the course of the whole of 2009, more than $46 *trillion*[1] in stocks were traded: four times more than the revenues of all the companies in the Fortune 500 put together.[2]

This furious velocity of trading is something fairly new. In the 1950s, the average share of common stock in an American company was held for about six years before being traded—consistent with the idea that stocks are a long-term investment. By the 2000s, the velocity of trading had increased roughly twelvefold. Instead of being held for six years, the same share of stock was traded after just six *months*.[3] The trend shows few signs of abating: stock market volumes have been doubling once every four or five years. With the advent of high-

FIGURE 11-1: AVERAGE TIME U.S. COMMON STOCK WAS HELD

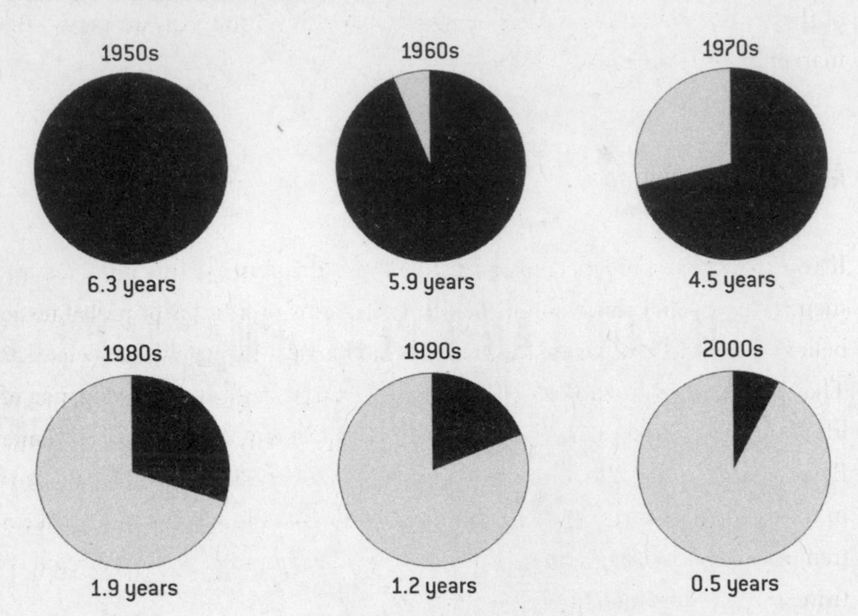

frequency trading, some stocks are now literally bought and sold in a New York microsecond.[4]

Economics 101 teaches that trading is rational only when it makes both parties better off. A baseball team with two good shortstops but no pitching trades one of them to a team with plenty of good arms but a shortstop who's batting .190. Or an investor who is getting ready to retire cashes out her stocks and trades them to another investor who is just getting his feet wet in the market.

But very little of the trading that occurs on Wall Street today conforms to this view. Most of it reflects true differences of opinion—contrasting predictions—about the future returns of a stock.* Never before in human history have so many predictions been made so quickly and for such high stakes.

Why so much trading occurs is one of the greatest mysteries in finance.[5]

* Some of it also represents market making: certain investment firms, like a well-stocked 7-Eleven, hold a lot of inventory and can be counted on to be open for business when no one else is, hoping to make a few pennies at a time for their trouble.

More and more people seem to think they can outpredict the collective wisdom of the market. Are these traders being rational? And if not, can we expect the market to settle on a rational price?

A Trip to Bayesland

If you follow the guidance provided by Bayes's theorem, as this book recommends, then you'll think about the future in terms of a series of probabilistic beliefs or forecasts. What are the chances that Barack Obama will be reelected? That Lindsay Lohan will be arrested again? That we will discover evidence of life on other another planet? That Rafael Nadal will win at Wimbledon? Some Bayesians assert[6] that the most sensible way to think about these probabilities is in terms of the betting line we would set. If you take the idea to its logical extreme, then in Bayesland we all walk around with giant sandwich boards advertising our odds on each of these bets:

FIGURE 11-2: BAYESIAN SANDWICH BOARD

TODAY'S PRICES	
• OBAMA WINS REELECTION	55%
• LOHAN GETS ARRESTED	99%
• STOCK MARKET CRASHES	10%
• LIFE ON MARS	2%
• NADAL WINS WIMBLEDON	30%

In Bayesland, when two people pass each other by and find that they have different forecasts, they are obliged to do one of two things. The first option is to come to a consensus and revise their forecasts to match. If my sandwich board says that Nadal has a 30 percent chance to win Wimbledon, and yours says that he has a 50 percent chance instead, then perhaps we both revise our estimates to 40 percent. But we don't necessarily have to meet in the middle; if you're more up on Lindsay Lohan gossip than I am, perhaps I capitulate to you and concede that your Lindsay Lohan forecast is better, thereby adopting it as my own belief. Either way, we walk away from the meeting with the same number in mind—a revised and, we hope, more accurate forecast about the probability of some real-world event.

But sometimes we won't agree. The law of the land says that we must then settle our differences by placing a bet on our forecasts. In Bayesland, you *must* make one of these two choices: come to a consensus or bet.* Otherwise, to a Bayesian, you are not really being rational. If after we have our little chat, you still think your forecast is better than mine, you should be happy to bet on it, since you stand to make money. If you don't, you should have taken my forecast and adopted it as your own.

Of course, this whole process would be incredibly inefficient. We'd have to maintain forecasts on thousands and thousands of events and keep a ledger of the hundreds of bets that we had outstanding at any given time. In the real world, this is the function that markets play. They allow us to make transactions at one fixed price, at a consensus price, rather than having to barter or bet on everything.[7]

The Bayesian Invisible Hand

In fact, free-market capitalism and Bayes' theorem come out of something of the same intellectual tradition. Adam Smith and Thomas Bayes were contemporaries, and both were educated in Scotland and were heavily influenced by the philosopher David Hume. Smith's "invisible hand" might be thought of as a Bayesian process, in which prices are gradually updated in response to changes in supply and demand, eventually reaching some equilibrium. Or, Bayesian reasoning might be thought of as an "invisible hand" wherein we gradually update and improve our beliefs as we debate our ideas, sometimes placing bets on them when we can't agree. Both are consensus-seeking processes that take advantage of the wisdom of crowds.

It might follow, then, that markets are an especially good way to make predictions. That's really what the stock market is: a series of predictions about the

---

* One could imagine that a small fudge factor might be allowed if our probability estimates were close but not exactly the same, since there is some inconvenience associated with betting.

future earnings and dividends of a company.[8] My view is that this notion is *mostly* right *most* of the time. I advocate the use of betting markets for forecasting economic variables like GDP, for instance. One might expect these markets to improve predictions for the simple reason that they force us to put our money where our mouth is, and create an incentive for our forecasts to be accurate.

Another viewpoint, the efficient-market hypothesis, makes this point much more forcefully: it holds that it is *impossible* under certain conditions to outpredict markets. This view, which was the orthodoxy in economics departments for several decades, has become unpopular given the recent bubbles and busts in the market, some of which seemed predictable after the fact. But, the theory is more robust than you might think.

And yet, a central premise of this book is that we must accept the fallibility of our judgment if we want to come to more accurate predictions. To the extent that markets are reflections of our collective judgment, they are fallible too. In fact, a market that makes perfect predictions is a logical impossibility.

Justin Wolfers, Prediction Markets Cop

If there really were a Bayesland, then Justin Wolfers, a fast-talking, ponytailed polymath who is among America's best young economists, would be its chief of police, writing a ticket anytime he observed someone refusing to bet on their forecasts. Wolfers challenged me to a dinner bet after I wrote on my blog that I thought Rick Santorum would win the Iowa caucus, bucking the prediction market Intrade (as well as my own predictive model), which still showed Mitt Romney ahead. In that case, I was willing to commit to the bet, which turned out well for me after Santorum won by literally just a few dozen votes after a weeks-long recount.* But there have been other times when I have been less willing to accept one of Wolfers' challenges. Presuming you are a betting man as I am, what good is a prediction if you aren't willing to put money on it?

Wolfers is from Australia, where he supported himself in college by run-

* I met Santorum for my *New York Times* story on the Iowa vote count dispute after the initial tally had shown Romney ahead. Santorum remembered my bet and jokingly asserted that the bet was my motivation for tracking him down. It may have provided some additional incentive.

ning numbers for a bookie in Sydney.[9] He now lives in Philadelphia, where he teaches at the Wharton School and writes for the *Freakonomics* blog. I visited Wolfers at his home, where he was an outstanding host, having ordered a full complement of hoagie sandwiches from Sarcone's to welcome me, my research assistant Arikia Millikan, and one of his most talented students, David Rothschild. But he was buttering me up for a roast.

Wolfers and Rothschild had been studying the behavior of prediction markets like Intrade, a sort of real-life version of Bayesland in which traders buy and sell shares of stock that represent real-world news predictions—everything from who will win the Academy Award for Best Picture to the chance of an Israeli air strike on Iran. Political events are especially popular subjects for betting. One stock, for instance, might represent the possibility that Hillary Clinton would win the Democratic nomination in 2008. The stock pays a dividend of $100 if the proposition turns out to be true (Clinton wins the nomination) but nothing otherwise. However, the traders can exchange their shares as much as they want until the outcome is determined. The market price for a share thus represents a consensus prediction about an outcome's likelihood. (At one market,[10] shares in Clinton stock crashed to $18 after she lost the Iowa caucuses, rebounded to $66 when she won the New Hampshire primary, and slowly drifted back down toward $0 as Obama outlasted her in the campaign.) Markets like these have a long tradition in politics, dating back to at least the 1892 presidential election, when stocks in Grover Cleveland and Benjamin Harrison were traded just steps away from the American Stock Exchange.[11]

"You should tell Nate about the comparison paper," Wolfers said to Rothschild a few minutes into lunch, a mischievous grin on his face.

"I did a paper for an academic journal that looked at de-biased Internet-based polling, comparing it to prediction markets in 2008, showing they were comparable," Rothschild volunteered.

"That's way too polite," Wolfers interrupted. "It was Intrade versus Nate."

"And Intrade won," Rothschild said.

Rothschild's paper, which was published in *Public Opinion Quarterly*,[12] compared the forecasts I made at FiveThirtyEight over the course of the 2008 election cycle with the predictions at Intrade. It concluded that, although FiveThirtyEight's forecasts had done fairly well, Intrade's were better.

The Benefits (and Limitations) of Group Forecasts

I have some quarrels with the paper's methodology. Intrade's forecasts beat FiveThirtyEight's only after Wolfers and Rothschild made certain adjustments to them after the fact; otherwise FiveThirtyEight won.[13] Perhaps more important, a new forecast at FiveThrityEight fairly often moved the Intrade price in the same direction, suggesting that the bettors there were piggybacking off it to some extent.

Nevertheless, there is strong empirical and theoretical evidence that there is a benefit in aggregating different forecasts. Across a number of disciplines, from macroeconomic forecasting to political polling, simply taking an average of everyone's forecast rather than relying on just one has been found to reduce forecast error,[14] often by about 15 or 20 percent.

But before you start averaging everything together, you should understand three things. First, while the aggregate forecast will essentially always be better than the typical individual's forecast, that doesn't necessarily mean it will be *good*. For instance, aggregate macroeconomic forecasts are much too crude to predict recessions more than a few months in advance. They are somewhat better than individual economists' forecasts, however.

Second, the most robust evidence indicates that this wisdom-of-crowds principle holds when forecasts are made *independently* before being averaged together. In a true betting market (including the stock market), people can and do react to one another's behavior. Under these conditions, where the crowd begins to behave more dynamically, group behavior becomes more complex.

Third, although the aggregate forecast is better than the typical individual's forecast, it does not necessarily hold that it is better than the *best* individual's forecast. Perhaps there is some polling firm, for instance, whose surveys are so accurate that it is better to use their polls and their polls alone rather than dilute them with numbers from their less-accurate peers.

When this property has been studied over the long run, however, the aggregate forecast has often beaten even the very best individual forecast. A study of the Blue Chip Economic Indicators survey, for instance, found that the aggregate forecast was better over a multiyear period than the forecasts issued by *any one* of the seventy economists that made up the panel.[15] Another study by Wolfers, looking at predictions of NFL football games, found that the consensus forecasts produced by betting markets were better than about 99.5 percent of those from individual handicappers.[16] And this is certainly true of political polling; models that treat any one poll as the Holy Grail are more prone to embarrassing failures.[17] Reducing error by 15 or 20 percent by combining forecasts may not sound all that impressive, but it's awfully hard to beat in a competitive market.

So I told Wolfers and Rothschild that I was ready to accept the principle behind their conclusion, if not all the details. After all, bettors at Intrade can use FiveThirtyEight's forecasts to make their predictions as well as whatever other information they deem to be relevant (like the forecasts issued by our competitors, some of which are also very good). Of course, the bettors could interpret that information in a biased fashion and get themselves into trouble. But it is not like the FiveThirtyEight forecasts—or anybody else's—are beyond reproach.

Wolfers seemed disappointed that I was willing to concede so much ground. If I wasn't sure I could beat Intrade, why not just join them and adopt their predictions as my own?

"I'm surprised by your reaction, actually," he told me. "If there's something else that should beat it and does beat it, what's the point of doing what you're doing?"

For one thing, I find making the forecasts intellectually interesting—and they help to produce traffic for my blog.

Also, while I accept the theoretical benefits of prediction markets, I don't know that political betting markets like Intrade are all that good right now—the standard of competition is fairly low. Intrade is becoming more popular, but it is still small potatoes compared with the stock market or Las Vegas. In the weeks leading up to the Super Tuesday primaries in March 2012, for instance, about $1.6 million in shares were traded there;[18] by contrast, $8 million is traded in the New York Stock Exchange in a single *second*. The biggest profit made by any one trader from his Super Tuesday bets was about $9,000, which is not enough to make a living, let alone to get rich. Meanwhile, Intrade is in a legal gray area and most of the people betting on American politics are from Europe or from other countries. There have also been some cases of market manipulation*[19] or blatant irrational pricing[20] there. And these markets haven't done very well at aggregating information in instances where there isn't much information worth aggregating, like in trying to guess the outcome of Supreme Court cases from the nebulous clues the justices provide to the public.

Could FiveThirtyEight and other good political forecasters beat Intrade if it were fully legal in the United States and its trading volumes were an order of magnitude or two higher? I'd think it would be difficult. Can they do so right now? My educated guess[21] is that some of us still can, if we select our bets carefully.[22]

Then again, a lot of smart people have failed miserably when they thought they could beat the market.

The Origin of Efficient-Market Hypothesis

In 1959, a twenty-year-old college student named Eugene Fama, bored with the Tufts University curriculum of romance languages and Voltaire, took a job working for a professor who ran a stock market forecasting service.[23] The job was a natural fit for him; Fama was a fierce competitor who had been the first in his family to go to college and who had been a star athlete at Boston's Malden Catholic High School despite standing at just five feet eight. He combed through data on past stock market returns looking for anything that could provide an investor with an advantage, frequently identifying statistical patterns that suggested the stock market was highly predictable and an investor could make a fortune by exploiting them. The professor almost always responded skeptically, advising Fama to wait and see how the strategies performed in the real world before starting to invest in them. Almost always, Fama's strategies failed.

Equally frustrated and fascinated by the experience, Fama abandoned his

* Because it is so cheap to bet on Intrade, but it is receiving an increasing amount of media attention and can generate more favorable press coverage for a campaign, it could conceivably be in the interest of a candidate or a "super PAC" to buy shares as a cheap form of advertising.

plans to become a high school teacher and instead enrolled at the University of Chicago's Graduate School of Business, where in 1965 he managed to get his Ph.D. thesis published. The paper had something of the flavor of the baseball statistician Bill James's pioneering work during the 1980s, leaning on a mix of statistics and sarcasm to claim that much of the conventional wisdom about how stocks behaved was pure baloney. Studying the returns of dozens of mutual funds in a ten-year period from 1950 to 1960, Fama found that funds that performed well in one year were no more likely to beat their competition the next time around.[24] Although he had been unable to beat the market, nobody else really could either:

> A superior analyst is one whose gains . . . are *consistently* greater than those of the market. Consistency is the crucial word here, since for any given short period of time . . . some people will do much better than the market and some will do much worse.
>
> Unfortunately, by this criterion, this author does not qualify as a superior analyst. There is some consolation, however . . . . [O]ther more market-tested institutions do not seem to qualify either.[25]

The paper, although it would later be cited more than 4,000 times,[26] at first received about as much attention as most things published by University of Chicago graduate students.[27] But it had laid the groundwork for efficient-market hypothesis. The central claim of the theory is that the movement of the stock market is unpredictable to any meaningful extent. Some investors inevitably perform better than others over short periods of time—just as some gamblers inevitably win at roulette on any given evening in Las Vegas. But, Fama claimed, they weren't able to make good enough predictions to beat the market over the long run.

Past Performance Is Not Indicative of Future Results

Very often, we fail to appreciate the limitations imposed by small sample sizes and mistake luck for skill when we look at how well someone's predictions have done. The reverse can occasionally also be true, such as in examining the batting averages of baseball players over short time spans: there is skill there, perhaps even quite a bit of it, but it is drowned out by noise.

In the stock market, the data on the performance of individual traders is noisy enough that it's very hard to tell whether they are any good at all. "Past performance is not indicative of future results" appears in mutual-fund brochures for a reason.

Suppose that in 2007 you wanted to invest in a mutual fund, one that focused mostly on large-capitalization American stocks like those that make up the Dow Jones Industrial Average or the S&P 500. You went to E*Trade, which offered literally hundreds of choices of such funds and all sorts of information about them, like the average return they had achieved over the prior five years. Surely you would have been better off investing in a fund like EVTMX (Eaton Vance Dividend Builder A), which had beaten the market by almost 10 percent annually from 2002 through 2006? Or if you were feeling more daring, JSVAX (Janus Contrarian T), which had invested in some unpopular stocks but had bettered the market by 9 percent annually during this period?

Actually, it wouldn't have made any difference. When I looked at how these mutual funds performed from 2002 through 2006, and compared it with how they performed over the next five years from 2007 through 2011, there was literally no correlation between them. EVTMX, the best-performing fund from 2002 through 2006, was only average over the next five years. And high-performing JSVAX was 3 percent *worse* per year than the market average. As Fama found, there was just no consistency in how well a fund did, even over five-year increments. Other studies have identified very modest correlations in mutual fund performance from year to year,[28] but it's so hard to tell them apart (figure 11-3)[29] that you're best off just selecting the one with the cheapest fees—or eschewing them entirely and investing in the market yourself.

The Misery of the Chartist

Fama reserved his harshest criticism, however, for what he called "chartists"—people who claim to be able to predict the direction of stock prices (as Fama

FIGURE 11-3: INCONSISTENT MUTUAL FUND PERFORMANCE

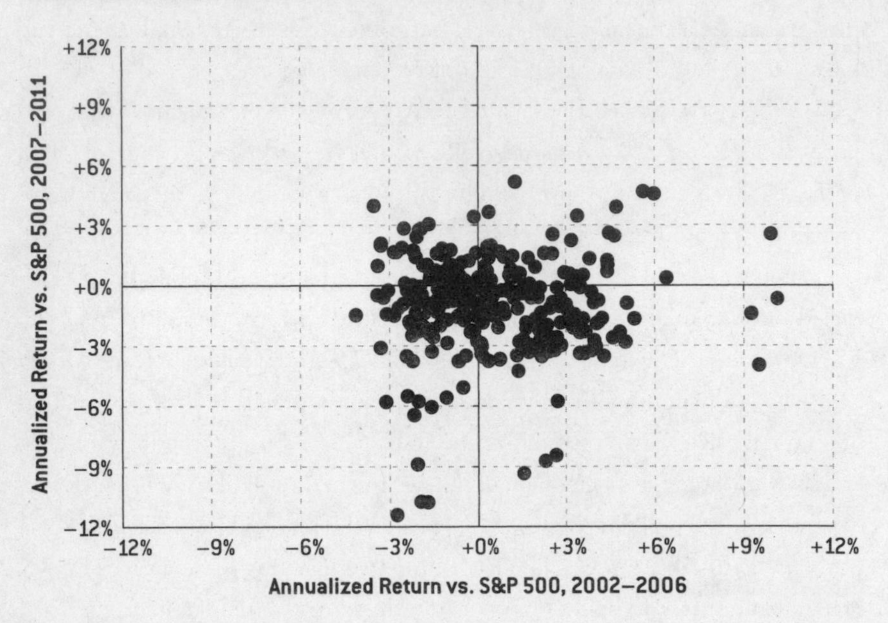

had tried and failed to do) solely on the basis of past statistical patterns, without worrying about whether the company had made a profit or a loss or whether it sold airplanes or hamburgers. (The more polite term for this activity is *technical analysis*.)

Perhaps we should have some sympathy for the poor chartist: distinguishing the noise from the signal is not always so easy. In figure 11-4, I have presented a series of six stock market charts. Four of them are fakes and were literally generated by telling my computer to flip a coin* (or rather, pick a random series of 1's and 0's). The other two are real and depict the actual movement of the Dow Jones Industrial Average over the first 1,000 trading days of the 1970s and 1980s, respectively. Can you tell which is which? It isn't so easy. (The answer is in the endnotes.[30]) Investors were looking at stock-price movements like these and were mistaking noise for a signal.

* This experiment re-creates a famous experiment by the Princeton economist Burton Malkiel, who had his class flip coins to generate these random charts, and then showed them to a technical analyst who insisted that he buy the stocks immediately.

FIGURE 11-4: RANDOM-WALK AND ACTUAL STOCK-MARKET CHARTS

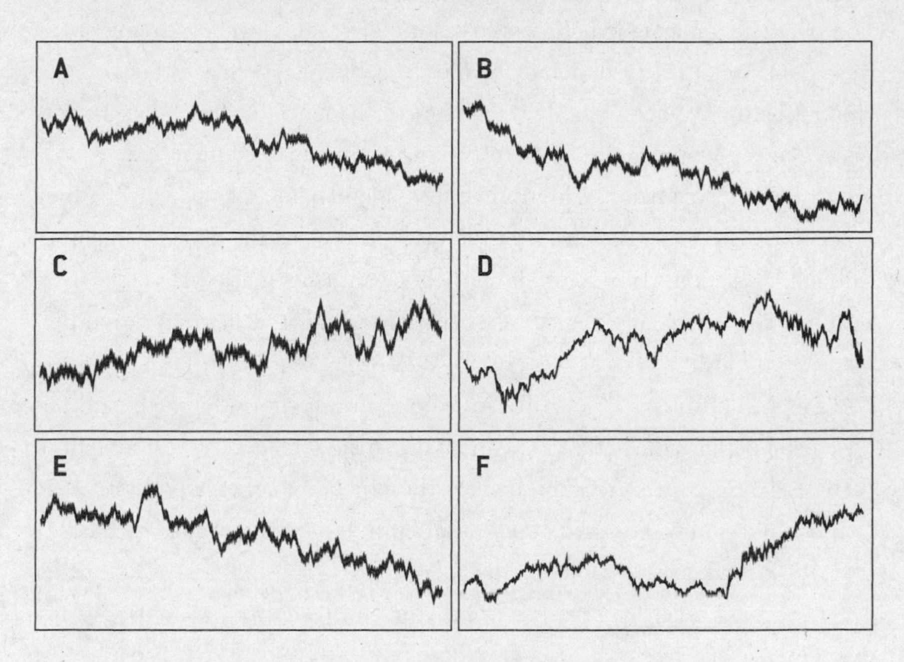

Three Forms of Efficient-Market Hypothesis

After looking at enough of this type of data, Fama refined his hypothesis to cover three distinct cases,[31] each one making a progressively bolder claim about the predictability of markets.

First, there is the **weak form** of efficient-market hypothesis. What this claims is that stock-market prices cannot be predicted from analyzing past statistical patterns alone. In other words, the chartist's techniques are bound to fail.

The **semistrong form** of efficient-market hypothesis takes things a step further. It argues that fundamental analysis—meaning, actually looking at publicly available information on a company's financial statements, its business model, macroeconomic conditions and so forth—is also bound to fail and will also not produce returns that consistently beat the market.

Finally, there is the **strong form** of efficient-market hypothesis, which claims that even private information—insider secrets—will quickly be incorporated into market prices and will not produce above-average returns. This version of efficient-market hypothesis is meant more as the logical extreme of the theory and is not believed literally by most proponents of efficient markets (including Fama.[32]) There is fairly unambiguous evidence, instead, that insiders make above-average returns. One disturbing example is that members of Congress, who often gain access to inside information about a company while they are lobbied and who also have some ability to influence the fate of companies through legislation, return a profit on their investments that beats market averages by 5 to 10 percent *per year*,[33] a remarkable rate that would make even Bernie Madoff blush.

But the debates over the weak form and semistrong forms of the hypothesis have been perhaps the hottest topic in all the social sciences. Almost nine hundred academic papers are published on the efficient-market hypothesis every year,[34] and it is now discussed almost as often in financial journals[35] as the theory of evolution is discussed in biological ones.[36]

Efficient-market hypothesis is sometimes mistaken for an excuse for the excesses of Wall Street; whatever else those guys are doing, it seems to assert, at least they're behaving rationally. A few proponents of the efficient-market hypothesis might interpret it in that way. But as the theory was originally drafted, it really makes just the opposite case: the stock market is fundamentally and profoundly *unpredictable*. When something is truly unpredictable, nobody from your hairdresser to the investment banker making $2 million per year is able to beat it consistently.

However, as powerful as the theory claims to be, it comes with a few qualifications. The most important is that it pertains to returns on a *risk-adjusted basis*. Suppose you pursue an investment strategy that entails a 10 percent chance of going broke every year. This is exceptionally foolish—if you followed the strategy over a twenty-year investment horizon, there's only a 12 percent chance that your money would live to tell about it. But if you are that ballsy, you deserve an excess profit. All versions of the efficient-market hypothesis allow for investors to make an above-average return provided it's proportionate to the additional risks they are taking on.

Another important qualification is that profits are measured net of the cost of trading. Investors incur transaction costs every time they trade a stock. These costs are fairly small in most circumstances—perhaps 0.25 percent of a trade.[37] But they accumulate the more often you trade and can be quite devastating to an overactive trader. This gives efficient-market hypothesis a bit of a buffer zone. Some investment strategies might be a tiny bit profitable in a world where trading was free. But in the real world, a trader needs to earn a large enough profit to cover this additional expense, in somewhat the same way that a winning poker player needs to beat the game by a large enough margin to cover the house's take.

A Statistical Test of Efficient-Market Hypothesis

Opponents of efficient-market hypothesis have two good ways to attempt to disprove it. One is to demonstrate that some investors are consistently beating the stock market. The other is more direct: illustrate predictability in the returns.

One simple way to refute the hypothesis would be to demonstrate that stock-price movements are correlated from one day to the next. If the stock market rises on Tuesday, does that mean it is also more likely to rise on Wednesday? If so, that means that an investor could potentially benefit through a simple strategy of buying stocks each day that the market rises and selling them or shorting them each time that it declines. Depending on how large the investor's transaction costs were, he might be able to beat the market in this way.

Suppose that we looked at the daily closing price of the Dow Jones Industrial Average in the 10 years between 1966 and 1975—the decade just after Fama had published his thesis. Over this period, the Dow moved in the same direction from day to day—a gain was followed by a gain or a loss by a loss—58 percent of the time. It switched directions just 42 percent of the time. That seems nonrandom and it is: a standard statistical test[38] would have claimed that there was only about a 1-in-7 quintillion possibility (1 chance in 7,000,000,000,000,000) that this resulted from chance alone.

But statistical significance does not always equate to practical significance. An investor could not have profited from this trend.

Suppose that an investor had observed this pattern for ten years—gains tended to be followed by gains and losses by losses. On the morning of January 2, 1976, he decided to invest $10,000 in an index fund[39] which tracked the Dow Jones Industrial Average. But he wasn't going to be a passive investor. Instead he'd pursue what he called a Manic Momentum strategy to exploit the pattern. Every time the stock market declined over the day, he would pull all his money out, avoiding what he anticipated would be another decline the next day. He'd hold his money out of the market until he observed a day that the market rose, and then he would put it all back in. He would pursue this strategy for ten years, until the last trading day of 1985, at which point he would cash out his holdings for good, surely assured of massive profits.

How much money would this investor have at the end of the ten-year period? If you ignore dividends, inflation, and transaction costs, his $10,000 investment in 1976 would have been worth about $25,000 ten years later using the Manic Momentum strategy. By contrast, an investor who had adopted a simple buy-and-hold strategy during the same decade—buy $10,000 in stocks on January 2, 1976, and hold them for ten years, making no changes in the interim—would have only about $18,000 at the end of the period. Manic Momentum seems to have worked! Our investor, using a very basic strategy that exploited a simple statistical relationship in past market prices, substantially beat the market average, seeming to disprove the efficient-market hypothesis in the process.

But there is a catch. We ignored this investor's transaction costs. This makes an enormous difference. Suppose that the investor had pursued the Manic Momentum strategy as before but that each time he cashes into or out of the market, he paid his broker a commission of 0.25 percent. Since this investor's strategy requires buying or selling shares hundreds of times during this period, these small costs will nickel-and-dime him to death. If you account for his transaction costs, in fact, the $10,000 investment in the Manic Momentum strategy would have been worth only about $1,100 ten years later, eliminating not only his profit but also almost all the money he put in originally. In this case, there is just a little bit of predictability in stock-market returns—but not nearly enough to make a profit from them, and so efficient-market hypothesis is not violated.

The other catch is that the pattern has since reversed itself. During the 2000s, the stock market *changed* direction from day to day about 54 percent of the time, just the opposite of the pattern from earlier decades. Had the investor

FIGURE 11-5: VALUE OF STOCK HOLDINGS WITH AND WITHOUT TRANSACTION COSTS

pursued his Manic Momentum strategy for ten years beginning in January 2000, his $10,000 investment would have been whittled down to $4,000 by the end of the decade even *before* considering transaction costs.[40] If you do consider transaction costs, the investor would have had just $141 left over by the end of the decade, having lost almost 99 percent of his capital.

In other words: do not try this stuff at home. Strategies like these resemble a high-stakes game of rock-paper-scissors at best,* and the high transaction costs

* In rock-paper-scissors, the equilibrium strategy is to pick at random between rock, paper, and scissors. This strategy ensures that, over the long run, you cannot possibly lose. Unfortunately, it also ensures that you cannot possibly win. No matter how long you play, the expected return from the strategy is zero.

If you think your opponent is behaving predictably, of course, you can deviate from this strategy and try to outguess him. Perhaps like Bart Simpson, your opponent always plays rock. In this case, you should always play paper. The catch is that in trying to out-guess your opponent, your own strategy will become predictable. Once Bart realizes that you're always playing paper, he can beat you anytime he wants by switching to scissors.

Much technical trading in the stock market probably obeys this sort of cat-and-mouse dynamic, with technical traders simply trying to outguess other technical traders. However, the patterns they are trading on can dissipate or even reverse themselves once other investors become aware of them. The result is a lot of money being passed around from trader to trader, but the pile slowly growing smaller as it is eaten away by transaction costs.

they entail will deprive you of any profit and eat into much of your principal. As Fama and his professor had discovered, stock-market strategies that seem too good to be true usually are. Like the historical patterns on the frequency of earthquakes, stock market data seems to occupy a sort of purgatory wherein it is not quite random but also not quite predictable. Here, however, matters are made worse because stock market data ultimately describes not some natural phenomenon but the collective actions of human beings. If you do detect a pattern, particularly an obvious-seeming one, the odds are that other investors will have found it as well, and the signal will begin to cancel out or even reverse itself.

Efficient Markets Meet Irrational Exuberance

A more significant challenge to the theory comes from a *sustained* increase in stock prices, such as occurred in technology stocks during the late 1990s and early 2000s. From late 1998 through early 2000, the NASDAQ composite index more than tripled in value before all those gains (and then some) were wiped out over the two years that followed.

FIGURE 11-6: NASDAQ COMPOSITE, 1990–2004

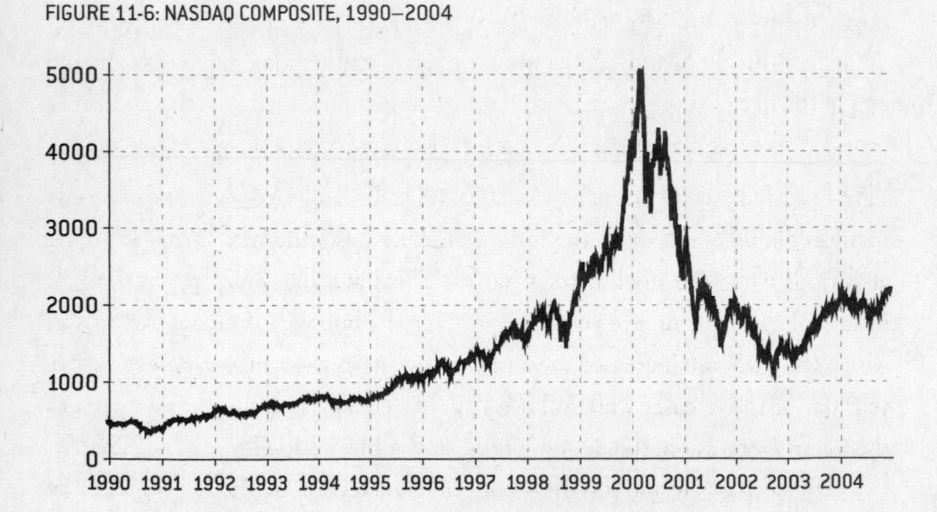

Some of the prices listed on the NASDAQ seemed to be plainly irrational. At one point during the dot-com boom, the market value of technology companies accounted for about 35 percent of the value of all stocks in the United States,[41] implying they would soon come to represent more than a third of private-sector profits. What's interesting is that the technology itself has in some ways exceeded our expectations. Can you imagine what an investor in 2000 would have done if you had shown her an iPad? And told her that, within ten years, she could use it to browse the Internet on an airplane flying 35,000 feet over Missouri and make a Skype call* to her family in Hong Kong? She would have bid Apple stock up to infinity.

Nevertheless, ten years later, in 2010, technology companies accounted for only about 7 percent of economic activity.[42] For every Apple, there were dozens of companies like Pets.com that went broke. Investors were behaving as though every company would be a winner, that they wouldn't have to outcompete each other, leading to an utterly unrealistic assumption about the potential profits available to the industry as a whole.

Still, some proponents of efficient-market hypothesis continue to resist the notion of bubbles. Fama, in what was otherwise a very friendly conversation, recoiled when I so much as mentioned the b-word. "That term has totally lost its meaning," he told me emphatically. "A bubble is something that has a predictable ending. If you can't tell you're in a bubble, it's not a bubble."

In order for a bubble to violate efficient-market hypothesis, it needs to be predictable in real-time. Some investors need to identify it as it is happening and then exploit it for a profit.

Identifying a bubble is of course much easier with the benefit of hindsight—but frankly, it does not seem all that challenging to do so in advance, as many economists did while the housing bubble was underway. Simply looking at periods when the stock market has increased at a rate much faster than its historical average can give you some inkling of a bubble. Of the eight times in which the S&P 500 increased in value by twice its long-term average over a five-year period,[43] five cases were followed by a severe and notorious crash, such as the Great Depression, the dot-com bust, or the Black Monday crash of 1987.[44]

A more accurate and sophisticated bubble-detection method is proposed by the Yale economist Robert J. Shiller, whose prescient work on the housing bub-

---

* Although this is technologically feasible, it will not make your flight attendant happy.

ble I discussed in chapter 1. Shiller is best known for his book *Irrational Exuberance*. Published right as the NASDAQ achieved its all-time high during the dot-com bubble, the book served as an antidote to others, such as *Dow 36,000*, *Dow 40,000* and *Dow 100,000*[45] that promised prices would keep going up, instead warning investors that stocks were badly overpriced on the basis of the fundamentals.

In theory, the value of a stock is a prediction of a company's future earnings and dividends. Although earnings may be hard to predict, you can look at what a company has made in the recent past (Shiller's formula uses the past ten years of earnings) and compare it with the value of the stock. This calculation—known as the P/E or price-to-earnings ratio—has gravitated toward a value of about 15 over the long run, meaning that the market price per share is generally about fifteen times larger than a company's annual profits.

There are exceptions in individual stocks, and sometimes they are justified. A company in an emerging industry (say, Facebook) might reasonably expect to make more in future years than in past ones. It therefore deserves a higher P/E ratio than a company in a declining industry (say, Blockbuster Video). Shiller, however, looked at the P/E ratio averaged across all companies in the S&P 500. In theory, over this broad average of businesses, the high P/E ratios for companies in emerging industries should be balanced out by those in declining ones and the market P/E ratio should be fairly constant across time.

But Shiller found that this had not been the case. At various times, the P/E ratio for all companies in the S&P 500 ranged everywhere from about 5 (in 1921) to 44 (when Shiller published his book in 2000). Shiller found that these anomalies had predictable-seeming consequences for investors. When the P/E ratio is 10, meaning that stocks are cheap compared with earnings, they have historically produced a real return[46] of about 9 percent per year, meaning that a $10,000 investment would be worth $22,000 ten years later. When the P/E ratio is 25, on the other hand, a $10,000 investment in the stock market has historically been worth just $12,000 ten years later. And when they are *very* high, above about 30—as they were in 1929 or 2000—the expected return has been negative.

However, these pricing patterns would not have been very easy to profit from unless you were very patient. They've become meaningful only in the long term, telling you almost nothing about what the market will be worth one month or one year later. Even looking several years in advance, they have only limited predictive power. Alan Greenspan first used the phrase "irrational exuberance" to describe technology stocks in December 1996,[47] at which point the P/E ratio of the S&P 500 was 28—not far from the previous record of 33 in 1929 in advance of Black Tuesday and the Great Depression. The NASDAQ was more richly valued still. But the peak of the bubble was still more than three years away. An investor with perfect foresight, who had bought the NASDAQ on the day that Greenspan made his speech, could have nearly quadrupled his money if he sold out at exactly the right time. Instead, it's really only at time horizons ten or twenty years out that these P/E ratios have allowed investors to make reliable predictions.

There is very little that is truly certain about the stock market,* and even this pattern could reflect some combination of signal and noise.[48] Still, Shiller's findings are backed by strong theory as well as strong empirical evidence, since his focus on P/E ratios ties back to the fundamentals of stock market valuation, making it more likely that they have evinced something real.

So how could stock prices be so predictable in the long run if they are so unpredictable in the short run? The answer lies in how traders behave in the competitive pressures that traders face—both from their rival firms and from their clients and their bosses.

Much of the theoretical appeal of efficient-market hypothesis is that errors in stock prices (like Bayesian beliefs) should correct themselves. Suppose you've observed that the stock price of MGM Resorts International, a large gambling company, increases by 10 percent every Friday, perhaps because traders are subconsciously looking forward to blowing their profits in Atlantic City over the weekend. On a particular Friday, MGM starts out priced at $100, so you expect it to rise to $110 by the end of the trading day. What should you do? You should

* Take, for instance, the oft-cited statistic that the stock market returns 7 percent annually after dividends and inflation. This is just a historical average. Reliable stock market data only goes back 120 years or so—not all that much data if you really want to know about the long run. Statistical tests suggest that the true long-run return—what we might expect over the *next* 120 years—could be anywhere from 3 percent to 10 percent instead of 7 percent. The answer to what economists call the "equity premium puzzle"—why stocks have returned so much more money than bonds in a way that is disproportionate to the risks they entail—may simply be that the returns stocks achieved in the twentieth century were anomalous, and the true long-run return is not as high as 7 percent.

FIGURE 11-7: TRAILING P/E RATIOS AND STOCK MARKET RETURNS

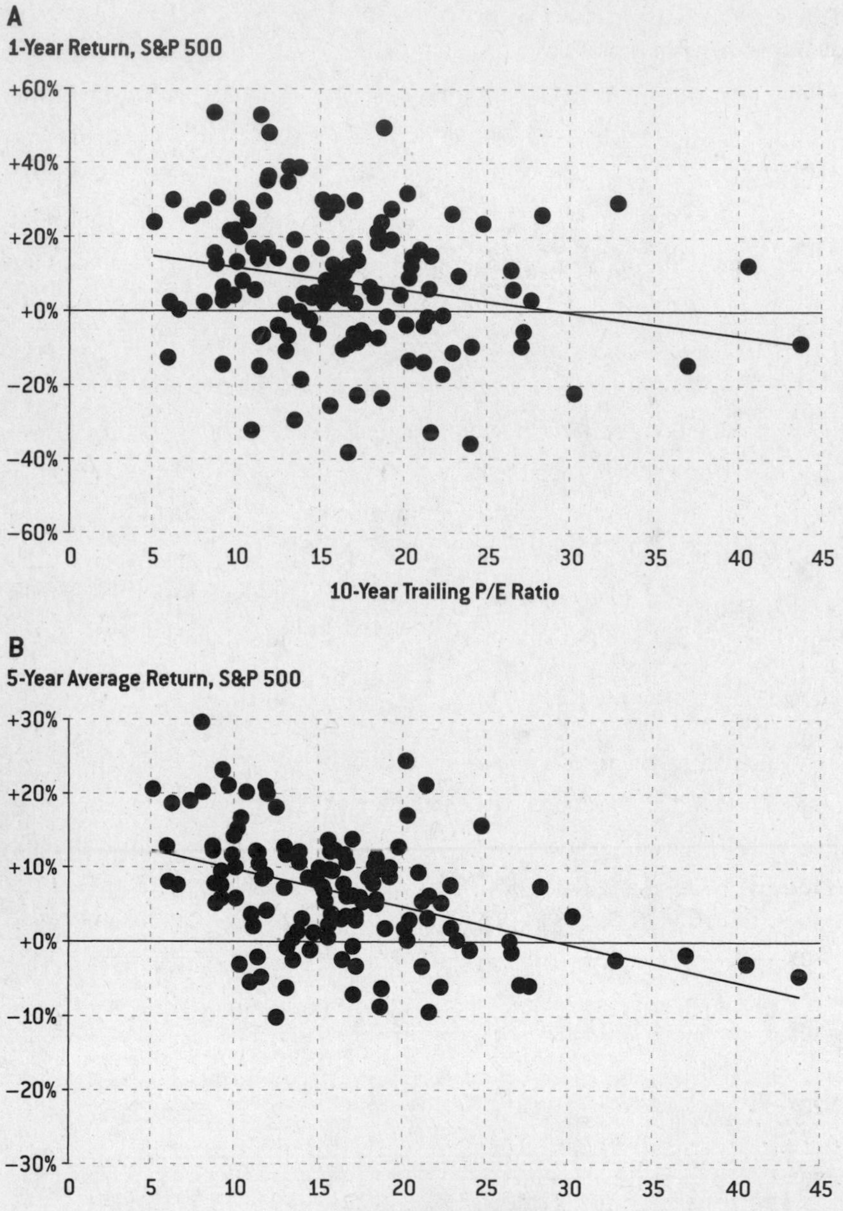

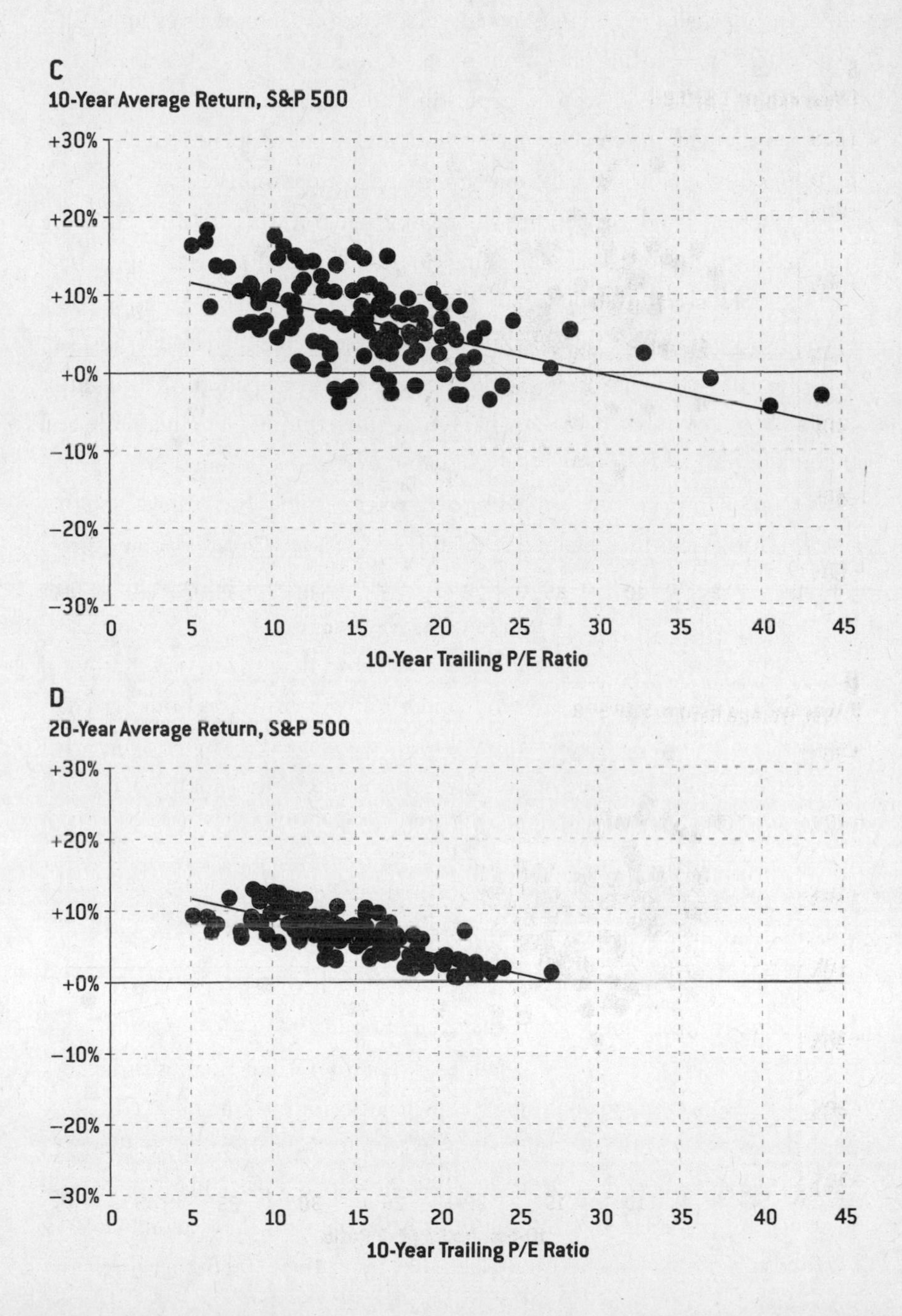
C
10-Year Average Return, S&P 500
+30%
+20%
+10%
+0%
−10%
−20%
−30%
0 5 10 15 20 25 30 35 40 45
10-Year Trailing P/E Ratio
D
20-Year Average Return, S&P 500
+30%
+20%
+10%
+0%
−10%
−20%
−30%
0 5 10 15 20 25 30 35 40 45
10-Year Trailing P/E Ratio

buy the stock, of course, expecting to make a quick profit. But when you buy the stock, its price goes up. A large enough trade[49] might send the price up to $102 from $100. There's still some profit there, so you buy the stock again and its price rises to $104. You keep doing this until the stock reaches its fair price of $110 and there are no more profits left. But look what happened: in the act of detecting this pricing anomaly, you have managed to eliminate it.

In the real world, the patterns will be nowhere near this obvious. There are millions of traders out there, including hundreds of analysts who concentrate on the gambling industry alone. How likely is it that you will have been the only one to have noticed that this stock always rises by 10 percent on Friday? Instead, you're usually left fighting over some scraps: a statistical pattern that might or might not be meaningful, that might or might not continue into the future and that might or might not be profitable enough to cover your transaction costs—and that other investors are competing with you to exploit. Nevertheless, all that competition means that the market should quickly adjust to large pricing errors and that the small ones may not be worth worrying about. At least that's how the theory goes.

But most traders, and *especially* the most active traders, are very short-term focused. They will arbitrage any profit opportunity that involves thinking a day, a month, or perhaps a year ahead, but they may not care very much about what happens beyond that. There may still be some predictability out there; it is just not within their job description to exploit it.

The Stampeding Herd

Henry Blodget first came to the nation's attention in 1998. After some meandering years split between freelance journalism and teaching English in Japan,[50] he had settled in at a job analyzing Internet stocks for the company CIBC Oppenheimer. As attention to Internet stocks increased, attention to Blodget's analysis did as well, and in December 1998 he issued a particularly bold call[51] in which he predicted that the shares of Amazon.com, then valued at $243, would rise to $400 within a year. In fact, they broke the $400 barrier barely *two weeks* later.[52]

Such was the mania of the time that this may even have been a self-fulfilling prophecy: Amazon stock jumped nearly 25 percent[53] within a few hours on the basis of Blodget's recommendation. The call helped catapult Blodget to fame and he took a multimillion-dollar job as an analyst for Merrill Lynch. Blodget has a particular gift[54] for distilling the zeitgeist of the market into coherent sentences. "What investors are buying," Blodget said of Internet stocks in 1998,[55] "is a particular vision of the future." His way with words and reputation for success led to ubiquitous television and radio appearances.

Blodget's call on Amazon still looks pretty good today: the shares he recommended at a price of $243 in 1998 would have traded as high as $1,300 by 2011 if measured on the same scale.[56] He urged investors to pay for value, concentrating on industry leaders like Amazon, Yahoo!, and eBay and noting that most of the smaller companies would "merge, go bankrupt, or quietly fade away."[57] In private correspondence, he trashed small companies with dubious business strategies—LifeMinders, Inc., 24/7 Media, and InfoSpace—all of which turned out to be pretty much worthless and lost 95 to 100 percent of their value.

The problem, the big problem, is that despite criticizing them privately, Blodget had recommended stocks like LifeMinders publicly, maintaining buy ratings on them and defending them on TV. Moreover, the discrepancies seemed to favor companies with whom Merrill did banking business.[58] Later charged with fraud by the Securities and Exchange Commission,[59] Blodget disputed some details of the case but eventually settled for a $4 million fine[60] and a lifetime ban from stock trading.

Blodget knows that whatever he says about Wall Street will be taken skeptically; a piece he once wrote for *Slate* magazine on the Martha Stewart trial had a 1,021-word disclosure statement attached to it.[61] He has had time, however, to digest the work of economists like Fama and Shiller and compare it with his actual experience as a Wall Street insider. He has also embarked on a new career as a journalist—Blodget is now the CEO of the highly successful blogging empire Business Insider. All this has given him a mature if somewhat jaded perspective on the lives of analysts and traders.

"If you talk to a lot of investment managers," Blodget told me, "the practical reality is they're thinking about the next week, possibly the next month or quarter. There isn't a time horizon; it's how you're doing *now*, relative to your competitors. You really only have ninety days to be right, and if you're wrong within ninety days, your clients begin to fire you. You get shamed in the media, and your performance goes to hell. Fundamentals do not help you with that."

Consider what would happen if a trader had read Shiller's book and accepted its basic premise that high P/E ratios signal an overvalued market. However, the trader cared only about the next ninety days. Historically, even when the P/E ratio in the market has been above 30—meaning that stock valuations are twice as high as they are ordinarily—the odds of a crash[62] over the next ninety days have been only about 4 percent.

If the trader had an unusually patient boss and got to look a whole year ahead, he'd find that the odds of a crash rose to about 19 percent (figure 11-8). This is about the same as the odds of losing a game of Russian roulette. The trader knows that he cannot play this game too many times before he gets burned. But what, realistically, are his alternatives?

FIGURE 11-8: HISTORICAL ODDS OF STOCK MARKET CRASH WITHIN ONE YEAR

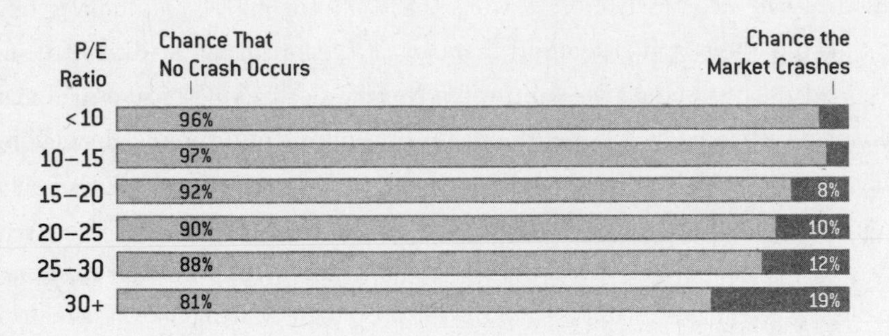

This trader must make a call—buy or sell. Then the market will crash or it will not. So there are four basic scenarios to consider. First, there are the two cases in which he turns out to have made the right bet:

- **The trader buys and the market rises.** In this case, it's business as usual. Everyone is happy when the stock market makes money. The trader gets a six-figure bonus and uses it to buy a new Lexus.
- **The trader sells and the market crashes.** If the trader anticipates a crash and

a crash occurs, he will look like a genius for betting on it when few others did. There's a chance that he'll get a significantly better job—as a partner at a hedge fund, for instance. Still, even geniuses aren't always in demand after the market crashes and capital is tight. More likely, this will translate into something along the lines of increased media exposure: a favorable write-up in the *Wall Street Journal*, a book deal, a couple of invitations to cool conferences, and so forth.

Which of these outcomes you prefer will depend significantly on your personality. The first is great for someone who enjoys the Wall Street life and likes to fit in with the crowd; the second for someone who enjoys being an iconoclast. It may be no coincidence that many of the successful investors profiled in Michael Lewis's *The Big Short*, who made money betting against mortgage-backed securities and other bubbly investments of the late 2000s, were social misfits to one degree or another.

But now consider what happens when the investor gets his bet *wrong*. This choice is much clearer.

- **The trader buys but the market crashes.** This is no fun: he's lost his firm a lot of money and there will be no big bonus and no new Lexus. But since he's stayed with the herd, most of his colleagues will have made the same mistake. Following the last three big crashes on Wall Street, employment at securities firms decreased by about 20 percent.[63] That means there is an 80 percent chance the trader keeps his job and comes out okay; the Lexus can wait until the next bull market.
- **The trader sells but the market rises.** This scenario, however, is a disaster. Not only will the trader have significantly underperformed his peers—he'll have done so after having stuck his neck out and screaming that they were fools. It is extremely likely that he will be fired. And he will not be well-liked, so his prospects for future employment will be dim. His career earnings potential will have been dramatically reduced.

If I'm this trader, a 20 percent chance of a crash would be nowhere near enough for me to take the sell side of the bet. Nor would a 50 percent chance.

I'd want a crash to be a near-certainty before I'd be ready to take the plunge, and I'd want everyone else to be in the sinking ship with me.

Indeed, the big brokerage firms tend to avoid standing out from the crowd, downgrading a stock only after its problems have become obvious.[64] In October 2001, fifteen of the seventeen analysts following Enron still had a "buy" or "strong buy" recommendation on the stock[65] even though it had already lost 50 percent of its value in the midst of the company's accounting scandal. Even if these firms know that the party is coming to an end, it may nevertheless be in their best interest to prolong it for as long as they can. "We thought it was the eighth inning, and it was the ninth," the hedge fund manager Stanley Druckenmiller told the *New York Times*[66] in April 2000 after his Quantum Fund had lost 22 percent of its value in just a few months. Druckenmiller knew that technology stocks were overvalued and were bound to decline—he just did not expect it to happen so soon.

In today's stock market, most trades are made with someone else's money (in Druckenmiller's case, mostly George Soros's). The 1990s and 2000s are sometimes thought of as the age of the day trader. But holdings by institutional investors like mutual funds, hedge funds, and pensions have increased at a much faster rate (figure 11-9). When Fama drafted his thesis in the 1960s, only about 15 percent of stocks were held by institutions rather than individuals.[67] By 2007, the percentage had risen to 68 percent.[68, 69]

These statistics represent a potential complication for efficient-market hypothesis: when it's not your own money on the line but someone else's, your incentives may change. Under some circumstances, in fact, it may be quite rational for traders to take positions that lose money for their firms and their investors if it allows them to stay with the herd and reduces their chance of getting fired.[70] There is significant theoretical and empirical evidence[71] for herding behavior among mutual funds and other institutional investors.[72] "The answer as to why bubbles form," Blodget told me, "is that it's in everybody's interest to keep markets going up."

Everything I've described up to this point could result from perfectly rational behavior on the part of individual participants in the market. It is almost as though the investors are responding *hyper*-rationally to their career incentives,

FIGURE 11-9: INDIVIDUAL AND INSTITUTIONAL INVESTOR TOTAL EQUITY HOLDINGS, UNITED STATES (ADJUSTED FOR INFLATION)[73]

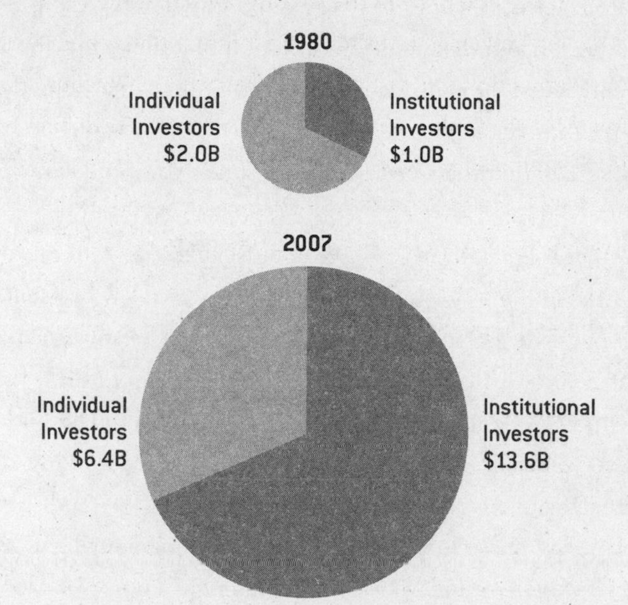

but not necessarily seeking to maximize their firm's trading profits. One conceit of economics is that markets as a whole can perform fairly rationally, even if many of the participants within them are irrational. But irrational behavior in the markets may result precisely because individuals are responding rationally according to their incentives. So long as most traders are judged on the basis of short-term performance, bubbles involving large deviations of stock prices from their long-term values are possible—and perhaps even inevitable.

Why We Herd

Herding can also result from deeper psychological reasons. Most of the time when we are making a major life decision, we're going to want some input from our family, neighbors, colleagues, and friends—and even from our competitors if they are willing to give it.

If I have a forecast that says Rafael Nadal has a 30 percent chance of winning Wimbledon and all the tennis fans I encounter say he has a 50 percent chance instead, I'd need to be very, very sure of myself to stick to my original position. Unless I had some unique information that they weren't privy to, or I were truly convinced that I had spent much more time analyzing the problem than they had, the odds are that my iconoclastic view is just going to lose money. The heuristic of "follow the crowd, especially when you don't know any better" usually works pretty well.

And yet, there are those times when we become too trusting of our neighbors—like in the 1980s "Just Say No" commercials, we do something because Everyone Else Is Doing it Too. Instead of our mistakes canceling one another out, which is the idea behind the wisdom of crowds,[74] they instead begin to reinforce one another and spiral out of control. The blind lead the blind and everyone falls off a cliff. This phenomenon occurs fairly rarely, but it can be quite disastrous when it does.

Sometimes we may also infer that the most confident-acting neighbor must be the best forecaster and follow his lead—whether he knows what he's doing. In 2008, for reasons that are still somewhat unclear, a rogue trader on Intrade started buying huge volumes of John McCain stock in the middle of the night when there was absolutely no news, while dumping huge volumes of Barack Obama stock.[75] Eventually the anomalies were corrected, but it took some time—often four to six hours—before prices fully rebounded to their previous values. Many traders were convinced that the rogue trader knew something they didn't—perhaps he had inside information about some impending scandal?

This is herding. And there's evidence that it's becoming more and more common in markets. The correlations in the price movements between different stocks and different types of assets are becoming sharper and sharper,[76] suggesting that everybody is investing in a little bit of everything and trying to exploit many of the same strategies. This is another of those Information Age risks: we share so much information that our independence is reduced. Instead, we seek out others who think just like us and brag about how many "friends" and "followers" we have.

In the market, prices may occasionally follow the lead of the worst investors. They are the ones making most of the trades.

Overconfidence and the Winner's Curse

A common experiment in economics classrooms, usually employed when the professor needs some extra lunch money, is to hold an auction wherein students submit bids on the number of pennies in a jar.[77] The student with the highest bid pays the professor and wins the pennies (or an equivalent amount in paper money if he doesn't like loose change). Almost invariably, the winning student will find that he has paid too much. Although some of the students' bids are too low and some are about right, it's the student who most overestimates the value of the coins in the jar who is obligated to pay for them; the worst forecaster takes the "prize." This is known as the "winner's curse."

The stock market has some of the same properties. Every now and then, the trader most willing to buy a stock will be the one who really does have some unique or proprietary insight about a company. But most traders are about average, and they're using mostly the same models with most of the same data. If they decide a stock is substantially undervalued when their peers disagree, most of the time it will be because they've placed too much confidence in their forecasting abilities and mistaken the noise in their model for a signal.

There is reason to suspect that of the various cognitive biases that investors suffer from, overconfidence is the most pernicious. Perhaps *the* central finding of behavioral economics is that most of us are overconfident when we make predictions. The stock market is no exception; a Duke University survey of corporate CFOs,[78] whom you might expect to be fairly sophisticated investors, found that they radically overestimated their ability to forecast the price of the S&P 500. They were constantly surprised by large movements in stock prices, despite the stock market's long history of behaving erratically over short time periods.

The economist Terrance Odean of the University of California at Berkeley constructed a model in which traders had this flaw and this flaw only: they were overconfident in estimating the value of their information. Otherwise, they were perfectly rational.[79] What Odean found was that overconfidence alone was enough to upset an otherwise rational market. Markets with overconfident traders will produce extremely high trading volumes, increased volatility, strange correlations in stock prices from day to day, and below-average returns for active traders—all the things that we observe in the real world.

Why It's Hard to Bust Bubbles

And yet, if the market is trending toward a bubble, efficient-market hypothesis would imply that some traders should step in to stop it, expecting to make enormous profits by selling the stock short. Eventually, the theory will be right: all bubbles burst. However, they can take a long time to pop.

The way to bet against an overvalued stock is to short it: you borrow shares at their current price with the promise of returning them at some point in the future based on their price at that time. If the stock goes down in value, you will make money on this trade. The problem comes if the stock goes up in value, in which case you will owe more money than you had borrowed originally. Say, for instance, that you had borrowed five hundred shares of the company InfoSpace on March 2, 1999, when they cost $27, promising to return them one year later. Borrowing these shares would have cost you about $13,400. One year later, however, InfoSpace was trading at $482 per share, meaning that you would be obligated to return about $240,000—almost *twenty times* the initial value of your investment. Although this bet would have turned out to be brilliant in the end—InfoSpace later traded for as little as $1.40 per share—you would have taken a bath and your ability to make future investments would be crippled. In fact, the losses from shorting a stock are theoretically *unlimited*.

In practice, the investor loaning you the shares can demand them back anytime she wants, as she assuredly will if she thinks you are a credit risk. But this also means she can quit anytime she's ahead, an enormous problem since overvalued stocks often become even more overvalued before reverting back to fairer prices. Moreover, since the investor loaning you the stocks knows that you may have to dig into your savings to pay her back, she will charge you a steep interest rate for the privilege. Bubbles can take months or years to deflate. As John Maynard Keynes said, "The market can stay irrational longer than you can stay solvent."

The Price Isn't Right

At other times, investors may not have the opportunity to short stocks at all. One somewhat infamous example, documented by the University of Chicago economists Richard Thaler and Owen Lamont,[80] is when the company 3Com spun off shares of its mobile phone subsidiary Palm into a separate stock offering. 3Com kept most of Palm's shares for itself, however, so a trader could also invest in Palm simply by buying 3Com stock. In particular, 3Com stockholders were guaranteed to receive three shares in Palm for every two shares in 3Com that they held. This seemed to imply Palm shares could trade at an absolute maximum of two-thirds the value of 3Com shares.

Palm, however, was a sexy stock at the time, whereas 3Com, although it had consistently earned a profit, had a stodgy reputation. Rather than being worth less than 3Com shares, Palm shares instead traded at a *higher* price for a period of several months. This should have allowed an investor, regardless of what he thought about Palm and 3Com, to make a guaranteed profit by buying 3Com shares and shorting Palm. On paper, it was a virtually no-risk arbitrage opportunity,[81] the equivalent of exchanging $1,000 for £600 British pounds at Heathrow Airport in London knowing you could exchange the £600 for $1,500 when you got off the flight in New York.

But shorting Palm proved to be very difficult. Few holders of Palm stock were willing to loan their shares out, and they had come to expect quite a premium for doing so: an interest rate of well over 100 percent per year.[82] This pattern was common during the dot-com bubble:[83] shorting dot-com stocks was prohibitively expensive when it wasn't literally impossible.

I met with Thaler after we both spoke at a conference in Las Vegas, where we ate an overpriced sushi dinner and observed the action on the Strip. Thaler, although a friend and colleague of Fama's, has been at the forefront of a discipline called behavioral economics that has been a thorn in the side of efficient-market hypothesis. Behavioral economics points out all the ways in which traders in the real-world are not as well-behaved as in the model.

"Efficient-market hypothesis has two components," Thaler told me between bites of *toro*. "One I call the No Free Lunch component, which is that you can't beat the market. Eugene Fama and I mostly agree about this component. The part he doesn't like to talk about is the Price Is Right component."

There is reasonably strong evidence for what Thaler calls No Free Lunch—it is difficult (although not literally impossible) for any investor to beat the market over the long-term. Theoretically appealing opportunities may be challenging to exploit in practice because of transaction costs, risks, and other constraints on trading. Statistical patterns that have been reliable in the past may prove to be ephemeral once investors discover them.

The second claim of efficient-market hypothesis, what Thaler refers to as the Price Is Right component, is more dubious. Examples like the discrepancy in pricing between Palm and 3Com stock simply could not have arisen if the price were right. You had the same commodity (the value of an interest in Palm) trading at two different and wildly divergent prices: at least one of them *must* have been wrong.

There are asymmetries in the market: bubbles are easier to detect than to burst. What this means is that the ultimatum we face in Bayesland—if you really think the market is going to crash, why aren't you willing to bet on it?—does not necessarily hold in the real world, where there are constraints on trading and on capital.

Noise in Financial Markets

There is a kind of symbiosis between the irrational traders and the skilled ones—just as, in a poker game, good players need some fish at the table to make the game profitable to play in. In the financial literature, these irrational traders are known as "noise traders." As the economist Fisher Black wrote in a 1986 essay simply called "Noise":

> Noise makes trading in financial markets possible, and thus allows us to observe prices for financial assets. [But] noise also causes markets to be somewhat inefficient. . . . Most generally, noise makes it very difficult to test either practical or academic theories about the way that financial or economic markets work. We are forced to act largely in the dark.[84]

Imagine there were no noise traders in the market. Everyone is betting on real information—signal. Prices are rational pretty much all the time, and the market is efficient.

But, if you think a market is efficient—efficient enough that you can't really beat it for a profit—then it would be irrational for you to place any trades. In fact, efficient-market hypothesis is intrinsically somewhat self-defeating. If all investors believed the theory—that they can't make any money from trading since the stock market is unbeatable—there would be no one left to make trades and therefore no market at all.

The paradox reminds me of an old joke among economists. One economist sees a $100 bill sitting on the street and reaches to grab it. "Don't bother," the other economist says. "If it were a real $100 bill, someone would already have picked it up." If everyone thought this way, of course, nobody would bother to pick up $100 bills until a naïve young lad who had never taken an economics course went about town scooping them up, then found out they were perfectly good and exchanged them for a new car.

The most workable solution to the paradox, identified by the Nobel Prize–winning economist Joseph Stiglitz and his colleague Sanford Grossman many years ago,[85] is to allow *some* investors to make *just a little* bit of profit: just enough to adequately compensate them for the effort they put in. This would not actually be all that difficult to accomplish in the real world. Although it might seem objectionable to you that securities analysts on Wall Street are compensated at $75 billion per year, this pales in comparison to the roughly *$17 trillion* in trades[86] that are made at the New York Stock Exchange alone. So long as they beat the market by 0.5 percent on their trades, they would be revenue-positive for their firms.

The equilibrium proposed by Stiglitz is one in which *some* minimal profits are available to *some* investors. Efficient-market hypothesis can't literally be true. Although some studies (like mine of mutual funds on E*Trade) seem to provide evidence for Fama's view that no investor can beat the market at all, others are more equivocal,[87] and a few[88] identify fairly tangible evidence of trading skill and excess profits. It probably isn't the mutual funds that are beating Wall Street; they follow too conventional a strategy and sink or swim together. But some hedge funds (not most) very probably beat the market,[89] and some proprietary trading desks at elite firms like Goldman Sachs almost certainly do. There also seems to be rather clear evidence of trading skill among options traders,[90] people who make bets on probabilistic assessments of how much a share price might move.* And while most individual, retail-level investors make common mistakes like trading too often and do *worse* than the market average, a select handful probably do beat the street.[91]

Buy High, Sell Low

You should not rush out and become an options trader. As the legendary investor Benjamin Graham advises, a little bit of knowledge can be a dangerous thing in the stock market.[92] After all, *any investor can do as well as the average investor with almost no effort*. All he needs to do is buy an index fund that tracks the average of the S&P 500.[93] In so doing he will come extremely close to replicating the average portfolio of every other trader, from Harvard MBAs to noise traders to George Soros's hedge fund manager. You have to be *really* good—or foolhardy—to turn that proposition down. In the stock market, the competition is fierce. The average trader, particularly in today's market, in which trading is dominated by institutional investors, is someone who will have ample credentials, a high IQ, and a fair amount of experience.

"Everybody thinks they have this supersmart mutual fund manager," Henry Blodget told me. "He went to Harvard and has been doing it for twenty-five years. How can he not be smart enough to beat the market? The answer is: Because there are nine million of him and they all have a fifty-million-dollar budget and computers that are collocated in the New York Stock Exchange. How could you possibly beat that?"

In practice, most everyday investors do not do even that well. Gallup and other polling organizations periodically survey Americans[94] on whether they think it is a good time to buy stocks. Historically, there has been a strong

* This is no surprise given how poor most of us—including most of us who invest for a living—are at estimating probabilities. The few who are good at it have the potential to clean up. However, most options traders receive a poor return, and it is a very risky activity on the whole.

relationship between these numbers and stock market performance—but the relationship runs in the exact *opposite* direction of what sound investment strategy would dictate. Americans tend to think it's a good time to buy when P/E ratios are inflated and stocks are overpriced. The highest figure that Gallup ever recorded in their survey was in January 2000, when a record high of 67 percent of Americans thought it was a good time to invest. Just two months later, the NASDAQ and other stock indices began to crash. Conversely, only 26 percent of Americans thought it was a good time to buy stocks in February 1990—but the S&P 500 almost quadrupled in value over the next ten years (figure 11-10).

FIGURE 11-10: PUBLIC SENTIMENT ABOUT STOCKS AND 10-YEAR ANNUAL RETURNS

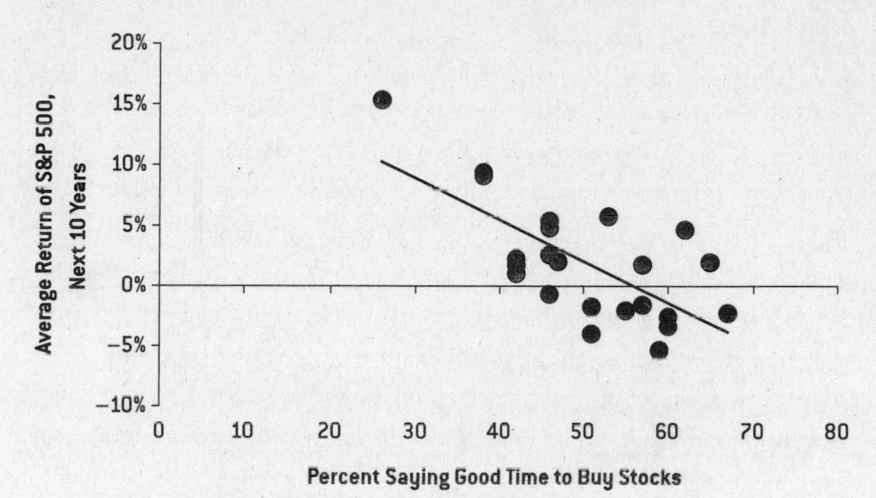

Most of us will have to fight these instincts. "Investors need to learn how to do exactly the reverse of what their fight-or-flight mechanism is telling them to do," Blodget told me. "When the market crashes, that is the time to get excited and put your money into it. It's not the time to get scared and pull money out. What you see instead is the more the market drops, the more money comes out of it. Normal investors are obliterated, because they continuously do exactly the wrong thing."

As Blodget says, these mistakes can be exceptionally costly for investors. Suppose that you had invested $10,000 in the S&P 500 in 1970, planning to cash it out forty years later upon your retirement in 2009. There were plenty of ups and downs during this period. But if you stuck with your investment through thick and thin, you would have made a profit of $63,000 when you retired, adjusted for inflation and not counting the original principal.[95] If instead you had "played it safe" by pulling your money out of the market every time it had fallen more than 25 percent from its previous peak, waiting until the market rebounded to 90 percent of its previous high before reinvesting, you would have just $18,000 in profit—a meager return of 2.6 percent per year.[96] Many investors, unfortunately, behave in exactly this fashion. Worse yet, they tend to make their initial investments at times when the market is overvalued, in which case they may struggle to make a positive return of any kind over the long term.

The next time the market is in a bubble, you will see signals like the flashing lights in a casino drawing you ever closer: the CNBC ticker full of green arrows . . . *Wall Street Journal* headlines about record returns . . . commercials for online brokerages that make a fortune seem only a mouse-click away. Avoiding buying during a bubble, or selling during a panic, requires deliberate and conscious effort. You need to have the presence of mind to ignore it. Otherwise you *will* make the same mistakes that everyone else is making.

Daniel Kahneman likens the problem to the Müller-Lyer illusion, a famous optical illusion involving two sets of arrows (figure 11-11). The arrows are exactly the same length. But in one case, the ends of the arrows outward, seem to signify expansion and boundless potential. In the other case, they point inward, making them seem self-contained and limited. The first case is analogous to how investors see the stock market when returns have been increasing; the second case is how they see it after a crash.

"There's no way that you can control yourself not to have that illusion," Kahneman told me. "You look at them, and one of the arrows is going to look longer than the other. But you can train yourself to recognize that this is a pattern that causes an illusion, and in that situation, I can't trust my impressions; I've got to use a ruler."

FIGURE 11-11: MÜLLER-LYER ILLUSION

The Other 10 Percent

The cognitive shortcuts that our mind takes—our heuristics—are what get investors into trouble. The idea that something going up will continue to go up couldn't be any more instinctive. It just happens to be completely wrong when it comes to the stock market.

Our instincts related to herding may be an even more fundamental problem. Oftentimes, it will absolutely be right to do what everyone else is doing, or at least to pay some attention to it. If you travel to a strange city and need to pick a restaurant for dinner, you probably want to select the one that has more customers, other things being equal. But occasionally it will backfire: you wind up at the tourist trap.

Likewise, when we were making forecasts in Bayesland, it usually behooved us to pay some attention to our neighbors and adjust our beliefs accordingly, rather than adopt the stubborn and often implausible notion that we knew better than everyone else.

I pay quite a bit of attention to what the consensus view is—what a market like Intrade is saying—when I make a forecast. It is never an absolute constraint. But the further I move away from that consensus, the stronger my evidence has to be before I come to the view that I have things right and everyone else has it wrong. This attitude, I think, will serve you very well most of the time. It implies that although you might occasionally be able to beat markets, it is not something you should count on doing every day; that is a sure sign of overconfidence.

But there are the exceptional cases. Fisher Black estimated that markets are basically rational 90 percent of the time. The other 10 percent of the time, the noise traders dominate—and they can go a little haywire.[97] One way to look at this is that markets are usually *very* right but occasionally *very* wrong. This, incidentally, is another reason why bubbles are hard to pop in the real world.

There might be a terrific opportunity to short a bubble or long a panic once every fifteen or twenty years when one comes along in your asset class. But it's very hard to make a steady career out of that, doing nothing for years at a time.

The Two-Track Market

Some theorists have proposed that we should think of the stock market as constituting two processes in one.[98] There is the signal track, the stock market of the 1950s that we read about in textbooks. This is the market that prevails in the long run, with investors making relatively few trades, and prices well tied down to fundamentals. It helps investors to plan for their retirement and helps companies capitalize themselves.

Then there is the fast track, the noise track, which is full of momentum trading, positive feedbacks, skewed incentives and herding behavior. Usually it is just a rock-paper-scissors game that does no real good to the broader economy—but also perhaps no real harm. It's just a bunch of sweaty traders passing money around.

However, these tracks happen to run along the same road, as though some city decided to hold a Formula 1 race but by some bureaucratic oversight forgot to close one lane to commuter traffic. Sometimes, like during the financial crisis, there is a big accident, and regular investors get run over.

This sort of duality, what the physicist Didier Sornette calls "the fight between order and disorder,"[99] is common in complex systems, which are those governed by the interaction of many separate individual parts. Complex systems like these can at once seem very predictable and very unpredictable. Earthquakes are very well described by a few simple laws (we have a very good idea of the long-run frequency of a magnitude 6.5 earthquake in Los Angeles). And yet they are essentially unpredictable from day to day. Another characteristic of these systems is that they periodically undergo violent and highly nonlinear*

* As Nassim Nicholas Taleb detailed in *The Black Swan* and as Fama also discussed in his thesis, the movement of stock prices does not follow a gentle bell-curve distribution. Instead, stock-price movements are characterized by very occasional but very large swings up or down. The distribution of stock-market crashes can also be modeled fairly well by a power-law distribution, which is the same function that governs the frequency of earthquakes.

phase changes from orderly to chaotic and back again. For Sornette and others who take highly mathematical views of the market, the presence of periodic bubbles seems more or less inevitable, an intrinsic property of the system.

I am partial toward this perspective. My view on trading markets (and toward free-market capitalism more generally) is the same as Winston Churchill's attitude toward democracy.[100] I think it's the worst economic system ever invented—except for all the other ones. Markets do a good job most of the time, but I don't think we'll ever be rid of bubbles.

But if we can't fully prevent the herd behavior that causes bubbles, can we at least hope to detect them while they are occurring? Say you accept Black's premise that the market is behaving irrationally 10 percent of the time. Can we know when we're in that 10 percent phase? Then we might hope to profit from bubbles. Or, less selfishly, we could create softer landings that lessened the need for abhorrent taxpayer bailouts.

Bubble detection does not seem so hopeless. I don't think we're ever going to bat 100 percent, or even 50 percent, but I think we can get somewhere. Some of the bubbles of recent years, particularly the housing bubble, were detected by enormous numbers of people well in advance. And tests like Shiller's P/E ratio have been quite reliable indicators of bubbles in the past.

We could try to legislate our way out of the problem, but that can get tricky. If greater regulation might be called for in some cases, constraints on short-selling—which make it harder to pop bubbles—are almost certainly counterproductive.

What's clear, however, is that we'll never detect a bubble if we start from the presumption that markets are infallible and the price is always right. Markets cover up some of our warts and balance out some of our flaws. And they certainly aren't easy to outpredict. But sometimes the price is wrong.

12

A CLIMATE OF HEALTHY SKEPTICISM

June 23, 1988, was an unusually hot day on Capitol Hill. The previous afternoon, temperatures hit 100 degrees at Washington's National Airport, the first time in decades they reached triple-digits so early in the summer.[1] The NASA climatologist James Hansen wiped his brow—the air-conditioning had inconveniently* ceased to function in the Senate Energy Committee's hearing room—and told the American people they should prepare for more of the same.

The greenhouse effect had long been accepted theory, predicted by scientists to warm the planet.[2] But for the first time, Hansen said, it had begun to produce an unmistakable signal in the temperature record: global temperatures had increased by about 0.4°C since the 1950s, and this couldn't be accounted for by natural variations. "The probability of a chance warming of that magnitude is about 1 percent," Hansen told Congress. "So with 99 percent confi-

* Senator Tim Wirth of Colorado later told PBS's *Frontline* that he and his colleagues had deliberately opened the windows in the hearing room the previous evening to prevent the air-conditioning from functioning effectively.

dence we can state that the warming trend during this time period is a real warming trend."[3]

Hansen predicted more frequent heat waves in Washington and in other cities like Omaha—already the change was "large enough to be noticeable to the average person." The models needed to be refined, he advised, but both the temperature trend and the reasons for it were clear. "It is time to stop waffling so much," Hansen said. "The evidence is pretty strong that the greenhouse effect is here."[4]

With nearly a quarter century having passed since Hansen's hearing, it is time to ask some of the same questions about global warming that we have of other fields in this book. How right, or how wrong, have the predictions about it been so far? What are scientists really agreed upon and where is there more debate? How much uncertainty is there in the forecasts, and how should we respond to it? Can something as complex as the climate system really be modeled well at all? Are climate scientists prone to the same problems, like overconfidence, that befall forecasters in other fields? How much have politics and other perverse incentives undermined the search for scientific truth? And can Bayesian reasoning be of any help in adjudicating the debate?

We should examine the evidence and articulate what might be thought of as *healthy* skepticism toward climate predictions. As you will see, this kind of skepticism does not resemble the type that is common in blogs or in political arguments over global warming.

The Noise and the Signal

Many of the examples in this book concern cases where forecasters mistake correlation for causation and noise for a signal. Up until about 1997, the conference of the winning Super Bowl team had been very strongly correlated with the direction of the stock market over the course of the next year. However, there was no credible causal mechanism behind the relationship, and if you had made investments on that basis you would have lost your shirt. The Super Bowl indicator was a false positive.

The reverse can sometimes also be true. Noisy data can obscure the signal, even when there is essentially no doubt that the signal exists. Take a relationship that few of us would dispute: if you consume more calories, you are more likely to become fat. Surely such a basic relationship would show up clearly in the statistical record?

I downloaded data from eighty-four countries for which estimates of both obesity rates and daily caloric consumption are publicly available.[5] Looked at in this way, the relationship seems surprisingly tenuous. The daily consumption in South Korea, which has a fairly meat-heavy diet, is about 3,070 calories per person per day, slightly above the world average. However, the obesity rate there is only about 3 percent. The Pacific island nation of Nauru, by contrast, consumes about as many calories as South Korea per day,[6] but the obsesity rate there is 79 percent. If you plot the eighty-four countries on a graph (figure 12-1) there seems to be only limited evidence of a connection between obesity and calorie consumption; it would not qualify as "statistically significant" by standard tests.*

There are, of course, many conflating factors that obscure the relationship. Certain countries have better genetics, or better exercise habits. And the data is rough: estimating how many calories an adult consumes in a day is challenging.[7] A researcher who took this statistical evidence too literally might incorrectly reject the connection between calorie consumption and obesity, a false negative.

It would be nice if we could just plug data into a statistical model, crunch the numbers, and take for granted that it was a good representation of the real world. Under some conditions, especially in data-rich fields like baseball, that assumption is fairly close to being correct. In many other cases, a failure to think carefully about causality will lead us up blind alleys.

There would be much reason to doubt claims about global warming were it not for their grounding in causality. The earth's climate goes through various warm and cold phases that play out over periods of years or decades or centuries. These cycles long predate the dawn of industrial civilization.

However, predictions are potentially much stronger when backed up by a

---

* As I discuss in chapter 6, the concept of "statistical significance" is very often problematic in practice. But to my knowledge there does not exist a community of "obesity skeptics" who cite statistics like these to justify a diet of Big Macs and Fritos.

FIGURE 12-1: CALORIE CONSUMPTION AND OBESITY RATES IN 84 COUNTRIES

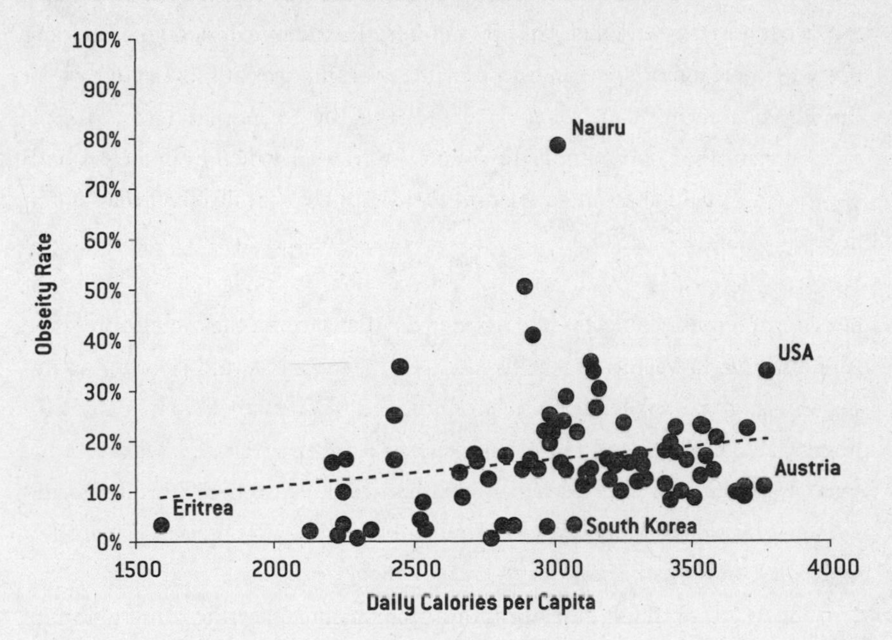

sound understanding of the root causes behind a phenomenon. We do have a good understanding of the cause of global warming: it is the greenhouse effect.

The Greenhouse Effect Is Here

In 1990, two years after Hansen's hearing, the United Nations' International Panel on Climate Change (IPCC) released more than a thousand pages of findings about the science of climate change in its First Assessment Report. Produced over several years by a team of hundreds of scientists from around the globe, the report went into voluminous detail on the potential changes in temperatures and ecosystems, and outlined a variety of strategies to mitigate these effects.

The IPCC's scientists classified just two findings as being absolutely certain, however. These findings did not rely on complex models, and they did not make highly specific predictions about the climate. Instead, they were based on relatively simple science that had been well-understood for more than 150 years and which is rarely debated even by self-described climate skeptics. They remain the most important scientific conclusions about climate change today.

The IPCC's first conclusion was simply that the greenhouse effect exists:

> There is a natural greenhouse effect that keeps the Earth warmer than it otherwise would be.[8]

The greenhouse effect is the process by which certain atmospheric gases—principally water vapor, carbon dioxide (CO_2), methane, and ozone—absorb solar energy that has been reflected from the earth's surface. Were it not for this process, about 30 percent[9] of the sun's energy would be reflected back out into space in the form of infrared radiation. That would leave the earth's temperatures much colder than they actually are: about 0° Fahrenheit or –18° Celsius[10] on average, or the same as a warm day on Mars.[11]

Conversely, if these gases become more plentiful in the atmosphere, a higher fraction of the sun's energy will be trapped and reflected back onto the surface, making temperatures much warmer. On Venus, which has a much thicker atmosphere consisting almost entirely of carbon dioxide, the average temperature is 460°C.[12] Some of that heat comes from Venus's proximity to the sun, but much of it is because of the greenhouse effect.[13]

There is no scenario in the foreseeable future under which the earth's climate will come to resemble that of Venus. However, the climate is fairly sensitive to changes in atmospheric composition, and human civilization thrives within a relatively narrow band of temperatures. The coldest world capital is Ulan Bator, Mongolia, where temperatures average about –1°C (or +30°F) over the course of the year;[14] the warmest is probably Kuwait City, Kuwait, where they average +27°C (+81°F).[15] Temperatures can be hotter or cooler during winter or summer or in sparsely populated areas,[16] but the temperature extremes are modest on an interplanetary scale. On Mercury, by contrast, which has little atmosphere to protect it, temperatures often vary between about –200°C and +400°C over the course of a single day.[17]

The IPCC's second conclusion made an elementary prediction based on the greenhouse effect: as the concentration of greenhouse gases increased in the atmosphere, the greenhouse effect and global temperatures would increase along with them:

> Emissions resulting from human activities are substantially increasing the atmospheric concentrations of the greenhouse gases carbon dioxide, methane, chlorofluorocarbons (CFCs) and nitrous oxide. These increases will enhance the greenhouse effect, resulting on average in additional warming of the Earth's surface. The main greenhouse gas, water vapor, will increase in response to global warming and further enhance it.

This IPCC finding makes several different assertions, each of which is worth considering in turn.

First, it claims that atmospheric concentrations of greenhouse gases like CO_2 are increasing, and as a result of human activity. This is a matter of simple observation. Many industrial processes, particularly the use of fossil fuels, produce CO_2 as a by-product.[18] Because CO_2 remains in the atmosphere for a long time, its concentrations have been rising: from about 315 parts per million (ppm) when CO_2 levels were first directly monitored at the Mauna Loa Observatory in Hawaii in 1959 to about 390 PPM as of 2011.[19]

The second claim, "these increases will enhance the greenhouse effect, resulting on average in additional warming of the Earth's surface," is essentially just a restatement of the IPCC's first conclusion that the greenhouse effect exists, phrased in the form of a prediction. The prediction relies on relatively simple chemical reactions that were identified in laboratory experiments many years ago. The greenhouse effect was first proposed by the French physicist Joseph Fourier in 1824 and is usually regarded as having been proved by the Irish physicist John Tyndall in 1859,[20] the same year that Charles Darwin published *On the Origin of Species*.

The third claim—that water vapor will also increase along with gases like CO_2, thereby enhancing the greenhouse effect—is modestly bolder. Water vapor, not CO_2, is the largest contributor to the greenhouse effect.[21] If there were an increase in CO_2 alone, there would still be some warming, but not as much as has been observed to date or as much as scientists predict going forward. But a basic thermodynamic principle known as the Clausius–Clapeyron relation, which was proposed and proved in the nineteenth century, holds that the atmosphere can retain more water vapor at warmer temperatures. Thus, as CO_2 and other long-lived greenhouse gases increase in concentration and warm the atmosphere, the amount of water vapor will increase as well, multiplying the effects of CO_2 and enhancing warming.

This Isn't Rocket Science

Scientists require a high burden of proof before they are willing to conclude that a hypothesis is incontrovertible. The greenhouse hypothesis has met this standard, which is why the original IPCC report singled it out from among hundreds of findings as the only thing that scientists were absolutely certain about. The science behind the greenhouse effect was simple enough to have been widely understood by the mid- to late nineteenth century, when the lightbulb and the telephone and the automobile were being invented—and not the atomic bomb or the iPhone or the Space Shuttle. The greenhouse effect isn't rocket science.

Indeed, predictions that industrial activity would eventually trigger global warming were made long before the IPCC—as early as 1897[22] by the Swedish chemist Svante Arrhenius, and at many other times[23] before the warming signal produced by the greenhouse signal had become clear enough to be distinguished from natural causes.

It now seems almost quaint to refer to the greenhouse effect. In the mid-1980s, the term *greenhouse effect* was about five times more common in English-language books[24] than the phrase *global warming*. But usage of *greenhouse effect* peaked in the early 1990s and has been in steady decline since. It is now used only about one-sixth as often as the term global warming, and one-tenth as often as the broader term *climate change*.[25]

This change has largely been initiated by climate scientists[26] as they seek to expand the predictive implications of the theory. However, the pullback from speaking about the *causes* of the change—the greenhouse effect—yields predictably misinformed beliefs about it.*

In January 2012, for instance, the *Wall Street Journal* published an editorial[27] entitled "No Need to Panic About Global Warming," which was signed by a set of sixteen scientists and advocates who might be considered global warming skeptics. Accompanying the editorial was a video produced by the *Wall Street Journal* that was captioned with the following phrase:

> A large number of scientists don't believe that carbon dioxide is causing global warming.

In fact, very few scientists doubt this—there is essentially no debate that greenhouse gases cause global warming. Among the "believers" in the theory was the physics professor William Happer of Princeton, who cosigned the editorial and who was interviewed for the video. "Most people like me believe that industrial emissions will cause warming," Happer said about two minutes into the video. Happer takes issue with some of the predictions of global warming's effects, but not with its cause.

I do not mean to suggest that you should just blindly accept a theory in the face of contradictory evidence. A theory is tested by means of its predictions, and the predictions made by climate scientists have gotten some things right and some things wrong. Temperature data is quite noisy. A warming trend might validate the greenhouse hypothesis *or* it might be caused by cyclical factors. A cessation in warming could undermine the theory *or* it might represent a case where the noise in the data had obscured the signal.

But even if you believe, as Bayesian reasoning would have it, that almost all scientific hypotheses should be thought of probabilistically, we should have a greater degree of confidence in a hypothesis backed up by strong and clear causal relationships. Newly discovered evidence that seems to militate against

---

* In this sense, the term *climate change* may be inferior to the more specific term *global warming*. *Climate change* creates the impression that any potential change in our environment—warming or cooling, more precipitation or less—is potentially consistent with the theory. In fact, some of these phenomena (like cooler temperatures) would contradict the predictions made by the theory under most circumstances.

the theory should nevertheless lower our estimate of its likelihood, but it should be weighed in the context of the other things we know (or think we do) about the planet and its climate.

Healthy skepticism needs to proceed from this basis. It needs to weigh the strength of new evidence against the overall strength of the theory, rather than rummaging through fact and theory alike for argumentative and ideological convenience, as is the cynical practice when debates become partisan and politicized.

Three Types of Climate Skepticism

It is hard to imagine a worse time and place to hold a global climate conference than Copenhagen in December, as the United Nations did in in 2009. During the winter solstice there, the days are short and dark—perhaps four hours of decent sunlight—and the temperatures are cold, with the wind whipping off the Øresund, the narrow strait that separates Denmark from Sweden.

Worse yet, the beer is expensive: the high taxes on alcohol and pretty much everything else in Denmark help to pay for a green-technology infrastructure that rivals almost anywhere in the world. Denmark consumes no more energy today than it did in the late 1960s,[28] in part because it is environmentally friendly and in part because of its low population growth. (By contrast, the United States' energy consumption has roughly doubled over the same period.[29]) The implicit message seemed to be that an energy-efficient future would be cold, dark, and expensive.

It is little wonder, then, that the mood at Copenhagen's Bella Center ranged far beyond skepticism and toward outright cynicism. I had gone to the conference, somewhat naively, seeking a rigorous scientific debate about global warming. What I found instead was politics, and the differences seemed irreconcilable.

Delegates from Tuvalu, a tiny, low-lying Pacific island nation that would be among the most vulnerable to rising sea levels, roamed the halls, loudly protesting what they thought to be woefully inadequate targets for greenhouse-gas reduction. Meanwhile, the large nations that account for the vast majority of greenhouse-gas emissions were nowhere near agreement.

President Obama had arrived at the conference empty-handed, having burned much of his political capital on his health-care bill and his stimulus package. Countries like China, India, and Brazil, which are more vulnerable than the United States to climate change impacts because of their geography but are reluctant to adopt commitments that might impair their economic growth, weren't quite sure where to stand. Russia, with its cold climate and its abundance of fossil-fuel resources, was a wild card. Canada, also cold and energy-abundant, was another, unlikely to push for any deal that the United States lacked the willpower to enact.[30] There was some semblance of a coalition among some of the wealthier nations in Europe, along with Australia, Japan, and many of the world's poorer countries in Africa and the Pacific.[31] But global warming is a problem wherein even if the politics are local, the science is not. CO_2 quickly circulates around the planet: emissions from a diesel truck in Qingdao will eventually affect the climate in Quito. Emissions-reductions targets therefore require near-unanimity, and not mere coalition-building, in order to be enacted successfully. That agreement seemed years if not decades away.

I was able to speak with a few scientists at the conference. One of them was Richard Rood, a soft-spoken North Carolinian who once led teams of scientists at NASA and who now teaches a course on climate policy to students at the University of Michigan.

"At NASA, I finally realized that the definition of rocket science is using relatively simple physics to solve complex problems," Rood told me. "The science part is relatively easy. The other parts—how do you develop policy, how do you respond in terms of public health—these are all relatively difficult problems because they don't have as well defined a cause-and-effect mechanism."

As I was speaking with Rood, we were periodically interrupted by announcements from the Bella Center's loudspeaker. "No consensus was found. Therefore I suspend this agenda item," said a French-sounding woman, mustering her best English. But Rood articulated the three types of skepticism that are pervasive in the debate about the future of climate.

One type of skepticism flows from self-interest. In 2011 alone, the fossil fuel industry spent about $300 million on lobbying activities (roughly double what they'd spent just five years earlier).[32, *] Some climate scientists I later spoke with for this chapter used conspiratorial language to describe their activities. But there is no reason to allege a conspiracy when an explanation based on rational self-interest will suffice: these companies have a financial incentive to preserve their position in the status quo, and they are within their First Amendment rights to defend it. What they say should not be mistaken for an attempt to make accurate predictions, however.

A second type of skepticism falls into the category of contrarianism. In any contentious debate, some people will find it advantageous to align themselves with the crowd, while a smaller number will come to see themselves as persecuted outsiders. This may especially hold in a field like climate science, where the data is noisy and the predictions are hard to experience in a visceral way. And it may be especially common in the United States, which is admirably independent-minded. "If you look at climate, if you look at ozone, if you look at cigarette smoking, there is always a community of people who are skeptical of the science-driven results," Rood told me.

Most importantly, there is scientific skepticism. "You'll find that some in the scientific community have valid concerns about one aspect of the science or the other," Rood said. "At some level, if you really want to move forward, we need to respect some of their points of view."

A Forecaster's Critique of Global Warming Forecasts

In climate science, this healthy skepticism is generally directed at the reliability of computer models used to forecast the climate's course. Scott Armstrong, a professor at the Wharton School at the University of Pennsylvania, is such a skeptic. He is also among the small group of people who have devoted their lives to studying forecasting. His book, *Principles of Forecasting*, should be considered canonical to anybody who is seriously interested in the field. I

---

* There is an alternative-energy lobby as well, although it is quite a bit smaller, spending about $30 million per year.

met with Armstrong in his office at Huntsman Hall in Philadelphia. He is seventy-four years old but has a healthy goatee and looks perhaps fifteen years younger.

In 2007, Armstrong challenged Al Gore to a bet. Armstrong posited that what he calls his "no-change" forecast—global temperatures would remain at their 2007 levels—would beat the IPCC's forecast, which predicted continued warming. Gore never accepted the bet, but Armstrong proceeded to publish the results without him. The bet was to be resolved monthly—whichever forecast was closer to the actual temperatures for that month won the round. Through January 2012, Armstrong's no-change forecast had prevailed over the IPCC's forecast of slow-but-steady warming in twenty-nine months out of forty-seven.[33]

FIGURE 12-2: ARMSTRONG–GORE BET

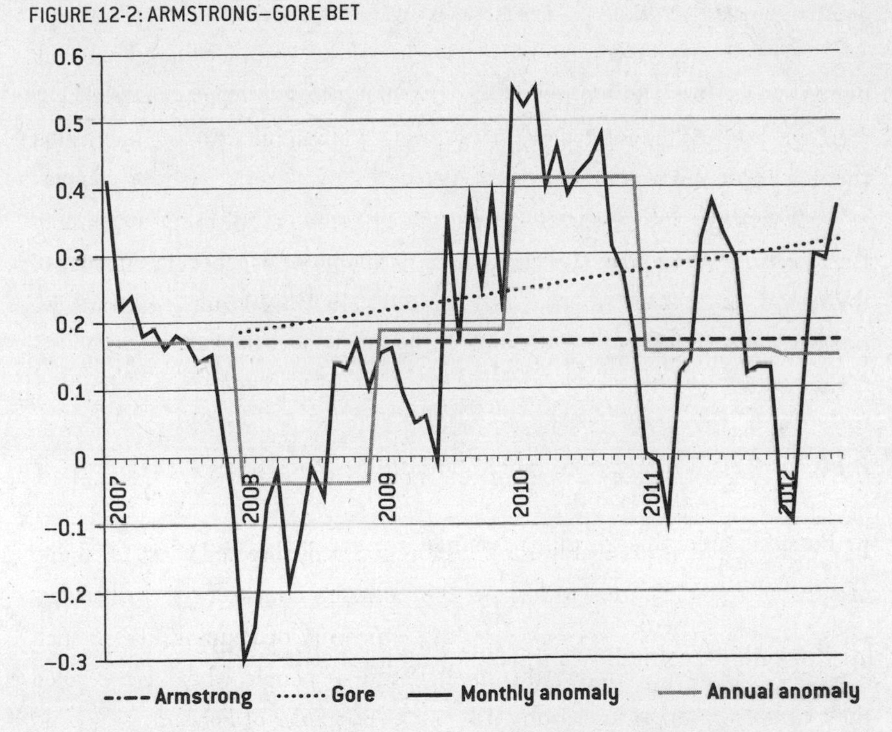

Armstrong told me that he does not doubt the science behind the greenhouse effect per se. "I mean there has been a little bit of warming," he told me. "But nobody is arguing about that over the last 150 years."

But Armstrong does have some grievances with the majority view on global warming.* In 2007, at about the same time he proposed his bet to Gore, Armstrong and his colleague Kesten Green subjected global warming forecasts to what they called an "audit."[34] The idea was to see how well global warming forecasts, especially those produced by the IPCC, abided by his forecasting principles.

The Armstrong and Green paper claimed to find the IPCC forecasts wanting; it suggested that they had failed to abide by seventy-two of eighty-nine forecasting principles. Eighty-nine forecasting principles[35] are probably too many.[36] Nevertheless, most of Armstrong's principles are good rules of thumb for forecasters, and when applied to global warming forecasts they can be simplified into what is essentially a three-pronged critique.

- First, Armstrong and Green contend that *agreement among forecasters is not related to accuracy*—and may reflect bias as much as anything else. "You don't vote," Armstrong told me. "That's not the way science progresses."
- Next, they say the *complexity of the global warming problem* makes forecasting a fool's errand. "There's been no case in history where we've had a complex thing with lots of variables and lots of uncertainty, where people have been able to make econometric models or any complex models work," Armstrong told me. "The more complex you make the model the worse the forecast gets."
- Finally, Armstrong and Green write that the forecasts *do not adequately account for the uncertainty intrinsic to the global warming problem*. In other words, they are potentially overconfident.

Complexity, uncertainty, and the value (or lack thereof) of consensus views are core themes of this book. Each claim deserves a full hearing.

All the Climate Scientists Agree on Some of the Findings

There is an unhealthy obsession with the term *consensus* as it is applied to global warming. Some who dissent from what they see as the consensus view

* Armstrong is an expert at the Heartland Institute, a conservative think tank that has opposed efforts to curb greenhouse emissions.

are proud to acknowledge it and label themselves as heretics.[37] Others, however, have sought strength in numbers, sometimes resorting to dubious techniques like circulating online petitions in an effort to demonstrate how much doubt there is about the theory.* Meanwhile, whenever any climate scientist publicly disagrees with any finding about global warming, they may claim that this demonstrates a lack of consensus about the theory.

Many of these debates turn on a misunderstanding of the term. In formal usage, consensus is not synonymous with unanimity—nor with having achieved a simple majority. Instead, consensus connotes *broad agreement after a process of deliberation*, during which time most members of a group coalesce around a particular idea or alternative. (Such as in: "We reached a consensus to get Chinese food for lunch, but Horatio decided to get pizza instead.")

A consensus-driven process, in fact, often represents an *alternative* to voting. Sometimes when a political party is trying to pick a presidential nominee, one candidate will perform so strongly in early-voting states like Iowa and New Hampshire that all the others drop out. Even though the candidate is far from having clinched the nomination mathematically, there may be no need for the other states to hold a meaningful vote if the candidate has demonstrated that he is acceptable to most key coalitions within the party. Such a candidate can be described as having won the nomination by consensus.

Science, at least ideally, is exactly this sort of deliberative process. Articles are published and conferences are held. Hypotheses are tested, findings are argued over; some survive the scrutiny better than others.

The IPCC is potentially a very good example of a consensus process. Their reports take years to produce and every finding is subject to a thorough—if somewhat byzantine and bureaucratic—review process. "By convention, every review remark has to be addressed," Rood told me. "If your drunk cousin wants to make a remark, it will be addressed."

The extent to which a process like the IPCC's can be expected to produce better predictions is more debatable, however. There is almost certainly some value in the idea that different members of a group can learn from one anoth-

* One such petition, which was claimed to have been signed by 15,000 scientists, later turned up names like Geri Halliwell, a.k.a. Ginger Spice of the Spice Girls, who had apparently given up her career as a pop star to pursue a degree in microbiology.

er's expertise. But this introduces the possibility of groupthink and herding. Some members of a group may be more influential because of their charisma or status and not necessarily because they have the better idea. Empirical studies of consensus-driven predictions have found mixed results, in contrast to a process wherein individual members of a group submit independent forecasts and those are averaged or aggregated together, which can almost always be counted on to improve predictive accuracy.[38]

The IPCC process may reduce the independence of climate forecasters. Although there are nominally about twenty different climate models used in the IPCC's forecast, they make many of the same assumptions and use some of the same computer code; the degree of overlap is significant enough that they represent the equivalent of just five or six independent models.[39] And however many models there are, the IPCC settles on just one forecast that is endorsed by the entire group.

Climate Scientists Are Skeptical About Computer Models

"It's critical to have a diversity of models," I was told by Kerry Emanuel, an MIT meteorologist who is one of the world's foremost theorists about hurricanes. "You do not want to put all your eggs in one basket."

One of the reasons this is so critical, Emanuel told me, is that in addition to the different assumptions these models employ, they also contain different bugs. "That's something nobody likes to talk about," he said. "Different models have different coding errors. You cannot assume that a model with millions and millions of lines of code, literally millions of instructions, that there isn't a mistake in there."

If you're used to thinking about the global warming debate as series of arguments between "skeptics" and "believers," you might presume that this argument emanates from a scientist on the skeptical side of the aisle. In fact, although Emanuel has described himself as conservative and Republican[40]—which is brave enough at MIT—he would probably not think of himself as a global warming skeptic. Instead, he is a member in good standing of the scientific establishment, having been elected to the National Academy of Sciences.

His 2006 book[41] presented a basically "consensus" (and extremely thoughtful and well-written) view on climate science.

Emanuel's concerns are actually quite common among the scientific community: climate scientists are in much broader agreement about some parts of the debate than others. A survey of climate scientists conducted in 2008[42] found that almost all (94 percent) were agreed that climate change is occurring now, and 84 percent were persuaded that it was the result of human activity. But there was much less agreement about the accuracy of climate computer models. The scientists held mixed views about the ability of these models to predict global temperatures, and generally skeptical ones about their capacity to model other potential effects of climate change. Just 19 percent, for instance, thought they did a good job of modeling what sea-rise levels will look like fifty years hence.

Results like these ought to be challenging to anyone who takes a caricatured view of climate science. They should cut against the notion that scientists are injudiciously applying models to make fantastical predictions about the climate; instead, the scientists have as much doubt about the models as many of their critics.[43] However, cinematographic representations of climate change, like Al Gore's *An Inconvenient Truth*, have sometimes been less cautious, portraying a polar bear clinging to life in the Arctic, or South Florida and Lower Manhattan flooding over.[44] Films like these are not necessarily a good representation of the scientific consensus. The issues that climate scientists actively debate are much more banal: for instance, how do we develop computer code to make a good representation of a cloud?

Climate Science and Complexity

Weather forecasters and climatologists often find themselves at odds;[45] a large number of meteorologists are either implicitly or explicitly critical of climate science.

Weather forecasters have endured decades of struggle to improve their forecasts, and they can still expect to receive angry e-mails whenever they get one wrong. It is challenging enough to predict the weather twenty-four hours in advance. So how can climate forecasters, who are applying somewhat analogous techniques, expect to predict what the climate will look like decades from now?

Some of the distinction, as in the case of the term *consensus*, is semantic. Climate refers to the long-term equilibriums that the planet achieves; weather describes short-term deviations from it.[46] Climate forecasters are not attempting to predict whether it will rain in Tulsa on November 22, 2062, although they are perhaps interested in whether it will be rainier on average throughout the Northern Hemisphere.

Meteorologists, nevertheless, have to wrestle with complexity:* the entire discipline of chaos theory developed out of what were essentially frustrated attempts to make weather forecasts. Climatologists have to deal with complexity as well: clouds, for instance, are small-scale phenomena that require a lot of computer power to model accurately, but they can have potentially profound effects on the feedback loops intrinsic to climate forecasts.[47]

The irony is that weather forecasting is one of the success stories in this book. Through hard work, and a melding of computer power with human judgment, weather forecasts have become much better than they were even a decade or two ago. Given that forecasters in most domains are prone to overconfidence, it is admirable that weather forecasters are hard on themselves and their forecasting peers. But the improvements they have made refute the idea that progress is hopeless in the face of complexity.

The improvements in weather forecasts are a result of two features of their discipline. First meteorologists get a lot of feedback—weather predictions play out daily, a reality check that helps keep them well-calibrated. This advantage is not available to climate forecasters and is one of the best reasons to be skeptical about their predictions, since they are made at scales that stretch out to as many as eighty or one hundred years in advance.

Meteorologists also benefit, however, from a strong understanding of the physics of the weather system, which is governed by relatively simple and easily observable laws. Climate forecasters potentially have the same advantage. We

---

* I use the term *complexity* in its plain-English meaning in this chapter meaning that *complex* is largely synonymous with *complicated*. There is also a particular scientific domain called complexity theory, which some scientists distinguish from chaos theory. We explore these very interesting ideas in chapter 5.

can observe clouds and we have a pretty good idea of how they behave; the challenge is more in translating that into mathematical terms.

One favorable example for climate forecasting comes from the success at forecasting the trajectories of some particularly big and important clouds—those that form hurricanes. Emanuel's office at MIT, designated as room 54-1814, is something of a challenge to find (I was assisted by an exceptional janitor who may as well have been the inspiration for *Good Will Hunting*). But it offers a clear view of the Charles River. It was easy to imagine a hurricane out in the distance: Would it careen toward Cambridge or blow out into the North Atlantic?

Emanuel articulated a distinction between two types of hurricane forecasts. One is purely statistical. "You have a long record of the phenomenon you're interested in. And you have a long record of what you consider to be viable predictors—like the wind in a large-scale flow of the atmosphere, or the temperature of the ocean, what have you," he said. "And without being particularly physical about it, you just use statistics to relate to what you're trying to predict to those predictors."

Imagine that a hurricane is sitting in the Gulf of Mexico. You could build a database of past hurricanes and look at their wind speed and their latitude and longitude and the ocean temperature and so forth, and identify those hurricanes that were most similar to this new storm. How had those other hurricanes behaved? What fraction struck populous areas like New Orleans and what fraction dissipated? You would not really need all that much meteorological knowledge to make such a forecast, just a good database.

Techniques like these can provide for crude but usable forecasts. In fact, up until about thirty years ago, purely statistical models were the primary way that the weather service forecasted hurricane trajectories.

Such techniques, however, are subject to diminishing returns. Hurricanes are not exactly rare, but severe storms hit the United States perhaps once every year on average. Whenever you have a large number of candidate variables applied to a rarely occurring phenomenon, there is the risk of overfitting your model and mistaking the noise in the past data for a signal.

There is an alternative, however, when you have some knowledge of the structure behind the system. This second type of model essentially creates a simulation of the physical mechanics of some portion of the universe. It takes much more work to build than a purely statistical method and requires a more solid understanding of the root causes of the phenomenon. But it is potentially more accurate. Models like these are now used to forecast hurricane tracks and they have been highly successful. As I reported in chapter 4, there has been roughly a threefold increase in the accuracy of hurricane track projections since the 1980s, and the location near New Orleans where Hurricane Katrina made landfall had been pinpointed well more than forty-eight hours in advance[48] (though not everyone chose to listen to the forecast). Statistically driven systems are now used as little more than the baseline to measure these more accurate forecasts against.

Beyond a Cookbook Approach to Forecasting

The criticisms that Armstrong and Green make about climate forecasts derive from their empirical study of disciplines like economics in which there are few such physical models available[49] and the causal relationships are poorly understood. Overly ambitious approaches toward forecasting have often failed in these fields, and so Armstrong and Green infer that they will fail in climate forecasting as well.

The goal of any predictive model is to capture *as much signal as possible and as little noise as possible*. Striking the right balance is not always so easy, and our ability to do so will be dictated by the strength of the theory and the quality and quantity of the data. In economic forecasting, the data is very poor and the theory is weak, hence Armstrong's argument that "the more complex you make the model the worse the forecast gets."

In climate forecasting, the situation is more equivocal: the theory about the greenhouse effect is strong, which supports more complicated models. However, temperature data is very noisy, which argues against them. Which consideration wins out? We can address this question empirically, by evaluating the success and failure of different predictive approaches in climate science. What matters most, as always, is how well the predictions do in the real world.

I would urge caution against reducing the forecasting process to a series of bumper-sticker slogans. Heuristics like Occam's razor ("other things being equal, a simpler explanation is better than a more complex one"[50]) sound sexy, but they are hard to apply. We have seen cases, as in the SIR models used to forecast disease outbreaks, where the assumptions of a model are simple and elegant—but where they are much too naïve to provide for very skillful forecasts. We have also seen cases, as in earthquake prediction, where unbelievably convoluted forecasting schemes that look great in the software package fail miserably in practice.

An admonition like "The more complex you make the model the worse the forecast gets" is equivalent to saying "Never add too much salt to the recipe." How much complexity—how much salt—did you begin with? If you want to get good at forecasting, you'll need to immerse yourself in the craft and trust your own taste buds.

Uncertainty in Climate Forecasts

Knowing the limitations of forecasting is half the battle, and on that score the climate forecasters do reasonably well. Climate scientists are keenly aware of uncertainty: variations on the term *uncertain* or *uncertainty* were used 159 times in just one of the three IPCC 1990 reports.[51] And there is a whole nomenclature that the IPCC authors have developed to convey how much agreement or certainty there is about a finding. For instance, the word "likely" taken alone is meant to imply at least a 66 percent chance of a prediction occurring when it appears in an IPCC report, while the phrase "virtually certain" implies 99 percent confidence or more.[52]

Still, it is one thing to be alert to uncertainty and another to actually estimate it properly. When it comes to something like political polling, we can rely on a robust database of historical evidence: if a candidate is ten points ahead in the polls with a month to go until an election, how often will she wind up winning? We can look through dozens of past elections to get an empirical answer to that.

The models that climate forecasters build cannot rely on that sort of technique. There is only one planet and forecasts about how its climate will evolve are made at intervals that leap decades into the future. Although climatologists might think carefully about uncertainty, *there is uncertainty about how much uncertainty there is.* Problems like these are challenging for forecasters in any discipline.

Nevertheless, it is possible to analyze the uncertainty in climate forecasts as having three component parts. I met with Gavin Schmidt, a NASA colleague of Hansen's and a somewhat sarcastic Londoner who is a coauthor of the blog RealClimate.org, at a pub near his office in Morningside Heights in New York.

Schmidt took out a cocktail napkin and drew a graph that looked something like what you see in figure 12-3, which illustrates the three distinct problems that climate scientists face. These different types of uncertainty become more or less prevalent over the course of a climate forecast.

First, there is what Schmidt calls **initial condition uncertainty**—the short-term factors that compete with the greenhouse signal and impact the way we experience the climate. The greenhouse effect is a long-term phenomenon, and it may be obscured by all types of events on a day-to-day or year-to-year basis.

The most obvious type of initial condition uncertainty is simply the weather; when it comes to forecasting the climate, it represents noise rather

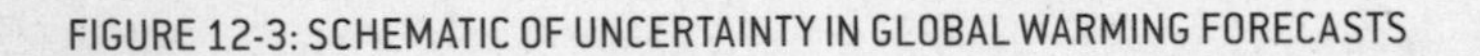
FIGURE 12-3: SCHEMATIC OF UNCERTAINTY IN GLOBAL WARMING FORECASTS

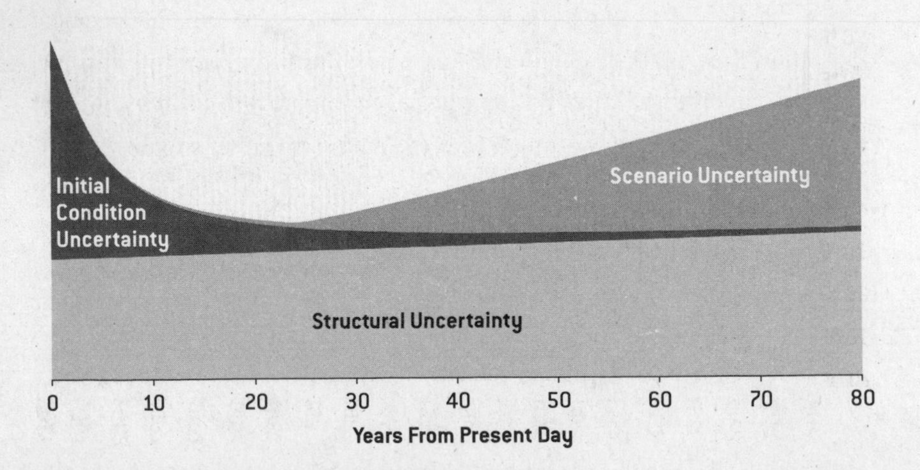

than signal. The current IPCC forecasts predict that temperatures might increase by 2°C (or about 4°F) over the course of the next century. That translates into an increase of just 0.2°C per decade, or 0.02°C per year. Such a signal is hard to perceive when temperatures can easily fluctuate by 15°C from day to night and perhaps 30°C from season to season in temperate latitudes.

In fact, just a few days before I met with Schmidt in 2011, there had been a freakish October snowstorm in New York and other parts of the Northeast. The snowfall, 1.3 inches in Central Park, set an October record there,[53] and was more severe in Connecticut, New Jersey, and Massachusetts, leaving millions of residents without power.[54]

Central Park happens to have a particularly good temperature record;[55] it dates back to 1869.[56] In figure 12-4, I have plotted the monthly average temperature for Central Park in the century encompassing 1912 through 2011. You will observe the seasons in the graphic; the temperature fluctuates substantially (but predictably enough) from warm to cool and back again—a little more so in some years than others. In comparison to the weather, the climate signal is barely noticeable. But it does exist: temperatures have increased by perhaps 4°F on average over the course of this one-hundred-year period in Central Park.

FIGURE 12-4: CENTRAL PARK (NEW YORK CITY) MONTHLY AVERAGE TEMPERATURES, 1912–2011, IN °F

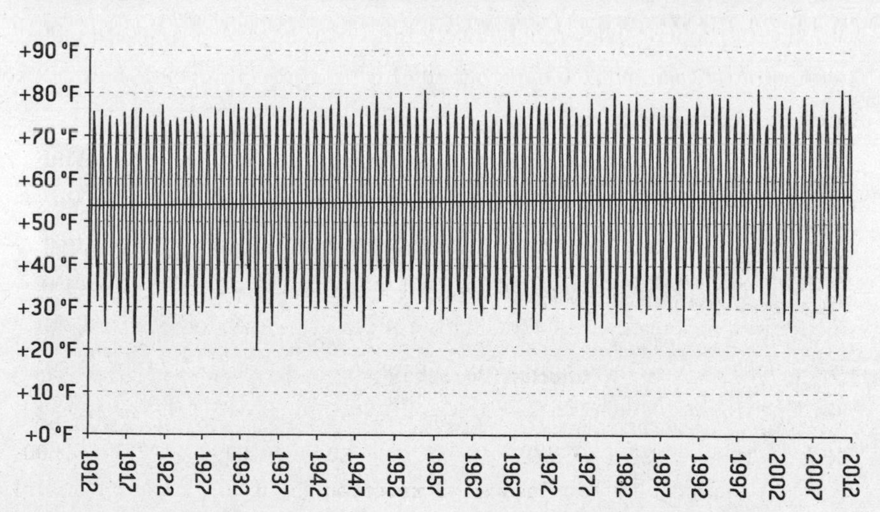

There are also periodic fluctuations that take hold at periods of a year to a decade at a time. One is dictated by what is called the ENSO cycle (the El Niño–Southern Oscillation). This cycle, which evolves over intervals of about three years at a time,[57] is instigated by temperature shifts in the waters of the tropical Pacific. El Niño years, when the cycle is in full force, produce warmer weather in much of the Northern Hemisphere, and probably reduce hurricane activity in the Gulf of Mexico.[58] La Niña years, when the Pacific is cool, do just the opposite. Beyond that, relatively little is understood about the ENSO cycle.

Another such medium-term process is the solar cycle. The sun gives off slightly more and slightly less radiation over cycles that last for about eleven years on average. (This is often measured through sunspots, the presence of which correlate with higher levels of solar activity.) But these cycles are somewhat irregular: Solar Cycle 24, for instance, which was expected to produce a maximum of solar activity (and therefore warmer temperatures) in 2012 or 2013, turned out to be somewhat delayed.[59] Occasionally, in fact, the sun can remain dormant for decades at a time; the Maunder Minimum, a period of about seventy years during the late seventeenth and early eighteenth centuries when there was very little sunspot activity, may have triggered cooler temperatures in Europe and North America.[60]

Finally, there are periodic interruptions from volcanoes, which blast sulfur—a gas that has an anti-greenhouse effect and tends to cool the planet—into the atmosphere. The eruption of Mount Pinatubo in 1991 reduced global temperatures by about 0.2°C for a period of two years, equivalent to a decade's worth of greenhouse warming.

The longer your time horizon, the less concern you might have about these medium-term effects. They can dominate the greenhouse signal over periods of a year to a decade at a time, but they tend to even out at periods beyond that.

Another type of uncertainty, however—what Schmidt calls **scenario uncertainty**—increases with time. This concerns the level of CO_2 and other greenhouse gases in the atmosphere. At near time horizons, atmospheric composition is quite predictable. The level of industrial activity is fairly constant, but CO_2 circulates quickly into the atmosphere and remains there for a long time. (Its chemical half-life has been estimated at about thirty years.[61]) Even if major industrialized countries agreed to immediate and substantial reductions in CO_2 emissions, it would take years to reduce the growth rate of CO_2 in the atmosphere, let alone to actually reverse it. "Neither you nor I will ever see a year in which carbon dioxide concentrations have gone down, not ever," Schmidt told me. "And not your children either."

Still, since climate models rely on specific assumptions about the amount of atmospheric CO_2, this can significantly complicate forecasts made for fifty or one hundred years out and affect them at the margin in the nearer term, depending on how political and economic decisions influence CO_2 emissions.

Last, there is the **structural uncertainty** in the models. This is the type of uncertainty that both climate scientists and their critics are rightly most worried about, because it is the most challenging to quantify. It concerns how well we understand the dynamics of the climate system and how well we can represent them mathematically. Structural uncertainty might increase slightly over time, and errors can be self-reinforcing in a model of a dynamic system like the climate.

Taken together, Schmidt told me, these three types of uncertainty tend to be at a minimum at a period of about twenty or twenty-five years in advance of a climate forecast. This is close enough that we know with reasonable certainty how much CO_2 there will be in the atmosphere—but far enough away that the effects of ENSO and volcanoes and the solar cycle should have evened out.

As it happens, the first IPCC report, published in 1990, falls right into this twenty-year sweet spot. So do some of the early forecasts made by James Hansen in the 1980s. It is time, in other words, to assess the accuracy of the forecasts. So how well did they do?

A Note on the Temperature Record

To measure the accuracy of a prediction, you first need a measuring stick—and climate scientists have quite a few choices. There are four major organizations that build estimates of global temperatures from thermometer readings at land and sea stations around the globe. These organizations include NASA[62] (which maintains its GISS[63] temperature record), NOAA[64] (the National Oceanic and Atmospheric Administration, which manages the National Weather Service), and the meteorological offices of the United Kingdom[65] and Japan.[66]

A more recent entrant into the temperature sweepstakes are observations from satellites. The most commonly used satellite records are from the University of Alabama at Huntsville and from a private company called Remote Sensing Systems.[67] The satellites these records rely on do not take the temperature directly—instead, they infer it by measuring microwave radiation. But the satellites' estimates of temperatures in the lower atmosphere[68] provide a reasonably good proxy for surface temperatures.[69]

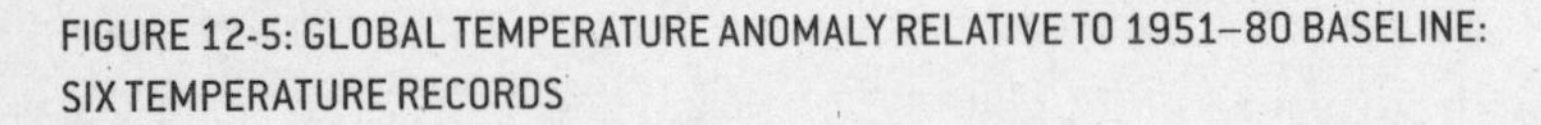

FIGURE 12-5: GLOBAL TEMPERATURE ANOMALY RELATIVE TO 1951–80 BASELINE: SIX TEMPERATURE RECORDS

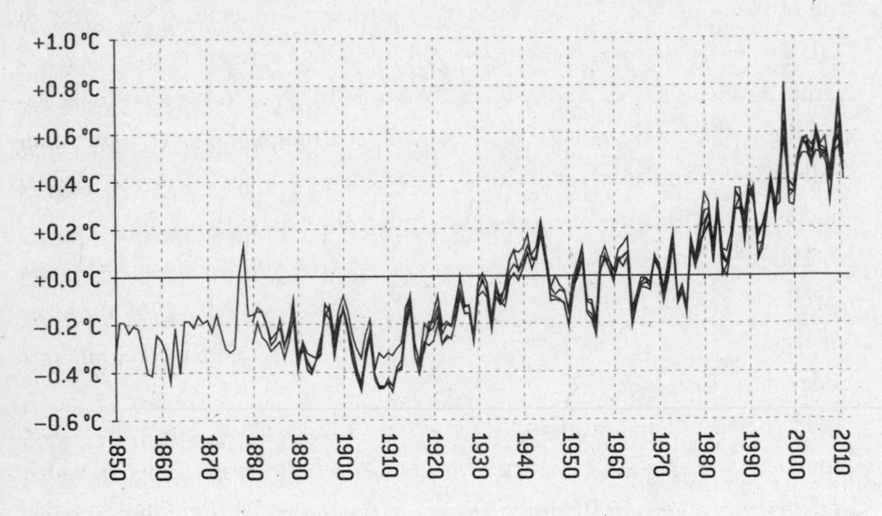

The temperature records also differ in how far they track the climate backward; the oldest are the observations from the UK's Met Office, which date back to 1850; the satellite records are the youngest and date from 1979. And the records are measured relative to different baselines—the NASA/GISS record is taken relative to average temperatures from 1951 through 1980, for instance, while NOAA's temperatures are measured relative to the average throughout the twentieth century. But this is easy to correct for,[70] and the goal of each system is to measure how much temperatures are rising or falling rather than what they are in any absolute sense.

Reassuringly, the differences between the various records are fairly modest[71] (figure 12-5). All six show both 1998 and 2010 as having been among the three warmest years on record, and all six show a clear long-term warming trend, especially since the 1950s when atmospheric CO_2 concentrations began to increase at a faster rate. For purposes of evaluating the climate forecasts, I've simply averaged the six temperate records together.

James Hansen's Predictions

One of the more forthright early efforts to forecast temperature rise came in 1981, when Hansen and six other scientists published a paper in the esteemed journal *Science*.[72] These predictions, which were based on relatively simple statistical estimates of the effects of CO_2 and other atmospheric gases rather than a fully fledged simulation model, have done quite well. In fact, they very slightly underestimated the amount of global warming observed through 2011.[73]

Hansen is better known, however, for his 1988 congressional testimony as well as a related 1988 paper[74] that he published in the *Journal of Geophysical Research*. This set of predictions did rely on a three-dimensional physical model of the atmosphere.

Hansen told Congress that Washington could expect to experience more frequent "hot summers." In his paper, he defined a hot summer as one in which average temperatures in Washington were in the top one-third of the summers observed from 1950 through 1980. He said that by the 1990s, Washington could expect to experience these summers 55 to 70 percent of the time, or roughly twice their 33 percent baseline rate.

In fact, Hansen's prediction proved to be highly prescient for Washington, DC. In the 1990s, six of the ten summers[75] qualified as hot (figure 12-6), right in line with his prediction. About the same rate persisted in the 2000s and Washington experienced a record heat wave in 2012.

FIGURE 12-6: HOT SUMMERS

BASELINE RATE: 33% **HANSEN PREDICTION (1988): 55%–70% OF THE TIME BY 1990S**				
City	Threshold	1990–1999	2000–2011	1990–2011
Washington, DC	86.2°F	60%	58%	59%
Omaha, NE	86.2°F	10%	42%	27%
New York, NY	81.4°F	80%	75%	77%
Memphis, TN	89.3°F	50%	67%	59%
Average		**50%**	**61%**	**56%**

In his paper, Hansen had also made these predictions for three other cities: Omaha, Memphis, and New York. These results were more mixed and go to illustrate the regional variability of the climate. Just 1 out of 10 summers in Omaha in the 1990s qualified as "hot" by Hansen's standard, well below the historical average rate of 33 percent. But 8 out of 10 summers in New York did, according to observations at LaGuardia Airport.

Overall, the predictions for the four cities were reasonably good, but were toward the lower end of Hansen's range. His global temperature predictions are harder to evaluate because they articulated a plethora of scenarios that relied on different assumptions, but they were also somewhat too high.[76] Even the most conservative scenario somewhat overestimated the warming experienced through 2011.

The IPCC's 1990 Predictions

The IPCC's 1990 forecasts represented the first true effort at international consensus predictions in the field and therefore received an especially large amount of attention. These predictions were less specific than Hansen's, although when they did go into detail they tended to get things mostly right. For instance, they predicted that land surfaces would warm more quickly than water surfaces, especially in the winter, and that there would be an especially substantial increase in temperature in the Arctic and other northerly latitudes. Both of these predictions have turned out to be correct.

The headline forecast, however, was that of the global temperature rise. Here, the IPCC's prediction left more to be desired.

The IPCC's temperature forecast, unlike Hansen's, took the form of a range of possible outcomes. At the high end of the range was a catastrophic temperature increase of 5°C over the course of the next one hundred years. At the low end was a more modest increase of 2°C per century, with a 3°C increase representing the most likely case.[77]

In fact, the actual temperature increase has been on a slower pace since the report was published (figure 12-7). Temperatures increased by an average of 0.015°C per year from the time the IPCC forecast was issued in 1990 through 2011, or at a rate of 1.5°C per century. This is about half the IPCC's most likely case, of 3°C warming per century, and also slightly less than the low end of their range at 2°C. The IPCC's 1990 forecast also overestimated the amount of sea-level rise.[78]

This represents a strike against the IPCC's forecasts, although we should consider one important qualification.

The IPCC forecasts were predicated on a "business-as-usual" case that assumed that there would be no success at all in mitigating carbon emissions.[79] This scenario implied that the amount of atmospheric CO_2 would increase to about four hundred parts per million (ppm) by 2010.[80] In fact, some limited efforts to reduce carbon emissions were made, especially in the European

FIGURE 12-7: ACTUAL GLOBAL TEMPERATURES, 1990–2011 VS. 1990 IPCC FORECAST RANGE
Anomaly vs. 1951–1980 Baseline

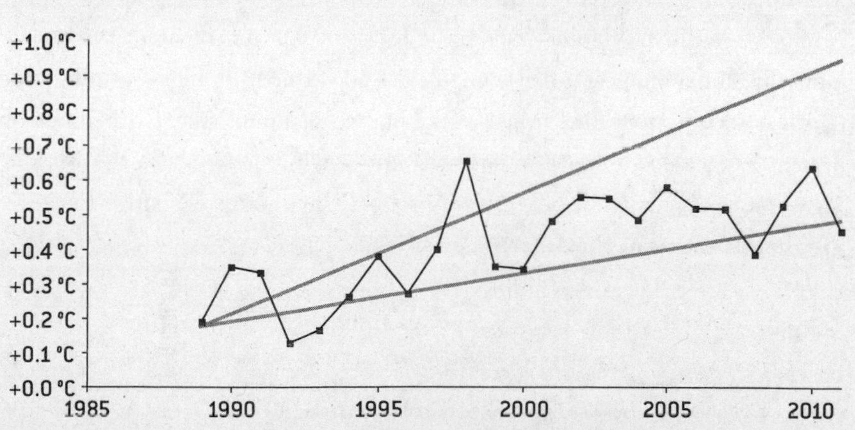

Union,[81] and this projection was somewhat too pessimistic; CO_2 levels had risen to about 390 ppm as of 2010.[82] In other words, the error in the forecast in part reflected scenario uncertainty—which turns more on political and economic questions than on scientific ones—and the IPCC's deliberately pessimistic assumptions about carbon mitigation efforts.*

Nevertheless, the IPCC later acknowledged their predictions had been too aggressive. When they issued their next forecast, in 1995, the range attached to their business-as-usual case had been revised considerably lower: warming at a rate of about 1.8°C per century.[83] This version of the forecasts has done quite well relative to the actual temperature trend.[84] Still, that represents a fairly dramatic shift. It is right to correct a forecast when you think it might be wrong rather than persist in a quixotic fight to the death for it. But this is evidence of the uncertainties inherent in predicting the climate.

The score you assign to these early forecasting efforts overall might depend on whether you are grading on a curve. The IPCC's forecast miss in 1990 is partly explained by scenario uncertainty. But this defense would be more persuasive if the IPCC had not substantially changed its forecast just five years later. On the other hand, their 1995 temperature forecasts have gotten things about right, and the relatively few specific predictions they made beyond global temperature rise (such as ice shrinkage in the Arctic[85]) have done quite well. If you hold forecasters to a high standard, the IPCC might deserve a low but not failing grade. If instead you have come to understand that the history of prediction is fraught with failure, they look more decent by comparison.

Uncertainty in forecasts is not necessarily a reason not to act—the Yale economist William Nordhaus has argued instead that it is precisely the uncertainty in climate forecasts that compels action,[86] since the high-warming scenarios could be quite bad. Meanwhile, our government spends hundreds of billions toward economic stimulus programs, or initiates wars in the Middle East, under the pretense of what are probably far more speculative forecasts than are pertinent in climate science.[87]

---

* If you scale back their warming estimates to reflect the smaller-than-assumed rate of CO_2 increase, you wind up with a revised projection of 1.4 °C to 3.6 °C in warming per century. The actual rate of increase, a pace of 1.5 °C per century since the report was published, falls within this range, albeit barely.

The Lessons of "Global Cooling"

Still, climate scientists put their credibility on the line every time they make a prediction. And in contrast to other fields in which poor predictions are quickly forgotten about, errors in forecasts about the climate are remembered for decades.

One common claim among climate critics is that there once had been predictions of global cooling and possibly a new ice age. Indeed, there were a few published articles that projected a cooling trend in the 1970s. They rested on a reasonable-enough theory: that the cooling trend produced by sulfur emissions would outweigh the warming trend produced by carbon emissions.

These predictions were refuted in the majority of the scientific literature.[88] This was less true in the news media. A *Newsweek* story in 1975 imagined that the River Thames and the Hudson River might freeze over and stated that there would be a "drastic decline" in food production[89]—implications drawn by the writer of the piece but not any of the scientists he spoke with.

If the media can draw false equivalences between "skeptics" and "believers" in the climate science debate, it can also sometimes cherry-pick the most outlandish climate change claims even when they have been repudiated by the bulk of a scientist's peers.

"The thing is, many people are going around talking as if they looked at the data. I guarantee that nobody ever has," Schmidt told me after New York's October 2011 snowstorm, which various media outlets portrayed as evidence either for or against global warming.

Schmidt received numerous calls from reporters asking him what October blizzards in New York implied about global warming. He told them he wasn't sure; the models didn't go into that kind of detail. But some of his colleagues were less cautious, and the more dramatic their claims, the more likely they were to be quoted in the newspaper.

The question of sulfur emissions, the basis for those global cooling forecasts in the 1970s, may help to explain why the IPCC's 1990 forecast went awry and why the panel substantially lowered their range of temperature predictions in 1995. The Mount Pinatubo eruption in 1991 burped sulfur into the atmosphere, and its effects were consistent with climate models.[90] But it nevertheless underscored that the interactions between different greenhouse gases can be challenging to model and can introduce error into the system.

Sulfur emissions from manmade sources peaked in the early 1970s before declining[91] (figure 12-8), partly because of policy like the Clean Air Act signed into law by President Nixon in 1970 to combat acid rain and air pollution. Some of the warming trend during the 1980s and 1990s probably reflected this decrease in sulfur, since SO_2 emissions counteract the greenhouse effect.

FIGURE 12-8: GLOBAL ANNUAL SULFUR EMISSIONS, 1900–2005

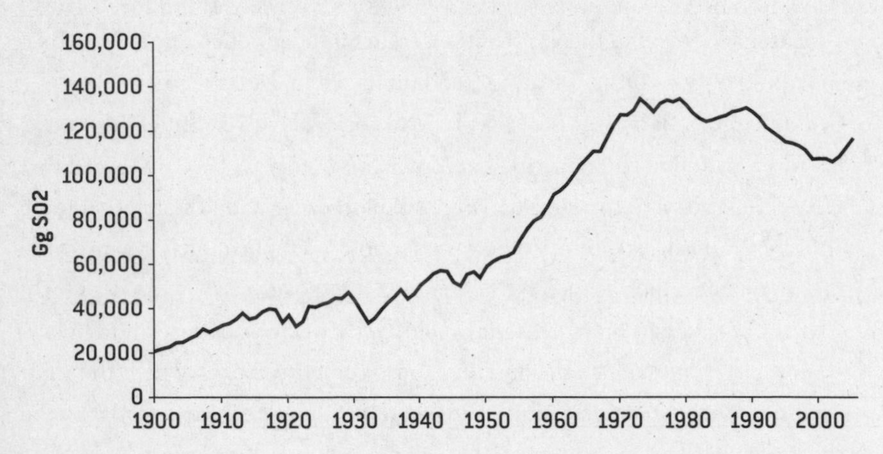

Since about 2000, however, sulfur emissions have increased again, largely as the result of increased industrial activity in China,[92] which has little environmental regulation and a lot of dirty coal-fired power plants. Although the negative contribution of sulfur emissions on global warming is not as strong as the positive contribution from carbon—otherwise those global cooling theories might have proved to be true!—this may have provided for something of a brake on warming.

A Simple Climate Forecast

So suppose that you have good reason to be skeptical of a forecast—for instance, because it purports to make fairly precise predictions about a very complex process like the climate, or because it would take years to verify the forecast's accuracy.

Sophomoric forecasters sometimes make the mistake of assuming that just because something is hard to model they may as well ignore it. Good forecasters always have a backup plan—a reasonable baseline case that they can default to if they have reason to worry their model is failing. (In a presidential election, your default prediction might be that the incumbent will win—that will do quite a bit better than just picking between the candidates at random.)

What is the baseline in the case of the climate? If the critique of global warming forecasts is that they are unrealistically complex, the alternative would be a simpler forecast, one grounded in strong theoretical assumptions but with fewer bells and whistles.

Suppose, for instance, that you had attempted to make a climate forecast based on an extremely simple statistical model: one that looked solely at CO_2 levels and temperatures, and extrapolated a prediction from these variables alone, ignoring sulfur and ENSO and sunspots and everything else. This wouldn't require a supercomputer; it could be calculated in a few microseconds on a laptop. How accurate would such a prediction have been?

In fact, it would have been very accurate—quite a bit better, actually, than the IPCC's forecast. If you had placed the temperature record from 1850 through 1989 into a simple linear regression equation, along with the level of CO_2 as measured in Antarctic ice cores[93] and at the Mauna Loa Observatory in Hawaii, it would have predicted a global temperature increase at the rate of 1.5°C per century from 1990 through today, exactly in line with the actual figure (figure 12-9).

Another technique, only slightly more complicated, would be to use estimates that were widely available at the time about the overall relationship between CO_2 and temperatures. The common currency of any global warming forecast is a value that represents the effect on temperatures from a doubling (that is, a 100 percent increase) in atmospheric CO_2. There has long been some agreement about this doubling value.[94] From forecasts like those made by the British engineer G. S. Callendar in 1938[95] that relied on simple chemical equations, to those produced by today's supercomputers, estimates have congregated[96] between 2°C and 3°C of warming from a doubling of CO_2.

FIGURE 12-9: ACTUAL GLOBAL TEMPERATURES. 1990–2011 VS. SIMPLE REGRESSION FORECAST
Anomaly vs. 1951–1980 Baseline

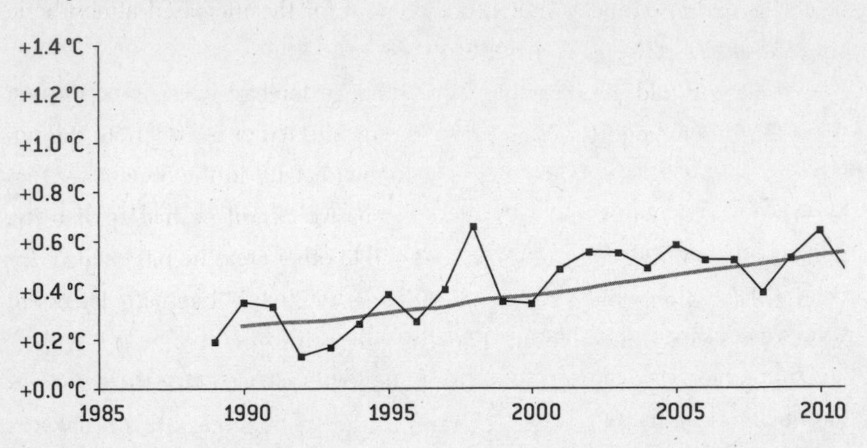

Given the actual rate of increase in atmospheric CO_2, that simple conversion would have implied temperature rise at a rate of between 1.1°C and 1.7°C per century from 1990 through the present day. The actual warming pace of 0.015°C per year or 1.5°C per century fits snugly within that interval.

James Hansen's 1981 forecasts, which relied on an approach much like this, did quite a bit better at predicting current tempertaures than his 1988 forecast, which relied on simulated models of the climate.

The Armstrong and Green critique of model complexity thus looks pretty good here. But the success of the more basic forecasting methods suggests that Armstrong's critique may have won the battle but not the war. He is asking some good questions about model complexity, and the fact that the simple models do pretty well in predicting the climate is one piece of evidence in favor of his position that simpler models are preferable. However, since the simple methods correctly predicted a temperature increase in line with the rise in CO_2, they are also evidence in favor of the greenhouse-effect hypothesis.

Armstrong's no-change forecast, by contrast, leaves some of the most basic scientific questions unanswered. The forecast used 2007 temperatures as its baseline, a year that was not exceptionally warm but which was nevertheless warmer than all but one year in the twentieth century. Is there a plausible hypothesis that explains why 2007 was warmer than 1987 or 1947 or 1907—other than through changes in atmospheric composition? One of the most tangible contributions of climate models, in fact, is that they find it impossible to replicate the *current* climate unless they account for the increased atmospheric concentration of CO_2 and other greenhouse gases.[97]

Armstrong told me he made the no-change forecast because he did not think there were good Bayesian priors for any alternative assumption; the no-change forecast, he has found, has been a good default in the other areas that he has studied. This would be a more persuasive case if he had applied the same rigor to climate forecasting that he had to other areas he has studied. Instead, as Armstrong told a congressional panel in 2011,[98] "I actually try not to learn a lot about climate change. I am the forecasting guy."

This book advises you to be wary of forecasters who say that the science is not very important to their jobs, or scientists who say that forecasting is not very important to their jobs! These activities are essentially and intimately related. A forecaster who says he doesn't care about the science is like the cook who says he doesn't care about food. What distinguishes science, and what makes a forecast scientific, is that it is concerned with the objective world. What makes forecasts fail is when our concern only extends as far as the method, maxim, or model.

An Inconvenient Truth About the Temperature Record

But if Armstrong's critique is so off the mark, what should we make of his proposed bet with Gore? It has not been a failed forecast at all; on the contrary, it has gone quite successfully. Since Armstrong made the bet in 2007, temperatures have varied considerably from month to month but not in any consistent pattern; 2011 was a slightly cooler year than 2007, for instance.

And this has been true for longer than four years: one inconvenient truth is that global temperatures did not increase at all in the decade between 2001 and 2011 (figure 12-10). In fact they declined, although imperceptibly.[99]

This type of framing can sometimes be made in bad faith. For instance, if you set the year 1998 as your starting point, which had record-high temperatures associated with the ENSO cycle, it will be easier to identify a cooling "trend." Conversely, the decadal "trend" from 2008 through 2018 will very probably be toward warming once it is calculated, since 2008 was a relatively

FIGURE 12-10: GLOBAL TEMPERATURES, 2001–2011
Anomaly vs. 1951–1980 Baseline

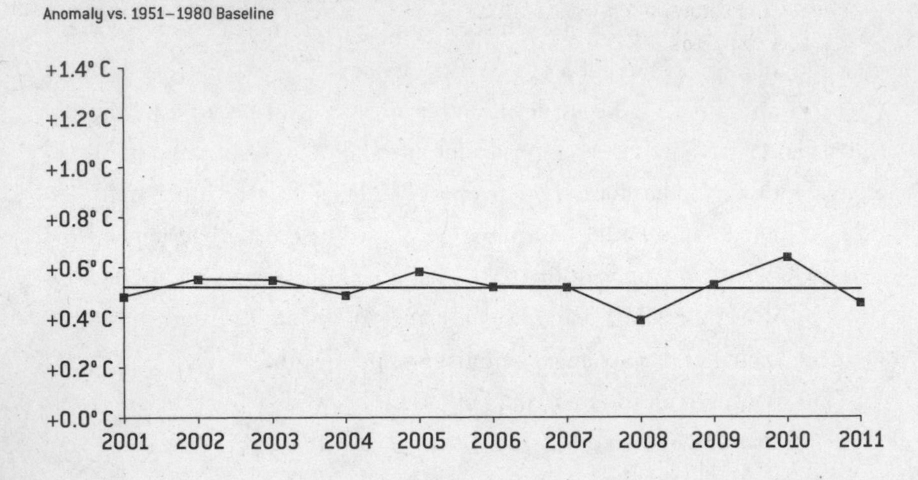

cool year. Statistics of this sort are akin to when the stadium scoreboard optimistically mentions that the shortstop has eight hits in his last nineteen at-bats against left-handed relief pitchers—ignoring the fact that he is batting .190 for the season.[100]

Yet global warming does not progress at a steady pace. Instead, the history of temperature rise is one of a clear long-term increase punctuated by periods of sideways or even negative trends. In addition to the decade between 2001 and 2011, for instance, there would have been little sign of warming between 1894 and 1913, or 1937 and 1956, or 1966 and 1977 (figure 12-11)—even though CO_2 concentrations were increasing all the while. This problem bears some resemblance to that faced by financial analysts: over the very long run, the stock market essentially always moves upward. But this tells you almost nothing about how it will behave in the next day, week, or year.

It might be possible to explain some of the recent sideways trend directly from the science; increased sulfur emissions in China might have played some role, for instance. And it might be remembered that although temperatures did not rise from 2001 through 2011, they were still much warmer than in any prior decade.

Nevertheless, this book encourages readers to think carefully about the signal and the noise and to seek out forecasts that couch their predictions in percentage or probabilistic terms. They are a more honest representation of the

FIGURE 12-11: GLOBAL TEMPERATURES, 1900–2011 WITH NEAR-TERM FLATLINES AND DOWNSHIFTS HIGHLIGHTED

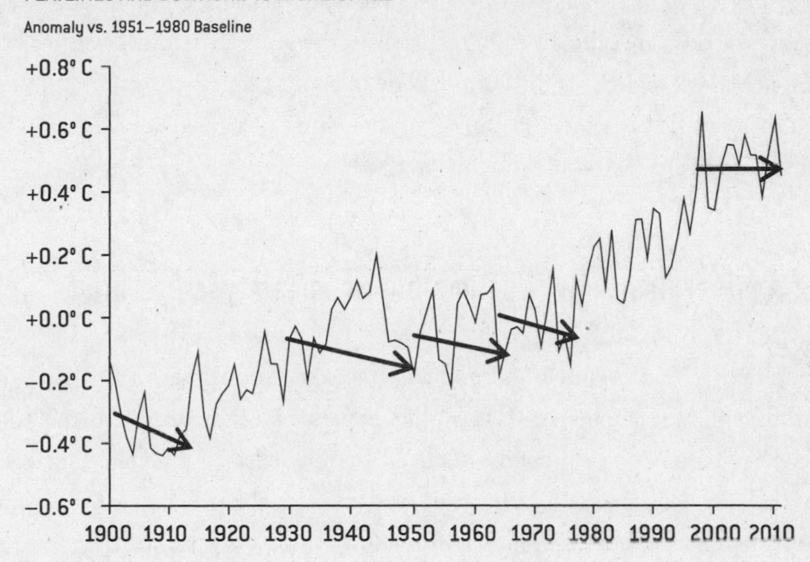

limits of our predictive abilities. When a prediction about a complex phenomenon is expressed with a great deal of confidence, it may be a sign that the forecaster has not thought through the problem carefully, has overfit his statistical model, or is more interested in making a name for himself than in getting at the truth.

Neither Armstrong nor Schmidt was willing to hedge very much on their predictions about the temperature trend. "We did some simulations from 1850 up to 2007," Armstrong told me. "When we looked one hundred years ahead it was virtually certain that I would win that bet."[101] Schmidt, meanwhile, was willing to offer attractive odds to anyone betting against his position that temperatures would continue to increase. "I could easily give you odds on the next decade being warmer than this decade," he told me. "You want 100-to-1 odds, I'd give it to you."

The statistical forecasting methods that I outlined earlier can be used to resolve the dispute—and they suggest that neither Armstrong nor Schmidt has it quite right. If you measure the temperature trend one decade at a time, it registers a warming trend about 75 percent of the time since 1900, but a cooling trend the other 25 percent of the time. As the growth rate of atmospheric CO_2

increases, creating a stronger greenhouse signal, periods of flat or cooling temperatures should become less frequent. Nevertheless, they are not impossible, nor are the odds anything like 100-to-1 against them. Instead, if you assume that CO_2 levels will increase at the current pace of about 2 ppm per year, the chance that there would be no net warming over the course of a given decade would be about 15 percent[102] according to this method.

Yet Another Reason Why Estimating Uncertainty Is Essential

Uncertainty is an essential and nonnegotiable part of a forecast. As we have found, sometimes an honest and accurate expression of the uncertainty is what has the potential to save property and lives. In other cases, as when trading stock options or wagering on an NBA team, you may be able to place bets on your ability to forecast the uncertainty accurately.

However, there is another reason to quantify the uncertainty carefully and explicitly. It is essential to scientific progress, especially under Bayes's theorem.

Suppose that in 2001, you had started out with a strong prior belief in the hypothesis that industrial carbon emissions would continue to cause a temperature rise. (In my view, such a belief would have been appropriate because of our strong causal understanding of the greenhouse effect and the empirical evidence for it up to that point.) Say you had attributed the chance of the global warming hypothesis's being true at 95 percent.

But then you observe some new evidence: over the next decade, from 2001 through 2011, global temperatures do not rise. In fact, they fall, although very slightly. Under Bayes's theorem, you should revise your estimate of the probability of the global warming hypothesis downward; the question is by how much.

If you had come to a proper estimate of the uncertainty in near-term temperature patterns, the downward revision would not be terribly steep. As we found, there is about a 15 percent chance that there will be no net warming over a decade even if the global warming hypothesis is true because of the variability in the climate. Conversely, if temperature changes are purely random and unpredictable, the chance of a cooling decade would be 50 percent since an increase and a decrease in temperatures are equally likely. Under Bayes's theorem (figure 12-12), a no-net-warming decade would cause you to revise downward your estimate of the global warming hypothesis's likelihood to 85 percent from 95 percent.

FIGURE 12-12: BAYES'S THEOREM—GLOBAL WARMING EXAMPLE

PRIOR PROBABILITY		
Initial estimate of how likely it is that global temperatures are increasing	x	95%
A NEW EVENT OCCURS: NO NET WARMING OVER 10 YEARS		
Probability of no net warming over 10 years if global warming hypothesis is correct	y	15%
Probability of no net warming over 10 years if global warming hypothesis is false	z	50%
POSTERIOR PROBABILITY		
Revised estimate of how likely it is that global warming is occurring, given no net temperature increase over 10 years	$\frac{xy}{xy + z(1-x)}$	85%

On the other hand, if you had asserted that there was just a 1 percent chance that temperatures would fail to increase over the decade, your theory is now in much worse shape because you are claiming that this was a more definitive test. Under Bayes's theorem, the probability you would attach to the global warming hypothesis has now dropped to just 28 percent.

When we advance more confident claims and they fail to come to fruition, this constitutes much more powerful evidence against our hypothesis. We can't really blame anyone for losing faith in our forecasts when this occurs; they are making the correct inference under Bayesian logic.

So what is the incentive to make more confident claims to begin with, especially when they are not really justified by the statistical evidence? There are all sorts of reasons that people may do this in practice. In the climate debate, it may be because these more confident claims can seem more persuasive—and they may be, but only if they are right. Attributing every weather anomaly to manmade climate change—other than the higher temperatures the global warming phenomenon is named for—is a high-stakes gamble, rooted more in politics than in science. There is little consensus about the ways that climate change might manifest itself other than through temperature increases and probably rising sea levels. Obviously, charging that every snowfall is evidence against the theory is just as ridiculous.

"We're in a Street Fight with These People"

The fundamental dilemma faced by climatologists is that global warming is a long-term problem that might require a near-term solution. Because carbon dioxide remains in the atmosphere for so long, decisions that we make about it today will affect the lives of future generations.

In a perfectly rational and benevolent world, this might not be so worrying. But our political and cultural institutions are not so well-devised to handle these problems—not when the United States Congress faces reelection every two years and when businesses are under pressure to meet earnings forecasts every quarter. Climate scientists have reacted to this challenge in a variety of ways, some involving themselves more in the political debate and others keeping it at arm's length.

Michael Mann, who is director of the Earth System Science Center at Penn State University, was once at the center of a controversy. "Climategate" concerned the hacking of a server at the Climatic Research Unit (CRU) at the University of East Anglia,[103] which produces the temperature record that the UK's Met Office uses. Skeptics alleged that Mann and other scientists had conspired to manipulate the CRU's temperature record.

The pertinent facts are that the scientists were cleared of wrongdoing by a panel of their peers,[104] and that the CRU's temperature record is quite consistent with the others[105]—but Mann and other scientists in the hacked e-mails demonstrated a clear concern with the public relations elements of how the science would be perceived. Mann is happy to tell you as much. I met with him on a crisp fall afternoon at his office at Penn State, where we chatted for about two hours.

Mann is exceptionally thoughtful about the science behind global warming. Like most other climatologists, he has little doubt about the theoretical mechanisms behind climate change, but he takes a skeptical view toward the predictions rendered by climate models.

"Any honest assessment of the science is going to recognize that there are things we understand pretty darn well and things that we sort of know," he told me. "But there are things that are uncertain and there are things we just have no idea about whatsoever."

"In my mind, one of the unfortunate consequences of this bad-faith public conversation we've been having is that we're wasting our time debating a proposition that is very much accepted within the scientific community, when we could be having a good-faith discussion about the uncertainties that do exist."

But Mann, who blogs along with Schmidt at RealClimate.org, sees himself as engaged in trench warfare against groups like the Heartland Institute. "We're in a street fight with these people," he told me, referring to a *Nature* editorial[106] that employed the phrase. The long-term goal of the street fight is to persuade the public and policy makers about the urgency (or lack thereof) of action to combat climate change. In a society accustomed to overconfident forecasters who mistake the confidence they express in a forecast for its veracity, expressions of uncertainty are not seen as a winning strategy by either side.

"Where you have to draw the line is to be very clear about where the uncertainties are, but to not have our statements be so laden in uncertainty that no one even listens to what we're saying," Mann told me. "It would be irresponsible for us as a community to not be speaking out. There are others who are happy to fill the void. And they're going to fill the void with disinformation."

The Difference Between Science and Politics

In practice, Mann's street fight is between "consensus" Web sites like RealClimate.org and "skeptical" ones like Watts Up With That,[107] and revolves around day-to-day scuffles about the latest journal article or weather pattern or political controversy. Both sides almost invariably stick up for others in their circle and refuse to yield ground. When you're a Jet, you're a Jet all the way.

I do not mean to suggest that the territory occupied by the two sides is symmetrical. In the scientific argument over global warming, the truth seems to be *mostly* on one side: the greenhouse effect almost certainly exists and will be exacerbated by manmade CO_2 emissions. This is very likely to make the planet warmer. The impacts of this are uncertain, but are weighted toward unfavorable outcomes.[108]

The street-fighter mentality, nevertheless, seems to be predicated on the notion that we are just on the verge of resolving our political problems, if only a few more people could be persuaded about the science. In fact, we are probably many years away. "There's a point when I come to the conclusion that we're going to have to figure out how to take the carbon out," Richard Rood told me in Copenhagen, anticipating that there was almost no way the 193 members of the United Nations would agree to mutually acceptable terms.

Meanwhile, the American public's confidence that global warming is occurring has decreased somewhat over the past several years.[109] And even if there were 100 percent agreement on the effects of climate change, some states and some countries would make out better than others in any plan to mitigate carbon emissions. "We have some very progressive Democratic governors in coal states," I was told by the governor of Washington, Christine Gregoire. "Boy, are they nervous about all this."

I don't know how to resolve these problems, which are not unique to the climate debate.[110] What I do know is that there is a fundamental difference between science and politics. In fact, I've come to view them more and more as opposites.

In science, progress is possible. In fact, if one believes in Bayes's theorem, scientific progress is inevitable as predictions are made and as beliefs are tested and refined.* The march toward scientific progress is not always straightforward, and some well-regarded (even "consensus") theories are later proved wrong—but either way science tends to move toward the truth.

In politics, by contrast, we seem to be growing ever further away from consensus. The amount of polarization between the two parties in the United States House, which had narrowed from the New Deal through the 1970s, had

* It may be hard to objectively measure scientific progress, but in the related realm of technology, the number of patents issued continues to increase at roughly exponential rates and is about double now what it was just a decade ago.

grown by 2011 to be the worst that it had been in at least a century.[111] Republicans have moved especially far away from the center,[112] although Democrats have to some extent too.

In science, one rarely sees *all* the data point toward one precise conclusion. Real data is noisy—even if the theory is perfect, the strength of the signal will vary. And under Bayes's theorem, no theory is perfect. Rather, it is a work in progress, always subject to further refinement and testing. This is what scientific skepticism is all about.

In politics, one is expected to give no quarter to his opponents. It is seen as a gaffe when one says something inconvenient—and true.[113] Partisans are expected to show equal conviction about a set of beliefs on a range of economic, social, and foreign policy issues that have little intrinsic relation to one another. As far as approximations of the world go, the platforms of the Democratic and Republican parties are about as crude as it gets.

It is precisely because the debate may continue for decades that climate scientists might do better to withdraw from the street fight and avoid crossing the Rubicon from science into politics. In science, dubious forecasts are more likely to be exposed—and the truth is more likely to prevail. In politics, a domain in which the truth enjoys no privileged status, it's anybody's guess.

The dysfunctional state of the American political system is the best reason to be pessimistic about our country's future. Our scientific and technological prowess is the best reason to be optimistic. We are an inventive people. The United States produces ridiculous numbers of patents,[114] has many of the world's best universities and research institutions, and our companies lead the market in fields ranging from pharmaceuticals to information technology. If I had a choice between a tournament of ideas and a political cage match, I know which fight I'd rather be engaging in—especially if I thought I had the right forecast.

13

WHAT YOU DON'T KNOW CAN HURT YOU

Franklin Delano Roosevelt said December 7 would live in infamy. The bombing of Pearl Harbor in 1941, the first foreign attack on American soil in more than a century,[1] was as shocking to the American psyche as the destruction of the World Trade Center sixty years later, and it transformed a vaguely menacing antagonist on the other side of the planet into a terrifying, palpable, existential threat. And yet, the attack on Pearl Harbor seemed quite predictable after the fact.

Many signals suggested an attack on Pearl Harbor was possible, and perhaps imminent. Diplomatic relations between the United States and Japan were in a state of rapid deterioration in November and December 1941. The Japanese desire for territorial expansion made the U.S. Pacific Fleet, which Roosevelt had moved to Pearl Harbor from San Diego precisely to deter such ambitions,[2] a natural flashpoint. Meanwhile, Japan's navy was repeatedly changing its call signs, an indication that its intentions were becoming more hostile, and there was increasing movement of Japanese troops and warships off the coasts of China and Southeast Asia.[3]

The most ominous signal of all was the silence. American intelligence officials had ingeniously succeeded in breaking PURPLE, the code that Japan used to encrypt its diplomatic messages, allowing us to decipher perhaps 97 percent of them.[4] Our attempts to decode Japanese military transmissions were less successful. But even if we could not understand the messages, we heard them and could trace their location. The steady stream of click-clacks from Japan's fleet of aircraft carriers ordinarily betrayed their whereabouts when they were out to sea.

From mid-November onward, however, there had been total radio silence; we had no idea where the carriers were. There were no global satellites in the 1940s, and only the primitive makings of radar. Air patrol reconnaissance missions were cost-prohibitive in the vast reaches of the Pacific and were carried out erratically at a distance of only three hundred or four hundred miles from the base.[5] The radio transmissions were our best means of detection, and without them an entire fleet of these ships, each of them the size of six football fields, had disappeared.

Many in the intelligence community concluded that the carriers were close to their home waters where they could rely on alternate means of communication.[6] The second possibility was that the fleet had ventured far into the Pacific, away from American naval installations.[7]

The Japanese carrier fleet, in fact, was en route to Hawaii. It had charted a precise course; like a quarterback perceiving the holes in his opponent's coverage, the fleet was maneuvering through the blind spots in our defenses. The ships initially traveled in a long, straight line toward the east-southeast, almost exactly bisecting our naval stations at the Midway Islands and in Dutch Harbor, Alaska. Then on December 4, after reaching longitude 165 west, they abruptly turned at a 45 degree angle toward Hawaii, where three days later they would launch the morning attack that killed almost 2,400 soldiers and sank four of the Navy's battleships.

The United States Congress declared war on Japan and its entry into World War II by a vote of 470 to 1 the next day.[8]

FIGURE 13-1: JAPANESE AIRCRAFT CARRIER COURSE TOWARD PEARL HARBOR

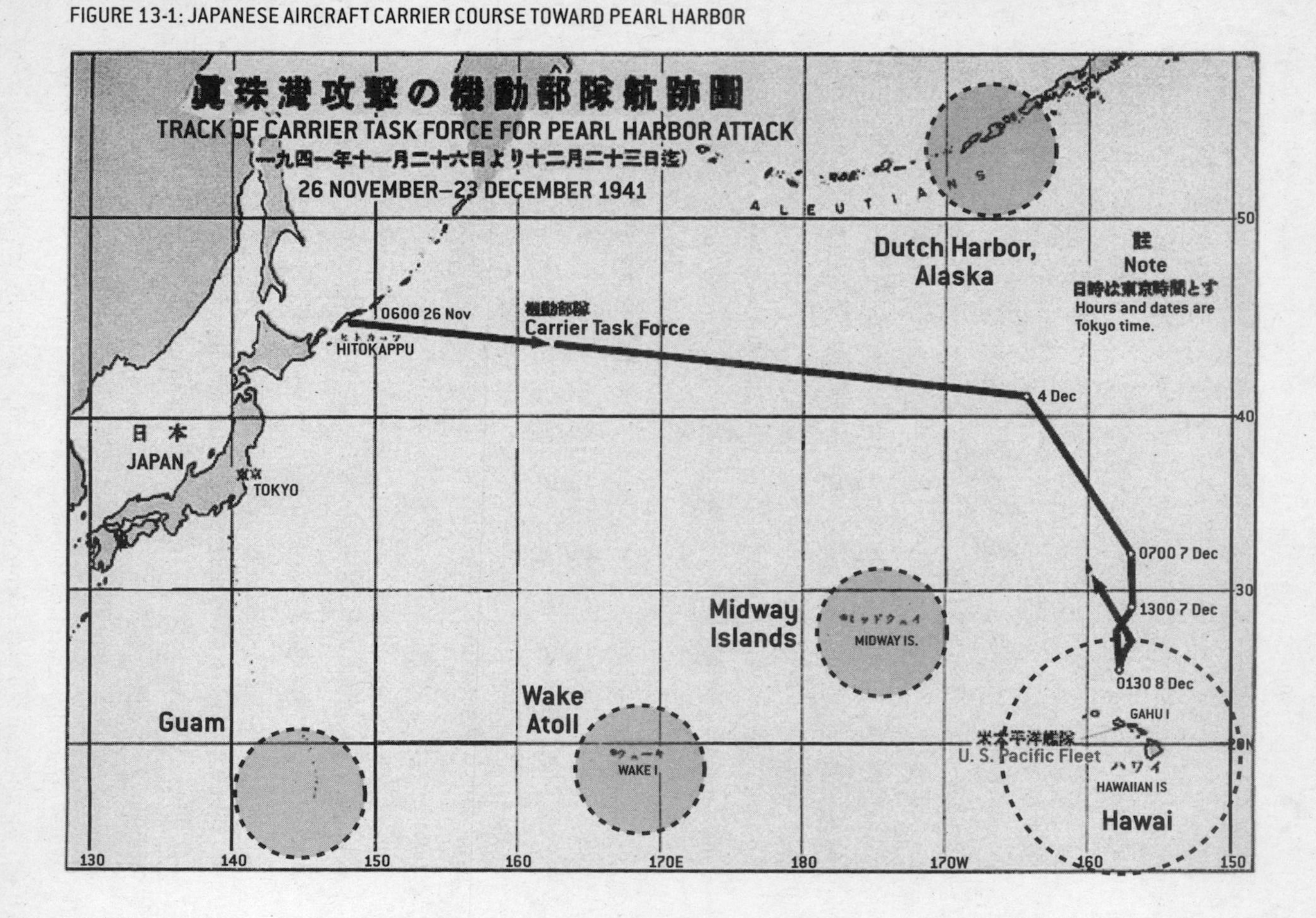

Signals, Signifying Nothing

When American Airlines Flight 77 collided with the limestone face of the Pentagon on September 11, 2001, killing fifty-nine innocent passengers and 125 of his colleagues, Donald Rumsfeld thought about Pearl Harbor.[9] He was eight years old on December 7, 1941, and was listening to his beloved Chicago Bears on the radio when the broadcast was interrupted by a bulletin with news of the surprise attack.[10]

I met Rumsfeld at his office in Washington in March 2012, having been warned that interviews with him could be hit or miss. Rumsfeld, at five-foot-seven and almost eighty years old, is not tremendously physically imposing, but he is intimidating in almost every other way. Born in Evanston, Illinois, he attended public high school and received an academic scholarship to Princeton, where he starred on the wrestling team and wrote his thesis on presidential powers before going off to serve in the Navy and then in Congress. His office walls are adorned with plaques and memorabilia from the four administrations he has served; he is the only person in American history to have had two separate stints as secretary of defense, first under President Ford from 1975 to 1977 and then, fully a quarter century later, under George W. Bush.

But Rumsfeld was in a good mood, having scrutinized the detailed outline for this book that I had given to his young and able chief of staff, Keith Urbahn.[11] I knew of Rumsfeld's interest in Pearl Harbor. He greeted me with a photocopy of the foreword to a remarkable book, Roberta Wohlstetter's 1962 *Pearl Harbor: Warning and Decision*, which outlined the myriad reasons why the Japanese attack had been such a surprise to our military and intelligence officers. Worse than being unprepared, we had mistaken our ignorance for knowledge and made ourselves more vulnerable as a result.

"In Pearl Harbor, what they prepared for were things that really didn't happen," Rumsfeld said. "They prepared for sabotage because they had so many Japanese descendants living in Hawaii. And so they stuck all the airplanes close together, so they could be protected. So of course the bombers came and they were enormously vulnerable, and they were destroyed."

In advance of Pearl Harbor, as Rumsfeld mentioned, we had a theory that sabotage—attack from within—was the most likely means by which our planes and ships would be attacked. The concern over sabotage was pervasive in Hawaii.[12] It was thought that the 80,000 Japanese nationals there might attack not just military bases but radio stations, pineapple farms, and dairy mills with little warning.* Any signals were interpreted in this context, logically or not, and we prepared for subterfuge.[13] We stacked our planes wingtip to wingtip, and our ships stern to bow, on the theory that it would be easier to monitor one big target than several smaller ones.

Meanwhile we theorized that, if Japan seemed to be mobilizing for an attack, it would be against Russia or perhaps against the Asian territorial possessions of the United Kingdom, Russia and the UK being countries that were already involved in the war. Why would the Japanese want to provoke the sleeping giant of the United States? We did not see that Japan believed our involvement in the war was inevitable,[14] and they wanted to strike us when we were least prepared and they could cause the most damage to our Navy. The imperial Japanese government of the time was not willing to abandon its hopes for territorial expansion. We had not seen the conflict through the enemy's eyes.

To Wohlstetter, a signal is a piece of evidence that tells us something useful about our enemy's intentions;[15] this book thinks of a signal as an indication of the underlying truth behind a statistical or predictive problem.† Wohlstetter's definition of *noise* is subtly different too. Whereas I tend to use noise to mean random patterns that might easily be mistaken for signals, Wohlstetter uses it to mean the sound produced by competing signals.[16] In the field of intelligence analysis, the absence of signals can signify something important (the absence of radio transmissions from Japan's carrier fleet signaled their move toward Hawaii) and the presence of *too many* signals can make it exceptionally challenging to discern meaning. They may drown one another out in an ear-splitting cacophony.

The next set of diagrams consist of a series of ten signals, each governed by an extremely simple and orderly mathematical function known as a sine wave.

---

* Although the term was not widely used at the time, the sorts of threats that planners were worried about might be thought of more along the lines of "terrorism" today.

† I chose the title of this book before I came across Wohlstetter's, but she pioneered the use of the metaphor fifty years ago.

In figure 13-2a, one of these signals is highlighted and is much more distinct than the others. After the fact of an attack or some other failure of prediction, this is how the world may look to us. We will see the signal: the paper trail, the pattern, the precursors. Following both Pearl Harbor and September 11, a significant minority of Americans asserted that the patterns were *so* clear that the government *must* have known about the attack, and therefore must have been complicit in planning or executing it.*

FIGURE 13-2A: COMPETING SIGNALS WITH ONE SIGNAL HIGHLIGHTED

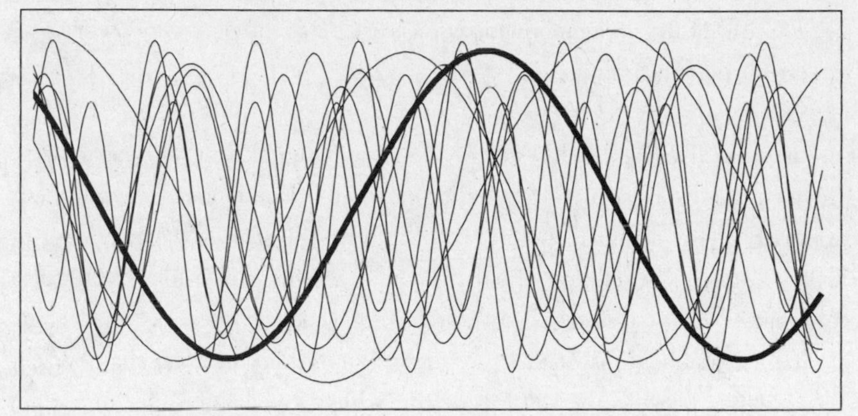

FIGURE 13-2B: COMPETING SIGNALS, UNDIFFERENTIATED

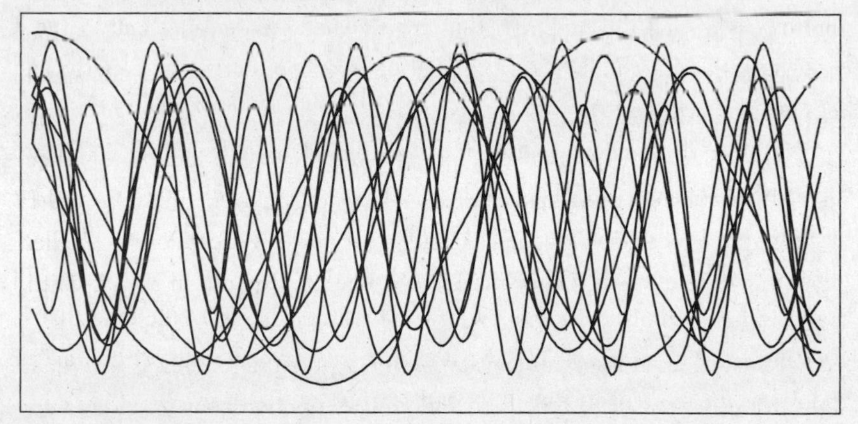

* A conspiracy theory might be thought of as the laziest form of signal analysis. As the Harvard professor H. L. "Skip" Gates says, "Conspiracy theories are an irresistible labor-saving device in the face of complexity."

But this is not usually how the patterns look to us in advance. Instead, they are more like figure 13-2b, an ugly mess of tangled spaghetti string. As Wohlstetter writes:[17]

> It is much easier *after* the event to sort the relevant from the irrelevant signals. After the event, of course, a signal is always crystal clear; we can now see what disaster it was signaling, since the disaster has occurred. But before the event it is obscure and pregnant with conflicting meanings. It comes to the observer embedded in an atmosphere of "noise," i.e., in the company of all sorts of information that is useless and irrelevant for predicting the particular disaster.

In cases like these, what matters is not our signal *detection* capabilities: provided that we have met some basic threshold of competence, we will have perceived plenty of signals before something on the scale of Pearl Harbor or September 11. The relevant signals will be somewhere in a file cabinet or a computer database. But so will a whole host of irrelevant ones. We need signal *analysis* capabilities to isolate the pertinent signals from the echo chamber.

Usually, we will have some views on which signals are more important and require our focus. It is good and necessary to have these views, up to a point. I've detailed the problems that ensue when we consider data without context. Rather than make useful predictions, we trip out on the patterns and get nowhere fast.

However, the context we provide can be biased and self-serving. As Cicero warned Shakespeare's Julius Caesar,[18] "Men may construe things, after their fashion / Clean from the purpose of the things themselves." We may focus on those signals which advance our preferred theory about the world, or might imply a more optimistic outcome. Or we may simply focus on the ones that fit with bureaucratic protocol, like the doctrine that sabotage rather than an air attack was the more likely threat to Pearl Harbor.

The Unfamiliar and the Improbable

Rumsfeld's favorite part of Wohlstetter's book is the foreword, composed by the Nobel Prize–winning economist Thomas Schelling, who was instrumental in translating John Nash's early work on game theory into national-security contexts. Schelling writes of our propensity to mistake the *unfamiliar* for the *improbable*:

> There is a tendency in our planning to confuse the unfamiliar with the improbable. The contingency we have not considered seriously looks strange; what looks strange is thought improbable; what is improbable need not be considered seriously.

Because of the United States' isolation from the European and Asian continents and the relatively good relations we have maintained with the rest of the Americas since the promulgation of the Monroe Doctrine, we have infrequently been the subject of foreign attack. The exceptions (September 11) and near-misses (the Cuban Missile Crisis) have therefore been exceptionally jarring to us. Before Pearl Harbor, the last foreign attack on American soil had been during the War of 1812.[19] Americans just do not live among the ruins of wars past, as people in Europe and Asia have throughout their history.

But our Hawaiian territory* sat in the middle of the Pacific Ocean: Honolulu is closer to Tokyo (3,860 miles) than to Washington, DC (4,825 miles). Because of its geographic position and the presence of our naval fleet, it made an obvious target to the Japanese. The unfamiliarity of attacks on American territory may have made us complacent about the threat.

Perhaps we went through a logical deduction something along these lines:

1. The United States is rarely attacked
2. Hawaii is a part of the United States
3. Therefore, Hawaii is unlikely to be attacked

---

* Hawaii did not become a state until 1959.

This is deeply flawed thinking. As I described in chapter 1, our predictions often fail when we venture "out of sample." The fact that the United States had rarely been attacked is an empirical observation, not an iron law. That Nebraska, say, had never been attacked by a foreign power gave no real evidentiary weight to the situation in Hawaii, given the latter's outlying geographic position in the Pacific and the precariousness of the war situation there.

But at least this flawed type of thinking would have involved some *thinking*. If we had gone through the thought process, perhaps we could have recognized how loose our assumptions were. Schelling suggests that our problems instead run deeper. *When a possibility is unfamiliar to us, we do not even think about it.* Instead we develop a sort of mind-blindness to it. In medicine this is called anosognosia:[20] part of the physiology of the condition prevents a patient from recognizing that they have the condition. Some Alzheimer's patients present in this way.

The predictive version of this syndrome requires us to do one of the things that goes most against our nature: admit to what we do not know.

Was 9/11 a Known Unknown?

> [T]here are known knowns; there are things we know we know. We also know there are known unknowns; that is to say we know there are some things we do not know. But there are also unknown unknowns—there are things we do not know we don't know.—Donald Rumsfeld[21]

Rumsfeld's famous line about "unknown unknowns," delivered in a 2002 press conference in response to a reporter's question about the presence of weapons of mass destruction in Iraq, is a corollary to Schelling's concern about mistaking the unfamiliar for the unlikely. If we ask ourselves a question and can come up with an exact answer, that is a known known. If we ask ourselves a question and can't come up with a very precise answer, that is a known unknown. An unknown unknown is when we haven't really thought to ask the question in the first place. "They are gaps in our knowledge, but gaps that we don't know exist," Rumsfeld writes in his 2011 memoir.[22]

The concept of the unknown unknown is sometimes misunderstood. It's common to see the term employed in formulations like this, to refer to a fairly specific (but hard-to-predict) threat:

> Nigeria is a good bet for a crisis in the not-too-distant future—an **unknown unknown** that poses the most profound implications for US and global security [emphasis added].[23]

This particular prophecy about the terrorist threat posed by Nigeria was rather prescient (it was written in 2006, three years before the Nigerian national Umar Farouk Abdulmutallab tried to detonate explosives hidden in his underwear while aboard a flight from Amsterdam to Detroit). However, it got the semantics wrong. Anytime you are able to enumerate a dangerous or unpredictable element, you are expressing a *known* unknown. To articulate what you don't know is a mark of progress.

Few things, as we have found, fall squarely into the binary categories of the predictable and the unpredictable. Even if you don't know to predict something with 100 percent certainty, you may be able to come up with an estimate or a forecast of the threat. It may be a sharp estimate or a crude one, an accurate forecast or an inaccurate one, a smart one or a dumb one.* But at least you are alert to the problem and you can usually get *somewhere*: we don't know exactly how much of a terrorist threat Nigeria may pose to us, for instance, but it is probably a bigger threat than Luxembourg.

The problem comes when, out of frustration that our knowledge of the world is imperfect, we fail to make a forecast at all. An unknown unknown is a contingency that *we have not even considered*. We have some kind of mental block against it, or our experience is inadequate to imagine it; it's as though it doesn't even exist.

This poses especially grave risks when we consider the signals we receive from terrorists. As before Pearl Harbor, there were many signals that pointed toward the September 11 attacks:

---

* You may even decide that the best estimate is that something is essentially random: you are certain that you are uncertain!

- There had been at least a dozen warnings[24] about the potential for aircraft to be used as weapons, including a 1994 threat by Algerian terrorists to crash a hijacked jet into the Eiffel Tower, and a 1998 plot by a group linked to Al Qaeda to crash an explosives-laden airplane into the World Trade Center.
- The World Trade Center had been targeted by terrorists before. The 1993 bombings by Ramzi Yousef and his co-conspirators, who had trained at Al Qaeda camps in Afghanistan, had killed six and were intended to bring the Twin Towers down.[25]
- Al Qaeda was known to be an exceptionally dangerous and inventive terrorist organization. It had a propensity for pulling off large-scale attacks, including the bombing of U.S. embassies in Kenya and Tanzania in 1998, which killed 224, and the bombing of the USS *Cole* in Yemen in 2000.[26]
- Secretary of State Condoleezza Rice had been warned in July 2001 about heightened Al Qaeda activity—and that the group was shifting its focus from foreign targets to the United States itself.[27] "It's my sixth sense," CIA chief George Tenet had said after seeing the intelligence. "But I feel it coming. This is going to be the big one."[28]
- An Islamic fundamentalist named Zacarias Moussaoui had been arrested on August 16, 2001, less than a month before the attacks, after an instructor at a flight training school in Minnesota reported he was behaving suspiciously.[29] Moussaoui, despite having barely more than 50 hours of training and having never flown solo, had sought training in a Boeing 747 simulator, an unusual request for someone who was nowhere near obtaining his pilot's license.[30]

It is much easier to identify the importance of these signals after the fact; our national security agencies have to sort through literally tens of thousands or even hundreds of thousands of potential warnings[31] to find useful nuggets of information. Most of them amount to nothing.

Still, the September 11 plot was exceptionally ambitious—and the terrorists were able to execute it with relatively few hitches. Nineteen terrorists had entered the air-transit system, successfully hijacking four planes. Three of the four planes hit their targets; United 93 failed to do so only because of the exceptional bravery of the passengers on board, who charged the cockpit after they learned what had happened to the other flights.* Not only had we failed to detect the plot, but it doesn't appear that we came all that close to doing so.

The 9/11 Commission Report identified four types of systemic failures that contributed to our inability to appreciate the importance of these signals, including failures of policy, capabilities, and management.[32] The most important category was failures of imagination.[33] The signals just weren't consistent with our familiar hypotheses about how terrorists behaved, and they went in one ear and out the other without our really registering them.

The North American Aerospace Defense Command (NORAD) had actually proposed running a war game in which a hijacked airliner crashed into the Pentagon. But the idea was dismissed as being "too unrealistic."[34] And in the unlikely event that such an attack were to occur, it was assumed, the plane would come from overseas and not from one of our domestic airports. (Ironically, this was the exact opposite of the mistake that we'd made before Pearl Harbor, where the possibility of an attack from abroad was dismissed because planners were concerned about sabotage.)

The possibility of a suicide attack may also have been hard to imagine. FAA policy was predicated on the idea that a hijacking would result in a tense standoff and perhaps a detour to some exotic airport in the Middle East. But it was assumed the terrorist would not want to destroy the plane, or to kill passengers other than as a negotiation tactic. Thus, cockpit doors were not tightly sealed and were often left entirely unlocked in practice.[35]

Yet suicide attacks had a rich history[36]—including, of course, the Japanese kamikaze pilots in World War II.[37] Moreover, suicide attacks had become much more common in the years immediately preceding September 11; one database of terrorist incidents[38] documented thirty-nine of them in 2000 alone, including the bombing of the USS *Cole* in Yemen, up from thirty-one in the 1980s.

However, World War II was a distant memory, and most of the suicide attacks had occurred in the Middle East or in Third World countries. The mental

* United 93 departed Newark Airport at 8:42 A.M., forty-one minutes after its scheduled departure time, hitting its cruising altitude just as the second plane hit the South Tower of the World Trade Center.

FIGURE 13-3: SUICIDE TERROR ATTACKS, 1979–2000

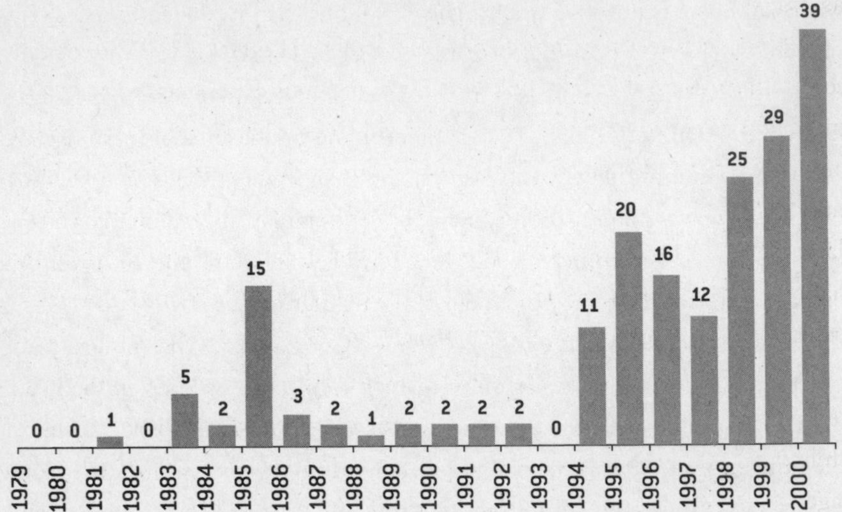

shortcut that Daniel Kahneman calls the availability heuristic[39]—we tend to overrate the likelihood of events that are nearer to us in time and space and underpredict the ones that aren't—may have clouded our judgment.

"You can reasonably predict behavior if people would prefer not to die," Rumsfeld told me. "But if people are just as happy dying, or feel that it's a privilege or that it achieves their goal, then they're going to behave in a very different way."

The Scale of September 11

The September 11 plot, then, was less a hypothesis that we evaluated and rejected as being unlikely—and more one that we had *failed to really consider* in the first place. It was too unfamiliar to us. Rumsfeld suggested in his memoir that September 11 was an unknown unknown.[40]

"The CIA would tell you that it was not a thought that was completely foreign to them," he told me. "But a lot of it came after the attack, in retrospect."

"I mean, it's fair to say Al Qaeda was a known unknown," added Urbahn, Rumsfeld's chief of staff. "But one of the things [Tenet] talks about in his book is that the scale was far different than any other attack. The fact that it was—*so big*."

Indeed, the sheer magnitude of the September 11 attack—2,977 innocent people killed—most differentiated it from our previous experience with terrorism. Even those who were the most concerned about what Al Qaeda was planning, like Tenet and Richard Clarke, the chairman of counterterrorism at the National Security Council, had trouble conceiving of its scale. The notes that Clarke had sent to Condoleezza Rice, for instance, implored her to imagine what might happen if hundreds of Americans died[41]—not thousands of Americans, as actually occurred.

Prior to September 11, the largest terror attack in a Western country* had been only about one-tenth as fatal, when in 1985 a Sikh extremist group hid a bomb in an Air India flight bound for Delhi from Montreal, killing 329 passengers. The destruction of Oklahoma City's Alfred P. Murrah federal building by the militant Timothy McVeigh in 1995, which killed 168, had been the largest attack on American soil.

But September 11 was not an outlier. Although the particulars of the events that day were not discerned in advance—and although they may have been very hard to predict—we had some reason to think that an attack on the scale of September 11 was possible.

The Mathematics of Terrorism: Why 9/11 Wasn't an Outlier

It might seem uncomfortable to think about terrorism in an abstract and mathematical fashion, as we are about to do here. To be clear, this is not a substitute for the sort of signal analysis that the intelligence community performs. But this type of thinking can help patch some of our blind spots and give us better estimates of the overall hazard that terrorism poses. We can better appreciate future risk if we analyze the existing data.

In 2008, I was invited to speak at a conference hosted by the Center for

---

* The largest terror attack in any country had been the arson of the Cinema Rex theater in Abadan, Iran, in 1978, which killed 422.

Strategic and International Studies (CSIS), a Washington-based foreign policy think tank. The timing couldn't have been much worse—it was just two weeks before the 2008 elections—but I was told that the panel involved national security and I suppose I felt it was my duty to attend.

The conference gathered experts from different domains in the hope that a group brainstorming session might stumble across one or two "outside-the-box" insights about how to predict and prevent terror attacks. There was a marketing executive from Coca-Cola on the panel, a police detective from New York, a man who created algorithms for the dating Web site eHarmony—and me. (There were also a number of experts whose work was more self-evidently related to terrorism: people who worked for the State Department, or the military, or for defense contractors around Washington, DC.)

I gave a short presentation describing my work predicting outcomes in baseball and in politics; it was received politely enough. But then it was time for the question-and-answer session. "That's all very nice, Nate," I was told. "But how in the hell is this applicable to terrorism?" (I'm paraphrasing, but only slightly.)

Frankly, the methods I presented at the conference probably weren't all that useful for national security analysis. Baseball and politics are data-rich fields that yield satisfying answers. Thousands of baseball games are played every year. Elections occur less frequently—and require somewhat more caution to forecast—but there are hundreds of polls released during every presidential campaign. All this data is publicly available, freely or cheaply.

Terrorism, seemingly, is not this way at all. The events that we are really concerned about, like September 11, occur only rarely. Moreover, terrorist groups try to conceal their intentions—Al Qaeda was known to be especially effective at this. Within the realm of terrorism, just as was the case immediately before Pearl Harbor, the absence of signals is sometimes more worrisome than the presence of them. If the CIA is able to hack into an Internet chat site thought to be used by radical groups, there may be quite a bit of loose and casual chatter at first, when organizations like Al Qaeda are simply hoping to find new and naïve recruits. But when an attack is being plotted, and the stakes are raised, the conversation usually goes offline.

At a microscopic level, then—at the level of individual terrorists, or individual terror schemes—there are unlikely to be any magic-bullet solutions to predicting attacks. Instead, intelligence requires sorting through the spaghetti strands of signals that I spoke about earlier. One expert I talked to at the CSIS conference provided another metaphor: detecting a terror plot is much more difficult than finding a needle in a haystack, he said, and more analogous to finding one particular needle in a pile full of needle parts.

Some problems that are highly unpredictable on a case-by-case basis may seem more orderly if we zoom out and take a macroscopic view. Outside insight into the mathematical properties of terrorism has proved more useful here.

Aaron Clauset, a professor at the University of Colorado with a background in physics and computer science, has published papers on the mathematics of everything from the evolution of whales[42] to the network dynamics of multiplayer role-playing games.[43] The intelligence community has a long-standing reputation for preferring alpha-male types. The thirty-something Clauset, whose catholic research interests might seem geeky to it, has met a mixture of encouragement and resistance when presenting his findings.

"Some people say it's a breath of fresh air," Clauset told me when we caught up on the phone. "That's a few people. Most people look at it and say, 'That's kind of weird. You want to use math?'"

Clauset's insight, however, is actually quite simple—or at least it seems that way with the benefit of hindsight. What his work found is that the mathematics of terrorism resemble those of another domain discussed in this book: earthquakes.

Imagine that you live in a seismically active area like California. Over a period of a couple of decades, you experience magnitude 4 earthquakes on a regular basis, magnitude 5 earthquakes perhaps a few times a year, and a handful of magnitude 6s. If you have a house that can withstand a magnitude 6 earthquake but not a magnitude 7, would it be right to conclude that you have nothing to worry about?

Of course not. According to the power-law distribution that these earthquakes obey, those magnitude 5s and magnitude 6s would have been a sign that larger earthquakes were possible—inevitable, in fact, given enough time. The big one is coming, eventually. You ought to have been prepared.

Terror attacks behave in something of the same way. The Lockerbie bombing and Oklahoma City were the equivalent of magnitude 7 earthquakes.

While destructive enough on their own, they also implied the potential for something much worse—something like the September 11 attacks, which might be thought of as a magnitude 8. It was not an outlier but instead part of the broader mathematical pattern.

Defining and Measuring Terrorism

To look at the statistics of terrorism, we first need to define exactly what it is. This can get a little tricky. Vladimir Lenin said that "the purpose of terrorism is to terrorize."[44] This is more insightful than it might seem: terrorists are not purely seeking to maximize their body count; instead, they want to maximize the amount of fear they inflict on a population so as to alter its behavior in some way. Death and destruction are just a means toward that end. "You may kill people to achieve that," Rumsfeld told me. "But that is not its purpose."

Still, there is a wide variety of violent behavior throughout the world, and so academics have sought a somewhat more precise definition to distinguish terrorism from its counterparts. One definition, employed by a widely used database of terrorist incidents,[45] requires that terrorist acts must be intentional, that they must entail actual or threatened violence, and that they must be carried out by "subnational actors" (meaning, not directly by sovereign governments themselves). The incidents, moreover, must be aimed at attaining a political, economic, social, or religious goal. And they must involve some element of intimidation or coercion—intended to induce fear in an audience beyond the immediate victims.

The type of terrorism that most explicitly meets these criteria and which is most familiar to us today has relatively modern roots. The UCLA political scientist David C. Rapoport dates it to 1979[46]—the year of the Iranian Revolution. He relates it to religious extremism, particularly among Islamist groups. This wave of terrorism is associated with a sharp rise in the number of attacks against Western countries and Western interests; from 1979 through 2000 the number of terror attacks against NATO countries rose almost threefold.

Most of the incidents, however, produced few fatalities, if any. From the Iranian Revolution through September 10, 2001, there were more than 4,000 attempted or successful terror attacks in NATO countries. But more than half the death toll had been caused by just seven of them. The three largest attacks—the Air India disaster, the Lockerbie bombing, and Oklahoma City—had accounted for more than 40 percent of the fatalities all by themselves.

This type of pattern—a very small number of cases causing a very large proportion of the total impact—is characteristic of a power-law distribution, the type of distribution that earthquakes obey. Clauset's insight was that terror attacks abide by a power-law distribution as well.

Suppose that we draw a graph (figure 13-4) plotting the frequency of terror attacks on one axis and their death tolls on the other. At first, this doesn't seem terribly useful. You can clearly see the power-law relationship: the number of attacks decreases very steeply with their frequency. But the slope is so steep that it seems to obscure any meaningful signal: you see a large number of very small attacks, and a small number of very large ones, with seemingly little room in between. The September 11 attacks look like an outlier.

FIGURE 13-4: TERROR ATTACK FREQUENCY BY DEATH TOLL IN NATO COUNTRIES, 1979–2009 (LINEAR SCALE)

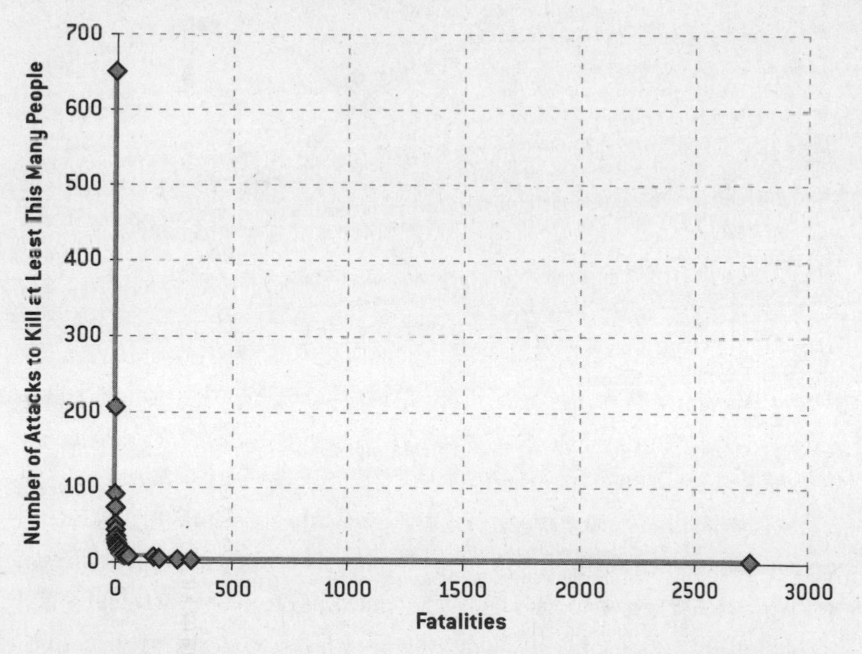

However, as was the case for earthquakes, the data is easier to comprehend when we plot it on a logarithmic scale (more specifically, as in figure 13-5, a double-logarithmic scale in which both the vertical and the horizontal axes are logarithmic). It's important to emphasize that I've done nothing to this data other than make it easier to visualize—it's still the same underlying information. But what had once seemed chaotic and random is now revealed to be rather orderly. When plotted on a double-logarithmic scale, the relationship between the frequency and the severity of terror attacks appears to be, more or less,[47] a straight line. This is, in fact, a fundamental characteristic of power-law relationships: when you plot them on a double-logarithmic scale, the pattern that emerges is as straight as an arrow.

FIGURE 13-5: TERROR ATTACK FREQUENCY BY DEATH TOLL, NATO COUNTRIES, 1979–2009 (LOGARITHMIC SCALE)

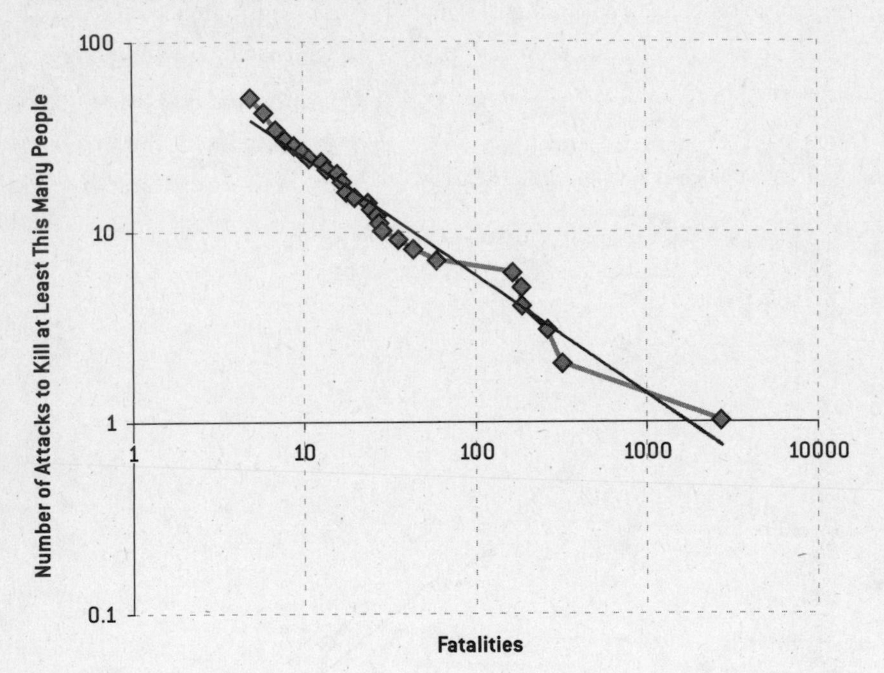

Power laws have some important properties when it comes to making predictions about the scale of future risks. In particular, they imply that disasters much worse than what society has experienced in the recent past are entirely possible, if infrequent. For instance, the terrorism power law predicts that a NATO country (not necessarily the United States) would experience a terror attack killing at least one hundred people about six times over the thirty-one-year period from 1979 through 2009. (This is close to the actual figure: there were actually seven such attacks during this period.) Likewise, it implies that an attack that killed 1,000 people would occur about once every twenty-two years. And it suggests that something on the scale of September 11,[48] which killed almost 3,000 people, would occur about once every forty years.

It's not that much of an accomplishment, however, to describe history in statistical terms. Sure, it's possible for a statistical model to accommodate an event like September 11 now that one has actually occurred. But what would Clauset's method have said about the possibility of such an attack *before* it happened?

September 11 certainly did shift the probabilities somewhat—just as the number of very large earthquakes in recent years implies that they are somewhat more common than we might have thought previously.[49] Nevertheless, even before it occurred, the power-law method would have concluded that an attack on the scale of September 11 was a clear possibility. If the power-law process is applied to data collected entirely *before* 9/11—everything from the

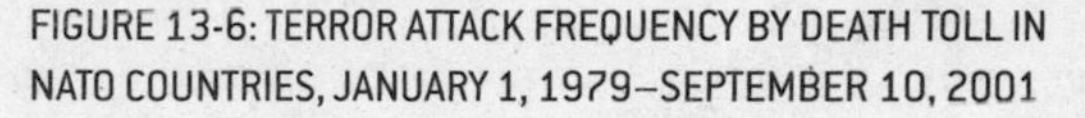
FIGURE 13-6: TERROR ATTACK FREQUENCY BY DEATH TOLL IN NATO COUNTRIES, JANUARY 1, 1979–SEPTEMBER 10, 2001

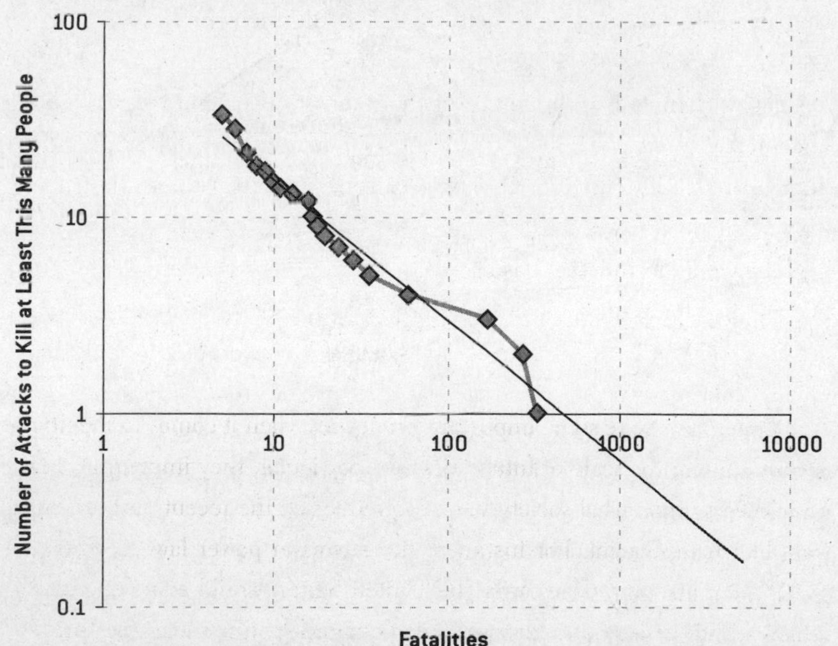

beginning of the modern wave of terrorism in 1979 through September 10, 2001—it implies that a September 11–scale attack would occur about once every eighty years in a NATO country, or roughly once in our lifetimes.[50]

This method does not tell us anything specific about exactly *where and when* the attack would occur. It is a long-term tendency, like the tendency toward earthquakes in California. And terror attacks, unlike earthquakes, can plausibly be prevented—this is an important qualification to Clauset's hypothesis.

What this data does suggest is that an attack on the scale of September 11 should not have been *unimaginable*. The power-law distribution demonstrates that events of a much larger scale than occurred in the past may plausibly occur in the future. Our lack of familiarity with them will be an especially poor guide to their likelihood.

Magnitude 9 Terrorism

But if the September 11 attacks were tantamount to a magnitude 8 earthquake, what about the potential for something larger still: the equivalent of a magnitude 9? Clauset's method gives us reason to believe that attacks that might kill tens of thousands or hundreds of thousands of people are a possibility to contemplate as well. The mechanism for such an attack is unpleasant but easy enough to identify. It would most likely involve weapons of mass destruction, particularly nuclear weapons.

The world, fortunately, hasn't had much experience with nuclear warfare. The atomic bombs dropped on Hiroshima and Nagasaki, Japan, in 1945 at the end of World War II killed about 200,000 people.[51] One estimate[52] holds that an equally powerful weapon, detonated at one of New York City's major ports, would kill on the order of 250,000 civilians. But nuclear technology has evolved since then. A larger and more modern weapon, blown up over midtown Manhattan, could kill as many as one million New Yorkers[53] while incinerating hundreds of billions of dollars in real estate. Simultaneous attacks on New York, Washington, Chicago, and Los Angeles could kill on the order of four million Americans, in line with a stated goal of Osama bin Laden.[54]

These estimates reflect—probably—worst-case scenarios. Because such an attack could be hundreds of times worse than 9/11, however, the question of its likelihood has been the subject of intense debate in the national security community.

One of the more pessimistic assessments comes from Graham Allison, a professor of political science at Harvard. Allison served in the administrations of both President Reagan and President Clinton, and his books and papers on the Cuban missile crisis have been cited thousands of times by other scholars.[55] So when Allison has something to say, his peers listen.

Allison came to a frightening conclusion in 2004: "A nuclear terrorist attack on America in the decade ahead is more likely than not,"[56] he wrote. Allison qualified his forecast by saying it assumed we remained "on the current path"—a world in which there are dangerous terrorist groups, vulnerable nuclear materials at many places around the world, and a lack of focus on the problem from U.S. policy makers.

We are more than a decade removed from the September 11 attacks. But when I spoke with Allison in 2010, he reaffirmed the gravity of threat he perceives. It's one that he takes quite literally, in fact. I called Allison from my desk at the *New York Times* office, a block from Times Square.[57] Allison told me that he'd be at least a little bit nervous in Times Square and wasn't sure if he'd be willing to work there every day.

Allison's probability estimate doesn't come from a statistical model. Instead it's "the basis on which [he]'d make bets."* Why does he see so much risk? "It's a knockoff of the old Sherlock Holmes version of motive, means, and opportunity," he told me.

The **motive** of terrorists, for Allison, is easy to discern. Osama bin Laden had said that he wanted to kill four million Americans, a number that could probably be achieved only through a nuclear attack. The modus operandi of Al Qaeda has been what Allison calls "spectaculars"—occasional but astonishing attacks that kill large numbers of innocent people. And the CIA had picked up Al Qaeda chatter about an "American Hiroshima" before the September 11 attacks.[58]

---

* Note that "the basis on which I'd make bets" is essentially a Bayesian prior possibility.

By **opportunity**, Allison means the ability of terrorist groups to smuggle a weapon into the United States. He has little doubt that this could happen. "How do crooks get into American cities every day?" Allison asked me. The United States has more than 3,700 ports, and receives more than six million cargo containers per year—but only 2 percent of them are physically inspected by customs agents.[59] "If you have any doubt, they could always hide it in a bale of marijuana," Allison half-joked.

So Allison is mostly focused on the **means**—the ability of a terrorist group to acquire a nuclear weapon. If we want to reduce the risk of a nuclear version of 9/11, controlling the means would be the way.

Experts believe there are about 20,000 nuclear warheads in the world today[60]—down from a peak of 65,000 in the 1980s. A threat could theoretically come from any of the nine countries that possess nuclear weapons today—even the United States has lost track of eleven of its nuclear weapons throughout its history[61]—and other countries may be trying to develop them. But Allison's concern stems primarily from two nuclear states: Russia and Pakistan.

In Allison's view, the risk has lessened some in the former country. In part because of successful programs like the one sponsored by senators Sam Nunn and Richard Lugar, there are no longer active nuclear weapons in the outlying states of the former Soviet Union. And in Russia itself, the number of nuclear weapons has declined to 11,000 today from a peak of 30,000 in 1985.

If the risk in Russia has been reduced, however, the threat posed by Pakistan has increased—perhaps markedly. "If you map weapons of mass destruction and terrorism, all the roads intersect in Pakistan," Allison told me.

Although Pakistan is ostensibly an ally of the United States, even the most generous interpretation would suggest that it represents a problem as well as a solution in the effort to contain terrorism. The country had initially been reluctant to cooperate with the United States after the September 11 attacks, and Pakistan's president later claimed that the U.S. had resorted to a threat to bomb it "back to the stone age" before it complied.[62] Osama bin Laden had been living in Abbottabad, Pakistan, for as many as six years[63] before he was killed. Meanwhile, Pakistan has roughly one hundred nuclear weapons and is building additional nuclear facilities and delivery systems at a rapid pace.[64] The country now ranks seventh in the world in the *Economist*'s Political Instability Index, up significantly from the recent past,[65] meaning that the risk of a coup d'état or a revolution is quite high. A new regime could be openly hostile to the United States. All the conditions that a terrorist might need to acquire a nuclear weapon could then be in place.

Because of the deteriorating situation in Pakistan, as well as the residual threat posed by other countries, Allison told me that he was still pretty much where he was in 2004—he envisions a better-than-even chance of a nuclear attack on the United States within the next decade unless there is a change in trajectory.

Allison also has his critics, including Michael Levi, whom I visited at his office at the Council on Foreign Relations in New York. Like Aaron Clauset, Levi has an eccentric background: he studied theoretical physics at Princeton and was a technical consultant to the popular television program *24*, which depicted a terrorist group attempting to detonate a nuclear weapon in Los Angeles.*

Levi thinks there is some risk of an attack. "When I first came here," he told me, "one of the first things I did was to draw the rings around Grand Central Station to see how a ten kiloton bomb would affect my apartment." But he thinks Allison's estimate is much too high, and questions several of his premises.

For one thing, Levi thinks that Allison takes the motive of terrorist groups too much for granted. It's not that Al Qaeda wouldn't have the aspiration to blow up Manhattan, he says. But groups and individuals have all sorts of aspirations that they make no effort to act on because they doubt their ability to achieve them. Terrorist groups, in Levi's view, place a high premium on the likelihood of their plots' succeeding. A failed attempt would blow the group's cover and invite unwanted attention from the United States and other governments. A failed attack could also harm the group's credibility with both existing members and potential new recruits. Terrorist organizations are fundamentally weak and unstable: as is supposedly true of new restaurants, 90 percent of

* They were stopped by Jack Bauer, of course.

them fail within the first year.[66] Their recruitment message relies, in large part, on convincing their members that they are able to deliver redemption from perceived injustices.[67] Al Qaeda, in particular, had a very high success rate on the attacks it attempted up to and including September 11, something that may help to explain its uncommon longevity. But if a terrorist group's ability to achieve that redemption is called into question, their recruits might go elsewhere.

A nuclear attack would be hard to pull off. It is not that terrorist groups are inherently averse to intricate planning—the September 11 attacks were in the works for five years. But the more complicated the plot, the more cooperation is required from a larger number of participants, each of whom carries the risk of defecting from the group or being detected by counterterrorism authorities. A nuclear attack would also require significant and highly specialized technical knowledge—far more than four terrorists learning how to fly a 767. The pool of nuclear physicists is small to begin with, and the number who could be trusted to maintain loyalty to a terrorist group is smaller still.[68] "If they found a guy who had an engineering degree from a university and put him in charge of the nuclear team," Levi told me, "I wouldn't expect much to come of that."

Finally, terrorists' goals are not necessarily to kill as many people as possible. Rather, it's to inflict fear and alter behavior. A nuclear attack would be incredibly terrifying, but it would not necessarily be one hundred or one thousand times more terrifying than September 11, even though it might kill that many more people. If the chance of success was low, it might not be the most effective means for terrorists to achieve their goals.

Others in the national security community, like Rumsfeld, are more concerned about a biological attack. A biological attack would require much less expertise than a nuclear one, but it might instill just as much fear. Moreover, it is a type of fear that might be less familiar to us. Particularly if the biological agent were contagious, like a reintroduction of the smallpox virus, it could remain an active concern for weeks or months, with schools and shopping malls shuttered, hospitals quarantined, and state borders closed. The remarkable resiliency that New York displayed just days after September 11 would be harder to achieve.

"Biological is different. It is also something that we're not comfortable with. People know how god-awful nuclear weapons are," Rumsfeld said. "Something that is contagious, and something that can go down through generations and can alter genes—the fear of a biological weapon is quite different than the fear of a nuclear weapon or even a chemical weapon."

The death toll that a biological attack might cause is hard to estimate—just as the spread of any transmissible disease is hard to predict until it occurs (as we found in chapter 7). Still, the worst-case scenarios are undoubtedly quite bad. A 2001 simulation called Dark Winter,[69] imagined that three million Americans might become infected with smallpox, and one million might die, if terrorists succeeded in spreading it with simultaneous hits on shopping malls in Oklahoma City, Philadelphia, and Atlanta.

Thinking Big About Terrorism

Clauset's method is agnostic on the means by which a magnitude 9 terrorist attack might occur—it just says that one is possible. Judging by the death tolls of attacks from 1979 through 2009, for instance, a power-law model like Clauset's could be taken to imply there is about a 10 percent chance of an attack that would kill at least 10,000 people in a NATO country over the next decade. There is a 3 percent chance of an attack that would kill 100,000, and a 0.6 percent chance of one that would kill one million or more.

These estimates need to be approached cautiously. There is still considerable uncertainty, especially about the potential for extremely large-scale events, and slightly different versions of this technique produce slightly different answers. However, there is one more instructive comparison to be made between terrorism and earthquakes.

The Gutenberg–Richter law dictates that, over the long term, the frequency of earthquakes is reduced about ten times for every one-point increase in magnitude. However, the energy released by earthquakes increases exponentially as a function of magnitude. In particular, for every one-point increase in magnitude, an earthquake's energy release increases by about thirty-two times. So a magnitude 6 earthquake releases around thirty-two times as much seismic energy as a magnitude 5, while a magnitude 7 is close to 1,000 times more powerful.

The force released by earthquakes scales up at a faster rate than their frequency decreases. If there are ten magnitude 6 earthquakes for every magnitude 7, the magnitude 7 tremor will account for considerably more damage[70] than all the magnitude 6s combined. Indeed, a mere handful of earthquakes are responsible for a very large fraction of their total seismic energy. In the one hundred years between 1906 and 2005, for instance, just three large earthquakes—the Chilean earthquake of 1960, the Alaskan earthquake of 1964, and the Great Sumatra Earthquake of 2004—accounted for almost half the total energy release of all earthquakes in the world over the entire century. So, seismologists and contingency planners are mostly concerned about very large earthquakes. A more modest earthquake in the wrong place at the wrong time can cause enormous damage (like the magnitude 7.0 earthquake in Haiti in 2010), but it's mostly the very high magnitude earthquakes that we have to worry about, even though they occur quite infrequently.

Consider again the case of terror attacks. The September 11 attacks alone killed more people—2,977, not counting the terrorists—than all other attacks in NATO countries over the thirty-year period between 1979 and 2009 combined (figure 13-7). A single nuclear or biological attack, meanwhile, might dwarf the fatality total of September 11.

So even if Levi is right that the odds of these attacks are extremely low, they represent the bulk of the hazard. The power-law method, for instance, estimates that the odds of an incident that would kill one million people—an atomic bomb detonated in Times Square—is only about 1 in 1,600 per year. But one million people dying once every 1,600 years represents 625 fatalities per year, considerably more than the roughly 180 people in NATO countries who have died on average in terror attacks since 1979. When it comes to terrorism, we need to think big, about the probability for very large magnitude events and how we might reduce it, even at the margin. Signals that point toward such large attacks should therefore receive a much higher strategic priority.

This mathematical argument for a focus on larger-scale threats cuts some-

FIGURE 13-7: SHARE OF FATALITIES FROM TERROR ATTACKS IN NATO COUNTRIES, 1979–2009

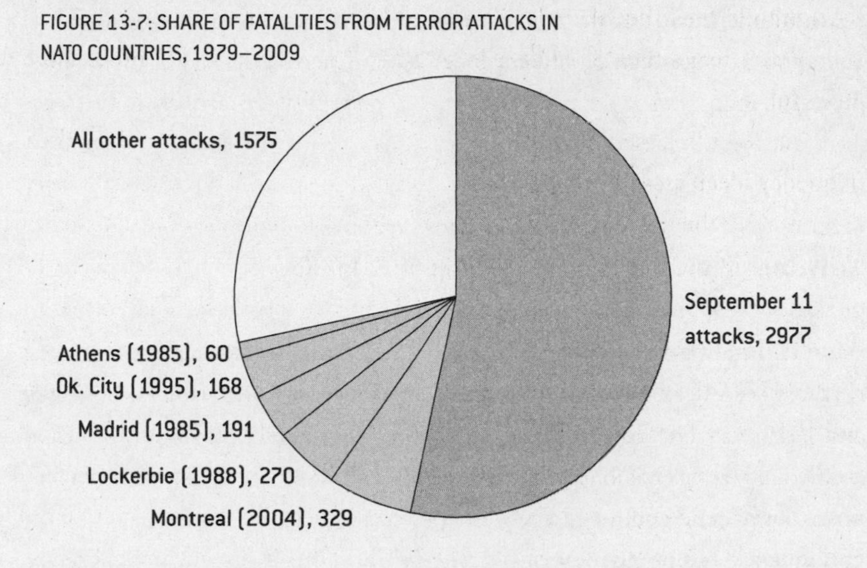

what against the day-to-day imperatives of those who are actively involved in homeland security. In 1982, the social scientists James Q. Wilson and George L. Kelling introduced what they called the "broken windows" theory of crime deterrence.[71] The idea was that by focusing on smaller-scale types of crime, like vandalism and misdemeanor drug offenses,[72] police could contribute to an overall climate of lawfulness and therefore prevent bigger crime. The empirical evidence for the merit of this theory is quite mixed.[73, 74]

However, the theory was very warmly embraced by police departments from Los Angeles to New York because it lowered the degree of difficulty for police departments and provided for much more attainable goals. It's much easier to bust a sixteen-year-old kid for smoking a joint than to solve an auto theft or prevent a murder. Everybody likes to live in a cleaner, safer neighborhood. But it's unclear whether the broken-windows theory is more than window dressing.

Likewise, the ever more cumbersome requirements for commercial flights fall into the category of what the security expert Bruce Schneier calls "security theater"[75]—they are more for show than to actually deter terrorists. It's by no means completely irrational to be worried about airport security; airplanes have been the subject of a large number of terror attacks in the past, and terrorism can have a copycat element.[76] Yet even accounting for crashes that had nothing to do with terrorism, only about one passenger for every twenty-five million was killed on an American commercial airliner during the decade of the 2000s.[77] Even if you fly twenty times per year, you are about twice as likely to be struck by lightning.

Why Don't Terrorists Blow Up Shopping Malls?

Mostly, these efforts are aimed at foiling dumb terrorists—and there are surely a few of those, like the underwear bomber. A clever terrorist would probably be able to outwit these efforts, however, or he would redirect his attention to more vulnerable targets like buses or trains. Really, he would need to look no further than the check-in counter, where there are just as many people gathered but much less security. Terrorists have already figured this out: a suicide bomber killed thirty-five at the international arrivals area of Moscow's Domodedovo Airport in 2011.[78]

For that matter, a virtually unlimited number of unsecured targets exist that have nothing to do with the transportation system. Why don't terrorists go shoot up a shopping mall?

One reason there aren't all that many terror attacks may be that there aren't all that many terrorists. It is very difficult to get a head count of terrorists, but one commonly cited estimate is that Al Qaeda had only about 500 to 1,000 operatives at its peak.[79] This figure includes hangers-on and wannabes, as well as the people engaged in all the nonviolent functions that groups like Al Qaeda must perform: some doofus has to reboot Al Qaeda's server when their network goes down. Kathleen Carley of Carnegie Mellon University, who studies the social networks of insurgent organizations, told me that even within what we think of as an extremist group, perhaps only 1 percent of members are true extremists. It's much easier to facilitate global jihad as Bin Laden's IT consultant than it is to walk into a crowded mall and blow yourself up.

Still, we should be careful when we ask these sorts of questions—we may, yet again, be confusing the unfamiliar for the improbable. The conundrum of why terrorists don't target shopping centers would seem ridiculous to someone in Israel, where it happens all the time.

The Israeli Approach to Terrorism Prevention

One of the obvious critiques of Clauset's power-law hypothesis is that, unlike in the case of something like earthquakes, terrorism can be stopped by human intervention.

Clauset's research suggests that the power-law distribution exists not in spite of the competition between terrorists and counterterrorism forces but perhaps *because* of it. The pattern of the points on the graph is dictated by the different counterterrorism strategies of the individual countries. There may be a sort of equilibrium that exists between terrorists and society, a balance that has been struck between freedom and security, although it may vary over time and place.[80] We must always accept a certain amount of risk from terrorism when we want to live in a free society, whether or not we want to admit it.

"It will ebb and flow," Rumsfeld told me. "We are probably the most vulnerable to terrorism. We being free people. Because that is our nature. We expect to be able to get up and walk out the door and send our children to school and not have to look around the corner and see if we're going to be killed by something or blown up. To the extent we alter our behavior as free people dramatically, they've won."

Although Israel is targeted by terrorists much more frequently than the United States, Israelis do not live in fear of terrorism. A 2012 survey of Israeli Jews found that only 16 percent described terrorism as their greatest fear[81]—no more than the number who said they were worried about Israel's education system.

No Israeli politician would say outright that he tolerates small-scale terrorism, but that's essentially what the country does. It tolerates it because the alternative—having everyone be paralyzed by fear—is incapacitating and in line with the terrorists' goals. A key element in the country's strategy is making life as normal as possible for people after an attack occurs. For instance, police typically try to clear the scene of an attack within four hours of a bomb going off,[82] letting everyone get back to work, errands, or even leisure. Small-scale terrorism is treated more like crime than an existential threat.

What Israel certainly does not tolerate is the potential for large-scale terrorism (as might be made more likely, for instance, by one of their neighbors acquiring weapons of mass destruction). There is some evidence that their approach is successful: Israel is the one country that has been able to bend Clauset's curve. If we plot the fatality tolls from terrorist incidents in Israel using the power-law method (figure 13-8), we find that there have been significantly fewer large-scale terror attacks than the power-law would predict; no incident since 1979 has killed more than two hundred people. The fact that Israel's power-law graph looks so distinct is evidence that our strategic choices do make some difference.

FIGURE 13-8: TERROR ATTACK FREQUENCY BY DEATH TOLL IN ISRAEL, 1979–2009

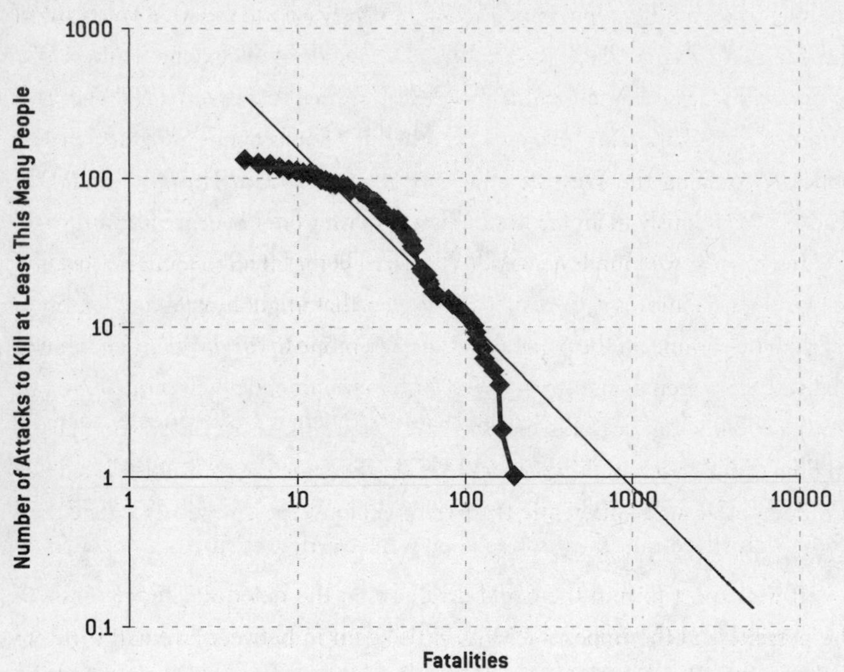

How to Read Terrorists' Signals

Whatever strategic choices we make, and whatever trade-off we are willing to accept between security and freedom, we must begin with the signal. Good intelligence is still our first line of defense against terror attacks.

One of the stated goals of the Iraq War was to prevent the further development of the country's weapons of mass destruction programs. Of course, there were virtually no WMDs there. Independent analyses of the decision to go into Iraq have generally concluded that the Bush White House did not pressure the intelligence community into coming up with faulty information—there was something of a consensus at the time in organizations like the CIA that Iraq was actively developing its WMD programs[83]—but it did misrepresent the intelligence to the American public in several important ways.[84]

Although this view has merit, I'm not sure that it's possible to make so crisp a distinction between what the Bush administration told the public, what they believed themselves, and what they learned from intelligence officials. In signal analysis, as with other types of prediction, it is very easy to see what you want in the mess of tangled data. The unreliable signals provided by sources like "Curveball"—an Iraqi national named Rafid Ahmed Alwan al-Janabi who later admitted[85] to fabricating evidence about Iraq's bioweapons program in the hopes of sparking the Western powers to topple Saddam Hussein—could be read too credulously in an environment where we were too eager for war.

In chess, as we found in chapter 9, human beings tend to focus on just one or two potential moves at the expense of others that might be superior. Computers, which examine all the possibilities, are less prone to this mistake, and moves that would be seen as signifying genius if they were made by Bobby Fischer or Garry Kasparov can often be identified in a few seconds on a laptop. Computers do not evaluate each of these moves to quite the same depth. But they manage the trade-off a little differently than we do. They cast a wide net rather than look for one perfect solution.

If we expect to find the world divided into the deterministic contours of the possible and the impossible, with little room in between, we will wind up with overconfident predictions on the one hand, or unknown unknowns on the other. We just aren't that good at prediction, and we need to hedge our bets rather than believe solely in one hypothesis, as Phil Tetlock's hedgehogs did.

Perhaps no field puts these skills to the test more than matters of national security. As Bruce Schneier notes,[86] the essence of a security problem is that it defines us by our weakest link. If you secure your front door with titanium plating, armed guards, and a phalanx of pit bulls, it doesn't do you any good if you also have a creaky side door that any two-bit burglar could break into. These threats are asymmetric; all of America's naval power in the Pacific didn't help it much when the Japanese fleet slipped right through the blind spots in our defenses and found that most of our arsenal was conveniently located in just one place, where it could easily be attacked. This is why events like Pearl Harbor and September 11 produce the sort of cognitive dissonance they do. Where our enemies will strike us is predictable: it's where we least expect them to.

Some of the type of thinking I encourage in this book can probably be helpful in the realm of national security analysis.[87] For instance, the Bayesian approach toward thinking about probability is more compatible with decision making under high uncertainty. It encourages us to hold a large number of hypotheses in our head at once, to think about them probabilistically, and to update them frequently when we come across new information that might be more or less consistent with them.

The closest we came to catching Al Qaeda red-handed in advance of September 11 was in the arrest of Zacarias Moussaoui, the Islamic extremist who had an unnatural interest in learning how to fly a 747. Were there innocent explanations for that? I suppose he could have been some sort of flyboy with a lot of time on his hands. But if we had attached some prior possibility, even a small one, to the hypothesis that "terrorists might hijack planes and crash them into buildings," our estimate of its likelihood would surely have increased substantially once we came across this intelligence. Instead, however, we failed to consider the hypothesis at all—it was an unknown unknown to us. As the 9/11 Commission concluded, "the system was not tuned to comprehend the potential significance of this information," and so Moussaoui's arrest contributed little to our ability to foil the plot.[88]

This is not to suggest that our intelligence officials have gotten everything wrong. If some blame must be given for the failure to detect the September 11 attacks, some credit must be given to both the Bush and Obama administrations for the lack of attacks since then—something I certainly would not have predicted eleven years ago. Like a baseball umpire, an intelligence analyst risks being blamed when something goes wrong but receives little notice when she does her job well. I do not consider this field to reflect an unmitigated failure of prediction as I do some of the others in this book; considering the challenges it faces, it may be one of the more successful ones.

Still, as the 9/11 Commission deduced, the most important source of failure in advance of the attacks was our lack of imagination. When we are making predictions, we need a balance between curiosity and skepticism.[89] They can be compatible. The more eagerly we commit to scrutinizing and testing our theories, the more readily we accept that our knowledge of the world is uncertain, the more willingly we acknowledge that perfect prediction is impossible, the less we will live in fear of our failures, and the more liberty we will have to let our minds flow freely. By knowing more about what we don't know, we may get a few more predictions right.

CONCLUSION

A Major League shortstop has some plays he can always make, some plays he can never make, and some he'll have to dive for. The diving plays are the most spectacular and will catch our attention. But they can lead to a myopic view of the shortstop's abilities.

The legendary shortstop Derek Jeter was a frequent subject of debate during the *Moneyball* era. Broadcasters and scouts noticed that Jeter seemed to make an especially large number of diving plays and concluded that he was an exceptional shortstop for that reason. Stat geeks crunched the numbers and detected a flaw in this thinking.[1] Although Jeter was a terrific athlete, he often got a slow jump on the ball and dove because he was making up for lost time. In fact, the numbers suggested that Jeter was a fairly poor defensive shortstop, despite having won five Gold Glove awards. The plays that Jeter had to dive for, a truly great defensive shortstop like Ozzie Smith might have made easily—perhaps receiving less credit for them because he made them look routine.

FIGURE C-1: SHORTSTOP DIVING RANGES

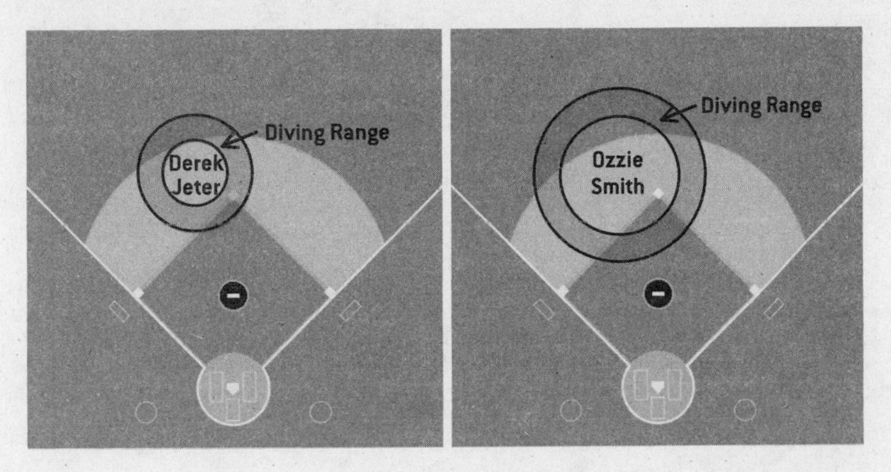

Whatever range of abilities we have acquired, there will always be tasks sitting right at the edge of them. If we judge ourselves by what is hardest for us, we may take for granted those things that we do easily and routinely.

One of the most spectacularly correct predictions in history was that of the English astronomer Edmund Halley, who in 1705 predicted that a great comet would return to the earth in 1758. Halley had many doubters, but the comet returned just in the nick of time.[2] Comets, which in antiquity were regarded as being wholly unpredictable omens from the gods,[3] are now seen as uncannily regular and predictable things.

Astronomers predict that Halley's Comet will next make its closest approach to the earth on July 28, 2061. By that time, many problems in the natural world that now vex our predictive abilities will have come within the range of our knowledge.

Nature's laws do not change very much. So long as the store of human knowledge continues to expand, as it has since Gutenberg's printing press, we will slowly come to a better understanding of nature's signals, if never all its secrets.

And yet if science and technology are the heroes of this book, there is the risk in the age of Big Data about becoming too starry-eyed about what they might accomplish.

There is no reason to conclude that the affairs of men are becoming more predictable. The opposite may well be true. The same sciences that uncover the laws of nature are making the organization of society more complex. Technology is completely changing the way we relate to one another. Because of the Internet, "the whole context, all the equations, all the dynamics of the propagation of information change," I was told by Tim Berners-Lee, who invented the World Wide Web in 1990.[4]

The volume of information is increasing exponentially. But relatively little of this information is useful—the signal-to-noise ratio may be waning. We need better ways of distinguishing the two.

This book is less about what we know than about the difference between what we know and what we think we know. It recommends a strategy so that we might close that gap. The strategy requires one giant leap and then some small steps forward. The leap is into the Bayesian way of thinking about prediction and probability.

Think Probabilistically

Bayes's theorem begins and ends with a probabilistic expression of the likelihood of a real-world event. It does not require you to believe that the world is intrinsically uncertain. It was invented in the days when the regularity of Newton's laws formed the dominant paradigm in science. It does require you to accept, however, that your subjective perceptions of the world are approximations of the truth.

This probabilistic element of the Bayesian way may seem uncomfortable at first. Unless we grew up playing cards or other games of chance, we were probably not encouraged to think in this way. Mathematics classrooms spend more time on abstract subjects like geometry and calculus than they do on probability and statistics. In many walks of life, expressions of uncertainty are mistaken for admissions of weakness.

When you first start to make these probability estimates, they may be quite poor. But there are two pieces of favorable news. First, these estimates are just a starting point: Bayes's theorem will have you revise and improve them as you encounter new information. Second, there is evidence that this is something we can learn to improve. The military, for instance, has sometimes trained soldiers in these techniques,[5] with reasonably good results.[6] There is also evidence that doctors think about medical diagnoses in a Bayesian manner.[7]

It is probably better to follow the lead of our doctors and our soldiers than our television pundits.

Our brains process information by means of approximation.[8] This is less an existential fact than a biological necessity: we perceive far more inputs than we can consciously consider, and we handle this problem by breaking them down into regularities and patterns.

Under high stress, the regularities of life will be stripped away. Studies of people who survived disasters like the September 11 attacks found that they could recall some minute details about their experiences and yet often felt almost wholly disconnected from their larger environments.[9] Under these circumstances, our first instincts and first approximations may be rather poor, often failing to recognize the gravity of the threat. Those who had been forced to make decisions under extreme stress before, like on the battlefield, were more likely to emerge as heroes, leading others to safety.[10]

Our brains simplify and approximate just as much in everyday life. With experience, the simplifications and approximations will be a useful guide and will constitute our working knowledge.[11] But they are not perfect, and we often do not realize how rough they are.

Consider the following set of seven statements, which are related to the idea of the efficient-market hypothesis and whether an individual investor can beat the stock market. Each statement is an approximation, but each builds on the last one to become slightly more accurate.

1. No investor can beat the stock market.
2. No investor can beat the stock market over the long run.
3. No investor can beat the stock market over the long run relative to his level of risk.

4. No investor can beat the stock market over the long run relative to his level of risk and accounting for his transaction costs.
5. No investor can beat the stock market over the long run relative to his level of risk and accounting for his transaction costs, unless he has inside information.
6. Few investors beat the stock market over the long run relative to their level of risk and accounting for their transaction costs, unless they have inside information.
7. It is hard to tell how many investors beat the stock market over the long run, because the data is very noisy, but we know that most cannot relative to their level of risk, since trading produces no net excess return but entails transaction costs, so unless you have inside information, you are probably better off investing in an index fund.

The first approximation—the unqualified statement that no investor can beat the stock market—seems to be extremely powerful. By the time we get to the last one, which is full of expressions of uncertainty, we have nothing that would fit on a bumper sticker. But it is also a more complete description of the objective world.

There is nothing wrong with an approximation here and there. If you encountered a stranger who knew nothing about the stock market, informing him that it is hard to beat, even in the crude terms of the first statement, would be a lot better than nothing.

The problem comes when we mistake the approximation for the reality. Ideologues like Phil Tetlock's hedgehogs behave in this way. The simpler statements seem more universal, more in testament to a greater truth or grander theory. Tetlock found, however, that his hedgehogs were very poor at making predictions. They leave out all the messy bits that make life real and predictions more accurate.

We have big brains, but we live in an incomprehensibly large universe. The virtue in thinking probabilistically is that you will force yourself to stop and smell the data—slow down, and consider the imperfections in your thinking. Over time, you should find that this makes your decision making better.

Know Where You're Coming From

Bayes's theorem requires us to state—explicitly—how likely we believe an event is to occur *before* we begin to weigh the evidence. It calls this estimate a prior belief.

Where should our prior beliefs come from? Ideally, we would like to build on our past experience or even better the collective experience of society. This is one of the helpful roles that markets can play. Markets are certainly not perfect, but the vast majority of the time, collective judgment will be better than ours alone. Markets form a good starting point to weigh new evidence against, particularly if you have not invested much time in studying a problem.

Of course, markets are not available in every case. It will often be necessary to pick something else as a default. Even common sense can serve as a Bayesian prior, a check against taking the output of a statistical model too credulously. (These models are approximations and often rather crude ones, even if they seem to promise mathematical precision.) Information becomes knowledge only when it's placed in context. Without it, we have no way to differentiate the signal from the noise, and our search for the truth might be swamped by false positives.

What isn't acceptable under Bayes's theorem is to pretend that you don't have *any* prior beliefs. You should work to reduce your biases, but to say you have none is a sign that you have many. To state your beliefs up front—to say "Here's where I'm coming from"[12]—is a way to operate in good faith and to recognize that you perceive reality through a subjective filter.

Try, and Err

This is perhaps the easiest Bayesian principle to apply: make a lot of forecasts. You may not want to stake your company or your livelihood on them, especially at first.* But it's the only way to get better.

*You may even want to hone your skills in fields like sports, which offer rich data sets and rich opportunities for testing your techniques. The NCAA bracket pool may be less of a diversion than you realize, especially if you take it seriously.

Bayes's theorem says we should update our forecasts any time we are presented with new information. A less literal version of this idea is simply trial and error. Companies that really "get" Big Data, like Google, aren't spending a lot of time in model land.* They're running thousands of experiments every year and testing their ideas on real customers.

Bayes's theorem encourages us to be disciplined about how we weigh new information. If our ideas are worthwhile, we ought to be willing to test them by establishing falsifiable hypotheses and subjecting them to a prediction. Most of the time, we do not appreciate how noisy the data is, and so our bias is to place too much weight on the newest data point. Political reporters often forget that there is a margin of error when polls are reported, and financial reporters don't always do a good job of conveying how imprecise most economic statistics are. It's often the outliers that make the news.

But we can have the opposite bias when we become too personally or professionally invested in a problem, failing to change our minds when the facts do. If an expert is one of Tetlock's hedgehogs, he may be too proud to change his forecast when the data is incongruous with his theory of the world. Partisans who expect every idea to fit on a bumper sticker will proceed through the various stages of grief before accepting that they have oversimplified reality.

The more often you are willing to test your ideas, the sooner you can begin to avoid these problems and learn from your mistakes. Staring at the ocean and waiting for a flash of insight is how ideas are generated in the movies. In the real world, they rarely come when you are standing in place.[13] Nor do the "big" ideas necessarily start out that way. It's more often with small, incremental, and sometimes even accidental steps that we make progress.

---

*A sign that you're spending too much time in model land is if you start to use the word "prediction" to refer to how your model fits the *past* data. As I describe in chapter 5, it is very easy to *overfit* a model, thinking you have captured the signal when you're just describing the noise. Sticking to the simple, commonsense definition of a prediction as something that applies strictly to a *future* event may reduce the risk of these errors.

Our Perceptions of Predictability

Prediction is difficult for us for the same reason that it is so important: it is where objective and subjective reality intersect. Distinguishing the signal from the noise requires both scientific knowledge and self-knowledge: the serenity to accept the things we cannot predict, the courage to predict the things we can, and the wisdom to know the difference.[14]

Our views on how predictable the world is have waxed and waned over the years. One simple measure of it is the number of times the words "predictable" and "unpredictable" are used in academic journals.[15] At the dawn of the twentieth century, the two words were used almost exactly as often as one another. The Great Depression and the Second World War catapulted "unpredictable" into the dominant position. As the world healed from these crises, "predictable" came back into fashion, its usage peaking in the 1970s. "Unpredictable" has been on the rise again in recent years.

FIGURE C-2: THE PERCEPTION OF PREDICTABILITY, 1900–2012

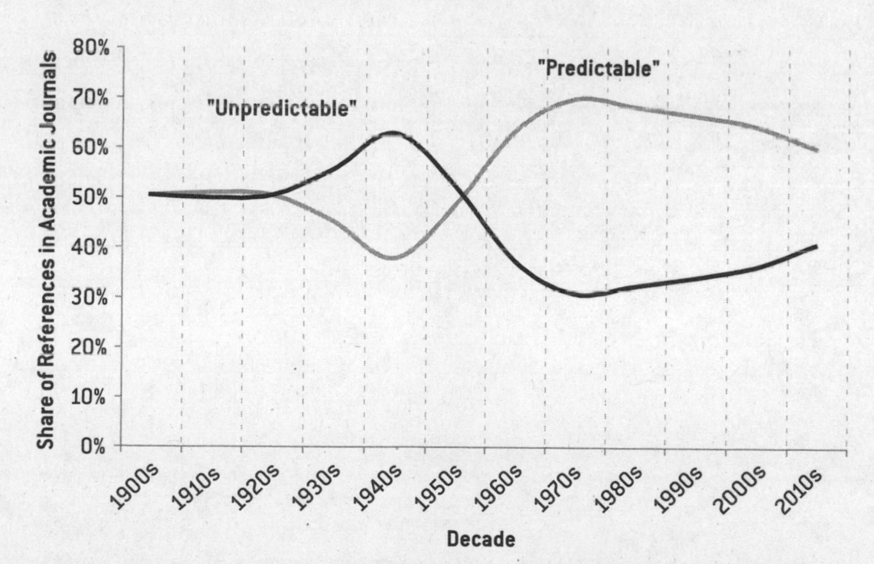

These perceptions about predictability are more affected by the fashions of the sciences[16] and the shortness of our memories—has anything really bad happened *recently*?—than by any real change in our forecasting skills. How good we think we are at prediction and how good we really are may even be inversely correlated. The 1950s, when the world was still shaken by the war and was seen as fairly unpredictable, was a time of more economic[17] and scientific[18] productivity than the 1970s, the decade when we thought we could predict everything, but couldn't.

These shifting attitudes have reverberated far beyond academic journals. If you drew the same chart based on the use of the words "predictable" and "unpredictable" in English-language *fiction*, it would look almost exactly the same as in figure C-2.[19] An unpredicted disaster, even if it has no direct effect on us, shakes our confidence that we are in control of our fate.

But our bias is to think we are better at prediction than we really are. The first twelve years of the new millennium have been rough, with one unpredicted disaster after another. May we arise from the ashes of these beaten but not bowed, a little more modest about our forecasting abilities, and a little less likely to repeat our mistakes.

ACKNOWLEDGMENTS

As the author Joseph Epstein has noted, it is a lot better to have written a book than to actually be writing one. Writing a book requires a tremendous amount of patience, organization, and discipline, qualities that I lack and that writing a blog do not very much encourage.

I was therefore highly dependent on many others who had those qualities in greater measure, and whose wisdom helped to shape the book in many large and small ways.

Thank you to my parents, Brian David Silver and Sally Thrun Silver, to whom this book is dedicated, and to my sister, Rebecca Silver.

Thank you to Virginia Smith for being a terrific editor in all respects. She, as Laura Stickney, Ann Godoff, and Scott Moyers, believed in the vision of the book. They made few compromises in producing a book that fulfilled that vision and yet tolerated many excuses when I needed more time to get it there.

Thank you to my literary agent, Sydelle Kramer, for helping me to conceive of and sell the project. Her advice was invariably the right kind: gentle enough, but never too gentle, on the many occasions when the book seemed at risk of running off the rails.

Thank you to my research assistant, Arikia Millikan, who provided boundless enthusiasm for the book, and whose influence is reflected in its keen interest in science and technology. Thank you to Julia Kamin, whose organizational skills helped point the way forward when the book was at a critical stage. Thank you to Jane Cavolina and Ellen Cavolina Porter, who produced high-quality transcriptions on a demanding schedule.

Thank you to Emily Votruba, Veronica Windholz, Kaitlyn Flynn, Amanda

Dewey, and John Sharp for turning the book around against an extremely tight production schedule, and for their understanding that "today" usually meant "tonight" and that "tonight" usually meant "5 in the morning."

Thank you to Robert Gauldin for his love and support. Thank you to Shashank Patel, Kim Balin, Bryan Joiner, Katie Halper, Jason MacLean, Maryam Saleh, and Jessica Klein for tolerating my rambling on about the book for hours at a time on the one hand or going into hiding for weeks at a time on the other.

Thank you to Micah Cohen at the *New York Times*, who assisted with this book in more ways than I can count.

Thank you to my bosses and colleagues at the *New York Times*, especially Megan Liberman, Jim Roberts, David Leonhardt, Lisa Tozzi, Gerry Mullany, Rick Berke, Dick Stevenson, Derek Willis, Matt Ericson, Greg Veis, and Hugo Lindgren, who trusted me to manage the demands of the book production cycle along with those of the news cycle. Thank you to Bill Keller, Gerry Marzorati, and Jill Abramson for bringing me into the *New York Times* family.

Thank you to John Sides, Andrew Gelman, Tom Schaller, Ed Kilgore, Renard Sexton, Brian McCabe, Hale Stewart, and Sean Quinn for their contributions to the FiveThirtyEight blog.

Thank you to Richard Thaler and Anil Kashyap, of the University of Chicago, for reviewing the chapters related to economics and finance. Thank you to David Carr, Kathy Gauldin, and Page Ashley for reminding me of the importance of finishing the book, and to Will Repko for helping to instill that work ethic that might get it there.

Thank you to Gary Huckabay, Brandon Adams, Rafe Furst, Kevin Goldstein, Keith Urbahn, Matthew Vogel, Rachel Hauser, Jennifer Bloch, Thom Shanker, Kyu-Young Lee, and Mark Goldstein for serving as connectors and facilitators at key points along the way.

Many people were polled on the title of this book. Thank you to Jonah Peretti, Andrea Harner, Kyle Roth, Jessi Pervola, Ruth Welte, Brent Silver, Richard Silver, Amanda Silver, Roie Lindegren, Len Lindegren, Zuben Jelveh, Douglas Jester, Justin Wolfers, J. Stephen Steppard, Robert Erikson, Katie Donalek, Helen Lee, Katha Pollitt, Jeffrey Toobin, David Roberts, Felix Salmon, Hillary Bok, Heather Hurlburt, Art Goldhammer, David Karol, Sara Robinson, Max Sawicky, Michael O'Hare, Marc Tracy, Daniel Davies, E. J. Graff, Paul Starr,

Russ Wellen, Jeffrey Hauser, Dana Goldstein, Suzy Khimm, Jonathan Zasloff, Avi Zenilman, James Galbraith, Greg Anrig, Paul Waldman, and Bob Kuttner for providing their advice.

This book is fairly scrupulous about citing the origin of its ideas, but some people I interviewed were more influential in determining its direction than might be inferred by the number of times that they appear in the text. This list includes Daniel Kahneman, Vasik Rajlich, Dr. Alexander "Sandy" McDonald, Roger Pielke Jr., John Rundle, Thomas Jordan, Irene Eckstrand, Phil Gordon, Chris Volinsky, Robert Bell, Tim Berners-Lee, Lisa Randall, Jay Rosen, Simon Jackman, Diane Lauderdale, Jeffrey Sachs, Howard Lederer, Rodney Brooks, Henry Abbott, and Bruce Bueno de Mesquita among others.

I hope to return all these favors someday. I will start by buying the first beer for anybody on this list, and the first three for anybody who should have been, but isn't.

—Nate Silver
Brooklyn, NY

NOTES

PREFACE TO THE PAPERBACK EDITION

1. Google Trends, November 3, 2014. http://www.google.com/trends/explore#q=big%20data.
2. NewsLibrary.com search engine, November 3, 2014.
3. NewsLibrary.com search engine, November 3, 2014.
4. Gartner, Inc., "Gartner's 2014 Hype Cycle for Emerging Technologies Maps the Journey to Digital Business," accessed November 3, 2014. http://www.gartner.com/newsroom/id/2819918.
5. This calculation is based on the final set that I ran prior to the election, which account for the various permutations of results in each of the fifty states.
6. Ian Schwartz, "George Will Predicts Romney Wins Big, 321–217," RealClearPolitics.com, accessed November 3, 2014. http://www.realclearpolitics.com/video/2012/11/04/george_will_predicts_romney_wins_big_321-217.html.
7. Michael Barone, "Barone: Going Out on a Limb: Romney Beats Obama, Handily," accessed November 3, 2014. http://www.washingtonexaminer.com/barone-going-out-on-a-limb-romney-beats-obama-handily/article/2512470.
8. Bureau of Labor Statistics' Occupational Employment Statistics Data, 2003 and 2013.

INTRODUCTION

1. The Industrial Revolution is variously described as starting anywhere from the mid-eighteenth to the early nineteenth centuries. I choose the year 1775 somewhat arbitrarily as it coincides with the invention of James Watt's steam engine and because it is a nice round number.
2. Steven Pinker, *The Better Angels of Our Nature: Why Violence Has Declined* (New York: Viking, Kindle edition, 2011); locations 3279–3282.
3. Much of the manuscript production took place at monasteries. Belgium often had among the highest rates of manuscript production per capita because of its abundant monasteries. Relieved of their need to produce manuscripts, a few of these monasteries instead began to shift their focus to producing their wonderful Trappist beers. So another of those unintended consequences: Gutenberg's invention, however indirectly, may be at least a little bit responsible for improving the quality of the world's beer.
4. Albania De la Mare, *Vespasiano da Bisticci Historian and Bookseller* (London: London University, 2007), p. 207.
5. Elizabeth Eisenstein, *The Printing Revolution in Early Modern Europe* (Cambridge, England: Cambridge University Press, 1993), p. 16.
6. That which has been is what will be,
   That which is done is what will be done,
   And there is nothing new under the sun.

Is there anything of which it may be said,
"See, this is new"?
It has already been in ancient times before us.
There is no remembrance of former things,
Nor will there be any remembrance of things that are to come
By those who will come after.

—Ecclesiastes 1:9-11 (New King James translation)
http://www.biblegateway.com/passage/?search=Ecclesiastes+1&version=NKJV

7. De la Mare, *Vespasiano da Bisticci Historian and Bookseller*, p. 207.
8. Eisenstein, *The Printing Revolution in Early Modern Europe*, p. 17.
9. Eltjo Burnigh and Jan Luiten Van Zanden, "Charting the 'Rise of the West': Manuscripts and Printed Books in Europe, a Long-Term Perspective from the Sixth Through Eighteenth Centuries," *Journal of Economic History*, vol. 69, issue 2; June 2009.
10. "Recognizing and Naming America," The Library of Congress, Washington, DC. http://www.loc.gov/rr/geogmap/waldexh.html.
11. Eisenstein, *The Printing Revolution in Early Modern Europe*, p. 209.
12. Louis Edward Inglebart, *Press Freedoms: A Descriptive Calendar of Concepts, Interpretations, Events, and Court Actions, from 4000 B.C. to the Present* (Westport, CT: Greenwood Publishing, 1987).
13. Renato Rosadlo, "The Cultural Impact of the Printed Word: A Review Article," in Andrew Shyrock, ed. *Comparative Studies in Society and History*, vol. 23, 1981, pp. 508–13. http://journals.cambridge.org/action/displayJournal?jid=CSS.
14. Eisenstein, *The Printing Revolution in Early Modern Europe*, p. 168.
15. Arthur Geoffrey Dickens, *Reformation and Society in Sixteenth Century Europe* (New York: Harcourt, Brace & World, 1970), p. 51. http://www.goodreads.com/book/show/3286085-reformation-and-society-in-sixteenth-century-europe.
16. Pinker, *The Better Angels of Our Nature*, Kindle locations 3279–3282.
17. "War and Violence on Decline in Modern Times," National Public Radio (transcript), December 7, 2011.http://www.npr.org/2011/12/07/143285836/war-and-violence-on-the-decline-in-modern-times.
18. Simon Augustine Blackmore, *The Riddles of Hamlet* (Stratford, England: Stratford and Company, 1917). http://www.shakespeare-online.com/plays/hamlet/divineprovidence.html.
19. Whose end is purposed by the mighty gods?
Yet Caesar shall go forth, for these predictions
Are to the world in general as to Caesar.

—William Shakespeare, *The Tragedy of Julius Caesar*, Act II, Scene II; ll. 27–29.

20. Douglas Harper, *Online Etymology Dictionary*.http://www.etymonline.com/index.php?term=forecast.
21. www.etymonline.com/index.php?term=predict.
22. One representative seventeenth-century text used the term *forecast* in this way:

> Men in all trades consider . . . where 'tis best to buy, and what things are likeliest to sell, and forecast in their own minds, what ways and methods are likeliest to make them thrive in their several occupations. John Kettlewell, *Five Discourses on So Many Very Important Points of Practical Religion* (A. and J. Churchill, 1696); http://books.google.com/books?id=ADo3AAAAMAAJ&dq

23. Not least because Calvinists and Protestants believed in predestination.
24. Max Weber, *The Protestant Ethic and the Spirit of Capitalism* (Abingdon, Oxon, England: Routledge Classics, 2001).

25. Eisenstein, *The Printing Revolution in Early Modern Europe*, p. 269.
26. J. Bradford DeLong, *Estimating World GDP, One Million* B.C.—*Present* (Berkeley, CA: University of California Press, 1988). http://econ161.berkeley.edu/TCEH/1998_Draft/World_GDP/Estimating_World_GDP.html.
27. Figure 1-2 is drawn from DeLong's estimates, although converted to 2010 U.S. dollars rather than 1990 U.S. dollars as per his original.
28. Google Books Ngram Viewer. http://books.google.com/ngrams/graph?content=information+age%2C+computer+age&year_start=1800&year_end=2000&corpus=0&smoothing=3.
29. Susan Hough, *Predicting the Unpredictable: The Tumultuous Science of Earthquake Prediction* (Princeton: Princeton University Press, Kindle edition, 2009), locations 862–869.
30. Robert M. Solow, "We'd Better Watch Out," *New York Times Book Review*, July 12, 1987. http://www.standupeconomist.com/pdf/misc/solow-computer-productivity.pdf.
31. "U.S. Business Cycle Expansions and Contractions," National Bureau of Economic Research, http://www.nber.org/cycles.html.
32. Although, as we explain in this book, economic statistics are much rougher than people realize.
33. Figures are adjusted to 2005 U.S. dollars.
34. I use the number of patent applications rather than patent grants for this metric because patent grants can be slowed by bureaucratic backlog. Among the only real bipartisan accomplishments of the 112th Congress was the passage of the America Invents Act in September 2011, which was passed by a an 89–9 majority in the Senate and sped patent applications.
35. For figures on U.S. research and development spending, see "U.S. and International Research and Development: Funds and Alliances," National Science Foundation. http://www.nsf.gov/statistics/seind02/c4/c4s4.htm.

    For patent applications, see "U.S. Patent Statistics Chart Calendar Years 1963–2011," U.S. Patent and Trade Office. http://www.uspto.gov/web/offices/ac/ido/oeip/taf/us_stat.htm.

    Note that the patent applications statistics in figure I-3 include those from U.S.-based inventors only as the U.S. Patent and Trade Office also processes many patent applications that originate from abroad.
36. "What Is Big Data?," IBM. http://www-01.ibm.com/software/data/bigdata/.
37. Chris Anderson, "The End of Theory: The Data Deluge Makes the Scientific Method Obsolete," *Wired* magazine, June 23, 2008. http://www.wired.com/science/discoveries/magazine/16-07/pb_theory.
38. Nate Silver, "Models Based on 'Fundamentals' Have Failed at Predicting Presidential Elections," *FiveThirtyEight*, *New York Times*, March 26, 2012. http://fivethirtyeight.blogs.nytimes.com/2012/03/26/models-based-on-fundamentals-have-failed-at-predicting-presidential-elections/.
39. John P. A. Ioannidis, "Why Most Published Research Findings Are False," *PLOS Medicine*, 2, 8 (August 2005), e124. http://www.plosmedicine.org/article/info:doi/10.1371/journal.pmed.0020124.
40. Brian Owens, "Reliability of 'New Drug Target' Claims Called into Question," *NewsBlog*, *Nature*, September 5, 2011. http://blogs.nature.com/news/2011/09/reliability_of_new_drug_target.html.
41. This estimate is from Robert Birge of Syracuse University. http://www.sizes.com/people/brain.htm.
42. Alvin Toffler, *Future Shock* (New York: Bantam Books, 1990), p. 362.
43. "The Polarization of the Congressional Parties," VoteView.com. http://voteview.com/political_polarization.asp.
44. Dan M. Kahan, et al., "The Polarizing Impact of Science Literacy and Numeracy on Perceived Climate Change Risks," *Nature Climate Change*, May 27, 2012. http://www.nature.com/nclimate/journal/vaop/ncurrent/full/nclimate1547.html.
45. Karl Popper, *The Logic of Scientific Discovery* (Abingdon, Oxon, England: Routledge Classics, 2007), p. 10.

CHAPTER 1: A CATASTROPHIC FAILURE OF PREDICTION

1. "S&P/Case-Shiller Home Price Index: Las Vegas, NV." http://ycharts.com/indicators/case_shiller_home_price_index_las_vegas.
2. Jeffrey M. Jones, "Trust in Government Remains Low," Gallup.com, September 18, 2008. http://www.gallup.com/poll/110458/trust-government-remains-low.aspx.
3. Although the polling on the bailout bills was somewhat ambiguous at the time they were passed, later statistical analyses suggested that members of Congress who had voted for the bailout were more likely to lose their seats. See for example: Nate Silver, "Health Care and Bailout Votes May Have Hurt Democrats," *FiveThirtyEight, New York Times*, November 16, 2011. http://fivethirtyeight.blogs.nytimes.com/2010/11/16/health-care-bailout-votes-may-have-hurt-democrats/.
4. As measured by comparison of actual to potential GDP. As of the fourth quarter of 2011, the output gap was about $778 billion, or slightly more than $2,500 per U.S. citizen. "Real Potential Gross Domestic Product," Congressional Budget Office, United States Congress. http://research.stlouisfed.org/fred2/data/GDPPOT.txt.
5. More technically, S&P's ratings consider only the likelihood of a default, in any amount, while the two other companies also consider a default's potential magnitude.
6. Anna Katherine Barnett-Hart, "The Story of the CDO Market Meltdown: An Empirical Analysis," thesis, Harvard University, p. 113. http://www.hks.harvard.edu/m-rcbg/students/dunlop/2009-CDOmeltdown.pdf.
7. Diane Vazza, Nicholas Kraemer and Evan Gunter, "2010 Annual U.S. Corporate Default Study and Rating Transitions," Standard & Poor's, March 30, 2011. http://www.standardandpoors.com/ratings/articles/en/us/?articleType=HTML&assetID=1245302234800.
8. S&P downgraded U.S. Treasuries to a AA-plus rating in 2011.
9. Mark Adelson, "Default, Transition, and Recovery: A Global Cross-Asset Report Card of Ratings Performance in Times of Stress," Standard & Poor's, June 8, 2010. http://www.standardandpoors.com/products-services/articles/en/us/?assetID=1245214438884.
10. Barnett-Hart, "The Story of the CDO Market Meltdown: An Empirical Analysis."
11. Most of the CDOs that did not default outright are nevertheless now close to worthless; more than 90 percent of the mortgage-backed securities issued during 2006 and 2007 have since been downgraded to below investment-grade (BlackRock Solutions, as of May 7, 2010; per a presentation provided to author by Anil Kashyap, University of Chicago).
12. "Testimony of Deven Sharma, President of Standard & Poor's, Before the Committee on Oversight and Government Reform," United States House of Representatives, October 22, 2008. http://oversight-archive.waxman.house.gov/documents/20081022125052.pdf.
13. This remains S&P's first line of defense today. When I contacted an S&P spokeswoman, Catherine Mathis, in September 2011, asking for an interview on their ratings of mortgage-backed debt, she responded in nearly identical language. "We were not alone, which you should point out," she wrote in an e-mail. "Many homeowners, financial institutions, policy makers, and investors did not expect the U.S. housing market to decline as quickly or as far as it did." S&P declined my request for a full interview. For additional evidence on this point, see Kristopher S. Gerardi, Andreas Lehnert, Shane M. Sherlund and Paul S. Willen, "Making Sense of the Subprime Crisis," Public Policy Discussion Papers No. 09-1, Federal Reserve Bank of Boston, December 22, 2008. http://www.bos.frb.org/economic/ppdp/2009/ppdp0901.htm.
14. Robert J. Shiller, *Irrational Exuberance*, (Princeton, NJ: Princeton University Press, 2000).
15. Dean Baker, "The Run-Up in Home Prices: Is It Real or Is It Another Bubble?" Center for Economic and Policy Research, August. 2002. http://www.cepr.net/index.php/publications/reports/the-run-up-in-home-prices-is-it-real-or-is-it-another-bubble/.
16. "In Come the Waves," *The Economist*, June 16, 2005. http://www.economist.com/node/4079027?story_id=4079027.
17. Paul Krugman, "That Hissing Sound," *New York Times*, August 8, 2005. http://www.nytimes.com/2005/08/08/opinion/08krugman.html.

18. Google "Insights for Search" beta; "housing bubble" (worldwide). http://www.google.com/insights/search/#q=housing%20bubble&cmpt=q.
19. Google "Insights for Search" beta; "housing bubble" (United States). http://www.google.com/insights/search/#q=housing+bubble&cmpt=q&geo=US.
20. Newslibrary.com search; United States sources only.
21. The volume of discussion in the news media forms a parallel to that of the stock market bubble in the late 1990s, references to which increased tenfold in news accounts between 1994 and 1999, peaking just the year before markets crashed.
22. Janet Morrissey, "A Corporate Sleuth Tries the Credit Rating Field," *New York Times*, February 26, 2011. http://www.nytimes.com/2011/02/27/business/27kroll.html?pagewanted=all.
23. Alex Veiga, "U.S. Foreclosure Rates Double," Associated Press, November 1, 2007. http://www.azcentral.com/realestate/articles/1101biz-foreclosures01-ON.html.
24. Elliot Blair Smith, "'Race to Bottom' at Moody's, S&P Secured Subprime's Boom, Bust," Bloomberg, September 25, 2008. http://www.bloomberg.com/apps/news?pid=newsarchive&sid=ax3vfya_Vtdo.
25. Hedge funds and other sophisticated investors generally do not have these requirements: they will do their own due diligence on a security, happy to bet against a ratings agency's forecast when warranted (as it often is). But public pension funds and other institutional investors like college endowments are stuck. Jane's retirement savings and Johnny's college scholarship are dependent on the accuracy of the ratings agencies, while a hedge fund leaves itself the flexibility to make the big short.
26. Per interview with Jules Kroll.
27. Another catch-22 is that historically an NRSRO was required to be in business for three years before it could start rating securities. But without any ratings revenues, there was no way for the business to get off the ground.
28. Moody's is the only one of the three major agencies that does most of its business through its credit ratings alone. S&P, by contrast, is a part of McGraw-Hill, a publishing company, making its financials harder to scrutinize.
29. Chicago Booth School, "Credit Rating Agencies and the Crisis."
30. Jonathan Katz, Emanuel Salinas and Consstantinos Stephanou, "Credit Rating Agencies: No Easy Regulatory Solutions," The World Bank Group's "Crisis Response," Note Number 8, October 2009. http://rru.worldbank.org/documents/CrisisResponse/Note8.pdf.
31. "Moody's Corporation Financials (NYSE:MCO)," Google Finance. http://www.google.com/finance?q=NYSE:MCO&fstype=ii.
32. Sam Jones, "Alphaville: Rating Cows," *Financial Times*, October 23, 2008. http://ftalphaville.ft.com/blog/2008/10/23/17359/rating-cows/.
33. Ibid.
34. "New CDO Evaluator Version 2.3 Masters 'CDO-Squared' Analysis, Increases Transparency in Market," Standard & Poor's press release, May 6, 2004. http://www.alacrastore.com/research/s-and-p-credit-research-New_CDO_Evaluator_Version_2_3_Masters_CDO_Squared_Analysis_Increases_Transparency_in_Market-443234.
35. Efraim Benmelech and Jennifer Dlugosz, "The Alchemy of CDO Credit Ratings," *Journal of Monetary Economics* 56; April 2009. http://www.economics.harvard.edu/faculty/benmelech/files/Alchemy.pdf.
36. S&P later conducted another simulation under a yet-more-stressful scenario—a housing decline severe enough to trigger a recession—and came to the same conclusion. The results were at first made public but have since been deleted from S&P's Web site.
37. Douglas Adams, *Mostly Harmless* (New York: Del Rey, 2000).
38. Benmelech and Dlugosz, "The Alchemy of CDO Credit Ratings."
39. Barnett-Hart, "The Story of the CDO Market Meltdown: An Empirical Analysis."
40. The 20 percent chance of default refers to the rate over a five-year period.
41. And it can get worse than that. These securities can also be combined into derivatives of one another, which are even more highly leveraged. For instance, five Alpha Pools of mortgage

debt could be combined into a Super Alpha Pool, which would pay you out unless all five of the underlying Alpha Pools defaulted. The odds of this happening are just one in 336 nonillion (a one followed by thirty zeroes) if the mortgages are perfectly uncorrelated with one another—but 1 in 20 if they are perfectly correlated, meaning that it is leveraged by a multiple of 16,777,215,999,999,900,000,000,000,000,000.

42. Ingo Fender and John Kiff, "CDO Rating Methodology: Some Thoughts on Model Risk and Its Implications," BIS Working Papers No. 163, November 2004.
43. Statement of Gary Witt, Former Managing Director, Moody's Investment Service, to Fiscal Crisis Inquiry Commission; June 2, 2010. http://fcic-static.law.stanford.edu/cdn_media/fcic-testimony/2010-0602-Witt.pdf.
44. Financial Crisis Inquiry Commission, *Financial Crisis Inquiry Commission Report: Final Report of the National Commission on the Causes of the Financial and Economic Crisis in the United States* (U.S. Government Printing Office, 2011), p. 121. http://www.gpo.gov/fdsys/pkg/GPO-FCIC/pdf/GPO-FCIC.pdf.
45. Frank H. Knight, *Risk, Uncertainty, and Profit* (New York: Riverside Press; 1921). http://www.programme-finance.com/teletudiant/Knight.%20Risk,%20Uncertainty%20and%20Profit.pdf.
46. In a Texas hold 'em game with one card to come.
47. Why is this the case? "For most of the U.S., land is abundant," Anil Kashyap explained in a note to me. "So the value of a house and a plot of land is about what the cost of construction would be. With all the technical progress in building, construction costs have fallen steadily. So expecting appreciation is doubtful. The big exception is places where there are building restrictions. One way to see this is to notice that Texas had almost no bubble. Why? Because there is no zoning and few natural limitations."
48. During the 1950s, consumers had extraordinarily rich balance sheets. For a variety of reasons—prudence brought on by the memory of the Great Depression, restrictions on prices and production of consumer goods during the war years, and a booming labor market that required all hands on deck—Americans had an unprecedented amount of savings. During World War II, Americans kept as much as 25 percent of their incomes, and savings rates remained elevated in the years that followed.
49. "Historical Census of Housing Tables," Housing and Household Economic Statistics Division, U.S. Census Bureau; last updated October 31, 2011. http://www.census.gov/hhes/www/housing/census/historic/owner.html.
50. David R. Morgan, John P. Pelissero, and Robert E. England, *Managing Urban America* (Washington, DC: CQ Press, 2007).
51. "Annual Statistics: 2005," Housing and Household Economic Statistics Division, U.S. Census Bureau; last updated October 31, 2011. http://www.census.gov/hhes/www/housing/hvs/annual05/ann05t12.html.
52. "Historical Income Tables—Families," Housing and Household Economic Statistics Division, U.S. Census Bureau; last updated August 26, 2008. http://www.webcitation.org/query?url=http%3A%2F%2Fwww.census.gov%2Fhhes%2Fwww%2Fincome%2Fhistinc%2Ff01AR.html&date=2009-04-12.
53. In fact, as Anil Kashyap pointed out to me, the zip codes that saw the biggest increase in subprime lending had declining employment, rising crime, and deteriorating fundamentals. See Atif Mian and Amir Sufi, "The Consequences of Mortgage Credit Expansion: Evidence from the U.S. Mortgage Default Crisis," *Quarterly Journal of Economics* 124, no. 4 (2009). http://qje.oxfordjournals.org/content/124/4/1449.short for additional detail.
54. David Leonhardt, "Be Warned: Mr. Bubble's Worried Again," *New York Times*, August. 21, 2005. http://www.nytimes.com/2005/08/21/business/yourmoney/21real.html?pagewanted=all.
55. Urban Land Price Index in "Japan Statistical Yearbook 2012," Statistical Research and Training Institute, MIC. http://www.stat.go.jp/english/data/nenkan/1431-17.htm.

56. Karl E. Case and Robert J. Shiller, "Is There a Bubble in the Housing Market?" Cowles Foundation for Research in Economics, Yale University, 2004. http://www.econ.yale.edu/~shiller/pubs/p1089.pdf.
57. Some economists I spoke with, like Jan Hatzius, dispute the Case–Schiller data, noting that the quality of housing data is questionable prior to the 1950s. But even using 1953 as a starting point—when the data quality significantly improves—there had been no overall increase in the value of homes between 1953 and 1996. The paper gains that homeowners may have thought they had achieved were no more than enough to pay for inflation.
58. "S&P/Case Shiller Home Price Index."
59. "New Private Housing Units Authorized by Building Permits," Census Bureau, United States Department of Commerce. http://research.stlouisfed.org/fred2/data/PERMIT.txt.
60. Alex Veiga, "U.S. Foreclosure Rates Double," Associated Press, November 1, 2007. http://www.azcentral.com/realestate/articles/1101biz-foreclosures01-ON.html.
61. "Crist Seeks $50M for Homebuyers," *South Florida Business Journal*, September 13, 2007. http://www.bizjournals.com/southflorida/stories/2007/09/10/daily44.html.
62. Vikas Bajaj, "Federal Regulators to Ease Rules on Fannie Mae and Freddie Mac," *New York Times*, February 28, 2008. http://www.nytimes.com/2008/02/28/business/28housing.html.
63. Survey of Professional Forecasters, November 2007. See table 5, in which the economists give a probabilistic forecast range for gross domestic product growth during 2008. The chance of a decline in GDP of 2 percent or more is listed at 0.22 percent, or about 1 chance in 500. In fact, GDP declined by 3.3 percent in 2008. http://www.phil.frb.org/research-and-data/real-time-center/survey-of-professional-forecasters/2007/spfq407.pdf.
64. Households between the 20th and 80th percentiles in the wealth distribution.
65. Edward N. Wolff, "Recent Trends in Household Wealth in the United States: Rising Debt and the Middle-Class Squeeze—an Update to 2007," Working Paper No. 589, Levy Economics Institute of Bard College, March 2010. http://www.levyinstitute.org/pubs/wp_589.pdf.
66. Atif R. Mian and Amir Sufi, "House Prices, Home Equity-Based Borrowing, and the U.S. Household Leverage Crisis," Chicago Booth Research Paper No. 09-20, May 2010. http://papers.ssrn.com/sol3/papers.cfm?abstract_id=1397607.
67. The 14 percent decline is after inflation.
68. Wolff, "Recent Trends in Household Wealth."
69. Binyamin Applebaum, "Gloom Grips Consumers, and It May Be Home Prices," *New York Times*, October 18, 2011. http://www.nytimes.com/2011/10/19/business/economic-outlook-in-us-follows-home-prices-downhill.html?ref=business.
70. Dean Baker, "The New York Times Discovers the Housing Wealth Effect;" *Beat the Press* blog, The Center for Economic and Policy Research, October 19, 2011. www.cepr.net/index.php/blogs/beat-the-press/the-new-york-times-discovers-the-housing-wealth-effect.
71. Figures are based on reports from the Federal Reserve Bank of New York, which show an average daily trading volume in the mortgage-backed securities market of $320 billion per day. Over 250 trading days, this works out to about $80 trillion in trades over the course of a year. James Vickery and Joshua Wright, "TBA Trading and Liquidity in the Agency MBS," Staff Report No. 468, Federal Reserve Bank of New York, August 2010. www.ny.frb.org/research/staff_reports/sr468.pdf.
72. This trading volume is also quite large as compared with the actual value of mortgage-backed securities, which was about $8 trillion.
73. Ilan Moscovitz, "How to Avoid the Next Lehman Brothers," The Motley Fool, June 22, 2010. http://www.fool.com/investing/general/2010/06/22/how-to-avoid-the-next-lehman-brothers.aspx.
74. Robin Blackburn, "The Subprime Crisis," *New Left Review* 50 (Mar.–Apr. 2008). http://www.newleftreview.org/?view=2715.

75. Niall Ferguson, "The Descent of Finance," *Harvard Business Review* (July–August 2009). http://hbr.org/hbr-main/resources/pdfs/comm/fmglobal/the-descent-of-finance.pdf.
76. David Miles, Bank of England, "Monetary Policy in Extraordinary Times," speech given to the Centre for Economic Policy Research and London Business School, February 23, 2011. http://www.bankofengland.co.uk/publications/Documents/speeches/2011/speech475.pdf
77. Investopedia staff, "Case Study: The Collapse of Lehman Brothers," Investopedia; April 2, 2009. http://www.investopedia.com/articles/economics/09/lehman-brothers-collapse.asp#axzz1bZ61K9wz.
78. George A. Akerlof, "The Market for 'Lemons': Quality Uncertainty and the Market Mechanism," *Quarterly Journal of Economics* 84, no. 3 (Aug. 1970). http://sws.bu.edu/ellisrp/EC387/Papers/1970Akerlof_Lemons_QJE.pdf.
79. "Lehman Brothers F1Q07 (Qtr End 2/28/07) Earnings Call Transcript," Seeking Alpha, Mar. 14, 2007. http://seekingalpha.com/article/29585-lehman-brothers-f1q07-qtr-end-2-28-07-earnings-call-transcript?part=qanda.
80. Investopedia staff, "Case Study: The Collapse of Lehman Brothers."
81. Abigail Field, "Lehman Report: Why the U.S. Balked at Bailing Out Lehman," *DailyFinance*, March 15, 2010. http://www.dailyfinance.com/2010/03/15/why-the-u-s-balked-at-bailout-out-lehman/
82. Summers was also secretary of the treasury under President Clinton.
83. This is my example, not Summers's, who instead gave me an example involving the price of wheat.
84. Although lemonade does not behave this way, economists have sometimes debated whether another tasty beverage—French wine—does. Over certain price ranges, increasing the price of a wine may increase demand because the customer sees high price as a signifier of superior quality. Eventually, though, even the most spendthrift oenophiles are priced out, so the positive feedback does not continue indefinitely.
85. Per interview with George Akerlof. "You may know what to pay for House A versus House B versus House C because you can say one has a kitchen with gadgets that is worth $500 more than House B, which has a kitchen with no gadgets. But you don't know what the price of a house should be."
86. Carmen M. Reinhart and Kenneth S. Rogoff, "The Aftermath of the Financial Crisis," Working Paper 14656, NBER Working Paper Series, National Bureau of Economic Research, January 2009. http://www.bresserpereira.org.br/terceiros/cursos/Rogoff.Aftermath_of_Financial_Crises.pdf.
87. Carmen M. Reinhart and Vincent R. Reinhart, "After the Fall," presentation at Federal Reserve Bank of Kansas City Jackson Hole Symposium, August 2010. http://www.kcfed.org/publicat/sympos/2010/reinhart-paper.pdf.
88. This is one reason why it may not be harmful—some studies have even claimed that it may be helpful—for a president to experience a recession early in his term. The American economy was in recession in 1982, for example, but recovered from it with spectacular 8 percent growth in 1983 and 6 percent growth in 1984, helping Ronald Reagan to a landslide victory for a second term. There is some evidence, in fact, that presidents may have enough influence on fiscal and monetary policy to help perpetuate these outcomes. Since 1948, the median rate of GDP growth is 2.7 percent in the first year of a president's term and 2.8 percent in the second year—but 4.2 percent in both the third and the fourth years.

    For additional discussion, see Larry Bartels, "The President's Fate May Hinge in 2009," *The Monkey Cage*, November 2, 2011. http://themonkeycage.org/blog/2011/11/02/the-presidents-fate-may-hinge-on-2009-2/.
89. Ezra Klein, "Financial Crisis and Stimulus: Could This Time Be Different?" *Washington Post*, October 8, 2011. http://www.washingtonpost.com/business/financial-crisis-and-stimulus-could-this-time-be-different/2011/10/04/gIQALuwdVL_story.html.

90. Christina Romer and Jared Bernstein, "The Job Impact of the American Recovery and Reinvestment Plan," January 9, 2009. http://www.economy.com/mark-zandi/documents/The_Job_Impact_of_the_American_Recovery_and_Reinvestment_Plan.pdf.
91. Paul Krugman, "Behind the Curve," *New York Times*, March 8, 2008. http://www.nytimes.com/2009/03/09/opinion/09krugman.html.
92. Peter Roff, "Economists: Stimulus Not Working, Obama Must Rein in Spending," *US News & World Report*, June 10, 2010. http://www.usnews.com/opinion/blogs/peter-roff/2010/06/10/economists-stimulus-not-working-obama-must-rein-in-spending.
93. For instance, the *Wall Street Journal* forecasting panel in January 2009 predicted that unemployment would rise to 8.5 percent by the end of the year. By contrast, the White House's "without stimulus" forecast predicted 9.0 percent, whereas their "with stimulus" forecast predicted 7.8 percent. Since the *Wall Street Journal* panel would not have known at the time about what magnitude of stimulus might be passed, it is not surprising that they wound up somewhere in between the two cases.

    However, the White House's "with stimulus" forecast claimed unemployment would actually fall between the first half and second half of 2009, something which no consensus forecast from the *Wall Street Journal* panel or the Survey of Professional Forecasters ever did. So, although the White House's forecast wasn't all that far removed from private-sector forecasts in anticipating where the unemployment rate would end up by the close of 2009, it did envision a different trajectory, anticipating that the curve would already be bending downward by that point in time.
94. Klein, "Financial Crisis and Stimulus: Could This Time Be Different?"
95. Lisa Mataloni, "Gross Domestic Product: Fourth Quarter 2008 (Advance)," Bureau of Economic Analysis, U.S. Department of Commerce, January. 30, 2009. http://www.bea.gov/newsreleases/national/gdp/2009/gdp408a.htm.
96. "Real Gross Domestic Product, 1 Decimal," Bureau of Economic Analysis, U.S. Department of Commerce. http://research.stlouisfed.org/fred2/data/GDPC1.txt
97. Specifically, the forecast for the unemployment rate a year in advance of the forecast date.
98. This is based on my analysis of data from the Survey of Professional Forecasters.
99. Antonio Spilimbergo, Steve Symansky, and Martin Schindler, "Fiscal Multipliers," International Monetary Fund Staff Position Note SPN/09/11, May 20, 2009. http://econ.tu.ac.th/class/archan/RANGSUN/EC%20460/EC%20460%20Readings/Global%20Issues/Global%20Financial%20Crisis%202007-2009/Academic%20Works%20By%20Instituion/IMF/IMF%20Staff%20Position%20Notes/Fiscal%20Multipliers.pdf.
100. "93% of Drivers Consider Themselves Above Average. Are You Above Average?" Cheap Car Insurance, Aug. 24, 2011. http://www.cheapcarinsurance.net/above-avarege-driver/.
101. *Financial Crisis Inquiry Commission Report*, 2011. http://www.gpo.gov/fdsys/pkg/GPO-FCIC/pdf/GPO-FCIC.pdf.

CHAPTER 2: ARE YOU SMARTER THAN A TELEVISION PUNDIT?

1. *The McLaughlin Group* transcript, Federal News Service, taped October 31, 2008. http://www.mclaughlin.com/transcript.htm?id=687.
2. "Iowa Electronic Markets," Henry B. Tippie College of Business, University of Iowa. http://iemweb.biz.uiowa.edu/pricehistory/PriceHistory_GetData.cfm.
3. *The McLaughlin Group* transcript, Federal News Service; taped November 7, 2008. http://www.mclaughlin.com/transcript.htm?id=688.
4. Nate Silver, "Debunking the Bradley Effect," *Newsweek*, October 20, 2008. http://www.thedailybeast.com/newsweek/2008/10/20/debunking-the-bradley-effect.html.
5. Academic studies of *The McLaughlin Group* have come to similar conclusions. See for example: Lee Sigelman, Jarol B. Manheim, and Susannah Pierce, "Inside Dopes? Pundits as Political Forecasters," *The International Journal of Press/Politics* 1, 1 (January 1996). http://hij.sagepub.com/content/1/1/33.abstract.

6. Predictions were not evaluated for one of three reasons: the panelist's answer was too vague to formulate a testable hypothesis, or the prediction concerned an event in the distant future, or the panelist dodged McLaughlin's question and did not issue a prediction at all.
7. Two of the less-frequent panelists, Clarence Page and Mort Zuckerman, had more promising results, while Crowley's were especially poor—but none of these trends were meaningful statistically.
8. Eugene Lyons, *Workers' Paradise Lost* (New York: Paperback Library, 1967).
9. Tetlock has since moved to the University of Pennsylvania.
10. An "expert," to Tetlock's thinking, is anyone who makes a living on their supposed prowess in a particular field—for instance, the *Washington Post*'s Moscow correspondent would be an expert on the USSR, as would a tenured Sovietologist at Berkeley.
11. Phillip E. Tetlock, *Expert Political Judgment* (Princeton, NJ: Princeton University Press, 2006), pp. 107–108.
12. The CIA's incorrect estimates of the Soviet Union's GDP came in part because they were looking at the country's impressive military and extrapolating backward from that to estimate the overall size of its economy. But the Soviet Union was spending a much higher percentage of its domestic product on its military than were the free economies of Europe and North America.
13. Abram Bergson, "How Big Was the Soviet GDP?" *Comparative Economic Studies*, March 22, 1997. http://web.archive.org/web/20110404205347/http://www.allbusiness.com/government/630097-1.html.
14. Some of this section is per my interviews with Bruce Bueno de Mesquita, the New York University political scientist.
15. Louis Menand, "Everybody's an Expert," *New Yorker*, December 6, 2005. http://www.newyorker.com/archive/2005/12/05/051205crbo_books1.
16. Dick Morris, "Bush Will Rebound from Katrina Missteps," *The Hill*, September 7, 2005.
17. Justin Gardner, "Dick Morris' Crazy Electoral Map," Donklephant.com, October 15, 2008. http://donklephant.com/2008/10/15/dick-morris-crazy-electoral-map/.
18. Dick Morris and Eileen McGann, "Goal: 100 House Seats," DickMorris.com, September 27, 2010. http://www.dickmorris.com/goal-100-house-seats/.
19. Dick Morris, "Krauthammer's 'Handicapping the 2012 Presidential Odds,' " DickMorris.com, April 25, 2011. http://www.dickmorris.com/comments-on-krauthammers-handicapping-the-2012-presidential-odds/.
20. Phillip E. Tetlock, *Expert Political Judgment*, p. 79.
21. James A. Barnes and Peter Bell, "Political Insiders Poll," *National Journal*, October 28, 2010. http://www.nationaljournal.com/magazine/political-insiders-poll-20101028?print=true.
22. *FiveThirtyEight's* forecasting model, which has no insider knowledge, called nine of the eleven races correctly—better than both the Democratic insiders (who called an average of 6.9 right) and the Republicans (8.4).
23. Note that these predictions weren't even internally consistent: if Democratic insiders expected to win almost all the marginal races, for instance, they ought to have predicted their party would perform well enough to hold the House.
24. Although race and gender were certainly key factors in the 2008 race, the media often defaulted to them as explanations when there were better hypotheses at hand. For instance, at some points in the 2008 primary campaign, it was noted that Obama was doing relatively poorly among Hispanic voters, the implication almost always being that Hispanic voters distrusted black candidates. In fact, I later found, Obama's lagging numbers among Hispanics had little do to with their race and everything to do with their income status. Clinton generally did better with working-class voters, and because many Hispanics were working-class, she also did better with that group. But controlling for income, there was essentially no difference in Clinton's performance among whites and Hispanics. And indeed, Obama had

little trouble winning Hispanic votes once he squared off against the Republican nominee, John McCain, taking about two-thirds of their votes.

25. "Election Results: House Big Board," *New York Times*, November 2, 2010. http://elections.nytimes.com/2010/results/house/big-board.
26. Nate Silver, "A Warning on the Accuracy of Primary Polls," *FiveThirtyEight*, *New York Times*, March 1, 2012. http://fivethirtyeight.blogs.nytimes.com/2012/03/01/a-warning-on-the-accuracy-of-primary-polls/.
27. Nate Silver, "Bill Buckner Strikes Again," *FiveThirtyEight*, *New York Times*; September 29, 2011. http://fivethirtyeight.blogs.nytimes.com/2011/09/29/bill-buckner-strikes-again/.
28. Otherwise, you should have assigned the congressman a 100 percent chance of victory instead.
29. Matthew Dickinson, "Nate Silver Is Not a Political Scientist," in Presidential Power: A Nonpartisan Analysis of Presidential Power, Blogs Dot Middlebury, November 1, 2010. http://blogs.middlebury.edu/presidentialpower/2010/11/01/nate-silver-is-not-a-political-scientist/.
30. Sam Wang, "A Weakness in FiveThirtyEight.com," Princeton Election Consortium, August 8, 2008. http://election.princeton.edu/2008/08/04/on-a-flaw-in-fivethirtyeightcom/.
31. Douglas A. Hibbs Jr., "Bread and Peace Voting in U.S. Presidential Elections," *Public Choice* 104 (January 10, 2000): pp. 149–180. http://www.douglas-hibbs.com/HibbsArticles/Public%20Choice%202000.pdf.
32. Hibbs's model predicted Al Gore to win 54.8 percent of the two-party vote (that is, excluding votes for third-party candidates) when Gore in fact won 50.3 percent—an error of 4.5 percent. His model claimed to have a standard error of about 2 points in predicting any one candidate's vote share (or about 4 points when predicting the margin between them). His forecast thus overestimated Gore's standing by 2.25 standard deviations, something that should occur only about 1 in 80 times according to the normal distribution.
33. James E. Campbell, "The Referendum That Didn't Happen: The Forecasts of the 2000 Presidential Election," *PS: Political Science & Politics* (March 2001). http://cas.buffalo.edu/classes/psc/fczagare/PSC%20504/Campbell.pdf.
34. Andrew Gelman and Gary King, "Why Are American Presidential Election Campaign Polls So Predictable?" *British Journal of Political Science* 23, no. 4 (October 1993). http://www.rochester.edu/College/faculty/mperess/ada2007/Gelman_King.pdf.
35. Nate Silver, "Models Based on 'Fundamentals' Have Failed at Predicting Presidential Elections," *FiveThirtyEight*, *New York Times*, March 26, 2012.
36. Between 1998 and 2008, the average poll for a U.S. Senate race conducted in the final three weeks of the campaign missed by 5.0 points, whereas the average poll for a U.S. House race missed by 5.8 points.
37. One mild criticism of Cook Political's methodology is that they classify too many races as Toss-Up, even where there is probably enough evidence to posit a modest advantage for one or another of the candidates. *FiveThirtyEight*'s methodology, which identifies a nominal favorite in all races no matter how slim his apparent advantage, correctly identified the winner in thirty-eight of the fifty races (76 percent) that Cook Political characterized as Toss-Up in 2010.
38. Between 1998 and 2010, there were seventeen instances in which Cook classified a race one way (favoring the Democrat, for instance) while the average of polls came to the opposite conclusion (perhaps finding a narrow advantage for the Republican). The Cook forecasts made the right call in thirteen of those seventeen cases.
39. Kapanke was later ousted from his State Senate seat in the Wisconsin recall elections of 2011.
40. Scott Schneider, "Democrats Unfairly Accuse Dan Kapanke of Ethics Violations," *La Crosse Conservative Examiner*, August 27, 2010. http://www.examiner.com/conservative-in-la-crosse/democrats-unfairly-accuse-dan-kapanke-of-ethics-violations.
41. Cook Political eventually moved its rating to Lean Democrat from Likely Democrat several

weeks later, but on account of the Democrats' deteriorating situation in the national political environment rather than anything that had happened during Kapanke's interview.

42. Paul E. Meehl, "When Shall We Use Our Heads Instead of the Formula," *Journal of Counseling Psychology* 4, no. 4 (1957), pp. 268–273. http://mcps.umn.edu/assets/pdf/2.10_Meehl.pdf.
43. Douglas Harper, *Online Etymology Dictionary.* http://www.etymonline.com/index.php?term=objective.

CHAPTER 3: ALL I CARE ABOUT IS W'S AND L'S

1. Nate Silver in Jonah Keri, et al., *Baseball Between the Numbers: Why Everything You Know About the Game Is Wrong* (New York: Basic Books, 2006).
2. Danny Knobler, "The Opposite of a 'Tools Guy,' Pedroia's Simply a Winner," CBSSports.com, November 18, 2008. http://www.cbssports.com/mlb/story/11116048.
3. Nate Silver, "Lies, Damned Lies: PECOTA Takes on Prospects, Wrap-up," BaseballProspectus.com, March 8, 2006. http://www.baseballprospectus.com/article.php?articleid=4841.
4. Law is also a former writer at *Baseball Prospectus.*
5. Keith Law, "May Rookies Struggling to Show They Belong," ESPN.com, May 12, 2007. http://insider.espn.go.com/mlb/insider/columns/story?columnist=law_keith&id=2859877.
6. Pedroia's day-by-day statistics are per Baseball-Reference.com.
7. Tommy Craggs, "Dustin Pedroia Comes out Swinging," *Boston Magazine,* April 2009. http://www.bostonmagazine.com/articles/dustin_pedroia/page5.
8. This calculation is based on the binomial distribution and assumes that a player gets five hundred at-bats.
9. Or, more precisely, the typical *hitter.* Pitchers follow different and more irregular aging patterns.
10. More particularly, a hitter's age-32 season seems to be about the point at which his skills begin to decline at an accelerated rate.
11. Jeff Sonas, "The Greatest Chess Player of All Time, Part II, Chessbase.com, April 28, 2004. http://www.chessbase.com/newsdetail.asp?newsid=2354.
12. Bruce Weinberg and David Galenson, "Creative Careers: The Life Cycles of Nobel Laureates in Economics," NBER Working Paper No. 11799, National Bureau of Economic Research, November 2005. http://www.econ.brown.edu/econ/sthesis/IanPapers/tcl.html.
13. Del Jones, "Does Age Matter When You're CEO?" *USA Today,* September 11, 2008. http://www.usatoday.com/money/companies/management/2008-08-12-obama-mccain-age-ceos_N.htm.
14. Gary Huckabay, "6-4-3," *Baseball Prospectus,* August 2, 2002. http://www.baseballprospectus.com/article.php?articleid=1581.
15. Arlo Lyle, "Baseball Prediction Using Ensemble Learning," thesis submitted to the University of Georgia, 2007. http://www.ai.uga.edu/IAI/Theses/lyle_arlo.pdf.
16. Bill James, "Whatever Happened to the Hall of Fame," *Fireside* (1995): p. 89.
17. Note, however, that a significant amount of the back-end processing for PECOTA occurred in a statistical language called STATA.
18. I chose World War II as a cutoff point because a number of developments occurred shortly after the war that made professional baseball the modern game that it is today: the breaking of the color barrier by Jackie Robinson (1947); the first televised World Series (1947); the movement of teams to the West Coast (1957); the introduction of night baseball, which had initially occurred as early as 1935 but gained more currency during the war as workers put in long hours at factories and needed some recreation at night.
19. Alan Schwarz, "The Great Debate," *Baseball America,* January. 7, 2005. http://www.baseballamerica.com/today/features/050107debate.html.
20. Per interview with Billy Beane.

21. Nate Silver, "What Tim Geithner Can Learn from Baseball," *Esquire*, March 11, 2009. http://www.esquire.com/features/data/mlb-player-salaries-0409.
22. As a result of my original agreement in 2003 and a subsequent agreement in 2009, Baseball Prospectus now fully owns and operates PECOTA. Beginning with the 2010 season, the PECOTA forecasts reflect certain changes, improvements, and departures from my original methodology. The methods I describe herein apply to the 2003–2009 version of PECOTA specifically.
23. Nate Silver, "PECOTA Takes on the Field," *Baseball Prospectus*, January 16, 2004. http://www.baseballprospectus.com/article.php?articleid=2515.
24. Nate Silver, "Lies, Damned Lies: Projection Reflection," *Baseball Prospectus*, October 11, 2006. http://www.baseballprospectus.com/article.php?articleid=5609.
25. Ibid.
26. Dave Van Dyck, "Computer Crashes White Sox," *Chicago Tribune*, March 11, 2007. http://articles.chicagotribune.com/2007-03-11/sports/0703110075_1_computer-paul-konerko-projections.
27. Steve Slowinski, "The Projection Rundown: The Basics on Marcels, ZIPS, CAIRO, Oliver, and the Rest," FanGraphs.com, February 16, 2011. http://www.fangraphs.com/library/index.php/the-projection-rundown-the-basics-on-marcels-zips-cairo-oliver-and-the-rest/.
28. Silver, "Lies, Damned Lies: PECOTA Takes on Prospects, Wrap-up."
29. There are many different versions of WARP and similar statistics. Naturally, I use the *Baseball Prospectus* version for these calculations.
30. Dave Cameron, "Win Values Explained: Part Six," FanGraphs.com, January 2, 2009. http://www.fangraphs.com/blogs/index.php/win-values-explained-part-six/.
31. Silver, "Lies, Damned Lies: PECOTA Takes on Prospects, Introduction," *Baseball Prospectus*, February 1, 2007. http://www.baseballprospectus.com/article.php?articleid=5836.
32. "All-Time Top 100 Prospects," *Baseball America*. http://www.baseballamerica.com/today/prospects/rankings/top-100-prospects/all-time.html.
33. "1997 Oakland Athletics Batting, Pitching, & Fielding Statistics," Baseball-Reference.com. http://www.baseball-reference.com/teams/OAK/1997.shtml
34. It should be noted, however, that with the average major league payroll now at about $100 million, a forecasting system that allowed a baseball team to spend its money 2 percent more efficiently would save them $2 million. Compared with the five-figure salaries that statistical analysts receive in major league front offices, that's quite a bargain.
35. "Detroit Tigers 11, Kansas City Athletics 4: Game Played on Tuesday, April 13, 1965 (D) at Municipal Stadium," Retrosheet.org. http://www.retrosheet.org/boxesetc/1965/B04130KC.11965.htm.
36. "John Sanders, Grand Island," Inductee 2002, Nebraska High School Sports Hall of Fame Foundation. http://www.nebhalloffame.org/2002/sanders.htm.
37. Steve Treder, "Cash in the Cradle: The Bonus Babies," *The Hardball Times*, November 1, 2004. http://www.hardballtimes.com/main/article/cash-in-the-cradle-the-bonus-babies/.
38. Mike Pesca, "The Man Who Made Baseball's Box Score a Hit," National Public Radio, July 30, 2009. http://www.npr.org/templates/story/story.php?storyId=106891539.
39. Why teams persisted in believing the contrary for so long is a good question. It may have had to do with the fact that a walk was traditionally perceived as a mistake made by the pitcher, rather than a skill exhibited by the batter. It may have been that a walk is perceived as too passive in a culture that prizes machismo. But the industry has grown wise to the value of OBP, which is now the category most highly correlated with salaries paid to free agents. So the A's can no longer exploit the inefficiency—after years of finishing near the top of the league in walks drawn, they placed just tenth out of fourteen American League teams in the category in 2009.
40. Ken C. Winters, "Adolescent Brain Development and Drug Abuse," Treatment Research Institute, November 2004. http://www.factsontap.org/docs/2004Nov_AdolescentBrain.pdf.

41. Per interview with John Sanders.
42. Players spend their first two major-league seasons subject to something called the reserve clause, which dictates that they may not sign with any other clubs. This means that the player has almost no leverage at all, and is usually signed for something close to the league minimum salary of $400,000. After that, he spends his third through sixth years subject to arbitration, in which both he and the team submit salary requests, and a three-person panel decides which is the more appropriate. But salaries awarded in arbitration are typically only about 60 percent of those secured by free agents with comparable skills, so the teams are getting these players at a substantial discount.
43. Nor is there any reason that you couldn't have scouts rate a player's mental tools along with his physical ones.
44. Jeremy Greenhouse, "Touching Bases: Best PITCHf/x Pitches of 2009," Baseball Analysts; March 4, 2010. http://baseballanalysts.com/archives/2010/03/top_pitchfx_pit.php.
45. Ibid.
46. "Baseball Hall of Fame Second Basemen," Baseball Almanac. http://www.baseball-almanac.com/hof/hofst2b.shtml.
47. Some of the conversation with James is also taken from a subsequent phone call I made to him.

CHAPTER 4: FOR YEARS YOU'VE BEEN TELLING US THAT RAIN IS GREEN

1. Forecaster Stewart, "Tropical Depression Twelve: ZCZC MIATCDAT2 ALL, TTAA00 KNHC DDHHMM," National Hurricane Center, National Weather Service, August. 23, 2005. http://www.nhc.noaa.gov/archive/2005/dis/al122005.discus.001.shtml?
2. Based on statistics from StormPulse.com from the 2000 through 2011 Atlantic Basin hurricane seasons. The exact figure for the percentage of tropical depressions that became hurricanes during this period was 43 percent, while 88 percent of tropical depressions became at least tropical storms.
3. Stewart, "Tropical Storm Katrina: ZCZC MIATCDAT2 ALL, TTAA00 KNHC DDHHMM," August 24, 2005. http://www.nhc.noaa.gov/archive/2005/dis/al122005.discus.005.shtml?
4. By convention, cyclones do not receive names until they become tropical storms with sustained wind speeds of at least 39 mph. They become hurricanes when their wind increases to 74 mph. So Tropical Depression Twelve was briefly Tropical Storm Katrina before becoming Hurricane Katrina.
5. Forecaster Knabb, "Hurricane Katrina: : ZCZC MIATCDAT2 ALL, TTAA00 KNHC DDHHMM," National Hurricane Center, National Weather Service; August 27, 2005. http://www.nhc.noaa.gov/archive/2005/dis/al122005.discus.016.shtml?
6. "Washing Away—Special Report from the Times-Picayune," *Times-Picayune*, June 23–27, 2002. http://www.nola.com/hurricane/content.ssf?/washingaway/index.html.
7. Ezra Boyd, "The Evacuation of New Orleans for Hurricane Katrina: A Synthesis of the Available Data," presentation for the National Evacuation Conference, February 5, 2010. http://www.nationalevacuationconference.org/files/presentations/day2/Boyd_Ezra.pdf.
8. "Survey of Hurricane Katrina Evacuees," *Washington Post*, Harvard University, and the Kaiser Family Foundation, September 2005. http://www.kff.org/newsmedia/upload/7401.pdf.
9. "The Weatherman," *Curb Your Enthusiasm*, season 4, episode 4, HBO, January. 25, 2004.
10. joesixpacker, "The Mitt Romney Weathervane," YouTube, December. 24, 2011. http://www.youtube.com/watch?v=PWPxzDd661M.
11. Willis I. Milham, *Meteorology. A Text-Book on the Weather, the Causes of Its Changes, and Weather Forecasting for the Student and General Reader* (New York: Macmillan, 1918).
12. Aristotle, *Meteorology*, translated by E. W. Webster. Internet Classics Archive. http://classics.mit.edu/Aristotle/meteorology.html.
13. Pierre-Simon Laplace, "A Philosophical Essay on Probabilities" (Cosmo Publications, 2007).
14. The uncertainty principle should not be confused with the *observer effect*, which is the idea

that the act of measuring a system (such as shooting a laser beam at a particle of light) necessarily disrupts it. The two beliefs are not inherently incompatible—but the uncertainty principle is a stronger statement and is not so satisfyingly intuitive. Indeed, Heisenberg believed his uncertainty principle to be quite counterintuitive. But basically the idea is that, beyond a certain degree of resolution, at the very moment we think we're able to pin down *exactly* where a particle is, the particle no longer behaves as a point of matter but instead as something much different: as a *wave* that is *moving*. About the most satisfying demonstration I have seen on this is from the MIT physicist Walter Lewin: Acorvettes, "Quantum Mechanics, the Uncertainty Principle, Light Particles," YouTube, August 4, 2007. http://www.youtube.com/watch?v=KT7xJ0tjB4A.

15. "London Weather," in Official London Guide, Visitlondon.com. http://www.visitlondon.com/travel/weather.
16. Some of Richardson's failure was the result of a fairly minor specification error, it was later revealed. If he had corrected for it, he would have produced a reasonably accurate forecast.
17. J. G. Charney, R. Fjörtoft, and J. von Neumann, "Numerical Integration of the Barotropic Vorticity Equation," *Tellus* 2 (1950): pp. 237–254. http://mathsci.ucd.ie/~plynch/eniac/CFvN-1950.pdf.
18. "Moore's Law," Intel Corporation, 2005. ftp://download.intel.com/museum/Moores_Law/Printed_Materials/Moores_Law_2pg.pdf.
19. Lorenz's paper was not originally published but instead delivered at a talk to the American Association for the Advancement of Science on December 29, 1972. However, it was later published in Lorenz's book *The Essence of Chaos* (Seattle: University of Washington Press, 1995). http://www.washington.edu/uwpress/search/books/LORESS.html.
20. Douglas Allchin, "Penicillin and Chance," SHiPS Resource Center. http://www1.umn.edu/ships/updates/fleming.htm.
21. Per interview with Richard Loft.
22. 5^5^5 is 298,023,223,876,953,000—about 298 quadrillion. But 5^6^5 is 931,322,574,615,479,000,000, or about 931 *quintillion*. That "small" mistake would have led us to overestimate the value we were trying to derive by a factor of 3,125.
23. Yes, calculus is actually useful for something.
24. NCAR is separate from this part of the bureaucracy, instead run by means of a nonprofit consortium of research universities with funding from the National Science Foundation; this is why it has nicer buildings.
25. "History of the National Weather Service," Public Affairs Office, National Weather Service, National Oceanic and Atmospheric Administration, United States Department of Commerce. http://www.weather.gov/pa/history/index.php.
26. "The Blizzard of 1888," Nebraska State Historical Society, last updated June 4, 2004. http://www.nebraskahistory.org/publish/markers/texts/blizzard_of_1888.htm.
27. The most junior weather forecasters might start out at grade 5 of the government's general pay scale, where salaries are about $27,000 per year before cost-of-living adjustments. The top salary for a government employee under the schedule is about $130,000 plus cost-of-living adjustments.
28. "National Weather Service: FY 2012 Budget Highlights," National Oceanic and Atmospheric Administration, United States Department of Commerce. http://www.corporateservices.noaa.gov/nbo/FY09_Rollout_Materials/NWS_One_Pager_FINAL.pdf.
29. "Weather Impact on USA Economy," *National Oceanic and Atmospheric Association Magazine*, Nov. 1, 2001. http://www.magazine.noaa.gov/stories/mag4.htm.
30. This is probably not entirely coincidental—weather forecasting is a 24/7 business, and everyone at the World Weather Building takes their rotation on the night shift. With the lack of sunlight and windows at the World Weather Building, I sometimes had the feeling of being in a submarine.
31. "HPC% Improvement to NCEP Models (1-Inch Day 1 QPF Forecast)," Hydro Meteorological

Prediction Center, National Oceanic and Atmosphéric Association. http://www.hpc.ncep.noaa.gov/images/hpcvrf/1inQPFImpann.gif.

32. "HPC Pct Improvement vs MOS (Max Temp MAE: Stations Adjusted >= 1 F)," Hydro Meteorological Prediction Center, National Oceanic and Atmospheric Association. http://www.hpc.ncep.noaa.gov/images/hpcvrf/max1.gif.
33. "Weather Fatalities," National Weather Service, National Oceanic and Atmospheric Association. http://www.nws.noaa.gov/om/hazstats/images/weather_fatalities.pdf.
34. "NHC Tropical Cyclone Forecast Verification," National Hurricane Center, National Weather Service, National Oceanic and Atmospheric Association; updated March 1, 2012. http://www.nhc.noaa.gov/verification/verify5.shtml.
35. Another type of competition is for the taxpayer's dollar. So long as the memory of Hurricane Katrina lingers—an event which, in addition to its human toll, established the precedent that the government is held accountable for its response to natural disasters—the Weather Service is probably immune from massive budget cuts. But concerns over budgets remain a constant paranoia in Camp Springs—the fear is that some bright bulb in Washington will get wind of how well the computers are doing, and decide that the human forecasters are redundant. President Obama's proposed 2013 budget for the Weather Service increased funding for weather satellites but cut it for basic operations and research.
36. Traffic estimates are per Alexa.com.
37. Although *ten* inches of snow sounds like a lot more precipitation than *one* inch of rain, they're in fact roughly equivalent, because snow molecules are much less dense. If ten inches of snow were to melt, they would typically leave about one inch of water.
38. Allan H. Murphy, "What Is a Good Forecast? An Essay on the Nature of Goodness in Weather Forecasting," *American Meteorological Society* 8 (June 1993): pp. 281–293. http://www.swpc.noaa.gov/forecast_verification/Assets/Bibliography/i1520-0434-008-02-0281.pdf.
39. "History for Lansing, MI: Friday January 13, 1978," Wunderground.com. http://www.wunderground.com/history/airport/KLAN/1978/1/13/DailyHistory.html?req_city=Lansing&req_state=MI&req_statename=Michigan.
40. Data courtesy of Eric Floehr, ForecastWatch.com.
41. In fact, predicting a 50 percent chance of rain is a fairly bold forecast, since it typically rains on only about 20 percent of the days at an average location in the United States.
42. At one point in advance of the 2012 presidential election, I told a group of executives I had been invited to speak with that I thought the best forecast at the time was that Barack Obama and Mitt Romney each had a 50 percent chance of winning. They demanded that I stop being so wishy-washy and give them a real answer.
43. The wet bias that Floehr identified refers to predicting precipitation more *often* than it occurs in reality. It does not necessarily imply that the forecasters predict too little precipitation conditional on rain occurring. In fact, Floehr has also found that the weather forecasters tend to *underestimate* precipitation associated with the most severe winter storms, storms like the "Snowpocalypse" in New York City in 2012.
44. Although calibration is a very important way to judge a forecast, it won't tell you everything. Over the long run, for instance, it rains about 20 percent of the days at a randomly chosen location in the United States. So you can have a *well-calibrated* forecast just by guessing that there is always a 20 percent chance of rain. However, this forecast has no real skill; you are just defaulting to climatology. The counterpart to calibration is what is called discrimination or resolution, which is a measure of how much you vary your forecasts from one case to the next. A forecaster who often predicts a 0 percent chance of rain, or a 100 percent chance, will score better on discrimination than one who always guesses somewhere in the middle.

    Good evaluations of forecasts account for both of these properties—either individually or with statistical measures like the Brier score that attempt to account for both properties at once.

    The reason I say calibration is the best measure of a forecast is pragmatic: most expert forecasters have no trouble with discrimination. In fact, they discriminate too much—that is, their forecasts are overconfident.

45. "Performance Characteristics and Biases of the Operational Forecast Models," National Weather Service Weather Forecast Office, Louisville, KY; National Weather Service, National Oceanic and Atmospheric Association; May 23, 2004. http://www.crh.noaa.gov/lmk/soo/docu/models.php.
46. Sarah Lichtenstein, Baruch Fischhoff, and Lawrence D. Phillips, "Calibration of Probabilities: The State of the Art to 1980," Decision Research, Perceptronics, for Office of Naval Research, 1986. http://www.dtic.mil/cgi-bin/GetTRDoc?AD=ADA101986.
47. J. Eric Bickel and Seong Dae Kim, "Verification of the Weather Channel Probability of Precipitation Forecasts," *American Meteorological Society* 136 (December 2008): pp. 4867–4881. http://faculty.engr.utexas.edu/bickel/Papers/TWC_Calibration.pdf.
48. J. D. Eggleston, "How Valid Are TV Weather Forecasts?" Freakonomics.com, Apr. 21, 2008. http://www.freakonomics.com/2008/04/21/how-valid-are-tv-weather-forecasts/comment-page-6/#comments.
49. Per interview with Max Mayfield.
50. Mayfield was born September 19, 1948, and was fifty-six when Katrina hit; he is in his midsixties now.
51. The cone of uncertainty is officially supposed to cover two-thirds of the potential landfall locations, although according to Max Mayfield, storms have remained within the cone somewhat more often than that in practice.
52. "Vermont Devastation Widespread, 3 Confirmed Dead, 1 Man Missing," BurlingtonFreePress.com, August 29, 2011. http://www.burlingtonfreepress.com/article/20110829/NEWS02/110829004/Vermont-devastation-widespread-3-confirmed-dead-1-man-missing.
53. Associated Press, "Hurricane Rita Bus Owner Found Guilty," *USA Today*, October 3, 2006. http://www.usatoday.com/news/nation/2006-10-03-rita-bus_x.htm.
54. "Hurricane Katrina Timeline," The Brookings Institution. http://www.brookings.edu/fp/projects/homeland/katrinatimeline.pdf.
55. Douglas Brinkley, "How New Orleans Drowned," *Vanity Fair*, June 2006. http://www.vanityfair.com/politics/features/2006/06/brinkley_excerpt200606.
56. Keith Elder, et al., "African Americans' Decisions Not to Evacuate New Orleans Before Hurricane Katrina: A Qualitative Study," *American Journal of Public Health* 97, supplement 1 (April 2007). http://www.ncbi.nlm.nih.gov/pmc/articles/PMC1854973/.
57. H. Gladwin and W. G. Peacock, "Warning and Evacuation: A Night for Hard Houses," in *Hurricane Andrew: Ethnicity, Gender, and the Sociology of Disasters* (Oxford, England: Routledge, 1997), pp. 52–74.
58. Brinkley, "How New Orleans Drowned."
59. "Hurricane Katrina Timeline," Brookings Institution.
60. "Houston Shelter Residents' Reports of Evacuation Orders and Their Own Evacuation Experiences," in "Experiences of Hurricane Katrina Evacuees in Houston Shelters: Implications for Future Planning," by Mollyann Brodie, Erin Weltzien, Drew Altman, Robert J. Blendon and John M. Benson, *American Journal of Public Health* 9, no. 8 (August2006): pp. 1402–1408. http://www.ncbi.nlm.nih.gov/pmc/articles/PMC1522113/table/t2/.
61. Amanda Ripley, *The Unthinkable* (New York: Random House, 2008). Kindle edition.

CHAPTER 5: DESPERATELY SEEKING SIGNAL

1. John Dollar, "The Man Who Predicted the Earthquake," *Guardian*, April 5, 2010. http://www.guardian.co.uk/world/2010/apr/05/laquila-earthquake-prediction-giampaolo-giuliani.
2. "Scientists in the Dock," *Economist*, September 17, 2011. http://www.economist.com/node/21529006.
3. Roger A. Pielke Jr., "Lessons of the L'Aquila Lawsuit," *Bridges* 31 (October 2011). http://sciencepolicy.colorado.edu/admin/publication_files/2011.36.pdf.
4. "Eyewitnesses: Italy Earthquake," BBC News, April 6, 2009. http://news.bbc.co.uk/2/hi/europe/7985248.stm.

5. Michael Taylor, "L'Aquila, Guiliani, and the Price of Earthquake Prediction," The Pattern Connection, July 7, 2010. http://patternizer.wordpress.com/2010/07/07/laquila-guiliani-and-the-price-of-earthquake-prediction/.
6. John Bingham, "L'Aquila Dogged by Earthquakes Through 800 Year History," *Telegraph*, April 6, 2009. http://www.telegraph.co.uk/news/worldnews/europe/italy/5113215/LAquila-dogged-by-earthquakes-through-800-year-history.html.
7. M. Stucchi, C. Meletti, A. Ravida, V. D'Amio, and A. Capera, "Historical Earthquakes and Seismic Hazard of the L'Aquila Area," *Progettazione Sismica* 1, no. 3 (2010): pp. 23–24.
8. Elisabeth Malkin, "Once Built on a Lake, Mexico City Now Runs Dry," *New York Times*, March 16, 2006. http://www.nytimes.com/2006/03/16/world/americas/16iht-mexico.html.
9. Nicola Nosengo, "Italian Earthquake Toll Highlights Poor Preparedness," *Nature* news blog, May 22, 2012. http://blogs.nature.com/news/2012/05/italian-earthquake-toll-highlights-poor-preparedness.html.
10. "Così Posso Prevedere I Terremoti In Abruzzo Ci Sono 5 Apparecchi," *La Repubblica*, April 6, 2009. Translated into English using Google Translate. http://www.repubblica.it/2009/04/sezioni/cronaca/terremoto-nord-roma/giulianigiampaolo/giulianigiampaolo.html.
11. Symon Hill, "Earthquakes and Bad Theology," Symon Hill's Blog, Ekklesia, January 17, 2010. http://www.ekklesia.co.uk/node/11032.
12. William Pike, "The Haiti and Lisbon Earthquakes: 'Why, God?' " *Encyclopedia Britannica* blog, January 19, 2010. http://www.britannica.com/blogs/2010/01/the-haiti-and-lisbon-earthquakes-why-god/.
13. Rick Brainard, "The 1755 Lisbon Earthquake," *18th Century History*, 2005. http://www.history1700s.com/articles/article1072.shtml.
14. Susan Hough, "Confusing Patterns with Coincidences," *New York Times*, April 11, 2009. http://www.nytimes.com/2009/04/12/opinion/12hough.html.
15. John Roach, "Can the Moon Cause Earthquakes?" National Geographic News, May 23, 2005. http://news.nationalgeographic.com/news/2005/05/0523_050523_moonquake.html.
16. Since 1900, the ten deadliest earthquakes worldwide have killed about 1.6 million people, versus 1.2 million for hurricanes. Matt Rosenberg, "Top 10 Deadliest World Hurricanes Since 1900," About.com. http://geography.about.com/od/physicalgeography/a/deadlyhurricane.htm; "Earthquakes with 1,000 or More Deaths Since 1900," United States Geological Service. http://earthquake.usgs.gov/earthquakes/world/world_deaths.php.
17. I speak here of the perception, not of the reality. In a literal sense, earthquakes are much more common than hurricanes, with several million occurring worldwide every year compared with only several dozen hurricanes. However, the vast majority of these are tiny and are undetectable unless you have a seismometer, whereas hurricanes almost always make news.
18. "Legends of Unusual Phenomena Before Earthquakes—Wisdom or Superstition?" in *Earthquakes and Animals—From Folk Legends to Science* (Hackensack, NJ: World Scientific Publishing, 2005). www.worldscibooks.com/etextbook/5382/5382_chap01.pdf.
19. Giuliani was given a very favorable write-up in the estimable UK newspaper the *Guardian*, for instance. Dollar, "The Man Who Predicted the Earthquake."
20. Ed Wilson, Don Drysdale, and Carrie Reinsimar, "CEPEC Keeps Eye on Earthquake Predictions," State of California Department of Conservation, October 23, 2009. http://www.consrv.ca.gov/index/news/Pages/CEPECKeepsEyeOnEarthquakePredictions.aspx.
21. R. A. Grant and T. Halliday, "Predicting the Unpredictable: Evidence of Pre-Seismic Anticipatory Behaviour in the Common Toad," *Journal of Zoology* 700 (January 25, 2010). http://image.guardian.co.uk/sys-files/Environment/documents/2010/03/30/toads.pdf.
22. One obvious problem with the paper: the timing of the toads' behavior coincides very well with the foreshocks to L'Aquila—much better than with the main quake. So if they were sensitive to earthquakes, it seems more likely that they were reacting to these foreshocks, rather than "predicting" the main earthquake. Of course, seismometers were able to detect the foreshocks as well.

23. Per Bak, *How Nature Works: The Science of Self-Organized Criticality* (New York: Springer, 1999). Kindle edition, location 1357.
24. "FAQs—Earthquake Myths," United States Geological Survey. http://earthquake.usgs.gov/learn/faq/?categoryID=6&faqID=13.
25. According to the USGS, there is a 95.4 percent chance of an earthquake of at least magnitude 6.75 hitting somewhere in the 100-kilometer radius surrounding San Francisco within the next 100 years, and a 97.6 percent chance for Los Angeles. The probability of at least one of the cities experiencing a major earthquake is 99.9 percent.
26. "2009 Earthquake Probability Mapping," United States Geological Survey. http://geohazards.usgs.gov/eqprob/2009/.
27. "Earthquake Facts and Statistics," United States Geological Survey. http://earthquake.usgs.gov/earthquakes/eqarchives/year/eqstats.php.
28. The exceptions are in wealthy and earthquake-rich regions like California, Japan, and Italy, where there are seismometers on every proverbial block.
29. 1964 marks about the point at which record keeping for medium-size earthquakes significantly improved.
30. "Composite Earthquake Catalog," Advanced National Seismic System, Northern California Earthquake Data Center. http://quake.geo.berkeley.edu/cnss/.
31. In a box measuring three degrees latitude by three degrees longitude centered around Tehran.
32. "Largest and Deadliest Earthquakes by Year: 1990–2011," United States Geological Survey. http://earthquake.usgs.gov/earthquakes/eqarchives/year/byyear.php.
33. "Corruption Perceptions Index 2011," Transparency.org. http://cpi.transparency.org/cpi2011/.
34. Kishor Jaiswal and David Wald, "An Empirical Model for Global Earthquake Fatality Estimation," *Earthquake Spectra* 26, no. 4 (November 2010). http://earthquake.usgs.gov/earthquakes/pager/prodandref/Jaiswal_&_Wald_(2010)_Empirical_Fatality_Model.pdf.
35. An empirically derived relationship known as Omori's Law dictates that the number of aftershocks is inversely proportional to the amount of time that has passed since the initial earthquake hit. In other words, aftershocks are more likely to occur immediately after an earthquake than days later, and more likely to occur days later than weeks after the fact.
36. This chart and the others cover a grid of 1 degree of latitude by 1 degree of longitude, with the epicenter of the main quake at the center of the box. The exception is figure 5-4d (for Reno, Nevada) since there was no main earthquake; Reno's city hall serves as the center of the box in that case.
37. Instituto Nazionale di Geofisica e Volcuanologia database. http://web.archive.org/web/20101114020542/http://cnt.rm.ingv.it/earthquakes_list.php.
38. The exact magnitude of the Tōhoku earthquake is debated; the database I used to generate the charts list it at magnitude 9.1, while other sources have it at 9.0.
39. Hough, "Predicting the Unpredictable," Kindle locations 1098–1099.
40. Ibid., Kindle locations 1596–1598.
41. Ibid., Kindle locations 1635–1636.
42. *Anchorage Daily News*, June 27, 1981, p. A-10.
43. Hough, "Predicting the Unpredictable," Kindle location 1706.
44. W. H. Bakun and A. G. Lindh, "The Parkfield, California, Earthquake Prediction Experiment," *Science* 229, no. 4714 (August 16, 1985). http://earthquake.usgs.gov/research/parkfield/bakunLindh85.html.
45. Hough, "Predicting the Unpredictable," Kindle locations 488–491.
46. Per interview with John Rundle.
47. "UCLA Geophysicist Warns 6.4 Quake to Hit LA by Sept. 5," Agence France-Presse via Space Daily.com, April 15, 2004. http://www.spacedaily.com/news/tectonics-04d.html.
48. P. Shebalin, V. Keilis-Borok, A. Gabrielov, I. Zaliapin, and D. Turcotte, "Short-term Earthquake Prediction by Reverse Analysis of Lithosphere Dynamics," *Tectonophysics* 413 (December 13, 2005). http://www.math.purdue.edu/~agabriel/RTP_Tect.pdf.

49. J. Douglas Zechar and Jiancang Zhuang, "Risk and Return: Evaluating Reverse Tracing of Precursors Earthquake Predictions," *Geophysical Journal International* (May 17, 2010). http://bemlar.ism.ac.jp/zhuang/pubs/zechar10.pdf.
50. Likewise, a Google News search returns no record of Keilis-Borok's predictions in 2003.
51. My sense from speaking with Dr. Keilis-Borok is that he was acting in good faith, but forecasters do not always do so. Sometimes, especially in economic forecasting, they'll even blatantly cheat by revising their method after the fact and claiming the new model would have predicted an event that they missed in real time. There is simply no substitute for having a clear public record of a prediction made before an event occurred.
52. Andrew Bridges, "Decade After Northridge, Earthquake Predictions Remain Elusive," Associated Press State & Local Wire, January 12, 2004.
53. "Information About the Keilis-Borok California Earthquake Prediction," United States Geological Survey. http://earthquake.usgs.gov/earthquakes/eqinthenews/2004/KB_prediction.php.
54. Zechar and Zhuang, "Risk and Return: Evaluating Reverse Tracing of Precursors Earthquake Predictions."
55. Arnaud Mignan, Geoffrey King, and David Bowman, "A Mathematical Formulation of Accelerating Moment Release Based on the Stress Accumulation Model," *Journal of Geophysical Research* 112, BO7308 (July 10, 2007). http://geology.fullerton.edu/dbowman/Site/Publications_files/Mignan_etal_JGR2007_1.pdf.
56. Arnaud Mignan, Geoffrey King, David Bowman, Robin Lacassin, and Renata Dmowska, "Seismic Activity in the Sumatra-Java Region Prior to the December 26, 2004 (Mw=9.0-9.3) and March 28, 2005 (Mw=8.7) Earthquakes," *Earth and Planetary Science Letters* 244 (March 13, 2006). http://esag.harvard.edu/dmowska/MignanKingBoLaDm_SumatAMR_EPSL06.pdf.
57. Specifically, the fit line in figure 5-6c is generated by a technique called Lowess regression. This technique is fine for many things and does not inherently lead to overfitting, but it requires you to set a smoothness parameter that will model anything from a very tight fit to a very loose one. In this case, obviously, I've chosen an implausibly tight fit.
58. For instance, if you apply the overfit curve to out-of-sample data—the circles from figure 5-5—it explains only about 40 percent of the variance in them. This substantial deterioration from in-sample to out-of-sample data is characteristic of an overfit model.
59. Freeman Dyson, "Turning Points: A Meeting with Enrico Fermi," *Nature* 427 (January 22, 2004). http://www.nature.com/nature/journal/v427/n6972/full/427297a.html.
60. Michael A. Babyak, "What You See May Not Be What You Get: A Brief, Nontechnical Introduction to Overfitting in Regression-Type Models," *Psychosomatic Medicine* 66 (February 19, 2004). http://os1.amc.nl/mediawiki/images/Babyak_-_overfitting.pdf.
61. M. Ragheb, "Fukushima Earthquake and Tsunami Station Blackout Accident." https://netfiles.uiuc.edu/mragheb/www/NPRE%20402%20ME%20405%20Nuclear%20Power%20Engineering/Fukushima%20Earthquake%20and%20Tsunami%20Station%20Blackout%20Accident.pdf.
62. Martin Fackler, "Tsunami Warnings, Written in Stone," *New York Times*, April 20, 2011. http://www.nytimes.com/2011/04/21/world/asia/21stones.html?pagewanted=all.
63. Specifically, this represents a 1-degree latitude by 1-degree longitude box, with 38.32 degrees north and 142.37 degrees east at the center.
64. Robert J. Geller, "Shake-up Time for Japanese Seismology," *Nature* 472, no. 7344 (April 28, 2011). http://kitosh.k.u-tokyo.ac.jp/uploader2/src/8.pdf.
65. Specifically, the chances of it are about 20 percent.
66. The odds of a .300 hitter going 0-for-5 are about 17 percent, assuming at-bats are independent of one another.
67. Earthsky.org staff, "Scientists Did Not Expect 9.0 Magnitude Earthquake in Japan," FastCompany.com, March 25, 2011. http://www.fastcompany.com/1742641/scientists-did-not-expect-90-magnitude-earthquake-in-japan.

68. Seth Stein and Emile A. Okal, "The Size of the 2011 Tohoku Earthquake Need Not Have Been a Surprise," *Eos Transactions American Geophysical Union* 92, no. 27 (July 5, 2011): p. 227. http://www.earth.northwestern.edu/people/seth/Texts/tohoku.pdf.
69. According to the ANSS catalog, there were twelve magnitude 7 earthquakes in the area measuring 10 degrees of latitude and 10 degrees of longitude in either direction from the epicenter of the Great Sumatra Earthquake of 2004, but none measuring at a magnitude 8 or greater during this period.
70. Like that of other earthquakes, the precise magnitude of the Great Sumatra Earthquake is disputed, with various estimates putting it between 9.0 and 9.3; I use a middle estimate of 9.2 here.
71. Geller, "Shake-up Time for Japanese Seismology."
72. SilentASMR, "2 Hours of Brown Noise (Read Description)," YouTube.com, February 25, 2012. http://ww.youtube.com/watch?v=0BfyKQaf0TU.
73. Livia Borghese, "Italian Scientists on Trial Over L'Aquila Earthquake," CNN World, September 20, 2011. http://articles.cnn.com/2011-09-20/world/world_europe_italy-quake-trial_1_geophysics-and-vulcanology-l-aquila-seismic-activity?_s=PM:EUROPE.
74. Thomas H. Jordan and Lucile M. Jones, "Operational Earthquake Forecasting: Some Thoughts on Why and How," *Seismological Research Letters* 81, 4 (July/August 2010). http://earthquake.usgs.gov/aboutus/nepec/meetings/10Nov_Pasadena/Jordan-Jones_SRL-81-4.pdf.
75. Alicia Chang, "Location a Major Factor in New Zealand Earthquake Devastation," *Washington Post*, February 22, 2011. http://www.washingtonpost.com/wp-dyn/content/article/2011/02/22/AR2011022205105.html.
76. Ya-Ting Leea, Donald L. Turcottea, James R. Holliday, Michael K. Sachs, John B. Rundlea, Chien-Chih Chen, and Kristy F. Tiampoe, "Results of the Regional Earthquake Likelihood Models (RELM) Test of Earthquake Forecasts in California," Proceedings of the National Academy of Sciences of the United States of America, September 26, 2011. http://www.pnas.org/content/early/2011/09/19/1113481108.abstract?sid=ea35f085-e352-42a8-8128-19149a05c795.

CHAPTER 6: HOW TO DROWN IN THREE FEET OF WATER

1. Christopher S. Rugaber, "Unexpected Jump in Unemployment Rate to 9.2% Stings Markets," *Denver Post*, July 9, 2011. http://www.denverpost.com/business/ci_18444012.
2. Christine Hauser, "Two Jobs Reports Point to a Higher Gain in June," *New York Times*, July 7, 2011. http://www.nytimes.com/2011/07/08/business/economy/data-point-to-growth-in-jobs-in-june.html.
3. Based on data from the Survey of Professional Forecasters, Federal Reserve Bank of Philadelphia. http://www.phil.frb.org/research-and-data/real-time-center/survey-of-professional-forecasters/anxious-index/.
4. Roger A. Pielke Jr., "Lessons of the L'Aquila Lawsuit," *Bridges* 31 (October 2011). http://sciencepolicy.colorado.edu/admin/publication_files/2011.36.pdf.
5. Teri Tomaszkiewicz, "Disaster Isn't Over When Media Leave: Discovering the Meaning of Memorial Day in North Dakota," *Milwaukee Journal Sentinel*, June 1, 1997.
6. Ashley Shelby, *Red River Rising: The Anatomy of a Flood and the Survival of an American City* (St. Paul: Borealis Books, 2004).
7. In fact, about two feet worth of sandbags were deployed at Grand Forks, meaning that its total protection was about 52 or 53 feet by the time the flood hit. But that still was not quite enough to prevent a 54-foot flood.
8. This figure can be derived from the margin of error, assuming the error is distributed normally.
9. Figure 6-1 is not drawn to scale.
10. Roger A. Pielke, Jr., "Who Decides? Forecasts and Responsibilities in the 1997 Red River Flood," *Applied Behavioral Science Review* 7, no. 2 (1999). http://128.138.136.233/admin/publication_files/resource-81-1999.16.pdf.
11. Pielke, "Who Decides? Forecasts and Responsibilities in the 1997 Red River Flood."

12. Alex Veiga, "U.S. Foreclosure Rates Double," Associated Press, November 1, 2007. http://www.azcentral.com/realestate/articles/1101biz-foreclosures01-ON.html.
13. Jonathan Stempel, "Countrywide Financial Plunges on Bankruptcy Fears," Reuters, August 16, 2007. http://uk.reuters.com/article/2007/08/16/countrywide-financial-idUKNOA62283620070816.
14. John B. Taylor, *Getting Off Track: How Government Actions and Interventions Caused, Prolonged, and Worsened the Financial Crisis* (Stanford, CA: Hoover Institution Press, Kindle edition, 2009), location 361.
15. Note, however, that the economists assigned a higher chance, about 20 percent, to a negative GDP reading in each of the four financial quarters.
16. In fact, this 1-in-500 estimate is a little generous, since it applied to any GDP reading below minus 2 percent, whereas the actual GDP figure of –3.3 percent was quite a bit lower than that. Although the economists did not explicitly quantify it, it can be inferred that they would have assigned only about a 1-in-2,000 chance to a GDP reading of –3.3 percent or worse.
17. Specifically, I've looked at the forecasts made each November about the GDP growth in the following year; for instance, the November 1996 forecast of GDP growth in 1997.
18. Michael P. Clements, "An Evaluation of the Survey of Professional Forecasters Probability Distribution of Expected Inflation and Output Growth," *Journal of Economic Literature*, November 22, 2002. http://www.icmacentre.ac.uk/pdf/seminar/clements2.pdf.
19. Based on the binomial distribution, the chance that a well-calibrated forecast would fall outside its 90 percent prediction interval six times in eighteen years is just 0.6 percent, or about 1 chance in 150.
20. This covers all releases of the Survey of Professional Forecasters from the fourth quarter of 1968 through the fourth quarter of 2010, excluding a few early cases where the economists were not asked to issue an annual forecast.
21. Prakash Loungani, "The Arcane Art of Predicting Recessions," *Financial Times* via International Monetary Fund, December 18, 2000. http://www.imf.org/external/np/vc/2000/121800.htm.
22. *Wall Street Journal* Forecasting panel, February 2009. http://online.wsj.com/article/SB123445757254678091.html.
23. Torsten Rieke, "Ganz oben in der Wall Street," *Handelsblatt*, October 19, 2005. http://www.handelsblatt.com/unternehmen/management/koepfe/ganz-oben-in-der-wall-street/2565624.html.
24. "Federal Reserve Economic Data," Economic Research, Federal Reserve Bank of St. Louis. http://research.stlouisfed.org/fred2/.
25. Lakshman Achuthan and Anirvan Benerji, *Beating the Business Cycle: How to Predict and Profit from Turning Points in the Economy* New York: Random House, 2004). Kindle edition, locations 1476–1477.
26. "U.S. Business Cycle Expansions and Contractions," National Bureau of Economic Research. http://www.nber.org/cycles.html.
27. Specifically, the stock market as measured by the S&P 500.
28. The original National Football League included the Pittsburgh Steelers, Baltimore Colts (now the Indianapolis Colts), and the Cleveland Browns (the original version of which became the Baltimore Ravens). The way the indicator is usually defined, these teams are counted as belonging to the original National Football Conference even though they have since moved to the American Football Conference, the successor of the American Football League. The fact that this slightly archaic definition is used is another tip-off that the indicator is contrived.
29. For instance, explaining the amount of stock market growth through a simple regression model that uses the conference affiliation of the Super Bowl winner and a constant term as its only inputs would yield this estimate.
30. "Powerball—Prizes and Odds," Multi-State Lottery Association. http://www.powerball.com/powerball/pb_prizes.asp.

31. Achuthan and Benerji, *Beating the Business Cycle*, Kindle location 1478.
32. Gene Sperling, "The Insider's Guide to Economic Forecasting," *Inc. Magazine*, August 1, 2003. http://www.inc.com/magazine/20030801/forecasting_pagen_3.html.
33. Ibid.
34. Douglas M. Woodham, "Are the Leading Indicators Signaling a Recession?" *Federal Reserve Bank of New York Review* (Autumn 1984). http://www.newyorkfed.org/research/quarterly_review/1984v9/v9n3article8.pdf.
35. By "real time," I mean based on the values of the Leading Economic Index as available to forecasters at the time, before revisions to the data and the composition of the index. See Francis X. Diebold and Glenn D. Rudebusch, "Forecasting Output with the Composite Leading Index: A Real-Time Analysis," *Journal of the American Statistical Association* 86, 415 (September 1991), pp. 603–610.
36. Mark J. Perry, "Consumer Confidence Is a Lagging Indicator: Expect Post-Recession Gloom Through 2010," Seeking Alpha, October 29, 2009. http://seekingalpha.com/article/169740-consumer-confidence-is-a-lagging-indicator-expect-post-recession-gloom-through-2010.
37. Robert Lucas, "Econometric Policy Evaluation: A Critique," and Karl Brunner and A. Meltzer, "The Phillips Curve and Labor Markets," Carnegie-Rochester Conference Series on Public Policy, American Elsevier, 1976, pp. 19–46. http://pareto.uab.es/mcreel/reading_course_2006_2007/lucas1976.pdf.
38. C.A.E. Goodhart, "Problems of Monetary Management: The U.K. Experience," *Papers in Monetary Economics*, Reserve Bank of Australia, 1975.
39. The term that economists use for this condition is exogeneity.
40. Job growth as measured by the percentage change in net nonfarm payrolls.
41. "Year" in this instance refers to the change from the second quarter of 2009 through the first quarter of 2010.
42. The National Bureau of Economic Research, "U.S. Business Cycle Expansions and Contractions."
43. "Japan: Gross Domestic Product, constant prices (National currency)," Global Insight and Nomura database via International Monetary Fund, last updated 2010. http://www.imf.org/external/pubs/ft/weo/2011/02/weodata/weorept.aspx?pr.x=38&pr.y=9&sy=1980&ey=2016&scsm=1&ssd=1&sort=country&ds=.&br=1&c=158&s=NGDP_R&grp=0&a=.
44. "Minutes of the Federal Open Market Committee;" Federal Reserve System, October 30–31, 2007. http://www.federalreserve.gov/monetarypolicy/files/fomcminutes20071031.pdf.
45. "Gauging the Uncertainty of the Economic Outlook from Historical Forecasting Errors," by David Reifschneider and Peter Tulip; Finance and Economics Discussion Series, Divisions of Research and Statistics and Monetary Affairs, Federal Reserve Board, November 19, 2007. http://www.federalreserve.gov/Pubs/FEDS/2007/200760/200760pap.pdf.
46. The government can do a better job of estimating incomes once it sees tax filings (Americans, as much as they might dislike taxes, are relatively honest about paying them). But income earned in January 2009 would not be reported to the IRS until April 15, 2010. Then the IRS might take a few more months to collect all the data and report it to the Bureau of Economic Analysis. So although this is highly useful information, it might become available only after a lag of eighteen months to two years—much too late to be of any use to forecasters. Nevertheless, the government continues to refine its estimates of indicators like GDP for years after the fact through what are known as benchmark revisions.
47. "Historical Data Files for the Real-Time Data Set: Real GNP/GDP (ROUTPUT)," Federal Reserve Bank of Philadelphia. http://www.philadelphiafed.org/research-and-data/real-time-center/real-time-data/data-files/ROUTPUT/.
48. Specifically, the 95 percent margin of error.
49. The opposite case, happily, can sometimes arise as well. The government initially reported negative growth in the third quarter of 1981. But the data now says that the economy grew at nearly a 5 percent clip.
50. Although economists often do not give enough attention to the distinction between real time

and revised data when they present their forecasts. Revisions tend to bring different economic indicators more in line with one another. But the data is much messier when it is revealed in real time; in the spring of 2012, for instance, some economic indicators (like personal income) were near-recessionary while others (like industrial production) were suggestive of strong growth. In a few years, the data from this period will probably look much cleaner and will tell a more consistent story; the personal income figures will be revised upward some or the industrial production numbers downward. But that is much too late for economists trying to make a forecast. Building a forecast model from revised data will lead you to overestimate the ease of the task of forecasting.

51. This is not just me being glib. One can plot the error made in annual GDP predictions by the Survey of Professional Forecasters against a time trend and find that there has been no overall improvement since 1968.
52. "U.S. Economy Tipping into Recession," Economic Cycle Research Institute, September 30, 2011. http://www.businesscycle.com/reports_indexes/reportsummarydetails/1091.
53. Chris Isidore, "Forecast Says Double-Dip Recession Is Imminent," *CNNMoney*; September 30, 2011. http://money.cnn.com/2011/09/30/news/economy/double_dip_recession/index.htm.
54. Economic Cycle Research Institute, "U.S. Economy Tipping into Recession," September 30, 2011. http://www.businesscycle.com/reports_indexes/reportsummarydetails/1091.
55. Achuthan and Benerji, *Beating the Business Cycle*, Kindle locations 192–194.
56. Chris Anderson, "The End of Theory: The Data Deluge Makes the Scientific Method Obsolete," *Wired* magazine, 16.07; June 23, 2008. http://www.wired.com/science/discoveries/magazine/16-07/pb_theory.
57. I don't publish economic forecasts, but I certainly would not claim that I was especially bullish at the time.
58. A variety of conceptually similar methods based on "leading indicators" were predicting considerable growth at the time, or at least very little chance of a recession. See Dwaine van Vuuren, "U.S. Recession—an Opposing View," *Advisor Perspectives*, January 3, 2012. http://www.advisorperspectives.com/newsletters12/US_Recession-An_Opposing_View.php. Although ECRI's methodology is a little opaque, it appears that their indices may have placed a lot of emphasis on commodities prices, which were declining in late 2011 after having been in something of a bubble.
59. From September 30, 2011 (S&P 500 closed at 1131.42) to March 30, 2012 (S&P closed at 1379.49).
60. Henry Blodget, "ECRI's Lakshman Achuthan: No, I'm Not Wrong—We're Still Headed for Recession," *Daily Ticker*, Yahoo! Finance; May 9, 2012. http://finance.yahoo.com/blogs/daily-ticker/ecri-lakshman-achuthan-no-m-not-wrong-still-145239368.html.
61. In the November forecasts made from 1968 through 2009 of the next year's GDP in the Survey of Professional Forecasters, the root mean square error (RMSE) for an individual economist's forecast was 2.27 points, while the RMSE of the aggregate forecast was 1.92 points. Thus, averaging the forecasts reduced error by about 18 percent.
62. Stephen K. McNees, "The Role of Judgment in Macroeconomic Forecasting Accuracy," *International Journal of Forecasting*, 6, no. 3, pp. 287–99, October 1990. http://www.sciencedirect.com/science/article/pii/016920709090056H.
63. About the only economist I am aware of who relies solely on statistical models without applying any adjustments to them is Ray C. Fair of Yale. I looked at the accuracy of the forecasts from Fair's model, which have been published regularly since 1984. They aren't bad in some cases: the GDP and inflation forecasts from Fair's model have been roughly as good as those of the typical judgmental forecaster. However, the model's unemployment forecasts have always been very poor, and its performance has been deteriorating recently as it considerably underestimated the magnitude of the recent recession while overstating the prospects for recovery. One problem with statistical models is that they tend to perform well until one of their assumptions is violated and they encounter a new situation, in which case they may produce very inaccurate forecasts. In this case, a global financial crisis would represent a new

situation to a model that had been "trained" on economic data from after World War II since none had occurred during that period.

64. For instance, if some economists were consistently better than others at forecasting GDP, you would expect those who made more accurate forecasts in even-numbered years (2000, 2002, 2004) to also make more accurate forecasts in odd-numbered ones (2001, 2003, 2005). But when I took the data from the Survey of Professional Forecasters and divided it up in this way, there was very little correlation between the two. Economists who made better forecasts in even-numbered years were only about average in odd-numbered ones, and vice versa.
65. Andy Bauer, Robert A. Eisenbeis, Daniel F. Waggoner, and Tao Zha, "Forecast Evaluation with Cross-Sectional Data: The Blue Chip Surveys," *Economic Review*, Federal Reserve Bank of Atlanta, 2003. http://www.frbatlanta.org/filelegacydocs/bauer_q203.pdf.
66. David Laster, Paul Bennett, and In Sun Geoum, "Rational Bias in Macroeconomic Forecasts," *Quarterly Journal of Economics*, 114, 1 (1999), pp. 293–318. http://www.newyorkfed.org/research/staff_reports/sr21.pdf.
67. Ibid.
68. David Reifschneider and Peter Tulip, "Gauging the Uncertainty of the Economic Outlook from Historical Forecasting Errors," Federal Reserve Board Financial and Economics Discussion Series #2007-60 (November 2007). http://www.federalreserve.gov/pubs/feds/2007/200760/200760abs.html.
69. Ibid.
70. In figure 6-6, the black lines indicating the historical average are calibrated to the mean forecast for 2012 GDP growth in the November 2011 Survey of Professional Forecasters, which was 2.5 percent.
71. Jeremy Kahn, "The Man Who Would Have Us Bet on Terrorism—Not to Mention Discard Democracy and Cryogenically Freeze Our Heads—May Have a Point (About the Betting, We Mean)," *Fortune* magazine via CNN Money, September 15, 2003. http://money.cnn.com/magazines/fortune/fortune_archive/2003/09/15/349149/index.htm.
72. Robin Hanson, *Futarchy: Vote Values, but Bet Beliefs* (Washington, DC: George Mason University), August 2000. http://hanson.gmu.edu/futarchy.html.
73. Felix Salmon, "Why the Correlation Bubble Isn't Going to Burst," Reuters, August 19, 2011. http://blogs.reuters.com/felix-salmon/2011/08/19/why-the-correlation-bubble-isnt-going-to-burst/.

CHAPTER 7. ROLE MODELS

1. The narrative in the first several paragraphs is based primarily on two accounts: a *New York Times Magazine* article dated September 5, 1976, and a more recent Slate.com article available here: Patrick Di Justo, "The Last Great Swine Flu Epidemic," Salon.com, April 28, 2009. http://www.salon.com/news/environment/feature/2009/04/28/1976_swine_flu.
2. A/Victoria was named for its apparent origins in the Australian state of Victoria, where Melbourne is located; it was the dominant strain of seasonal flu in the mid-1970s.
3. Jeffery K. Taubenberger and David M. Morens, "1918 Influenza: The Mother of All Pandemics," *Emerging Infectious Diseases*, 12, 1 (January 2006). http://www.webcitation.org/5kCUlGdKu.
4. John Barr, "The Site of Origin of the 1918 Influenza Pandemic and Its Public Health Implications," *Journal of Translational Medicine*, 2, 3 (January 2004).
5. Among many other examples, see Jane E. Brody, "Influenza Virus Continues to Keep Scientists Guessing," *New York Times*, July 23, 1976. http://query.nytimes.com/mem/archive/pdf?res=F30E16FB3E5B167493C1AB178CD85F428785F9.
6. Di Justo, "The Last Great Swine Flu Epidemic."
7. Harold M. Schmeck Jr., "Flu Experts Soon to Rule on Need of New Vaccine," *New York Times*, March 21, 1976. http://select.nytimes.com/gst/abstract.html?res=F40711FC355E157493C3AB1788D85F428785F9.

8. Di Justo, "The Last Great Swine Flu Epidemic."
9. *New York Times*, April 23, 1976.
10. $180 million was the projected cost of the vaccination program itself. But Democrats—sensing Ford's urgency and desperation—also managed to attach more than a billion dollars in unrelated social welfare spending to the bill.
11. Although there is certainly much less flu in the summer than the winter, part of the reason for the difference is that people are not looking for the flu in the summer and those with flulike symptoms may be diagnosed with other illnesses, I was told by Dr. Alex Ozonoff of Harvard University.
12. Boyce Rensberger, "U.S. Aide Doubts a Heavy Flu Toll," *New York Times*, July 2, 1976. http://select.nytimes.com/gst/abstract.html?res=F30614F83F5B167493C0A9178CD85F428785F9.
13. *New York Times*, June 9, 1976.
14. *New York Times*, July 20, 1976.
15. *New York Times*, June 8, 1976.
16. These are really worth watching and look like something out of a John Waters movie. See for example "1976 Swine Flu Propaganda," by tarot1984; YouTube.com; April 27, 2009. http://www.youtube.com/watch?v=ASibLqwVbsk.
17. Di Justo, "The Last Great Swine Flu Epidemic."
18. Harold M. Schmeck, "Swine Flu Program I Halted in 9 States as 3 Die After Shots; Deaths Occur in Pittsburgh," *New York Times*, October 13, 1976. http://select.nytimes.com/gst/abstract.html?res=F00910F63F5A167493C1A8178BD95F428785F9.
19. In 1976, about 3,000 people aged sixty-five and older would have died every day in the United States.
20. The *New York Times*, for instance, argued the following in an October 14 editorial:

    It is conceivable that the 14 elderly people who are reported to have died soon after receiving the vaccination died of other causes. Government officials in charge of the program claim that it is all a coincidence, and point out that old people drop dead every day. The American people have even become familiar with a new statistic: Among every 100,000 people 65 to 75 years old, there will be nine or ten deaths in every 24-hour period under most normal circumstances.

    Even using the official statistic, it is disconcerting that three elderly people in one clinic in Pittsburgh, all vaccinated within the same hour, should die within a few hours thereafter. This tragedy could occur by chance, but the fact remains that it is extremely improbable that such a group of deaths should take place in such a peculiar cluster by pure coincidence.

    Although this logic is superficially persuasive, it suffers from a common statistical fallacy. The fallacy is that, although the odds of three *particular* elderly people dying on the same *particular* day after having been vaccinated at the same *particular* clinic are surely fairly long, the odds that *some* group of three elderly people would die at *some* clinic on *some* day are much shorter.

    Assuming that about 40 percent of elderly Americans were vaccinated within the first 11 days of the program, then about 9 million people aged 65 and older would have received the vaccine in early October 1976. Assuming that there were 5,000 clinics nationwide, this would have been 164 vaccinations per clinic per day. A person aged 65 or older has about a 1-in-7,000 chance of dying on any particular day; the odds of at least three such people dying on the same day from among a group of 164 patients are indeed very long, about 480,000 to one against. However, under our assumptions, there were 55,000 opportunities for this "extremely improbable" event to occur—5,000 clinics, multiplied by 11 days. The odds of this coincidence occurring *somewhere* in America, therefore, were much shorter—only about 8 to 1 against.
21. Di Justo, "The Last Great Swine Flu Epidemic."

22. David Evans, Simon Cauchemez, and Frederick G. Hayden, "Prepandemic Immunization for Novel Influenza Viruses: 'Swine Flu' Vaccine, Guillain-Barré Syndrome, and the Detection of Rare Severe Adverse Events," *Journal of Infectious Diseases*, 200, no. 3 (2009), pp. 321–328. www.journals.uchicago.edu/doi/pdf/10.1086/603560.
23. Kimberly Kindy, "Officials Are Urged to Heed Lessons of 1976 Flu Outbreak," *Washington Post*, May 9, 2009. www.washingtonpost.com/wp-dyn/content/article/2009/05/08/AR2009050802050.html.
24. *New York Times*, December 30, 1976.
25. There were sporadic press accounts of other cases of the swine flu, such as one in Wisconsin, but they were never confirmed by the CDC and, at the very least, never progressed beyond a single documented victim.
26. Peter Doshi, Figure 3 in "Trends in Recorded Influenza Mortality: United States, 1900–2004," *American Journal of Public Health*, 98, no. 5, May 2008. http://www.ncbi.nlm.nih.gov/pmc/articles/PMC2374803/figure/f3/.
27. Indeed, as Carter's margin over Ford was only 2 points nationwide, it's possible that fiasco cost him the election. Then again, so could a lot of things.
28. Harold M. Schmeck, "U.S. Discloses Shortage of Swine Flu Vaccine for Children 3 to 17," *New York Times*, November 16, 1978. http://select.nytimes.com/gst/abstract.html?res=F70E17F9395B167493C4A8178AD95F428785F9.
29. *New York Times*, May 19, 1979.
30. Fortunately these were very mild years for the virus; had there been a major pandemic later in the 1970s, it would have been very hard to persuade people to receive vaccinations, and there could have been thousands of preventable deaths as a result.
31. *New York Times*, September 5, 1976.
32. Much less commonly, the flu can be carried by other mammalian species like whales and horses.
33. Indeed, because of the role that animals play in the origination of influenza, it would probably be impossible to completely eradicate it for any length of time (as we have done with some other diseases, like smallpox) unless we also eradicated it in birds and pigs; see also News Staff: "Avian Flu Research Sheds Light on Swine Flu—and Why Influenza A Can Never Be Eradicated," *Scientific Blogging, Science 2.0*, May 1, 2009. http://www.science20.com/news_articles/avian_flu_research_sheds_light_swine_flu_and_why_influenza_can_never_be_eradicated.
34. John R. Moore, "Swine Productions: A Global Perspective," Alltech Inc., Engormix.com, accessed on May 20, 2012. http://en.engormix.com/MA-pig-industry/articles/swine-production-global-perspective_124.htm.
35. "Food Statistics > Pork Consumption per Capita (Most Recent) by Country," NationMaster.com, accessed May 20, 2012. http://www.nationmaster.com/graph/foo_por_con_per_cap-food-pork-consumption-per-capita.
36. "Disease and Terror," *Newsweek*, April 29, 2009. http://www.thedailybeast.com/newsweek/2009/04/30/disease-and-terror.html.
37. Although there are alternate theories that H1N1 indeed originated in Asia; see for example Donald G. McNeil Jr., "In New Theory, Swine Flu Started in Asia, Not Mexico," *New York Times*, June 23, 2009. http://www.nytimes.com/2009/06/24/health/24flu.html.
38. Tom Blackwell, "Flu Death Toll in Mexico Could Be Lower Than First Thought," *National Post*, April 29, 2009. http://web.archive.org/web/20100523224652/http://www.nationalpost.com/news/story.html?id=1547114.
39. Jo Tuckman and Robert Booth, "Four-Year-Old Could Hold Key in Search for Source of Swine Flu Outbreak," *The Guardian*, April 27, 2009. http://www.guardian.co.uk/world/2009/apr/27/swine-flu-search-outbreak-source.
40. Keith Bradsher, "Assessing the Danger of New Flu," *New York Times*, April 27, 2009. http://www.nytimes.com/2009/04/28/health/28hong.html?scp=35&sq=h1n1&st=nyt.

41. "Tracking Swine Flu Cases Worldwide," *New York Times*, April 23, 2011. http://www.nytimes.com/interactive/2009/04/27/us/20090427-flu-update-graphic.html.
42. "Report to the President on U.S. Preparations for 2009-H1N1 Influenza, President's Council of Advisors on Science and Technology, Executive Office of the President, August 7, 2009. http://www.whitehouse.gov/assets/documents/PCAST_H1N1_Report.pdf.
43. Carl Bialik, "Swine Flu Count Plagued by Flawed Data," *Wall Street Journal*, January 23, 2010. http://online.wsj.com/article/SB10001424052748704509704575019313343580460.html.
44. This may have been in part because the H1N1 vaccine seemed to provide some measure of protection from the seasonal strains of the flu, which for the first time in many years produced no discernible peak in January and February as it normally does.
45. Stephen Davies, "The Great Horse-Manure Crisis of 1894," *The Freeman*, 54, no. 7, September 2004. http://www.thefreemanonline.org/columns/our-economic-past-the-great-horse-manure-crisis-of-1894/.
46. Sir William Petty, "An Essay Concerning the Multiplication of Mankind," 1682.
47. Tomas Frejka, "World Population Projections: A Concise History," Center for Policy Studies, Working Papers Number 66, March 1981. http://pdf.usaid.gov/pdf_docs/PNAAR555.pdf.
48. Haya El Nasser, "World Population Hits 7 Billion," *USA Today*, October 31, 2011.
49. Ronald Bailey, "Seven Billion People Today—Malthusians Still Wrong (and Always Will Be)," Reason.com, October 31, 2011. http://reason.com/blog/2011/10/31/seven-billion-people-today-mal.
50. Frejka, "World Population Projections."
51. "U.S. HIV and AIDS Cases Reported Through December 1999," HIV/AIDS Surveillance Report, 11, no. 2, U.S. Department of Health and Human Services, Centers for Disease Control and Prevention. http://www.cdc.gov/hiv/topics/surveillance/resources/reports/pdf/hasr1102.pdf.
52. James M. Hyman and E. Ann Stanley, "Using Mathematical Models to Understand the AIDS Epidemic," *Mathematical Biosciences* 90, pp. 415–473, 1988. http://math.lanl.gov/~mac/papers/bio/HS88.pdf.
53. The version I applied here was to log-transform both the year variable and the AIDS-cases variable, then calculate the exponent via regression analysis. The 95 percent confidence interval on the exponent ran from about 2.2 to 3.7 by this method, with a most likely value of about 2.9. When applied ten years into the future, those relatively modest-seeming differences turn into an exceptionally broad range of possible outcomes.
54. Richard Carter and Kamini N. Mendis, table 4 in "Evolutionary and Historical Aspects of the Burden of Malaria," *Clinical Microbiology Reviews*, 15, no. 4, pp. 564–594, October 2002.
55. Note that R_0 is often given as a range in the literature; I present the midpoint of that range for ease of reading. Sources: David L. Smith, F. Ellis McKenzie, Robert W. Snow, and Simon I. Hay, malaria: "Revisiting the Basic Reproductive Number for Malaria and Its Implications for Malaria Control," *PLoS Biology*, 5, no. 3, March 2007. http://www.ncbi.nlm.nih.gov/pmc/articles/PMC1802755/; ebola: G. Chowell, N. W. Hengartner, C. Castillo-Chavez, P. W. Fenimore, and J. M. Hyman, "The Basic Reproductive Number of Ebola and the Effects of Public Health Measures: The Cases of Congo and Uganda," *Journal of Theoretical Biology*, 229, no. 1, pp. 119–126, July 7, 2004. math.lanl.gov/~gchowell/publications/ebolaJTB.pdf; 1918 flu: Marc Lipsitch, Christina Mills, and James Robins, "Estimates of the Basic Reproductive Number for 1918 Pandemic Influenza in the United States: Implications for Policy," Global Health Security Initiative, 2005. www.ghsi.ca/documents/Lipsitch_et_al_Submitted%2020050916.pdf; 2009 flu and seasonal flu: Todd Neale, "2009 Swine Flu More Transmissible Than Seasonal Flu," *MedPage Today*, May 11, 2009. http://www.medpagetoday.com/InfectiousDisease/SwineFlu/14154; HIV/AIDS: R. M. Anderson and R. M. May, "Population Biology of Infectious Diseases: Part I," *Nature*, 280, pp. 361–367, August 2, 1979; SARS: J. Wallinga and P. Teunis, "Different Epidemic Curves for Severe Acute Respiratory Syndrome Reveal Similar Impacts of Control Measures," *American Journal of Epidemiology*, 160, no. 6,

pp. 509–516, 2004; others: "History and Epidemiology of Global Smallpox Eradication" in "Smallpox: Disease, Prevention, and Intervention" Centers for Disease Control and Prevention. http://www.bt.cdc.gov/agent/smallpox/training/overview/pdf/eradicationhistory.pdf.

56. "Acquired Immunodeficiency Syndrome (AIDS) Weekly Surveillance Report," Centers for Disease Control, December 31, 1984. http://www.cdc.gov/hiv/topics/surveillance/resources/reports/pdf/surveillance84.pdf.
57. Gregory M. Herek and John P. Capitanio, "AIDS Stigma and Sexual Prejudice," *American Behavioral Scientist*, 42, pp. 1126–1143, 1999. http://psychology.ucdavis.edu/rainbow/html/abs99_sp.pdf.
58. Marc Lacey and Elisabeth Malkin, "First Flu Death Provides Clues to Mexico Toll," *New York Times*, April 30, 2009. http://www.nytimes.com/2009/05/01/health/01oaxaca.html?scp=28&sq=h1n1&st=nyt.
59. Jo Tuckman and Robert Booth, "Four-Year-Old Could Hold Key in Search for Source of Swine Flu Outbreak," *The Guardian*, April 27, 2009. http://www.guardian.co.uk/world/2009/apr/27/swine-flu-search-outbreak-source.
60. CNN/Time/ORC International poll of Iowa Republican voters, December 21–27, 2011. http://i2.cdn.turner.com/cnn/2011/images/12/28/topstate3.pdf.
61. Selzer & Company poll of Iowa Republican voters, *Des Moines Register*, December 27–30, 2011. http://www.desmoinesregister.com/assets/pdf/FullTopLineResults.pdf.
62. Marshall L. Fisher, Janice H. Hammon, Walter R. Obermeyer, and Ananth Raman, "Making Supply Meet Demand in an Uncertain World," *Harvard Business Review*, May 1994. http://hbr.org/1994/05/making-supply-meet-demand-in-an-uncertain-world/ar/1.
63. I encountered a similar case when consulting for a major Hollywood movie studio in 2009. This studio, which was very sophisticated and data-driven in many ways, had a belief that a certain weekend—let's say it was the first weekend in October—was a particularly good time to release a big film even though it didn't correspond to anything especially important on the calendar. What had happened is that the studio had a film that performed unexpectedly well in that time slot one year, mostly because it was just a very good movie. The studio, however, attributed some of its success to the timing of its release. Every subsequent year, then, they would save one of their better releases for this same weekend in October and then market the hell out of it. Of course, a good and well-marketed film normally will do pretty well whenever it is released. Nevertheless, this fulfilled their prediction that early October was a good time to release a film and entrenched their belief further.
64. This calculation is based on the number of children who are classified as autistic and therefore are eligible for special education programs in public schools under the federal IDEAS Act. "Table 45. Children 3 to 21 Years Old Served Under Individuals with Disabilities Education Act, Part B, by Type of Disability: Selected Years, 1976–77 through 2008–09," *Digest of Educational Statistics*, National Center for Education Statistics, Institute of Education Sciences, 2010. http://nces.ed.gov/programs/digest/d10/tables/dt10_045.asp.
65. Per NewsLibrary.com. In figure 7-4, figures for articles discussing autism are adjusted to reflect the overall volume of articles in the NewsLibrary.com database in a given year, and then indexed so as to be on the same scale as the number of public schoolchildren receiving special education programs for autism under the IDEAS Act.
66. Tomohisa Yamashita, Kiyoshi Izumi, Koichi Kurumatani, "Effective Information Sharing Based on Mass User Support for Reduction of Traffic Congestion," presented at the New England Complex Systems Institute's Fifth International Conference on Complex Systems, May 16–21, 2004. http://www.necsi.edu/events/iccs/openconf/author/papers/f190.pdf.
67. Hyejin Youn, Hawoong Jeong, and Michael T. Gastner, "The Price of Anarchy in Transportation Networks: Efficiency and Optimality Control," *Physical Review Letters*, 101, August 2008. http://arxiv.org/pdf/0712.1598.pdf.
68. Hanna Kokko, "Useful Ways of Being Wrong," *Journal of Evolutionary Biology*, 18 (2005), pp. 1155–1157. http://www.anu.edu.au/BoZo/kokko/Publ/Wrong.pdf.

69. W. O. Kermack and A. G. McKendrick, "A Contribution to the Mathematical Theory of Epidemics," *Proceedings of the Royal Society* A, 115 (1927), pp. 700–721.
70. H-H. M. Truong, et al., "Increases in Sexually Transmitted Infections and Sexual Risk Behaviour Without a Concurrent Increase in HIV Incidence Among Men Who Have Sex with Men in San Francisco: A Suggestion of HIV Serosorting?," *Sexually Transmitted Infections*, 82, 6 (2006), pp. 461–466.
71. "Condom fatigue" is the idea that gay men had simply grown sick and tired of being told that they needed to use a condom every time they had sex.
72. Thomas H. Maugh II, "Experts Fear Resurgence of HIV Infection," *Los Angeles Times*, July 8, 2000. http://articles.latimes.com/2000/jul/08/news/mn-49552.
73. MSM is the preferred term in the medical literature in this context; it is more precise than terms like *homosexual* and particularly *gay*, which often refer to sexual identity rather than sexual behavior. Some men who identify as straight (or bisexual) nevertheless have sex with other men; and some men who identify as gay may nevertheless have sex with women, or may be celibate.
74. San Francisco Department of Public Health.
75. Christopher S. Hall and Gail Bolan, "Syphilis and HIV," HIV InSite Knowledge Base Chapter, University of California San Francisco; June 2006. http://hivinsite.ucsf.edu/InSite?page=kb-05-01-04.
76. H-H M. Truong et al., "Increases in Sexually Transmitted Infections and Sexual Risk Behaviour Without a Concurrent Increase in HIV Incidence Among Men Who Have Sex with Men in San Francisco: A Suggestion of HIV Serosorting?"
77. Fengyi Jin, et al., "Per-Contact Probability of HIV Transmission in Homosexual Men in Sydney in the Era of HAART," *AIDS*, 24, pp. 907–913, 2010. http://www.who.int/hiv/events/artprevention/jin_per.pdf.
78. Much of the research suggested that it was HIV-positive men who were driving the trend: most HIV-positive men would much prefer to have sex with other HIV-positive partners, particularly if they are not planning to use a condom. The advent of the Internet, as well as various types of support networks in the offline world, has made that much easier to do.
79. Larry Green, "Measles on Rise Nationwide; Chicago Worst Hit," *Los Angeles Times*, August 5, 1989. http://articles.latimes.com/1989-08-05/news/mn-469_1_chicago-health.
80. Justin Lessler et al., "Transmissibility of Swine Flu at Fort Dix, 1976," *Journal of the Royal Society Interface*, 4, no. 15, pp. 755–762, August 2007. http://rsif.royalsocietypublishing.org/content/4/15/755.full.
81. Ibid.
82. "Keep it sophisticatedly simple" was a phrase used by the late economist Arnold Zellner.
83. "Healthy Hand Washing Survey 2011," Bradley Corp. http://www.bradleycorp.com/handwashing/survey.jsp.
84. http://www.altpenis.com/penis_news/20060710032108data_trunc_sys.shtml.
85. "An Agent-Based Approach to HIV/AIDS Epidemic Modeling: A Case Study of Papua New Guinea, thesis, Massachusetts Institute of Technology, 2006. http://dspace.mit.edu/handle/1721.1/34528.
86. Shan Mei, et al., "Complex Agent Networks Explaining the HIV Epidemic Among Homosexual Men in Amsterdam," *Mathematics and Computers in Simulation*, 80, no. 5, January 2010. http://portal.acm.org/citation.cfm?id=1743988.
87. Donald G. McNeil Jr., "Predicting Flu with the Aid of (George) Washington," *New York Times*, May 3, 2009. http://www.nytimes.com/2009/05/04/health/04model.html?hp.
88. Michael A. Babyak, "What You See May Not Be What You Get: A Brief, Nontechnical Introduction to Overfitting in Regression-Type Models," *Statistical Corner, Psychosomatic Medicine*, 66 (2004), pp. 411–421.
89. Even if a prediction model is just a sort of thought experiment that is years away from producing useful results, it can still help us understand the scope of a problem. The Drake equation,

a formula that provides a framework for predicting the number of intelligent extraterrestrial species in the galaxy, is not likely to yield highly useful and verifiable predictions in the span of our lifetimes—nor, probably, in the span of human civilization. The uncertainties are too great. Too many of its parameters are not known to within an order of magnitude; depending on which values you plug in, it can yield answers anywhere from that we are all alone in the universe to that there are billions and billions of extraterrestrial species. However, the Drake equation has nevertheless been a highly useful lens for astronomers to think about life, the universe, and everything.

90. George E. P. Box and Norman R. Draper, *Empirical Model-Building and Response Surfaces* (New York: Wiley, 1987), p. 424.
91. "Norbert Wiener," Wikiquote.org. http://en.wikiquote.org/wiki/Norbert_Wiener.

CHAPTER 8: LESS AND LESS AND LESS WRONG

1. Roland Lazenby, *The Show: The Inside Story of the Spectacular Los Angeles Lakers in the Words of Those Who Lived It* (New York: McGraw-Hill Professional, 2006).
2. Mark Heisler, "The Times' Rankings: Top to Bottom/NBA," *Los Angeles Times*, November 7, 1999.
3. Tom Spousta, "Pro Basketball: Trail Blazers Have Had Some Success Containing O'Neal," *New York Times*, May 20, 2000. http://www.nytimes.com/2000/05/20/sports/pro-basketball-trail-blazers-have-had-some-success-containing-o-neal.html?scp=2&sq=lakers+portland&st=nyt.
4. "Blazer Blowout Shows Need for 'Sheed," Associated Press; May 22, 2000. http://web.archive.org/web/20041226093339/http://sportsmed.starwave.com/nba/2000/20000522/recap/porlal.html.
5. Tom Spousta, "Pro Basketball: Game 2 Was a Blur as Lakers Lost Focus," *New York Times*, May 24, 2000. http://www.nytimes.com/2000/05/24/sports/pro-basketball-game-2-was-a-blur-as-lakers-lost-focus.html?scp=3&sq=lakers+portland&st=nyt.
6. Tom Spousta, "Pro Basketball: Lakers Rally and Get Back on Track," *New York Times*, May 27, 2012. http://www.nytimes.com/2000/05/27/sports/pro-basketball-lakers-rally-and-get-back-on-track.html?scp=14&sq=lakers+portland&st=nyt.
7. Tom Spousta, "Pro Basketball: Everything Comes Up Roses for the Lakers," *New York Times*, May 29, 2000. http://www.nytimes.com/2000/05/29/sports/pro-basketball-everything-comes-up-roses-for-the-lakers.html?scp=16&sq=lakers+portland&st=nyt.
8. "Seventh Heaven: Blazers Send Series Back to L.A. for Game 7," Associated Press via *Sports Illustrated*, June 3, 2000. http://sportsillustrated.cnn.com/basketball/nba/2000/playoffs/news/2000/06/02/lakers_blazers_gm6_ap/.
9. That is, $300,000 from winning his $200,000 bet on Portland at 3-to-2 odds, less the $80,000 that Voulgaris originally bet on the Lakers.
10. Tom Spousta, "Pro Basketball: Trail Blazers Follow Plan to the Bitter End," *New York Times*, June 7, 2000. http://www.nytimes.com/2000/06/05/sports/pro-basketball-trail-blazers-follow-plan-to-the-bitter-end.html?scp=28&sq=lakers+portland&st=nyt.
11. Per play-by-play data downloaded from Basketballvalue.com. http://basketballvalue.com/downloads.php.
12. This is based on a logistic regression analysis I conducted of all games played in the 2009–2010 NBA regular season, where the independent variable is the score margin between the home team and the away team with fourteen minutes left to play in the game, and the dependent variable is whether or not the home team ultimately won. The regression model yields a value of .056 when the scoring margin is –16; that is, the home team has a 5.6 percent chance of victory when trailing by sixteen points, which translates into odds 17-to-1 against. I round down slightly to 15 to 1 because a team trailing by sixteen points at home will usually be inferior to its opponent, whereas the Lakers and Blazers were more evenly matched.

13. Voulgaris's odds of winning his bet at the start of the evening were about 50 percent: a 60 percent chance that the Lakers beat the Blazers in Game 7 multiplied by what I've estimated to be an 83 percent chance that the Lakers would beat the Pacers if advancing to the final. By that point in the game, the Lakers' odds of winning the Championship were down to about 5 percent: a 6 percent chance of coming back to beat the Blazers, multiplied by an 83 percent chance of beating the Pacers.
14. Miranda Hitti, "Testosterone Ups Home Field Advantage," *WebMD Health News*, June 21, 2006. http://www.webmd.com/fitness-exercise/news/20060621/testosterone-ups-home-field-advantage.
15. Most sports leagues hold their drafts in reverse order of finish: the team with the worst record is the first to pick. In the NBA, a sport where a single superstar talent can make an exceptional amount of difference, the league holds a draft lottery so as to discourage teams from tanking their games at the end of the season to improve their drafting position. Nevertheless, the worse a team does, the more Ping-Pong balls it gets in the lottery, and so teams will often play something other than their best basketball in these scenarios.
16. This asymmetry would not exist to the same extent if basketball teams were more focused on individual defensive statistics. But offense is relatively easy to measure, and defense is relatively hard; some teams don't even try to measure individual defensive performance at all. A player who scores a basket will therefore gain more market value than the man defending will lose by conceding one.
17. "2001–02 Cleveland Cavaliers Schedule and Results," Basketball-Reference.com. http://www.basketball-reference.com/teams/CLE/2002_games.html.
18. On average, a team will go either over or under the total five games in a row about five times per season. That works out to 150 such streaks per season between the thirty NBA teams combined.
19. D. R. Bellhouse, "The Reverend Thomas Bayes FRS: A Biography to Celebrate the Tercentenary of His Birth," *Statistical Science*, 19, 1, pp. 3–43; 2004. http://www2.isye.gatech.edu/~brani/isyebayes/bank/bayesbiog.pdf.
20. Bayes may also have been an Arian, meaning someone who followed the teachings of the early Christian leader Arias and who regarded Jesus Christ as the divine son of God rather than (as most Christians then and now believe) a direct manifestation of God.
21. Thomas Bayes, "Divine Benevolence: Or an Attempt to Prove That the Principal End of the Divine Providence and Government Is the Happiness of His Creatures." http://archive.org/details/DivineBenevolenceOrAnAttemptToProveThatThe.
22. Ibid.
23. Ibid.
24. The Late Rev. Mr. Bayes, Communicated by Mr. Price, in a Letter to John Canton, M. A. and F. R. S., "An Essay Towards Solving a Problem in the Doctrine of Chances," *Philosophical Transactions of the Royal Society of London*, 53, pp. 370–418; 1763. http://www.stat.ucla.edu/history/essay.pdf.
25. Donald A. Gillies, "Was Bayes a Bayesian?," *Historia Mathematica*, 14, no. 4, pp. 325–346, November 1987. http://www.sciencedirect.com/science/article/pii/0315086087900656.
26. David Hume, "Cause and Effect" in *An Enquiry Concerning Human Understanding (1772)* (Hackett Publishing Company, 1993). http://www.marxists.org/reference/subject/philosophy/works/en/hume.htm.
27. Some Christians regard Bayesian probability as more compatible with their worldview. Under Bayes's theorem, if you assign a 100 percent prior probability to the hypothesis that a Christian God exists, then no amount of worldly evidence will shake you from that conviction. It is plausible that Bayes was aware of this property; in introducing Bayes's essay, Richard Price mentioned that he thought Bayes's theorem helped to confirm "the existence of the Diety."

    For further discussion, see Steve Bishop, "Christian Mathematicians—Bayes, God &

Math: Thinking Christianly About Mathematics," *Education*, March 22, 2012. http://godandmath.com/2012/03/22/christian-mathematicians-bayes/.

28. "Fundamental Atheism," Free Atheist Church. https://sites.google.com/site/freeatheistchurch/fundamental-atheism.
29. Sharon Bertsch McGrayne, *The Theory That Would Not Die: How Bayes' Rule Cracked the Enigma Code, Hunted Down Russian Submarines, and Emerged Triumphant from Two Centuries of Controversy* (New Haven, CT: Yale University Press, Kindle edition), 427–436.
30. E. O. Lovett, "The Great Inequality of Jupiter and Saturn," *Astronomical Journal*, 15, 351 (1895), pp. 113–127.
31. McGrayne, *The Theory That Would Not Die*, Kindle location 19.
32. Pierre-Simon Laplace, "A Philosophical Essay on Probabilities" (1902), pp. 6–8.
33. Bret Schulte, "How Common Are Cheating Spouses?" *U.S. News & World Report*, March 27, 2008. http://www.usnews.com/news/national/articles/2008/03/27/how-common-are-cheating-spouses.
34. "Breast Cancer Risk by Age," Breast Cancer Centers for Disease Control and Prevention, last updated August 13, 2010. http://www.cdc.gov/cancer/breast/statistics/age.htm.
35. "Understanding Breast Exam Results—False Negative–False Positive Results," RealAge.com. http://www.realage.com/womens-health/breast-exam-results.
36. S. Eva Singletary, Geoffrey L. Robb, and Gabriel N. Hortobagyi, "Advanced Therapy of Breast Disease," B. C. Decker, May 30, 2004.
37. Gina Kolata, "Panel Urges Mammograms at 50, Not 40," *New York Times*, November 16, 2009. http://www.nytimes.com/2009/11/17/health/17cancer.html.
38. Dan M. Kahan, et al., "The Polarizing Impact of Science Literacy and Numeracy on Perceived Climate Change Risks," *Nature Climate Change*, May 27, 2012. See Supplementary Information: http://www.nature.com/nclimate/journal/vaop/ncurrent/extref/nclimate1547-s1.pdf.
39. Twenty-five thousand days prior to September 11, 2001, would take us back to 1942.
40. John P. A. Ioannidis, "Why Most Published Research Findings Are False," *PLOS Medicine*, 2, e124, August 2005. http://www.plosmedicine.org/article/info:doi/10.1371/journal.pmed.0020124.
41. Brian Owens, "Reliability of 'New Drug Target' Claims Called into Question," *NewsBlog, Nature*, September 5, 2011. http://blogs.nature.com/news/2011/09/reliability_of_new_drug_target.html.
42. McGrayne, *The Theory That Would Not Die*, Kindle location 46.
43. Paul D. Stolley, "When Genius Errs: R. A. Fisher and the Lung Cancer Controversy," *American Journal of Epidemiology*, 133, 5, 1991. http://www.epidemiology.ch/history/PDF%20bg/Stolley%20PD%201991%20when%20genius%20errs%20-%20RA%20fisher%20and%20the%20lung%20cancer.pdf.
44. Alan Agresti and David B. Hitchcock, "Bayesian Inference for Categorical Data Analysis," *Statistical Methods & Applications*, 14 (2005), pp. 297–330. http://www.stat.ufl.edu/~aa/articles/agresti_hitchcock_2005.pdf.
45. John Aldrich, "R. A. Fisher on Bayes and Bayes' Theorem," *Bayesian Analysis*, 3, no. 1 (2008), pp. 161–170. http://ba.stat.cmu.edu/journal/2008/vol03/issue01/aldrich.pdf.
46. McGrayne, *The Theory That Would Not Die*, Kindle location 48.
47. Tore Schweder, "Fisherian or Bayesian Methods of Integrating Diverse Statistical Information?" *Fisheries Research*, 37, 1–3 (August 1998), pp. 61–75. http://www.sciencedirect.com/science/article/pii/S0165783698001271.
48. 2008 New Hampshire Democratic Primary polls via RealClearPolitics.com. http://www.realclearpolitics.com/epolls/2008/president/nh/new_hampshire_democratic_primary-194.html.
49. Nate Silver, "Rasmussen Polls Were Biased and Inaccurate; Quinnipiac, SurveyUSA Performed Strongly," *FiveThirtyEight*, *New York Times*, November 4, 2010. http://fivethirtyeight

.blogs.nytimes.com/2010/11/04/rasmussen-polls-were-biased-and-inaccurate-quinnipiac-surveyusa-performed-strongly/.

50. R. A. Grant and T. Halliday, "Predicting the Unpredictable: Evidence of Pre-Seismic Anticipatory Behaviour in the Common Toad," *Journal of Zoology*, 700, January 25, 2010. http://image.guardian.co.uk/sys-files/Environment/documents/2010/03/30/toads.pdf.
51. "Hate Group Formation Associated with Big-Box Stores," *ScienceNewsline.com*, April 11, 2012. http://www.sciencenewsline.com/psychology/2012041121000031.html.
52. Aldrich, "R. A. Fisher on Bayes and Bayes' Theorem."
53. McGrayne, *The Theory That Would Not Die*, Kindle location 111.
54. Sir Ronald A. Fisher, "Smoking: The Cancer Controversy," Oliver and Boyd. http://www.york.ac.uk/depts/maths/histstat/smoking.htm.
55. Jean Marston, "Smoking Gun," *NewScientist*, no. 2646, March 8, 2008. http://www.newscientist.com/article/mg19726460.900-smoking-gun.html.
56. McGrayne, *The Theory That Would Not Die*, Kindle location 113.
57. Stolley, "When Genius Errs."
58. Ibid.
59. Jo Tuckman and Robert Booth, "Four-Year-Old Could Hold Key in Search for Source of Swine Flu Outbreak," *The Guardian;* April 27, 2009. http://www.guardian.co.uk/world/2009/apr/27/swine-flu-search-outbreak-source
60. McGrayne, *The Theory That Would Not Die*, Kindle location 7.
61. Raymond S. Nickerson, "Null Hypothesis Significance Testing: A Review of an Old and Continuing Controversy," *Psychological Methods*, 5, 2 (2000), pp. 241–301. http://203.64.159.11/richman/plogxx/gallery/17/%E9%AB%98%E7%B5%B1%E5%A0%B1%E5%91%8A.pdf.
62. Andrew Gelman and Cosma Tohilla Shalizi, "Philosophy and the Practice of Bayesian Statistics," *British Journal of Mathematical and Statistical Psychology*, pp. 1–31, January 11, 2012. http://www.stat.columbia.edu/~gelman/research/published/philosophy.pdf.
63. Although there are several different formulations of the steps in the scientific method, this version is mostly drawn from "APPENDIX E: Introduction to the Scientific Method," University of Rochester. http://teacher.pas.rochester.edu/phy_labs/appendixe/appendixe.html.
64. Thomas S. Kuhn, *The Structure of Scientific Revolutions* (Chicago: University of Chicago Press, Kindle edition).
65. Jacob Cohen, "The Earth Is Round (p<.05)," *American Psychologist*, 49, 12 (December 1994), pp. 997–1003. http://ist-socrates.berkeley.edu/~maccoun/PP279_Cohen1.pdf.
66. Jeff Gill, "The Insignificance of Null Hypothesis Significance Testing," *Political Research Quarterly*, 52, 3 (September 1999), pp. 647–674. http://www.artsci.wustl.edu/~jgill/papers/hypo.pdf.
67. David R. Anderson, Kenneth P. Burnham, and William L. Thompson, "Null Hypothesis Testing: Problems, Prevalence, and an Alternative," *Journal of Wildlife Management*, 64, 4 (2000), pp. 912–923. http://cat.inist.fr/%3FaModele%3DafficheN%26cpsidt%3D792848.
68. William M. Briggs, "It Is Time to Stop Teaching Frequentism to Non-Statisticians," arXiv.org, January 13, 2012. http://arxiv.org/pdf/1201.2590.pdf.
69. David H. Krantz, "The Null Hypothesis Testing Controversy in Psychology," *Journal of the American Statistical Association*, 44, no. 448 (December 1999). http://www.jstor.org/discover/10.2307/2669949?uid=3739832&uid=2&uid=4&uid=3739256&sid=47698905120317.

CHAPTER 9. RAGE AGAINST THE MACHINES

1. "Poe Invents the Modern Detective Story," National Historic Site Philadelphia, National Park Service, U.S. Department of the Interior. http://www.nps.gov/edal/forteachers/upload/detective.pdf.
2. Nick Eaton, "Gallup: Bill Gates Is America's Fifth-Most Admired Man," *Seattle Post-Intelligencer*, December 27, 2010. http://blog.seattlepi.com/microsoft/2010/12/27/gallup-bill-gates-is-americas-fifth-most-admired-man/.

3. Joann Pan, "Apple Tops Fortune's 'Most Admired' List for Fifth Year in a Row," *Mashable*, March 2, 2012. http://mashable.com/2012/03/02/apple-tops-fortunes-most-admired-list-five-years-straight-video/.
4. David Kravets, "Stock-Picking Robot 'Marl' Is a Fraud, SEC Says," Threat Level, *Wired*, April 23, 2012. http://www.wired.com/threatlevel/2012/04/stock-picking-robot/.
5. "What Is the Stock Trading Robot 'MARL'?," Squidoo.com. http://www.squidoo.com/StockTradingRobotMARL.
6. *Philadelphia Inquirer*, "Computer Predicts Odds of Life, Death," *Orlando Sentinel*, July 9, 1992. http://articles.orlandosentinel.com/1992-07-09/news/9207090066_1_apache-system-critical-care-critical-care.
7. Nick Montfort, *Twisty Little Passages: An Approach to Interactive Fiction* (Boston: MIT Press, 2005), p. 76.
8. Claude E. Shannon, "Programming a Computer for Playing Chess," *Philosophical Magazine*, Series 7, 41, 314, March 1950. http://archive.computerhistory.org/projects/ chess/related_materials/software/2-0%20and%202-1.Programming_a_computer_for_playing_chess.shannon/2-0%20and%202-1.Programming_a_computer_for_playing_chess.shannon.062303002.pdf.
9. William G. Chase and Herbert A. Simon, "The Mind's Eye in Chess" in *Visual Information Processing* (New York: Academic Press, 1973).
10. Douglas Harper, *Online Etymology Dictionary*. http://www.etymonline.com/index.php?term=eureka.
11. Amos Tversky and Daniel Kahneman, "Judgement Under Uncertainty: Heuristics and Biases," *Science*, 185 (September 27, 1974), pp. 1124–1131. http://www.econ.yale.edu/~nordhaus/homepage/documents/tversky_kahn_science.pdf.
12. Lauren Himiak, "Bear Safety Tips," National & States Parks, About.com. http://usparks.about.com/od/backcountry/a/Bear-Safety.htm.
13. billwall, "Who Is the Strongest Chess Player?" Chess.com, October 27, 2008. http://www.chess.com/article/view/who-is-the-strongest-chess-player.
14. Feng-hsiung Hsu, Thomas Anantharaman, Murray Campbell, and Andreas Nowatzyk, "A Grandmaster Chess Machine," *Scientific American*, October 1990. http://www.disi.unige.it/person/DelzannoG/AI2/hsu.html.
15. Ibid.
16. "The Chip vs. the Chessmaster," Nova (documentary), March 26, 1991.
17. Garry Kasparov, "'The Chess Master and the Computer," *New York Review of Books*, February 11, 2010. http://www.nybooks.com/articles/archives/2010/feb/11/the-chess-master-and-the-computer/.
18. "Frequently Asked Questions: Deep Blue;" IBM Research via Internet Archive WayBack Machine beta. http://web.archive.org/web/20071028124110/http://www.research.ibm.com/deepblue/meet/html/d.3.3a.shtml#difficult.
19. Chess Opening Explorer, chessgames.com. http://www.chessgames.com/perl/explorer.
20. Murray Campbell, A. Joseph Hoane Jr., and Feng-hsiung Hsu, "Deep Blue," sjeng.org, August 1, 2001. http://sjeng.org/ftp/deepblue.pdf.
21. IBM Research, "Frequently Asked Questions: Deep Blue."
22. "1, Nf3 d5, 2. g3 Bg4" Chess Opening Explorer, chessgames.com. http://www.chessgames.com/perl/explorer?node=1959282&move=3&moves=Nf3.d5.g3.Bg4&nodes=74.77705.124843.1959282.
23. Trading a bishop for a knight early in the game, as Deep Blue threatened to do, may not be a good trade because bishops are more valuable when a player still has both of them on the board. If you have just one bishop, your opponent can move with relative impunity on the half of the squares that the remaining bishop can't physically cover. In other words, you'd rather have one knight and both bishops than two knights and one bishop.
24. Position Search, chessgames.com. http://www.chessgames.com/perl/chess.pl?node=1967201.

25. Adriaan D. de Groot, *Thought and Choice in Chess* (Amsterdam, Holland: Amsterdam University Press, Amsterdam Academic Archive, 2008).
26. Ibid.
27. Shannon, "Programming a Computer for Playing Chess."
28. Uly, January 23, 2010 (2:52 P.M.), comment on "computer eval – winning chances" by ppipper, on Rybka Chess Community Forum. http://rybkaforum.net/cgi-bin/rybkaforum/topic_show.pl?tid=15144.
29. "Kasparov vs. Deep Blue, Game 1, May 3, 1997," Chess Corner. http://www.chesscorner.com/games/deepblue/dblue1.htm.
30. Robert Byrne, "In Late Flourish, a Human Outcalculates a Calculator," *New York Times*, May 4, 1997. http://www.nytimes.com/1997/05/04/nyregion/in-late-flourish-a-human-outcalculates-a-calculator.html?scp=3&sq=kasparov&st=nyt.
31. deka, "Analysis by Rybka 3 14ply," February 26, 2010. http://web.zone.ee/chessanalysis/study%20on%20chess%20strength.pdf.
32. Frederic Friedel, "Garry Kasparov vs. Deep Blue," ChessBase.com, May 1997. http://www.chessbase.com/columns/column.asp?pid=146.
33. Ibid.
34. Ibid.
35. "Deep Blue: Overview," IBM100 Icons of Progress, IBM. http://www.research.ibm.com/deepblue/games/game2/html/move34b.shtml.
36. Which was the correct move is still debated. The other computers of the day—less advanced than Deep Blue—found Qb6, the queen advancement, to be the right play, and by a somewhat large margin. But the very fact that Deep Blue had deviated from the computer play had been what distinguished its move. When I set up the position on a contemporary computer engine called Rybka, it also picked Qb6, but found the position to be much closer, the difference amounting to not more than about three-tenths of a pawn worth of strength. The difference is small enough that it's easy to imagine Deep Blue, a somewhat idiosyncratic engine that was specifically tailored to match up with Kasparov, picking the alternative line.
37. Maurice Ashley, Patrick Wolff, and Yasser Seirawan, "Game 2, black 36 . . . axb5," IBM Research real-time text commentary, May 11, 2007. http://web.archive.org/web/20080614011112/http://www.research.ibm.com/deepblue/games/game2/html/move36b.shtml.
38. The actual match itself was played upstairs, in a specially designed television studio on the thirty-fifth floor of the Equitable Center, with no spectators permitted.
39. Bruce Weber, "Computer Defeats Kasparov, Stunning the Chess Experts," *New York Times*, May 5, 1997. http://www.nytimes.com/1997/05/05/nyregion/computer-defeats-kasparov-stunning-the-chess-experts.html?scp=3&sq=kasparov&st=nyt.
40. Bruce Weber, "Wary Kasparov and Deep Blue Draw Game 3," *New York Times*, May 7, 1997. http://www.nytimes.com/1997/05/07/nyregion/wary-kasparov-and-deep-blue-draw-game-3.html?scp=1&sq=kasparov+hand+of+god&st=nyt.
41. Frederic Friedel, "Garry Kasparov vs. Deep Blue," *Multimedia Report*, ChessBase Magazine 58. http://www.chessbase.com/columns/column.asp?pid=146.
42. Bruce Weber, "Swift and Slashing, Computer Topples Kasparov," *New York Times*, May 12, 1997. http://www.nytimes.com/1997/05/12/nyregion/swift-and-slashing-computer-topples-kasparov.html?scp=3&sq=kasparov&st=nyt.
43. This metaphor is borrowed from Bill Wyman, a music critic for the *Chicago Reader*, who ranked it as the greatest moment in rock history. Bill Wyman, "The 100 Greatest Moments in Rock History," *Chicago Reader*, September 28, 1995. http://www.chicagoreader.com/chicago/the-100-greatest-moments-in-rock-history/Content?oid=888578.
44. Campbell, Hoane Jr., and Feng-hsiung, "Deep Blue."
45. Larry Page, "PageRank: Bringing Order to the Web," Stanford Digital Library Project, August

18, 1997. http://web.archive.org/web/20020506051802/www-diglib.stanford.edu/cgi-bin/WP/get/SIDL-WP-1997-0072?1.

46. "How Search Works," by Google via YouTube, March 4, 2010. http://www.youtube.com/watch?v=BNHR6IQJGZs.
47. Per interview with Vasik Rajlich.
48. "Amateurs beat GMs in PAL / CSS Freestyle," ChessBase News. http://www.chessbase.com/newsdetail.asp?newsid=2467.
49. Kasparov, "The Chess Master and the Computer."

CHAPTER 10. THE POKER BUBBLE

1. "Chris Moneymaker Ranking History" in The Mob Poker Database, thehendonmob.com. http://pokerdb.thehendonmob.com/player_graphs/chris_moneymaker_18826.
2. I first played in one of the smaller events at the World Series of Poker in 2005, although I did not participate in the $10,000 main event of the tournament until 2009.
3. The catch was that you had to play a certain number of hands at the site before you could cash out any winnings.
4. The example that follows is idealized, in the sense that we will be applying a relatively formal and rigorous mathematical process to consider a fairly typical poker hand, alongside the more impressionistic view that a player might have at the table. In a real poker game, players must make decisions much more quickly, both for reasons of etiquette and because spending too much time contemplating a hand would itself reveal information to the opponents. However, the thought process described here is what all poker players are striving for, whether they realize it or not. The question is who can come to the closest approximation of it under the pressures that a real game represents.
5. If the opponent raised with a middling hand, such as a pair of nines on the river, he would essentially be doing so as a bluff since he wouldn't expect us to call with a weaker hand.
6. Nate Silver, "Sanity Check: 88 Hand" twoplustwo.com; May 14, 2012. http://forumserver.twoplustwo.com/56/medium-stakes-pl-nl/sanity-check-88-hand-1199549/.
7. G4mblers, "Biggest Pot in TV Poker History—Tom Dwan vs Phil Ivey Over 1.1 Million," YouTube; January 28, 2010. http://www.youtube.com/watch?v=GnxFohpljqM.
8. "About Tom Dwan;" PokerListings.com. http://www.pokerlistings.com/poker-player_tom-dwan.
9. locke, "Isildur1 & the Poker Economy," PokerTableRatings.com, December 11, 2009. http://www.pokertableratings.com/blog/2009/12/isildur1-the-poker-economy/.
10. "Player Profile: durrrr;" Highstakes DataBase. http://www.highstakesdb.com/profiles/durrrr.aspx.
11. PokerListings.com, "About Tom Dwan."
12. The quotes included in this book are from an interview I conducted by phone with Dwan in May 2012, but I had met him in person on previous occasions.
13. Bill Chen and Jerrod Ankenman, "The Mathematics of Poker," Conjelco, November 30, 2006.
14. Darse Billings, "Algorithms and Assessment in Computer Poker," thesis submitted to Department of Computing Science, University of Alberta; 2006. http://www.cs.virginia.edu/~evans/poker/readings/billings-ch1.pdf.
15. Tommy Angelo, "Elements of Poker," Tommy Angelo Operations, Kindle edition, p. 209, December 13, 2007.
16. Robert Koch, *Living Life the 80/20 Way* (Boston: Nicholas Brealey Publishing, 2004).
17. My analysis was limited to players who played in games featuring a big blind of $2 or higher; this is about the cheapest increment at which enough money could potentially be made from the game to sustain the existence of some professional players.

18. Specifically, I split the sample data into even- and odd-numbered months; if a player is truly skilled, he should be winning in the even-numbered months as well as the odd-numbered ones. I then applied a regression analysis to predict a player's win rate (measured as big blinds won per one hundred hands) from one half of the sample to the other; the results of the regression are taken to be tantamount to the player's long term success rate. The variables in the regression were a player's win rate, multiplied by the natural logarithm of the number of hands who he played, along with a variable indicating how many hands he played of the ones he was dealt. Players who were too tight or too loose were considerably less likely to replicate their success from one period to the next, holding other factors equal—in fact, this was often a better predictor of a player's out-of-sample win rate than the past win rate itself, unless he had played a very large number of hands.
19. Online poker tables typically consisted of ten seats; those in bricks-and-mortar casinos often contain nine seats instead.
20. In the online poker data, the rake was equivalent to about $57 per one hundred hands, according to my estimate. The figure would potentially be quite similar at a bricks-and-mortar casino. The Bellagio, for instance, usually charges players a fee of $6 per half hour. To get in one hundred hands of no-limit poker in a casino environment, a player would typically need to spend about four hours at the table (it is a fairly slow game). That means she'd need to pay the time charge eight times, for $48 total. Add in the customary $1 tip to the dealer when she wins a hand, and you get to roughly the same $57 figure. No adjustment was made for the benefits—and the hidden costs—of the free cocktails in Las Vegas.
21. "Listen, here's the thing, if you can't spot the sucker in your first half hour at the table, then you are the sucker," Mike McDermott in *Rounders* via monologuedb.com. http://www.monologuedb.com/dramatic-male-monologues/rounders-mike-mcdermott/.
22. "Guy Laliberte's Accounts on Full Tilt Poker Down Millions of Dollars in 2008," *PokerKingBlog.com*, January 2, 2009. http://www.pokerkingblog.com/2009/01/02/guy-laliberte-the-engine-of-the-high-stakes-economy-on-full-tilt-poker/.
23. Games tended to be a bit fishier in the winter, when more people were staying indoors and were logged on to their computers, and in the summer when the World Series of Poker was being played, than they were in the spring or the fall.
24. James McManus, "Full Tilt Boogie: The UIGEA and You," Grantland.com, December 8, 2011. http://www.grantland.com/story/_/id/7333093/uigea-you.
25. Rocco Havel, "Taking Stock of the UIGEA," *Tight Poker*, April 16, 2008. http://www.tightpoker.com/news/taking-stock-of-the-uigea-487/.
26. Branon Adams, "The Poker Economy," *Bluff Magazine*, November, 2006. http://www.bluffmagazine.com/magazine/The-Poker-Economy-Brandon-Adams-584.htm.
27. Nate Silver, "After 'Black Friday,' American Poker Faces Cloudy Future," *FiveThirtyEight*, *New York Times*, April 20, 2011. http://fivethirtyeight.blogs.nytimes.com/2011/04/20/after-black-friday-american-poker-faces-cloudy-future/.
28. Based on a minimum of fifty at-bats for the player in both April and May.
29. Contrast this with a sport like tennis, where the structure of the game is such that even modest differences in skill level manifest themselves very quickly. The very best players in the world, like Rafael Nadal and Novak Djokovic, win only about 55 percent of the points they play—barely more than half. However, hundreds of points are played over the course of a single match, while baseball players might get just four or five at-bats. It would be silly to conclude that Nadal is better at tennis than Josh Hamilton is at baseball. But Nadal almost always wins while Hamilton has plenty of nights when he goes 0-for-4. In tennis, you reach the long run much more quickly.
30. This estimate is based on my own history, as well as the statistics from other limit hold 'em players in my databases.
31. Chen and Ankenman, "The Mathematics of Poker;" pp. 40–43.

32. Although this calculation is made more complicated by the fact that the distribution of long-term win rates among poker players almost certainly does not follow a bell curve (normal distribution). Instead, per our application of the Pareto principle of prediction, it is left-skewed, with a "fat tail" of losing players.
33. Martin Harris, "Polaris 2.0 Defeats Stoxpoker Team in Man-Machine Poker Championship Rematch," *PokerNews.com*, July 10, 2008. http://www.pokernews.com/news/2008/07/man-machine-II-poker-championship-polaris-defeats-stoxpoker-.htm.
34. "Poker Services;" Poker Royalty. http://pokerroyalty.com/poker-business.php.
35. "Annual per Capita Lottery Sales, by Educational Attainment," bp0.blogger.com. http://bp0.blogger.com/_bYktpmgngXA/RclJid4kTxI/AAAAAAAAAHU/PnDE3 Orpqc/s1600-h/Compound_Charts3.JPG.
36. Angelo, "Elements of Poker," Kindle location 2.
37. Ibid., Kindle location 55.

CHAPTER 11. IF YOU CAN'T BEAT 'EM . . .

1. "Stocks Traded, Total Value (% of GDP)," World Bank World Development Indicators. http://data.worldbank.org/indicator/CM.MKT.TRAD.GD.ZS.
2. "Fortune 500;" CNN Money. http://money.cnn.com/magazines/fortune/fortune500/2009/full_list/101_200.html.
3. "Stocks Traded, Turnover Ratio (%),"World Bank World Development Indicators. http://data.worldbank.org/indicator/CM.MKT.TRNR/countries.
4. Adrianne Jeffries, "High-Frequency Trading Approaches the Speed of Light," BetaBeat.com, February 17, 2012. http://www.betabeat.com/2012/02/17/high-frequency-trading-approaches-the-speed-of-light/.
5. Terrance Odean, "Do Investors Trade Too Much?" *American Economic Review*, 89, no. 5 (December 1999), pp. 1279–1298. http://web.ku.edu/~finpko/myssi/FIN938/Odean_Do%20Investors%20Trade%20Too%20Much_AER_1999.pdf.
6. Bruno de Finetti, "La Prévision: Ses Lois Logiques, Ses Sources Subjectives," *Annales de l'Institut Henri Poincaré*, 7 (1937).
7. Markets provide us with valuable information whether or not we participate in them directly. An economist might hate bananas, refusing to buy bananas at the supermarket at any price, but she still might be interested in knowing what they cost, to help her calculate the inflation rate. Or an orchard might be interested to know what the price was to decide whether it should plant more banana trees.

   Economists call this property "price discovery," and it's one of the key advantages of a free-market economy; the common price for a good or service in the market provides valuable information about the supply and demand for it, and the price can rise or fall in response. If prices are set by a central authority instead, it's much more cumbersome to determine which goods should be produced.
8. William F. Sharpe, *Investments* (Englewood Cliffs, NJ: Prentice-Hall, 1978).
9. Caren Chesler, "A Bettor World," *The American*, American Enterprise Institute, May/June 2007. http://www.american.com/archive/2007/may-june-magazine-contents/a-bettor-world.
10. Iowa Electronic Markets. http://iemweb.biz.uiowa.edu/pricehistory/pricehistory_SelectContract.cfm?market_ID=214.
11. Per interview with Justin Wolfers and David Rothschild.
12. David Rothschild, "Forecasting Elections: Comparing Prediction Markets, Polls, and Their Biases," *Public Opinion Quarterly*, 73, no. 5 (2009), pp. 895–916. http://assets.wharton.upenn.edu/~rothscdm/RothschildPOQ2009.pdf.
13. Intrade has suffered from some systematic—and perhaps predictable—biases. In particular, research by Wolfers and others has shown it suffers from something called favorite-longshot bias, which means that bettors tend to overprice low-probability events. For instance, events

that Intrade might suggest had a 1 in 10 chance of occurring in fact occur only 1 in 20 times over the long run, and events that the market implied had a 1 in 30 chance of occurring might in fact happen just 1 time in 100. Correcting for this bias was enough for Intrade to outperform *FiveThirtyEight* in 2008, according to Rothschild's paper, while *FiveThirtyEight* was quite a bit better without this adjustment.

14. Forecast error in this context would typically be measured by root-mean squared error (RMSE).
15. Andy Bauer, Robert A. Eisenbeis, Daniel F. Waggoner, and Tao Zha, "Forecast Evaluation with Cross-Sectional Data: The Blue Chip Surveys," *Economic Review*, Federal Reserve Bank of Atlanta, 2003. http://www.frbatlanta.org/filelegacydocs/bauer_q203.pdf.
16. Emile Servan-Schreiber, Justin Wolfers, David M. Pennock, and Brian Galebach, "Prediction Markets: Does Money Matter?" *Electronic Markets*, 14, no. 31 (September 2004). http://hcs.ucla.edu/lake-arrowhead-2007/Paper6_Watkins.pdf.
17. Several models of the 2010 U.S. House elections, for instance, relied on the Gallup poll and the Gallup poll alone. Although Gallup is usually a very good polling firm, its surveys were way off that year and implied a much larger gain for the Republicans than they actually achieved—perhaps 80 or 90 seats rather than the 63 they actually gained. However, taking the polling average would have gotten you quite close to the right result.
18. "Super Tuesday 2012 on Intrade;" Intrade.com, March 8, 2012. http://www.intrade.com/v4/reports/historic/2012-03-07-super-tuesday-2012/.
19. Nate Silver, "Intrade Betting Is Suspicious," *FiveThirtyEight*, September 24, 2008. http://www.fivethirtyeight.com/2008/09/intrade-betting-is-suspcious.html.
20. Nate Silver, "Evidence of Irrationality at Intrade," "Live Coverage: Alabama and Mississippi Primaries;" *FiveThirtyEight*, *New York Times*, March 13, 2012. http://fivethirtyeight.blogs.nytimes.com/2012/03/13/live-coverage-alabama-and-mississippi-primaries/#evidence-of-irrationality-at-intrade.
21. I haven't actually bet on Intrade, as I would consider it a conflict of interest.
22. More specifically, my hope (or belief) is that my subjective probability estimates (my Bayesian priors) might be better than Intrade's on average. I consider it less likely that this would be true of the forecast models that we run, since those models make certain simplifying assumptions and since it is important to think carefully about whether your model is flawed or whether the consensus is flawed when there is a divergence between them. (More often than not, the model is flawed.) If you follow the FiveThirtyEight blog, you may sometimes see me writing that you should bet against the FiveThirtyEight model for this reason.
23. Eugene F. Fama, "My Life in Finance," *Annual Review of Financial Economics*, 3 (2011), pp. 1–15. http://faculty.chicagobooth.edu/brian.barry/igm/fama_mylifeinfinance.pdf.
24. Eugene F. Fama, "The Behavior of Stock-Market Prices," *Journal of Business*, 38, no. 1 (January 1965), pp. 34–105. http://stevereads.com/papers_to_read/the_behavior_of_stock_market_prices.pdf.
25. Ibid., p. 40.
26. Google Scholar search. http://scholar.google.com/scholar?q=BEHAVIOR++OF+STOCK-MARKET++PRICES&hl=en&btnG=Search&as_sdt=1%2C33&as_sdtp=on.
27. According to a search of the Google News archive, Fama's name was not mentioned in the mainstream press until it appeared in a 1971 *New York Times* article. Marylin Bender, "Chicago School Foes to the Head of the Class," *New York Times*, May 23, 1971. http://query.nytimes.com/mem/archive/pdf?res=F00614F8355F127A93C1AB178ED85F458785F9.
28. William F. Sharpe, "Mutual Fund Performance," *Journal of Business*, 39, 1 (January 1966), part 2: Supplement on Security Prices, pp. 119–138. http://finance.martinsewell.com/fund-performance/Sharpe1966.pdf.

29. The sample consists of all mutual funds listed as balanced large-capitalization American equities funds by E*Trade's mutual funds screener as of May 1, 2012, excluding index funds. Funds also had to have reported results for each of the ten years from 2002 through 2006. There may be a slight bias introduced in that some funds that performed badly from 2002 through 2011 may no longer be offered to investors.
30. Charts A, B, C, and E are fake. Chart D depicts the actual movement of the Dow over the first 1,000 days of the 1970s, and chart F depicts the actual movement of the Dow over the first 1,000 days of the 1980s. Congratulations if you guessed correctly! Send your résumé and cover letter to Mad Money, 900 Sylvan Ave., Englewood Cliffs, NJ 07632.
31. Eugene F. Fama, "Efficient Capital Markets: A Review of Theory and Empirical Work," *Journal of Finance*, 25, 2 (1970), pp. 383–417.
32. Per interview with Eugene Fama.
33. Alan J. Ziobrowski, Ping Cheng, James W. Boyd, and Brigitte J. Ziobrowski, "Abnormal Returns from the Common Stock Investments of the U.S. Senate," *Journal of Financial and Quantiative Analysis*, 39, no. 4 (December 2004). http://www.walkerd.people.cofc.edu/400/Sobel/P-04.%20Ziobrowski%20-%20Abnormal%20Returns%20US%20Senate.pdf.
34. Google Scholar search. http://scholar.google.com/scholar?hl=en&q=%22efficient+markets%22&as_sdt=0%2C33&as_ylo=1992&as_vis=0.
35. Google Scholar search. http://scholar.google.com/scholar?hl=en&q=%22efficient+markets+hypothesis%22&btnG=Search&as_sdt=1%2C33&as_ylo=2000&as_vis=0.
36. Google Scholar search. http://scholar.google.com/scholar?as_q=&num=10&as_epq=theory+of+evolution&as_oq=&as_eq=&as_occt=any&as_sauthors=&as_publication=&as_ylo=1992&as_yhi=&as_sdt=1&as_subj=bio&as_sdtf=&as_sdts=33&btnG=Search+Scholar&hl=en.
37. John Aidan Byrne, "Elkins/McSherry—Global Transaction Costs Decline Despite High Frequency Trading," *Institutional Investor*, November 1, 2010. http://www.institutionalinvestor.com/Popups/PrintArticle.aspx?ArticleID=2705777.
38. Specifically, a linear regression of the sign of that day's stock price (1 indicating a positive movement and –1 indicating a negative movement) on the previous day's stock price. The trend is also highly statistically significant if you take a regression of the percentage change in the stock price on the previous day's percentage change. Note, however, that standard forms of regression analysis assume that errors are normally distributed, whereas the stock market does not obey a normal distribution. Economists like Fama think this is a problem when applying standard statistical tests to analyze patterns in stock prices.
39. Index funds would not have been widely available in 1976; the analysis assumes that the investor's returns would track that of the Dow Jones Industrial Average.
40. This is much worse than the market-average return. Although the 2000s were a poor decade for stocks, a buy-and-hold investor would have had about $9,000 rather than $4,000 left over by the end of the period.
41. Carlota Perez, "The Double Bubble at the Turn of the Century: Technological Roots and Structural Implications," *Cambridge Journal of Economics*, 33 (2009), pp. 779–805. http://www.relooney.info/Cambridge-GFC_14.pdf.
42. Based on a comparison of revenues from technology companies in the Fortune 500 to revenues for all companies in the Fortune 500 as of 2010. The companies I count as technology companies are Amazon.com, Apple, Avaya, Booz Allen Hamilton, Cisco Systems, Cognizant Technology Solutions, Computer Sciences, Corning, Dell, eBay, EMC, Google, Harris, Hewlett-Packard, IBM, Liberty Media, Microsoft, Motorola Solutions, NCR, Oracle, Pitney Bowes, Qualcomm, SAIC, Symantec, Western Digital, Xerox, and Yahoo!
43. Specifically, a 90 percent increase in value over the trailing five years, adjusted for dividends and inflation. This would correspond to a 14 percent annualized increase, twice the long-run rate of 7 percent.

44.

CASES IN WHICH THE S&P 500 INCREASED BY 90 PERCENT OVER A FIVE-YEAR PERIOD

Year(s)	Crash?	Description
1881–83	No	Although stocks lost 36 percent of their nominal value between June 1881 and January 1885, much of this is accounted for by the severe deflation that was gripping the country at the time. This should probably not be considered a crash, although returns were below-average even on an inflation-adjusted basis.
1901	**Yes**	**Panic of 1901.** Initially confined to a relatively narrow industry, railroad stocks, which happened to be very popular with small investors, the downturn eventually spread to the broader market with the S&P 500 losing more than a quarter of its value from 1901 to 1903.
1925–30	**Yes**	**Great Depression.** Stock prices declined by almost 80 percent over a three-year period from 1929–1932 amid the worst financial crisis in history.
1937	**Yes**	**Recession of 1937–38.** Stocks lost more than 40 percent of their value over a 14-month period as the economy went into a severe double-dip recession.
1954–59	No	The early 1950s were an extremely good time to buy stocks as investors benefited from America's postwar prosperity boom. Toward the end of the 1950s was somewhat less favorable, as stock prices were choppy throughout the 1960s.
1986–87	**Yes**	**Black Monday** came on October 19, 1987, with the Dow declining by almost 23 percent in a single day. An investor who persevered through the period would have seen very good returns in the 1990s.
1989	No	An investor buying in 1989 would have experienced the robust returns of the 1990s without the trauma of Black Monday. A good time to buy stocks.
1997–2000	**Yes**	**Dot-com crash.** Someone buying the S&P 500 at the top of the market in August 2000 and selling it at the bottom in February 2003 would have lost almost half his money. An investor with a portfolio rich in technology stocks would have done even worse.

45. The original title of *Dow 100,000* was apparently the more modest *Dow 30,000* before its publisher realized that it was going to be one-upped. I infer this because the original description of the book at Amazon.com refers to it as "*Dow 30,000,*" emphasizing its prediction that the Dow would increase to 30,000 by 2010. The book also predicted that the Dow would

rise to 100,000 by 2020, giving it its new title. See http://www.amazon.com/Dow-100-000-Fact-Fiction/dp/0735201374 retrieved November 25, 2011.

46. By "real return," I refer to the share price plus dividends but less inflation. I assume that dividends are automatically re-invested in the stock index rather than held.
47. Alan Greenspan, "The Challenge of Central Banking in a Democratic Society," Remarks at the Annual Dinner and Francis Boyern Lecture of The American Enterprise Institute for Public Policy Research, Washington, DC, December 5, 1996. http://www.federalreserve.gov/boarddocs/speeches/1996/19961205.htm.
48. Charts like those in figure 11-7 look to be extremely rich with data, but they are somewhat deceptive. They include a data point to represent every year of stock-market returns: one of the circles in the chart, for instance, represents how stocks behaved in the twenty years between 1960 and 1980. Another circle represents how they behaved from 1961 to 1981. The problem is that those periods overlap with one another and therefore are double-counting the same data. If we're looking at how stock prices perform in twenty years at a time, we don't really have that much data to work with. Shiller's P/E ratio can first be calculated in 1881. Start in 1881 and count upward twenty years at a time . . . you get to 1901, 1921, 1941, 1961, 1981, and 2001. Then you run out of time after just six data points.
49. You should, of course, make as large a trade as you can if you are certain about what the stock is going to do.
50. "Henry Blodget's Risky Bet on the Future of News," *Bloomberg Businessweek*, July 8, 2010. http://www.businessweek.com/print/magazine/content/10_29/b4187058885002.htm.
51. Dan Mitchell and Scott Martin, "Amazon Up 46 Points; Report 'Clarified,'" CNET News, December 16, 1998. http://news.cnet.com/2100-1017-219176.html.
52. Amazon.com Inc. (AMZN) Historical Prices; Yahoo! Finance. http://finance.yahoo.com/q/hp?s=AMZN&a=00&b=1&c=1997&d=11&e=25&f=2011&g=d.
53. Amazon.com shares rose to an intraday high of $302 on December 16, 1988, after beginning the day at $243, before closing at $289.
54. Out of the more than one hundred people I interviewed for this book, Blodget is one of just two or three for whom you'd be happy to publish the interview transcript almost word for word.
55. *Denver Post*, April 16, 1998.
56. The price of a share of Amazon is now superficially six times cheaper than it was in 1998 because of stock splits.
57. Zinta Lundborg, "Report Card: Henry Blodget," *The Street*, June 27, 2000. http://www.thestreet.com/markets/analystrankings/977502.html.
58. "Vested Interest;" PBS *Now*; May 31, 2002. http://www.pbs.org/now/politics/wallstreet.html.
59. *Securities and Exchange Commission*, 450 Fifth Street, N.W. Washington, DC 20549, *Plaintiff,—against—Henry Mckelvey Blodget, Defendant*," United States District Court, Southern District of New York, April 28, 2003. http://www.sec.gov/litigation/complaints/comp18115b.htm.
60. "The Securities and Exchange Commission, NASD and the New York Stock Exchange Permanently Bar Henry Blodget from the Securities Industry and Require $4 Million Payment;" U.S. Securities and Exchange Commission, April 28, 2003. http://www.sec.gov/news/press/2003-56.htm.
61. David Carr, "Not Need to Know but Nice to Know," *MediaTalk, New York Times*, November 24, 2003. http://www.nytimes.com/2003/11/24/business/mediatalk-not-need-to-know-but-nice-to-know.html?ref=henryblodget.
62. Crash is defined here as a 20 percent decline in stock prices, net of dividends and inflation.
63. "Securities Industry Employment 2Q 2010;" *Securities Industry and Financial Markets Association Research Report*, 5, no. 13. http://www.cdfa.net/cdfa/cdfaweb.nsf/fbaad5956b2928b086256efa005c5f78/7b5325c9447d35518625777b004cfb5f/$FILE/SecuritiesIndustry_Employment_20100810_SIFMA.pdf.
64. There is also some evidence that the analysts know privately when a stock is a loser much

sooner than they are willing to say publicly, and will tip their firm's investment clients to this before they inform the general public. See for instance Jeffrey A. Buss, T. Clifton Green, and Narasimhan Jegadeesh, "Buy-Side Trades and Sell-Side Recommendations: Interactions and Information Content," Emory University, January 2010. http://www.bus.emory.edu/cgreen/docs/busse,green,jegadeesh_wp2010.pdf.

65. Sorin Sorescu and Avanidhar Subrahmanyam, "The Cross-Section of Analyst Recommendations," *Recent Work*, Anderson Graduate School of Management, UC Los Angeles, January 9, 2004. http://escholarship.org/uc/item/76x8k0cc;jsessionid=5ACA605CE152E3724AB2754A1E35FC6A#page-3.
66. Floyd Norris, "Another Technology Victim; Top Soros Fund Manager Says He 'Overplayed' Hand," *New York Times*, April 29, 2000. http://www.nytimes.com/2000/04/29/business/another-technology-victim-top-soros-fund-manager-says-he-overplayed-hand.html?pagewanted=2&src=pm.
67. John C. Bogle, "Individual Investor, R.I.P.," *Wall Street Journal*, October 3, 2005.
68. Jonathan Lewellen, "Institutional Investors and the Limits of Arbitrage," *Journal of Financial Economics*, 102 (2011), pp. 62–80. http://mba.tuck.dartmouth.edu/pages/faculty/jon.lewellen/docs/Institutions.pdf.
69. Individual investors, because they tend to make smaller trades, account for an even smaller fraction of the *trading* that takes place. One recent study (Alicia Davis Evans, "A Requiem for the Retail Investor?" *Virginia Law Review*, May 14, 2009. http://www.virginialawreview.org/content/pdfs/95/1105.pdf) holds that individual investors account for just 2 percent of market volume on the New York Stock Exchange.
70. Say that a trader has the opportunity to bet $1 million of his firm's capital on a short position in the market. He figures there is a 55 percent chance the market will crash over the next year and a 45 percent chance that it will improve. This is a pretty good bet. If the analyst has calculated the odds right, the expected return from it is $100,000.

    But the incentives may look nothing like this for the trader himself. There is a high probability that he will be fired if he gets the bet wrong. Short-sellers aren't popular on Wall Street, especially when they're underperforming their peers. He may benefit to some extent if he gets the bet right, but not enough to outweigh this. Better to wait until the crash is unavoidable and all traders fall down together.
71. David S. Scharfstein and Jeremy C. Stein, "Herd Behavior and Investment," *American Economic Review*, 80, no. 3 (June 1990), pp. 465–479. http://ws1.ad.economics.harvard.edu/faculty/stein/files/AER-1990.pdf.
72. Russ Wermers, "Mutual Fund Herding and the Impact on Stock Prices," *Journal of Finance*, 7, 2 (April 1999), pp. 581–622. http://www.rhsmith.umd.edu/faculty/rwermers/herding.pdf.
73. Total equities holdings for 2007 are estimated from the World Bank. (http://databank.worldbank.org/ddp/html-jsp/QuickViewReport.jsp?RowAxis=WDI_Series~&ColAxis=WDI_Time~&PageAxis=WDI_Ctry~&PageAxisCaption=Country~&RowAxisCaption=Series~&ColAxisCaption=Time~&NEW_REPORT_SCALE=1&NEW_REPORT_PRECISION=0&newReport=yes&ROW_COUNT=1&COLUMN_COUNT=51&PAGE_COUNT=1&COMMA_SEP=true). For 1980, they are taken from data on the market capitalization within the New York Stock Exchange ("Annual reported volume, turnover rate, reported trades [mils. of shares]," http://www.nyxdata.com/nysedata/asp/factbook/viewer_edition.asp?mode=table&key=2206&category=4) and multiplied by the ratio of all U.S. stock holdings to NYSE market capitalization as of 1988, per World Bank data. All figures are adjusted to 2007 dollars.
74. James Surowiecki, *The Wisdom of Crowds* (New York: Random House, 2004).
75. Silver, "Intrade Betting Is Suspicious."
76. Marko Kolanovic, Davide Silvestrini, Tony SK Lee, and Michiro Naito, "Rise of Cross-Asset Correlations," *Global Equity Derivatives and Delta One Strategy*, J.P. Morgan, May 16, 2011. http://www.cboe.com/Institutional/JPMCrossAssetCorrelations.pdf.
77. Richard H. Thaler, "Anomalies: The Winner's Curse," *Journal of Economic Perspectives*, 2,

no. 1 (1998), pp. 191–202. http://econ.ucdenver.edu/Beckman/Econ%204001/thaler-winner's%20curse.pdf.

78. Daniel Kahneman, *Thinking, Fast and Slow* (New York: Farrar, Straus and Giroux, 2011), pp. 261–262.
79. Odean, "Do Investors Trade Too Much?"
80. Owen A. Lamont and Richard H. Thaler, "Can the Market Add and Subtract? Mispricing in Tech Stock Carve-outs," *Journal of Political Economy*, 111, 2 (2003), pp. 227–268. http://faculty.chicagobooth.edu/john.cochrane/teaching/Empirical_Asset_Pricing/lamont%20and%20thaler%20add%20and%20subtract%20jpe.pdf.
81. Virtually no risk but not literally no risk; there was a small amount of risk from the fact that the IRS might reject the spin-off.
82. Lamont and Thaler, "Can the Market Add and Subtract? Mispricing in Tech Stock Carve-outs."
83. José Scheinkman and Wei Xiong, "Overconfidence and Speculative Bubbles," *Journal of Political Economy*, 111, 6 (2003), pp. 1183–1220. http://web.ku.edu/~finpko/myssi/FIN938/Schienkman%20%26%20Xiong.volume-return.JPE_2003.pdf.
84. Fisher Black, "Noise," *Journal of Finance*, 41, 3 (1986).
85. Sanford J. Grossman and Joseph E. Stiglitz, "On the Impossibility of Informationally Efficient Markets," *American Economic Review*, 70, 3 (June 1980), pp. 393–408. http://www.math.ku.dk/kurser/2003-1/invfin/GrossmanStiglitz.pdf.
86. Edgar Ortega, "NYSE Loses Market Share and NASDAQ Isn't the Winner (Update3)," Bloomberg, June 24, 2009. http://www.bloomberg.com/apps/news?pid=newsarchive&sid=amB3bwJD1mLM.
87. Kent Daniel, Mark Grinblatt, Sheridan Titman, and Russ Wermers, "Measuring Mutual Fund Performance with Characteristic-Based Benchmarks," *Journal of Finance*, 52 (1997), pp. 1035–1058.
88. Franklin R. Edwards and Mustafa Onur Caglaya, "Hedge Fund Performance and Manager Skill," *Journal of Futures Markets*, 21, 11 (November 2001), pp. 1003–1028.
89. Ardian Harri and B. Wade Brorsen, "Performance Persistence and the Source of Returns for Hedge Funds," Oklahoma State University, Agricultural Economics Working Paper, July 5, 2002. http://www.hedgefundprofiler.com/Documents/166.pdf.
90. Rob Bauer, Mathijs Cosemans, and Piet Eichholtz, "Option Trading and Individual Investor Performance," *Journal of Banking & Finance*, 33 (2009), pp. 731–746. http://arno.unimaas.nl/show.cgi?fid=15657.
91. Joshua D. Coval, David A. Hirshleifer, and Tyler Shumway, "Can Individual Investors Beat the Market?" School of Finance, Harvard University, Working Paper No. 04-025/Negotiation, Organization and Markets, Harvard University, Working Paper No. 02-45; Sept. 2005. http://my.psychologytoday.com/files/attachments/5123/sept-2005-distributed-version.pdf.
92. Benjamin Graham and Jason Zweig, *The Intelligent Investor* (New York: Harper Collins, rev. ed., Kindle edition, 2009).
93. The S&P 500 is almost certainly preferable for these purposes to the Dow Jones Industrial Average, both because it includes a much broader and more diversified array of stocks and because it is weighted according to the market capitalization of each stock rather than their prices, meaning that it does a much better job of replicating the portfolio of the average investor.
94. "Investing," PollingReport.com. http://www.pollingreport.com/invest.htm.
95. Calculations are based on Robert Shiller's stock market data and assume dividends are reinvested. They ignore transaction costs.
96. This is not a cherry-picked example. The average annual return for an investor who pursued this strategy would have been 2.8 percent since 1900.
97. Black, "Noise."
98. Didier Sornette, *Why Stock Markets Crash: Critical Events in Complex Financial Systems* (Princeton, NJ: Princeton University Press, Kindle edition, 2005), location 3045.

99. Ibid.
100. "Quotation #24926 from Classic Quotes;" Quotations Page. http://www.quotationspage.com/quote/24926.html.

CHAPTER 12. A CLIMATE OF HEALTHY SKEPTICISM

1. "History for Washington, DC: Wednesday, June 22, 1988," Wunderground.com. http://www.wunderground.com/history/airport/KDCA/1988/6/22/DailyHistory.html?req_city=Ronald+Reagan+Washington+National&req_state=DC&req_statename=District+of+Columbia.
2. Kerry A. Emanuel, "Advance Written Testimony," Hearing on Climate Change: Examining the Processes Used to Create Science and Policy, House Committee on Science, Space and Technology, U.S. House of Representatives, March 31, 2011. http://science.house.gov/sites/republicans.science.house.gov/files/documents/hearings/Emanuel%20testimony.pdf.
3. James E. Hansen, "The Greenhouse Effect: Impacts on Current Global Temperature and Regional Heat Waves," statement made before the United States Senate Committee on Energy and Natural Resources, June 23, 1988. http://image.guardian.co.uk/sys-files/Environment/documents/2008/06/23/ClimateChangeHearing1988.pdf.
4. Philip Shabecoff, "Global Warming Has Begun, Expert Tells Senate," *New York Times*, June 24, 1988. http://www.nytimes.com/1988/06/24/us/global-warming-has-begun-expert-tells-senate.html?pagewanted=all&src=pm.
5. For most countries, statistics on obesity come from the World Health Organization's Global Database on Body Mass Index. http://apps.who.int/bmi/index.jsp. Data on caloric consumption is from the Food and Agriculture Organization of the United Nations. http://www.fao.org/fileadmin/templates/ess/documents/food_security_statistics/FoodConsumptionNutrients_en.xls.
6. "Nauru: General Data of the Country," Populstat.info. http://www.populstat.info/Oceania/naurug.htm.
7. One common technique requires adults to dutifully record everything they eat over a period of weeks, and trusts them to do so honestly when there is a stigma attached to overeating (and more so in some countries than others).
8. J. T. Houghton, G. J. Jenkins, and J. J. Ephraums, "Report Prepared for Intergovernmental Panel on Climate Change by Working Group I," *Climate Change: The IPCC Scientific Assessment* (Cambridge: Cambridge University Press, 1990), p. XI.
9. David R. Williams, "Earth Fact Sheet," NASA Goddard Space Flight Center, last updated November 17, 2010. http://nssdc.gsfc.nasa.gov/planetary/factsheet/earthfact.html.
10. Yochanan Kushnir, "The Climate System," Columbia University. http://eesc.columbia.edu/courses/ees/climate/lectures/radiation/.
11. "What Is the Typical Temperature on Mars?" Astronomy Cafe. http://www.astronomycafe.net/qadir/q2681.html.
12. Jerry Coffey, "Temperature of Venus," *Universe Today*, May 15, 2008. http://www.universetoday.com/14306/temperature-of-venus/.
13. Venus's temperatures are much warmer on average than Mercury, which has little atmosphere and whose temperatures vary from –200°C to +400°C over the course of a typical day.
14. Matt Rosenberg, "Coldest Capital Cities: Is Ottawa the Coldest Capital City?" About.com. http://geography.about.com/od/physicalgeography/a/coldcapital.htm.
15. "Kuwait City Climate," World-Climates.com. http://www.world-climates.com/city-climate-kuwait-city-kuwait-asia/.
16. High temperatures average 116°F in August in Kuwait; lows average –17° F in Ulan Bator in January.
17. "Mercury Statistics," Windows to the Universe. http://www.windows2universe.org/mercury/statistics.html.
18. "Human-Related Sources and Sinks of Carbon Dioxide" in *Climate Change—Greenhouse*

Gas Emissions, Environmental Protection Agency. http://www.epa.gov/climatechange/emissions/co2_human.html.

19. "Full Mauna Loa CO_2 Record" in *Trends in Atmospheric Carbon Dioxide*, Earth System Research Laboratory, National Oceanic & Atmospheric Administration Research, U.S. Department of Commerce. http://www.esrl.noaa.gov/gmd/ccgg/trends/#mlo_full.
20. Isaac M. Held and Brian J. Soden, "Water Vapor Feedback and Global Warming," *Annual Review of Energy and the Environment*, 25 (November 2000), pp. 441–475. http://www.annualreviews.org/doi/abs/10.1146%2Fannurev.energy.25.1.441.
21. Gavin Schmidt, "Water Vapor: Feedback or Forcing?" RealClimate.com, April 6, 2005. http://www.realclimate.org/index.php?p=142.
22. Kerry A. Emanuel, "Advance Written Testimony."
23. J. H. Mercer, "West Antarctic Ice Sheet and CO2 Greenhouse Effect: A Threat of Disaster," *Nature*, 271 (January 1978), pp. 321–325. http://stuff.mit.edu/~heimbach/papers_glaciology/nature_mercer_1978_wais.pdf.
24. Google Books' Ngram Viewer. http://books.google.com/ngrams/graph?content=greenhouse+effect%2Cglobal+warming%2Cclimate+change&year_start=1960&year_end=2010&corpus=0&smoothing=3.
25. Broadly the same trends are present in academic journals.
26. Erik Conway, "What's in a Name? Global Warming vs. Climate Change," NASA.gov. http://www.nasa.gov/topics/earth/features/climate_by_any_other_name.html.
27. "No Need to Panic About Global Warming;" *Wall Street Journal*, January 26, 2012. http://online.wsj.com/article/SB10001424052970204301404577171531838421366.html?mod=WSJ_Opinion_LEADTop.
28. "Denmark Energy Use (kt of oil equivalent)," World Bank data via Google Public Data, last updated March 30, 2012. http://www.google.com/publicdata/explore?ds=d5bncppjof8f9_&met_y=eg_use_pcap_kg_oe&idim=country:DNK&dl=en&hl=en&q=denmark+energy+consumption#!ctype=l&strail=false&bcs=d&nselm=h&met_y=eg_use_comm_kt_oe&scale_y=lin&ind_y=false&rdim=region&idim=country:DNK&ifdim=region&hl=en&dl=en.
29. "United States Energy Use (kt of oil equivalent)," World Bank data via Google Public Data, last updated March 30, 2012. http://www.google.com/publicdata/explore?ds=d5bncppjof8f9_&met_y=eg_use_pcap_kg_oe&idim=country:USA&dl=en&hl=en&q=denmark+energy+consumption#!ctype=l&strail=false&bcs=d&nselm=h&met_y=eg_use_comm_kt_oe&scale_y=lin&ind_y=false&rdim=region&idim=country:DNK:USA&ifdim=region&hl=en&dl=en.
30. "FAQ: Copenhagen Conference 2009;" CBCNews.ca, December 8, 2009. http://www.cbc.ca/news/world/story/2009/12/01/f-copenhagen-summit.html.
31. Nate Silver, "Despite Protests, Some Reason for Optimism in Copenhagen," *FiveThirtyEight.com*, December 9, 2009. http://www.fivethirtyeight.com/2009/12/despite-protests-some-reasons-for.html.
32. "Energy/Natural Resources: Lobbying, 2011," OpenSecrets.org. http://www.opensecrets.org/industries/lobbying.php?ind=E.
33. "The Climate Bet," theclimatebet.com. http://www.theclimatebet.com/gore.png.
34. Kesten C. Green and J. Scott Armstrong, "Global Warming: Forecasts by Scientists Verses Scientific Forecasts," *Energy & Environment*, 18, 7+8 (2007). http://www.forecastingprinciples.com/files/WarmAudit31.pdf.
35. Armstrong articulates 139 on his Web site (rather than 89), although not all could be applied to the IPCC forecasts. J. Scott Armstrong, "Standards and Practices for Forecasting," in *Principles of Forecasting: A Handbook for Researchers and Practitioners* (New York: Kluwer Academic Publishers, June 17, 2001). http://forecastingprinciples.com/files/standardshort.pdf.
36. One problem this introduces is that some of the rules verge on being self-contradictory. For instance, one of Armstrong's principles holds that forecasters should "use all important variables" while another touts the virtue of simplicity in forecasting methods. Indeed, the trade-

off between parsimony and comprehensiveness when building a forecast model is an important dilemma. I'm less certain that much is accomplished by attempting to boil it down to a series of conflicting mandates rather than thinking about the problem more holistically. It also seems unlikely to me, given the number of rules that Armstrong suggests, that very many forecasts of any kind would receive a high grade according to his forecast audit.

37. Nicholas Dawidoff, "The Civil Heretic," *New York Times Magazine*, March 25, 2009. http://www.nytimes.com/2009/03/29/magazine/29Dyson-t.html?pagewanted=all.
38. J. Scott Armstrong, "Combining Forecasts," in *Principles of Forecasting: A Handbook for Researchers and Practitioners* (New York: Kluwer Academic Publishers, June 17, 2001). http://repository.upenn.edu/cgi/viewcontent.cgi?article=1005&context=marketing_papers.
39. Per interview with Chris Forest of Penn State University.
40. Neela Banerjee, "Scientist Proves Conservatism and Belief in Climate Change Aren't Incompatible," *Los Angeles Times*, January 5, 2011. http://articles.latimes.com/2011/jan/05/nation/la-na-scientist-climate-20110105.
41. Kerry Emanuel, *What We Know About Climate Change* (Boston: MIT Press, 2007). http://www.amazon.com/About-Climate-Change-Boston-Review/dp/0262050897.
42. Dennis Bray and Hans von Storch, "CliSci2008: A Survey of the Perspectives of Climate Scientists Concerning Climate Science and Climate Change," Institute for Coastal Research, 2008. http://coast.gkss.de/staff/storch/pdf/CliSci2008.pdf.
43. And these doubts are not just expressed anonymously; the scientists are exceptionally careful, in the IPCC reports, to designate exactly which findings they have a great deal of confidence about and which they see as more speculative.
44. Ronarld Bailey, "An Inconvenient Truth: Gore as Climate Exaggerator," Reason.com, June 16, 2006. http://reason.com/archives/2006/06/16/an-inconvenient-truth.
45. Leslie Kaufman, "Among Weathercasters, Doubt on Warming," *New York Times*, March 29, 2010. http://www.nytimes.com/2010/03/30/science/earth/30warming.html?pagewanted=all.
46. "What's the Difference Between Weather and Climate?," NASA, February 1, 2005. http://www.nasa.gov/mission_pages/noaa-n/climate/climate_weather.html.
47. Anthony Del Genio, "Clouds and Climate Change: The Thick and Thin of It," Goddard Institute for Space Studies, NASA, December 2000. http://www.giss.nasa.gov/research/briefs/delgenio_03/.
48. "KATRINA Graphics Archive," National Hurricane Center, National Weather Service. http://www.nhc.noaa.gov/archive/2005/KATRINA_graphics.shtml.
49. Gavin Schmidt, "Green and Armstrong's Scientific Forecast," RealClimate.org, July 20, 2007. http://www.realclimate.org/index.php/archives/2007/07/green-and-armstrongs-scientific-forecast/.
50. "Occam's Razor;" Wikipedia.org. http://en.wikipedia.org/wiki/Occam's_razor.
51. John Theodore Houghton, G. J. Jenkins, J. J. Ephraums, eds. *Climate Change: The IPCC Scientific Assessment* (Cambridge: Cambridge University Press, 1990). http://www.ipcc.ch/ipccreports/far/wg_I/ipcc_far_wg_I_full_report.pdf.
52. "1.6: The IPCC Assessments of Climate Change and Uncertainties" in *Contribution of Working Group I to the Fourth Assessment Report of the Intergovernmental Panel on Climate Change*; 2007. http://www.ipcc.ch/publications_and_data/ar4/wg1/en/ch1s1-6.html.
53. "New York Snow: Central Park Sets the October Record from Noreaster," Associated Press via *Huffington Post*, October 29, 2011. http://www.huffingtonpost.com/2011/10/29/new-york-snow-noreaster_n_1065378.html.
54. Anne Barnard and Sarah Maslin Nir, "Cleaning Up After Natures Plays a Trick," *New York Times*, October 30, 2011. http://www.nytimes.com/2011/10/31/nyregion/october-snowstorm-sows-havoc-on-northeastern-states.html?pagewanted=all.
55. This makes Central Park's temperature record among the oldest in the United States. The oldest in the world is probably from the English Midlands, which has been updated continuously since 1659.
56. "Average Monthly & Annual Temperatures at Central Park," Eastern Regional Headquarters,

National Weather Service. http://www.erh.noaa.gov/okx/climate/records/monthannualtemp.html.

57. From peak to peak or trough to trough; the cycle goes from trough to peak in about half that time, or eighteen months.
58. Mark C. Bove, et al., "Effect of El Niño on U.S. Landfalling Hurricanes, Revisited," *Bulletin of the American Meteorological Society*, 79, 11 (1998). http://www.aoml.noaa.gov/hrd/Landsea/elnino/.
59. Victoria Jaggard, "Sun Headed into Hibernation, Solar Studies Predict," *National Geographic News*, June 14, 2011. http://news.nationalgeographic.com/news/2011/06/110614-sun-hibernation-solar-cycle-sunspots-space-science/.
60. Sarah Ineson, et al., "Solar Forcing of Winter Climate Variability in the Northern Hemisphere," *Nature Geoscience* 4 (October 9, 2011), pp. 753–757. http://www.nature.com/ngeo/journal/vaop/ncurrent/full/ngeo1282.html.
61. Berrien Moore II and B. H. Braswell, "The Lifetime of Excess Atmospheric Carbon Dioxide," *Global Biogeochemical Cycles*, 8, 1 (1994), pp. 23–38. http://www.agu.org/pubs/crossref/1994/93GB03392.shtml.
62. "Global Land-Ocean Temperature Index in 0.01 Degrees Celsius Base period: 1951–1980," Goddard Institute of Space Studies, NASA. http://data.giss.nasa.gov/gistemp/tabledata_v3/GLB.Ts+dSST.txt.
63. GISS stands for the Goddard Institute of Space Studies. Some scientists express a preference for the NASA/GISS record because it does a better job of accounting for the Arctic and a few other areas where temperature stations are sparse. This is potentially important because there has been more warming in the Arctic than in any other part of the globe.
64. Global Temperature Anomalies, National Atmospheric and Oceanic Association. ftp://ftp.ncdc.noaa.gov/pub/data/anomalies/annual.land_ocean.90S.90N.df_1901-2000mean.dat.
65. Climatic Research Unit, School of Environmental Sciences, University of East Anglia. http://www.cru.uea.ac.uk/cru/data/temperature/hadcrut3gl.txt.
66. Japan Meteorological Agency. http://www.data.kishou.go.jp/climate/cpdinfo/temp/list/an_wld.html.
67. Note that the two satellite records use some of the same underlying data.
68. Some analyses have mistakenly used the satellite temperature records for the upper atmosphere rather than the lower atmosphere. The upper atmosphere is not necessarily predicted to warm—and in fact, may actually cool—under the greenhouse effect.
69. If the satellite technique is slightly less precise since it relies on inference, it does provide some advantages over measuring temperatures from thermometer readings, as the traditional sources do. In particular, these measurements are not subject to the so-called "heat island effect," which is the tendency of downtown business centers to show higher temperatures because of the materials used in tall buildings, which often reflect heat and thereby leave the surrounding areas somewhat warmer. Studies suggest that the impact of the heat-island effect is small, and the station-based temperature records make efforts to correct for it. Nevertheless, having the satellite measurements in addition to station-based temperature measurements provides for some redundancy.
70. For instance, one can adjust the temperature records to the same scale by looking at the years in which they overlapped with one another.
71. Their correlations (where 1 represents an exact match and 0 represents no relationship at all) are all .90 or higher.
72. J. Hansen, et al., "Climate Impact of Increasing Atmospheric Carbon Dioxide," *Science*, 213, 4511(August 28, 1981). http://thedgw.org/definitionsOut/..%5Cdocs%5CHansen_climate_impact_of_increasing_co2.pdf.
73. Geert Jan van Oldenborgh and Rein Haarsma, "Evaluating a 1981 Temperature Projection," RealClimate.org, April 2, 2012. http://www.realclimate.org/index.php/archives/2012/04/evaluating-a-1981-temperature-projection/.
74. J. Hansen, et al., "Global Climate Changes as Forecast by Goddard Institute for Space Stud-

ies Three-Dimensional Model," *Journal of Geophysical Research*, 93, D8 (August 20, 1988), pp. 9341–9364. http://pubs.giss.nasa.gov/abs/ha02700w.html.

75. I use the meteorological definition of summer—the calendar months of June, July, and August—rather than the astronomical definition in which summer does not begin until about June 21.
76. This observation comes mainly from my own evaluation of Hansen's forecasts, but see also Steve McIntyre, "Thoughts on Hansen et al. 1988," *Climate Audit*, January 16, 2008. http://climateaudit.org/2008/01/16/thoughts-on-hansen-et-al-1988/.
77. The charts that accompanied the IPCC report showed a roughly linear increase in temperature. Thus, although we know that temperatures are subject to substantial yearly fluctuations, the IPCC forecast implied how much they are supposed to increase on average: by between 0.02°C and 0.05°C per year.
78. Roger Pielke Jr., "Verification of IPCC Sea Level Rise Forecasts 1990, 1995, 2001," *Prometheus*, January 15, 2008. http://cstpr.colorado.edu/prometheus/archives/climate_change/001323verification_of_ipcc.html.
79. "Policymakers' Summary," in *Climate Change: The IPCC Scientific Assessment* (Cambridge: Cambridge University Press, 1990), p. XVIII.
80. Ibid., figure 5, p. XIX.
81. "EU Greenhouse Gas Emissions: More Than Half Way to the 20 % Target by 2020," European Environment Agency, April 13, 2011. http://www.eea.europa.eu/pressroom/newsreleases/eu-greenhouse-gas-emissions-more.
82. Earth System Research Laboratory, "Full Mauna Loa CO_2 Record."
83. See section 2.7 in "IPCC Second Assessment: Climate Changes 1995," Intergovernmental Panel on Climate Change, p. 5. It refers to a "best estimate" of a 2°C increase in global mean surface temperatures in the 110 years between 1990 and 2100, which works out to approximately 1.8°C per 100 years. The note also expresses a range of projections between 0.9°C and 2.7°C in warming per century. So, even the high end of the IPCC's 1995 temperature range posited a (slightly) lower rate of warming than its best estimate in 1990. http://www.ipcc.ch/pdf/climate-changes-1995/ipcc-2nd-assessment/2nd-assessment-en.pdf.
84. Pielke, Jr., "Verification of IPCC Temperature Forecasts 1990, 1995, 2001, and 2007. http://cstpr.colorado.edu/prometheus/archives/climate_change/001319verification_of_ipcc.html.
85. Julienne Stroeve, Marika M. Holland, Walt Meier, Ted Scambos, and Mark Serreze, "Arctic Sea Ice Decline: Faster Than Forecast," *Geophysical Research Letters*, 34, 2007. http://www.ualberta.ca/~eec/Stroeve2007.pdf.
86. William Nordhaus, "The Challenge of Global Warming: Economic Models and Environmental Policy," 2007. http://nordhaus.econ.yale.edu/dice_mss_072407_all.pdf.
87. Richard B. Rood, Maria Carmen Lemos, and Donald E. Anderson, "Climate Projections: From Useful to Usability," University of Michigan, December 15, 2010. http://www.google.com/url?sa=t&rct=j&q=&esrc=s&source=web&cd=2&cts=1330695376711&ved=0CCsQFjAB&url=http%3A%2F%2Fclimateknowledge.org%2Fopenclimate%2Fdoclink%2F20101211_Projections_Usability_AGU_2010.ppt&ei=xcxQT7HYNoPg0QHNoMDqDQ&usg=AFQjCNH0X_mGc24M3bWTlVusJeltaZm_bA&sig2=bfPJPgcKUTj6czOP-WnegQ.
88. Thomas C. Peterson, William M. Connolley, and John Fleck, "The Myth of the 1970s Global Cooling Scientific Consensus," *Bulletin of the American Meteorological Society*, September 2008. http://scienceblogs.com/stoat/Myth-1970-Global-Cooling-BAMS-2008.pdf.
89. Peter Gwynne, "The Cooling World," *Newsweek*, April 28, 1975. http://denisdutton.com/newsweek_coolingworld.pdf.
90. Brian J. Soden, Richard T. Wetherald, Georgiy L. Stenchikov, and Alan Robock, "Global Cooling After the Eruption of Mount Pinatubo: A Test of Climate Feedback by Water Vapor," *Science*, 296; April 26, 2002. http://climate.envsci.rutgers.edu/pdf/SodenPinatubo.pdf.
91. S. J. Smith, et al., "Anthropogenic Sulfur Dioxide Emissions: 1850–2005," *Atmospheric Chemistry and Physics*, 11 (February 9, 2011), pp. 1101–1116. http://www.atmos-chem-phys.net/11/1101/2011/acp-11-1101-2011.pdf.

92. sfalke, "Country SO2 Emissions," Community Initiative for Emissions Research and Applications, October 5, 2010. http://ciera-air.org/wiki/country-so2-emissions.
93. For Antarctic Ice Core CO_2 measurements, see J.-M. Barnola, D. Raynaud, and C. Lorius, "Historical Carbon Dioxide Record from the Vostok Ice Core," Carbon Dioxide Information Analysis Center. http://cdiac.ornl.gov/trends/co2/vostok.html.
94. For a review of studies on the CO_2 doubling value from 1980 to 1995, see Kavita Kacholia and Ruth A. Reck, "Comparison of Global Climate Change Simulations for CO_2-Induced Warming: An Intercomparison of 108 Temperature Change Projections Published Between 1980 and 1995," *Climactic Change*, 35, 1 (1997), pp. 53–69. http://www.springerlink.com/content/g65v456v8621247m/.

    For a similar review for studies conducted prior to 1980, see Ruth A. Reck, "Introduction to the Proceedings of the Workshop on the Responsible Interpretation of Atmospheric Models and Related Data," General Motors Research Publication GMR-3800, 1981.
95. G. S. Callendar, "The Artificial Production of Carbon Dioxide and Its Influence on Climate," *Quarterly Journal of the Royal Meteorological Society*, 64 (1938), pp. 223–240.
96. Kacholia and Reck, "Comparison of Global Climate Change Simulations for CO_2-Induced Warming."
97. "How Reliable Are Climate Models?," *Skeptical Science*. http://www.skepticalscience.com/climate-models-intermediate.htm.
98. "Climate Change: Examining the Processes Used to Create Science and Policy," hearing before the Committee on Science, Space and Technology, U.S. House of Representatives, March 31, 2011. http://www.gpo.gov/fdsys/pkg/CHRG-112hhrg65306/pdf/CHRG-112hhrg65306.pdf.
99. The success of forecasting methods that date back to the 1990s instead depend on that especially large amount of warming that occurred during the 1990s.
100. Voros McCracken, "13 for His Last 24: Tomfoolery with Multiple Endpoints," *Primate Studies*, Baseball Think Factory, March 20, 2001. http://www.baseballthinkfactory.org/primate_studies/discussion/mccracken_2001-03-20_0/.
101. Armstrong was willing to acknowledge there was some chance his no-change forecast would go badly in the near term. He told me that he figured the chances of winning the bet with Gore—which looked at temperature increases over the next decade rather than the next century—were about 70 percent.
102. This estimate is based on the error term on the coefficient associated with temperature increase in the regression model. It assumes that the exact amount of carbon dioxide is known and that CO_2 will continue to increase at the same annual rate that it did between 2002 and 2011. In practice, the model underestimates the error slightly—and therefore somewhat underestimates the chance of a cooling decade—because the exact amount of CO_2 is an unknown, as well as because of any specification uncertainty in the model.
103. "Climatic Research Unit E-Mail Controversy;" Wikipedia.org. http://en.wikipedia.org/wiki/Climatic_Research_Unit_email_controversy.
104. Henry Chu, "Panel Clears Researchers in 'Climategate' Controversy," *Los Angeles Times*, April 15, 2010. http://articles.latimes.com/2010/apr/15/world/la-fg-climate-data15-2010apr15.
105. Including those from satellite records processed by private companies.
106. "Climate of Fear;" editorial in *Nature*, 464, 141 (March 11, 2010). http://www.nature.com/nature/journal/v464/n7286/full/464141a.html.
107. The site is run by the meteorologist Anthony Watts and derives its name accordingly.
108. This especially holds if we like the way things are now. In this sense, conservatism—if defined as preservation of the status quo—argues more strongly for action to mitigate climate change than liberalism does.
109. Lydia Saad, "In U.S., Global Warming Views Steady Despite Warm Winter," Gallup.com, March 30, 2012. http://www.gallup.com/poll/153608/global-warming-views-steady-despite-warm-winter.aspx.
110. The reluctance to tackle burgeoning national debt loads in the United States and other Western countries is another consequence of our short-term thinking.

111. Voteview, "An Update on Political Polarization (Through 2011)—Part II," VoteView.com. http://voteview.com/blog/?p=309.
112. Thomas E. Mann and Norman J. Ornstein, "Let's Just Say It: The Republicans Are the Problem," *Washington Post*, April 27, 2012. http://www.washingtonpost.com/opinions/lets-just-say-it-the-republicans-are-the-problem/2012/04/27/gIQAxCVUlT_story.html.
113. Michael Kinsley, "The Gaffer Speaks," The *Times* of London, April 23, 1988.
114. "Patents by Country, State, and Year; Utility Patents (December 2011)" Patent Technology Monitoring Team, U.S. Patent and Trademark Office. http://www.uspto.gov/web/offices/ac/ido/oeip/taf/cst_utl.htm.

CHAPTER 13: WHAT YOU DON'T KNOW CAN HURT YOU

1. Since the War of 1812. "A Sunday in December, Chapter 5: Fighting the Good Fight;" *Los Angeles Times*, December 3, 1991. http://articles.latimes.com/1991-12-03/news/wr-753_1_pearl-harbor/3.
2. Mark R. Peattie and David C. Evans, *Kaigun: Strategy, Tactics, and Technology in the Imperial Japanese Navy 1887–1941*; (Bethesda, MD: Naval Institute Press; 1997.
3. "Roberta Wohlstetter, *Pearl Harbor: Warning and Decision*" (Stanford, CA: Stanford University Press, 1962), p. 385.
4. Wohlstetter, *Pearl Harbor*, p. 173.
5. Ibid., pp. 12–13.
6. Ibid., p. 385.
7. Some analysts did think that the carriers were on the move but on a far more southerly route, toward the Marshall Islands.
8. The vote was 82–0 in the U.S. Senate and 388–1 in the U.S. House. Frank L. Kluckhorn, "U.S. Declares War, Pacific Battle Widens," *New York Times*; December 9, 1941.
9. Donald Rumsfeld, *Known and Unknown: A Memoir* (New York: Sentinel, 2011), Kindle edition, locations 6147–6148.
10. Ibid., Kindle locations 814–816.
11. Urbahn was the first to break the news of Osama bin Laden's death in May 2011, "Osama bin Laden Death First Revealed on Twitter;" *Daily Mirror*, May 2, 2011. http://www.mirror.co.uk/news/uk-news/osama-bin-laden-death-first-179280.
12. Associated Press, September 10, 1940 (as printed in *Tuscaloosa News*, September 11, 1941).
13. Wohlstetter, *Pearl Harbor*, p. 291.
14. Saburō Kurusu, "Historical Inevitability of the War of Greater East Asia,"Foreign Broadcast Intelligence Service, Tokyo; November 26, 1942. http://www.ibiblio.org/pha/policy/1942/421126a.html.
15. Wohlstetter, *Pearl Harbor*, pp. 1–2.
16. Ibid., p. 3.
17. Ibid., p. 387. Emphasis in original.
18. William Shakespeare, *The Tragedy of Julius Caesar* (1599), Act I, Scene III.
19. This would exclude an extremely brief incursion across the Rio Grande by Mexican troops during the Mexican-American War in 1846 and a raid on the town of Columbus, New Mexico, by Pancho Villa in 1916.
20. Errol Morris, "The Anosognosic's Dilemma: Something's Wrong but You'll Never Know What It Is (Part 1),"; *Opinionator, New York Times*, June 20, 2010. http://opinionator.blogs.nytimes.com/2010/06/20/the-anosognosics-dilemma-1/.
21. "DoD News Briefing—Secretary Rumsfeld and Gen. Myers;" News Transcript, U.S. Department of Defense; February 12, 2002. http://www.defense.gov/transcripts/transcript.aspx?transcriptid=2636.
22. Rumsfeld, *Known and Unknown*, Kindle location 196.
23. Harlan Ullman, "Known and Unknown Dangers (Terrorism)," *National Interest*, March 22, 2006.
24. "Report of the Joint Inquiry into the Terrorist Attacks of September 11, 2001," U.S. House

Permanent Select Committee on Intelligence and the Senate Select Committee on Intelligence; 107th Congress, 2nd Session; December 2002, pp. 209–214.

25. "Statement by J. Gilmore Childers, Esq., Orrick, Herrington & Sutcliffe LLP New York City, New York, and Henry J. DePippo, Esq., Nixon Hargrave Devans & Doyle Rochester, New York," before the Senate Judiciary Committee Subcommittee on Technology, Terrorism, and Government Information, Foreign Terrorists in America: Five Years After the World Trade Center, February 24, 1998. http://web.archive.org/web/20071227065444/http://judiciary.senate.gov/oldsite/childers.htm.
26. "Major Terrorist Acts Suspected of or Inspired by al-Qaeda;" InfoPlease.com. http://www.infoplease.com/ipa/A0884893.html.
27. "Two Months Before 9/11, an Urgent Warning to Rice;" *Washington Post*, October 1, 2006. http://www.washingtonpost.com/wp-dyn/content/article/2006/09/30/AR2006093000282.html.
28. Ibid.
29. Moussaoui, who had overstayed his visa, was arrested on immigration charges.
30. National Commission on Terrorist Attacks, *The 9/11 Commission Report: Final Report of the National Commission on Terrorist Attacks upon the United States* (New York: Norton Trade E-Books, 2011), Kindle edition), location 6914.
31. Ibid., Kindle location 9243.
32. Ibid., Kindle location 9092.
33. Some of this resulted because of the banalities of the national security system; bureaucracy might be thought of as the opposite of imagination. In Pearl Harbor, some of the signals of the impeding Japanese attack were detected by the Army and others by the Navy; some were detected in Washington and others in Hawaii. However, the information was not necessarily shared, so no one decision maker had much insight. As Schelling writes in Wohlstetter's book:

    > Surprise, when it happens to a government, is likely to be a complicated, diffuse, bureaucratic thing. It includes neglect of responsibility but also responsibility so poorly defined . . . that action gets lost. It includes gaps in intelligence, but also intelligence that, like a string of pearls too precious to wear, is too sensitive to give to those who need it. It includes the alarm that fails to work, but also the alarm that has gone off so often it has been disconnected. . . . It includes the contingencies that occur to no one, but also those that everyone assumes someone else is taking care of.

    Likewise, in advance of September 11, some key pieces of information were held by the FBI and some by the CIA, some by the State Department and some by the Department of Defense. Rumsfeld told me that many of George Tenet's revelations, for instance, he learned of for the first time only after reading Tenet's book. Meanwhile, the Bush administration had just taken over from the Clinton administration and there were the usual political chores to worry about; Rumsfeld, for instance, spent much of his first nine months in office trying to fend off budget cuts.
34. *The 9/11 Commission Report*, Kindle location 9253.
35. Ibid., Kindle locations 2907–2910.
36. Bruce Schneier, *Beyond Fear: Thinking Sensibly About Security in an Uncertain World* (New York: Springer, 2003), Kindle locations 951–952.
37. Note, however, that the kamikaze strategy was largely employed toward the end of the war, when Japan had started to lose badly; it was not in use at Pearl Harbor.
38. Global Terrorism Database, National Consortium for the Study of Terrorism and Responses to Terrorism, U.S. Department of Homeland Security, University of Maryland. http://www.start.umd.edu/gtd/search/Results.aspx?page=2&casualties_type=b&casualties_max=&start_yearonly=1979&end_yearonly=2000&dtp2=all&sAttack=1&count=100&expanded=no&charttype=line&chart=overtime&ob=GTDID&od=desc#results-table.

39. Amos Tversky and Daniel Kahneman, "Availability: A Heuristic for Judging Frequency and Probability," *Cognitive Psychology*, 5, 2 (Setepmber 1973), pp. 207–232. http://www.sciencedirect.com/science/article/pii/0010028573900339.
40. "Nineteen hijackers using commercial airliners as guided missiles to incinerate three thousand men, women, and children was perhaps the most horrific single unknown unknown America has experienced." Rumsfeld, *Known and Unknown*, Kindle locations 196–198.
41. *The 9/11 Commission Report*, Kindle locations 9198–9199.
42. Aaron Clauset, "Macroevolution of Whales and the Dynamics of Morphological Disparities," 2010 GSA Denver Annual Meeting, October 31, 2010. http://scholar.google.com/citations?view_op=view_citation&hl=en&user=e7VI_HcAAAAJ&sortby=pubdate&citation_for_view=e7VI_HcAAAAJ:qxL8FJ1GzNcC.
43. Winter Mason and Aaron Clauset, "Friends FTW! Friendship and Competition in Halo: Reach," *Arxiv*, March 3, 2012. http://scholar.google.com/citations?view_op=view_citation&hl=en&user=e7VI_HcAAAAJ&sortby=pubdate&citation_for_view=e7VI_HcAAAAJ:e5wmG9Sq2KIC.
44. Brig. S. S. Chandel, "Philosophy of Terrorism in Kashmir," Institute of Peace and Conflict Studies, Terrorism Articles, Number 480, March 2001. http://www.ipcs.org/article/terrorism/philosophy-of-terrorism-in-kashmir-480.html.
45. Global Terrorism Database. http://www.start.umd.edu/gtd/downloads/Codebook.pdf.
46. David C. Rapoport, "The Four Waves of Modern Terrorism," *Anthropoetics*, 8, 1 (June 5, 2006). http://www.international.ucla.edu/media/files/Rapoport-Four-Waves-of-Modern-Terrorism.pdf.
47. There is one slight "trick" that I've used here, and which Clauset also uses in his published work. Terror attacks producing very small numbers of fatalities—in this case, under five deaths—do not fit the data quite as cleanly and are eliminated from the analysis. Ordinarily, you would not want to do something like this; it's bad practice to throw out data unless you have a very good reason to do so. But in this case, it does not make much practical difference, since as numerous as small terror attacks are, they are responsible for a very small fraction of the death toll. Moreover, there may be modest biases in the coverage of the database—larger incidents are almost certain to be included, whereas those that kill just one or two people might not be. And there is some debate about whether terrorists themselves should be included in the death toll in the event of suicide attacks—RAND does include them—which has relatively more effect when the overall number of casualties is small.
48. In this case, an attack on the "scale of 9/11" refers to one that kills at least 2,749 people—the death toll at the World Trade Center site. The overall death toll on 9/11 was slightly higher—close to 3,000 people—but the RAND database and most others classify the attacks on the World Trade Center, the Pentagon, and United Flight 93 as having been separate (although related) attacks. This does not make much difference, however: an attack killing at least 3,000 people would be expected to occur about once every 44 years, rather than once every 41.
49. Peter M. Shearer and Phillip B. Stark, "Global Risk of Big Earthquakes Has Not Recently Increased," *PNAS*, December19, 2011. http://www.pnas.org/content/early/2011/12/12/1118525109.abstract.
50. In a 2012 paper that Clauset sent to me and which he submitted to the *Annals of Applied Statistics*, he went through a similar calculation with a more refined technique and put the risk of a 9/11-scale attack at between 11 percent and 35 percent in the 33 years between 1968 and 2001. This implies a slightly lower but still very tangible risk of a 9/11-scale attack, on the order of once per 130 years.
51. "How Many People Died as a Result of Atomic Bombings?," Frequently Asked Questions, Radiation Effects Research Foundation. http://www.rerf.or.jp/general/qa_e/qa1.html.
52. Ira Helfand, Lachlan Forrow, and Jaya Tiwari, "Nuclear Terrorism," *British Medical Journal*, 324, 7333 (February 9, 2002), pp. 356–359. https://www.ncbi.nlm.nih.gov/pmc/articles/PMC1122278/.
53. James Hoge, "'Nuclear Terrorism': Counting Down to the New Armageddon," *New York Times*, September 5, 2004. http://www.nuclearterror.org/nyt.htm.

54. Stewart Stogel, "Bin Laden's Goal: Kill 4 Million Americans," NewsMax.com, July 14, 2004. http://archive.newsmax.com/archives/articles/2004/7/14/215350.shtml.
55. Per Google Scholar search. http://scholar.google.com/scholar?hl=en&q=graham+allison&btnG=&as_sdt=1%2C33&as_sdtp=.
56. Graham Allison, *Nuclear Terrorism: The Ultimate Preventable Catastrophe* (New York: Times Books, 2004), Kindle edition, location 300.
57. Times Square is, in fact, named after the newspaper, as anyone who works there will be happy to point out.
58. Allison, *Nuclear Terrorism*, Kindle location 112.
59. J. F. Frittelli, et al., *Port and Maritime Security: Background and Issues* (New York. Novinka Books, 2003).
60. "Status of World Nuclear Forces;" Federation of American Scientists. http://www.fas.org/programs/ssp/nukes/nuclearweapons/nukestatus.html.
61. Allison, *Nuclear Terrorism*, Kindle location 3258.
62. Suzanne Goldenberg, "Bush Threatened to Bomb Pakistan, Says Musharraf," *The Guardian*, September 21, 2006. http://www.guardian.co.uk/world/2006/sep/22/pakistan.usa.
63. Jay Newton-Small, "Bin Laden May Have Lived at Abbottabad Compound for Six Years," *Swampland, Time*, May 3, 2011. http://swampland.time.com/2011/05/03/bin-laden-may-have-lived-at-abbottabad-compound-for-six-years/.
64. David Albright and Paul Brannan, "Pakistan Doubling Rate of Making Nuclear Weapons: Time for Pakistan to Reverse Course," Institute for Science and International Security, May 16, 2011. http://www.isis-online.org/isis-reports/detail/pakistan-doubling-rate-of-making-nuclear-weapons-time-for-pakistan-to-rever/.
65. "The Political Instability Index;" ViewsWire, Economist Intelligence Unit, *The Economist*. http://viewswire.eiu.com/site_info.asp?info_name=social_unrest_table&page=noads&rf=0.
66. Randy Borum, "Psychology of Terrorism," *Encyclopedia of Peace Psychology* (New York: Springer Science, 2010), p. 62. http://worlddefensereview.com/docs/PsychologyofTerrorism0707.pdf.
67. Mohammed M. Hafez, "Suicide Terrorism in Iraq: A Preliminary Assessment of the Quantitative Data and Documentary Evidence," *Studies in Conflict & Terrorism*, 29, 6 (September 2006), pp. 531–559. https://www.ncjrs.gov/app/publications/Abstract.aspx?id=237341.
68. Bribery or coercion of nuclear scientists is another concern. The United States funds a program called the Nuclear Cities Initiative, which helps nuclear scientists in the former USSR find other gainful lines of employment—rather than falling into the wrong hands.
69. "Dark Winter Exercise Overview;" Center for Biosecurity, University of Pittsburg Medical Center, June 22–23, 2001. http://www.upmc-biosecurity.org/website/events/2001_darkwinter/index.html.
70. I am using the term *damage* a bit loosely, since the damage from earthquakes on a human scale does not necessarily bear a one-to-one relationship to their energy release.
71. James Q. Wilson and George L. Kelling, "Broken Windows," *The Atlantic*, March 1982. http://www.manhattan-institute.org/pdf/_atlantic_monthly-broken_windows.pdf.
72. Bernard E. Harcourt and Jens Ludwig, "Reefer Madness: Broken Windows Policing and Misdemeanor Marijuana Arrests in New York City, 1989–2000," Criminology and Public Policy, University of Chicago Law & Economics, Olin Working Paper No. 317/University of Chicago, Public Law Working Paper No. 142; 2007. http://papers.ssrn.com/sol3/papers.cfm?abstract_id=948753.
73. Kees Keizer, Siegwart Lindenberg, and Linda Steg, "The Spreading of Disorder," *Science*, 322, 5908 (December 2008), pp. 1681–1685. http://www.sciencemag.org/content/322/5908/1681.abstract.
74. Bernard E. Harcourt and Jens Ludwig, "Broken Windows: New Evidence from New York City and a Five-City Social Experiment," *University of Chicago Law Review*, 73 (2006). http://lawreview.uchicago.edu/sites/lawreview.uchicago.edu/files/uploads/73.1/73_1_Harcourt_Ludwig.pdf.

75. Bruce Schneier, "Beyond Security Theater," Schneier on Security, November 13, 2009. http://www.schneier.com/blog/archives/2009/11/beyond_security.html.
76. Ibid., Kindle location 1035.
77. Nate Silver, "Crunching the Risk Numbers," *Wall Street Journal*, January 8, 2010. http://Online.wsj.com/article/SB10001424052748703481004574646963713065116.html.
78. Russian Authorities: Terrorist Bombing at Moscow Airport Kills 35;" CNN Wire; January 24, 2011. http://articles.cnn.com/2011-01-24/world/russia.airport.explosion_1_suicide-bomber-moscow-police-moscow-during-rush-hour?_s=PM:WORLD.
79. Ken Silverstein, "The Al Qaeda Clubhouse: Members Lacking," *Harper's* magazine, July 5, 2006. http://www.harpers.org/archive/2006/07/sb-al-qaeda-new-members-badly-needed-1151963690.
80. Aaron Clauset, Maxwell Young, and Kristian Skrede Gleditsch, "On the Frequency of Severe Terrorist Events," *Journal of Conflict Resolution*, 51, 1 (February 2007), pp. 58–87. http://www.cabdyn.ox.ac.uk/complexity_PDFs/CABDyN%20Seminars%202007_2008/Frequency%20Events_Gleditsch.pdf.
81. *Jerusalem Post* poll by TNS/Teleseker of 500 Jewish Israelis, January 23–24, 2012. http://thejerusalemreport.files.wordpress.com/2012/02/poll-new.jpg.
82. David Weisburd, Tal Jonathan, and Simon Perry, "The Israeli Model for Policing Terrorism: Goals, Strategies, and Open Questions," *Criminal Justice and Behavior*, 36, 12 (December 2009), pp. 1259–1278. http://scholar.googleusercontent.com/scholar?q=cache:ydYnY99dbqwJ:scholar.google.com/&hl=en&as_sdt=0,33.
83. "Iraq: What Did Congress Know, and When?," FactCheck.org, November 19, 2005. http://www.factcheck.org/iraq_what_did_congress_know_and_when.html.
84. "Report of the Select Committee on Intelligence on Postwar Findings About Iraq's WMD Programs and Links to Terrorism and How They Compare with Prewar Assessments;" U.S. Senate, 109th Congress, 2nd Session; September 8, 2006. http://intelligence.senate.gov/phaseiiaccuracy.pdf.
85. Martin Chulov and Helen Pidd, "Defector Admits to WMD Lies That Triggered Iraq War," *The Guardian*, February 15, 2011. http://www.guardian.co.uk/world/2011/feb/15/defector-admits-wmd-lies-iraq-war.
86. Schneier, "Beyond Security Theater," Kindle locations 1321–1322.
87. Harvey E. Lapan and Todd Sandler, "Terrorism and Signalling," *European Journal of Political Economy*, 9, 3 (August 1993), pp. 383–397;
88. *The 9/11 Commission Report*, Kindle locations 9286–9287.
89. Michael A. Babyak, "What You See May Not Be What You Get: A Brief, Nontechnical Introduction to Overfitting in Regression-Type Models," *Psychosomatic Medicine*, 66 (2004), pp. 411–.421; 2004. http://os1.amc.nl/mediawiki/images/Babyak_-_overfitting.pdf.

CONCLUSION

1. Brian Cartwright, "That Great Derek Jeter Conspiracy," *FanGraphs*, January 17, 2009. http://www.fangraphs.com/blogs/index.php/the-great-derek-jeter-conspiracy/.
2. Halley's Comet was first sited on Christmas Day in 1758. See Peter Lancaster Brown, *Halley and His Comet* (Suffolk, England: Blandford Press, 1985).
3. Mary Frances Williams, "The *Sidus Iulium*, the Divinity of Men, and the Golden Age in Virgil's *Aeneid*," *Leeds International Classical Studies*, vol. 2, issue 1, 2003. http://lics.leeds.ac.uk/2003/200301.pdf
4. The exact date of the invention of the World Wide Web is disputed but in 1990 Berners-Lee established the first successful connection between an HTTP client and the Internet. The set of hypertext documents called the World Wide Web are not to be confused with the Internet, the network by which the World Wide Web is accessed, which as everyone knows was invented by Al Gore.
5. Glenn Gunzelmann and Kevin A. Gluck, "Knowledge Tracing for Complex Training Applications: Beyond Bayesian Mastery Estimates," Air Force Research Laboratory,; Proceedings

of the Thirteenth Conference on Behavior Representation in Modeling and Simulation, 2004, pp. 383–84. http://act-r.psy.cmu.edu/papers/710/gunzelmann_gluck-2004.pdf.

6. Sarah Lichtenstein and Baruch Fischhoff, "Training for Calibration," prepared for U.S. Army Research Institute for the Behavioral and Social Sciences, ARI Technical Report TR-78-A32; November 1978. http://www.dtic.mil/cgi-bin/GetTRDoc?AD=ADA069703.
7. Christopher J. Gill, Lora Sabin and Christopher H. Schmidt, "Why Clinicians Are Natural Bayesians," *British Medical Journal*, vol. 330; May 7, 2005. http://www.ncbi.nlm.nih.gov/pmc/articles/PMC557240/.
8. Tomasso Poggio and Federico Girosi, "A Theory of Networks for Approximation and Learning," Massachusetts Institute of Technology Artificial Intelligence Laboratory and Center for Biological Information Processing, Whitaker College, A.I. Memo 1140, C.B.I.P. Paper 31, July 1989. http://www.dtic.mil/cgi-bin/GetTRDoc?AD=ADA212359.
9. Amanda Ripley, *The Unthinkable* (New York: Random House, Kindle edition), location 337–360.
10. Ibid., Kindle location 3688–98.
11. Joel Mokyr, *The Gifts of Athena; Historical Origins of the Knowledge Economy* (Princeton: Princeton University Press, Kindle Edition), location 160–162.
12. Jay Rosen, "The View from Nowhere: Questions and Answers," *Jay Rosen's Press Think*, November 10, 2010. http://pressthink.org/2010/11/the-view-from-nowhere-questions-and-answers/.
13. This is just a personal reflection—not an empirical observation—but I am being somewhat literal about this point. When I was working on this book and came across a thorny problem that I couldn't quite resolve, I found it much more productive to walk around, subjecting my brain to random inputs, than to stare at my computer screen or sit in a coffee shop. One of the advantages of living in New York is that it offers 24/7 access to the spontaneous behavior of eight million human beings who might jog your mind or your memory.
14. This is derived from Reinhold Niebuhr's Serenity Prayer. http://www.cptryon.org/prayer/special/serenity.html
15. The data shown in figure C-2 is based on searches conducted of the JSTOR catalog of print journals. I searched for cases in which either the word "predictable" or "unpredictable" appeared in the journal article at least once (but not both words in the same article), breaking down the results by the decade of publication. The percentages reflected in figure C-2 represent the number of uses of "predictable" and "unpredictable," respectively, relative to their total number.
16. Michel Foucault, *The Order of Things* (New York: Vintage, 1994).
17. Global per capita GDP growth averaged 3.4 percent per year in the 1950s but 2.6 percent in the 1970s. See J. Bradford DeLong, *Estimating World GDP, One Million B.C.—Present*; (Berkeley: University of California, 1988). http://econ161.berkeley.edu/TCEH/1998_Draft/World_GDP/Estimating_World_GDP.html.
18. The number of patent applications filed with the U.S. Patent and Trade Office rose by 18 percent over the decade of the 1950s but just 1 percent in the 1970s. See "U.S. Patent Activity Calendar Years 1790 to the Present," U.S. Patent and Trade Office. http://www.uspto.gov/web/offices/ac/ido/oeip/taf/h_counts.htm.
19. Google Books' Ngram Viewer. http://books.google.com/ngrams/graph?content=predictable%2Cunpredictable&year_start=1800&year_end=2000&corpus=4&smoothing=3.

INDEX

Page numbers in *italics* refer to figures. Page numbers beginning with 459 refer to endnotes.